T0313825

The Climate City

The Climate City

Edited by Martin Powell

WILEY Blackwell

This edition first published 2022
© 2022 John Wiley & Sons Ltd

All rights reserved. No part of this publication may be reproduced, stored in a retrieval system, or transmitted, in any form or by any means, electronic, mechanical, photocopying, recording or otherwise, except as permitted by law. Advice on how to obtain permission to reuse material from this title is available at http://www.wiley.com/go/permissions.

The right of Martin Powell to be identified as the author of this work has been asserted in accordance with law.

Registered Offices
John Wiley & Sons, Inc., 111 River Street, Hoboken, NJ 07030, USA
John Wiley & Sons Ltd, The Atrium, Southern Gate, Chichester, West Sussex, PO19 8SQ, UK

Editorial Office
9600 Garsington Road, Oxford, OX4 2DQ, UK

For details of our global editorial offices, customer services, and more information about Wiley products visit us at www.wiley.com.

Wiley also publishes its books in a variety of electronic formats and by print-on-demand. Some content that appears in standard print versions of this book may not be available in other formats.

Limit of Liability/Disclaimer of Warranty
In view of ongoing research, equipment modifications, changes in governmental regulations, and the constant flow of information relating to the use of experimental reagents, equipment, and devices, the reader is urged to review and evaluate the information provided in the package insert or instructions for each chemical, piece of equipment, reagent, or device for, among other things, any changes in the instructions or indication of usage and for added warnings and precautions. While the publisher and author have used their best efforts in preparing this work, they make no representations or warranties with respect to the accuracy or completeness of the contents of this work and specifically disclaim all warranties, including without limitation any implied warranties of merchantability or fitness for a particular purpose. No warranty may be created or extended by sales representatives, written sales materials or promotional statements for this work. The fact that an organization, website, or product is referred to in this work as a citation and/or potential source of further information does not mean that the publisher and author endorse the information or services the organization, website, or product may provide or recommendations it may make. This work is sold with the understanding that the publisher is not engaged in rendering professional services. The advice and strategies contained herein may not be suitable for your situation. You should consult with a specialist where appropriate. Further, readers should be aware that websites listed in this work may have changed or disappeared between when this work was written and when it is read. Neither the publisher nor author shall be liable for any loss of profit or any other commercial damages, including but not limited to special, incidental, consequential, or other damages.

Library of Congress Cataloging-in-Publication data applied for
Names: Powell, Martin, 1970- editor.
Title: The climate city / [edited by] Martin Powell.
Description: Hoboken, NJ, USA : Wiley-Blackwell, 2022. | Includes
    bibliographical references and index.
Identifiers: LCCN 2021041545 (print) | LCCN 2021041546 (ebook) | ISBN
    9781119746270 (hardback) | ISBN 9781119746300 (pdf) | ISBN 9781119746317
    (epub) | ISBN 9781119746294 (ebook)
Subjects: LCSH: Environmental protection. | Sustainable urban development. |
    Environmental policy. | Urban pollution.
Classification: LCC TD170 .C588 2022 (print) | LCC TD170 (ebook) | DDC
    363.7--dc23
LC record available at https://lccn.loc.gov/2021041545
LC ebook record available at https://lccn.loc.gov/2021041546

Cover Design by Wiley
Cover Image: © WATG (Wimberly, Allison, Tong & Goo) and Pixelflakes

Set in 9.5/12.5pt STIXTwoText by Integra software Pvt. Ltd, Pondicherry, India
Printed and bound by CPI Group (UK) Ltd, Croydon, CR0 4YY

C9781119746270_290422

Contents

Acknowledgements

For their support and encouragement, my love and thanks to Caroline, Jacob, Tessa, Albert, Annie, and Bluebell and Millie. To Daphne and Josephine for their original inspiration. Thanks to my Mum and Dad and my brother Andrew and my sister, Kathryn. Thanks to Pat and Randall. All the people along my career who have given me huge support which led me to this moment. These include Malcolm Horner, Professor George Korfiatis, Alan Swinger, Nicola Suozzo, Paul Craddock, Alistair Kirk, Peter Bishop, Frank Lee, Isabel Dedring, Rit Aggarwala, Jay Carson, Boris Johnson, Sir Simon Milton, Roland Busch, Alex Stuebler, Pedro Miranda, Savvas Verdis, Barbara Humpton, Camille Johnson, and Anthony Casciano.

A special mention to Cathe Reams, who helps us all to smile, and helped me look at the chapters in a particular sequence, and my fabulous niece, Livvie Hackland, who was drafted in to help me link the chapters and whose work ethic and writing are both spectacular.

Special thanks to Amalie Østergaard for her support from the City of Copenhagen and to Jakob Geiger for his support in The Resourceful City.

A big thank you to people I met along the way who inspired me to engage deeper with this topic: Samantha Heath for her tenacity, Mike Bloomberg for his targeted impact, and Al Gore, whose pursuit of detail, kind manner, and relentless ability to focus action in the right places at the right time (and his movie!) were a true inspiration to me and I have attempted to echo this approach in the book.

Annie Chiu, Paul Sayer, Todd Green, Skyler Van Valkenburgh, Amy Odum, Julie Musk, Janane Sivakumar and Hema Krishnamoourthy at Wiley, who calmly and professionally made this happen.

Each of my children had ideas for the cover, which led me to use the image of Fleet Street as it could be reimagined. I want to thank WATG for supplying it (https://www.watg.com/watg-unveils-innovative-green-block-to-help-make-london-the-worlds-first-national-park-city/). Founded in Honolulu in 1945 by George "Pete" Wimberly, WATG is one of the world's leading hospitality and destination design firms. Independent to this day and with a profound respect for heritage, its team of strategists, master planners, architects, landscape architects, and interior designers have created more than 400 built projects in 170 countries.

Finally, a big thank you to the 40 authors, each of whom agreed without any persuasion to provide their amazing chapters. You have been, and continue to be, great friends and great human beings. Your dedication to a better environment and a better future has shone through. It has made the book feel hopeful that something even better is ahead of us.

Authors Biographies

Peter Boyd

The Ambitious City – This chapter presents a case for combining high *ambition* with high *clarity* of definition for what we mean by "Net-Zero", highlighting the need to combine this high ambition with an appreciation and embrace of systems thinking, given the unique, complex, and intertwined nature of each city's challenges and opportunities.

Peter Boyd is Lecturer at the Yale School of the Environment, Lecturer in the Practice of Management at the School of Management, and Resident Fellow at the Center for Business and the Environment. Outside Yale, he is a director of REDD.plus, a digital platform to bring UN-registered REDD + forest carbon credits to a new cross-sector world of purchasers who want to achieve Paris-Agreement-compliant carbon neutrality as they transition to net-zero. Working with these and other partners, he is Founder of Time4Good, helping leaders and teams connect purpose to maximum positive impact.

He is former COO of Sir Richard Branson's Carbon War Room; former Chair of the Energy Efficiency Deployment Office for the UK Department of Energy & Climate Change; and former Project Lead for the B Team's "Net- Zero" initiative, focused on business encouragement, of an ambitious Paris Agreement at COP21. Following his first job with McKinsey & Co., his private-sector experience included over ten jobs in 12 years at the Virgin Group, including CEO of Virgin Mobile South Africa. Peter is originally from Edinburgh, Scotland; studied Philosophy, Politics and Economics at the University of Oxford; and now lives in Connecticut, where he serves as Chair of Sustainable Westport.

Martin Powell

The Civilized City – This chapter looks at the transformation of cities through time and how we can apply those learnings for a better future. Local solutions will tackle the global climate crisis.

Martin Powell was Environmental Advisor to the former Mayor of London, Boris Johnson, responsible for policy in water, waste, air quality, energy, climate change mitigation and adaptation, and biodiversity. He was also director for the design and delivery of the city's environmental programmes.

As Managing Director of Cambridge Management & Research, Martin worked for the Energy Saving Trust and the Institute for Sustainability and was Special Advisor to the C40 Cities Group chaired by Michael R. Bloomberg during his time as Mayor of New York. An engineer, he built his career working with organizations to structure their projects and programmes. Martin is also a trustee at Heart of the City, a charity that supports SMEs in London to tackle issues including climate action.

He has held several roles at Siemens including Global Head of Urban Development and is currently Head of Sustainability and Environmental Initiatives at Siemens Inc., with a focus on financing climate action.

Austin Williams

The Emerging City – Some countries don't need western lectures on sustainable development; they simply need to be allowed to develop. This chapter is about development without prefixes, full stop. This is Malawi's story.

Austin Williams is a senior lecturer in Professional Practice in Architecture at Kingston School of Art in London and Honorary Research Fellow at XJTLU University, Suzhou, China. He is director of the Future Cities Project, and the author of *China's Urban Revolution: Understanding Chinese Ecocities* (Bloomsbury, 2017) and *New Chinese Architecture: Twenty Women Building the Future* (Thames and Hudson, 2019).

Austin founded the mantownhuman manifesto featured in Penguin Classics' *100 Artists' Manifestos*. He has spoken at a wide range of conferences, from New York to Ningbo, from Hawaii to Hong Kong, and is a regular media commentator on development, environmentalism, and China. He has written for magazines as diverse as *Nature, Wired, Top Gear, Wallpaper, Times Literary Supplement*, and *The Economist*. He has directed over 200 short documentaries for NBS TV and authored and illustrated the *Shortcuts* design guides. For more information see WeChat/Twitter: Future_Cities andwww.futurecities.org.uk.

Patricia Holly Purcell

The Sustainable City – This chapter sets out the UN global frameworks for tackling climate change and the SDGs, how local governments feature in these international agendas, and the role of cities in advancing solutions to some of the greatest challenges of our time.

Patricia Holly Purcell has more than 15 years' experience in the public and private sectors leading global sustainability initiatives for multinational corporations and the UN. She is currently a private sector specialist for the United Nations Development Programme (UNDP), focused on unlocking and leveraging private sector engagement and investment to support increased ambition of countries' climate change goals and delivering opportunities to scale solutions across key sectors of the economy, including energy, ecosystems, health, and agriculture, among others.

Patricia is co-founder and Chair of the OECD Expert Group on Investing in the SDGs in Cities. Before joining UNDP, she served as Senior Strategic Advisor and Head of Partnerships to the UN Global Compact in New York. Previously, she was Senior Advisor to the UN Under Secretary-General and Executive Director of UN-Habitat, based in their Nairobi headquarters, where she led the Agency's strategic policy and programmatic initiatives, including creation of a global multilateral trust fund for sustainable development in conjunction with the World Bank. Prior to this, she served as Technical and Strategic Adviser to the Special Representative to the UN Secretary-General on Disaster Risk Reduction, based in Geneva.

Before joining the UN, Patricia was the founding director of Commercial Sustainability for the London-based Willis Group, a global insurance broker covering 180 countries. She began her career as a London-based financial journalist with an emphasis on climate change, writing for *The Economist, The Financial Times, The Times,* and *The Guardian.* She holds a Master's degree in Public Policy and Management from the University of York with a focus on the nexus between climate change and inequality, and is currently pursuing a PhD. She is originally from New York City and presently lives in Barcelona.

Amanda Eichel and Kerem Yilmaz

The Vocal City – Recognizing cities and the voice of cities in the 2015 Paris Agreement was the culmination of a nearly 30-year effort from advocates, city networks, and cooperative initiatives. There are, however, limits to what cities and the community that supports them can do alone – the voice of cities must better connect with the capabilities, skills, and learnings from other levels of government, as well as outside perspectives, to deliver action that both is locally appropriate and ensures climate friendly outcomes.

Amanda Eichel was the former Executive Director of the Global Covenant of Mayors for Climate and Energy based in Brussels. Previously, Amanda led efforts to grow the C40 Cities Climate Leadership Group under the chairmanship of Michael Bloomberg, where she built regional and programmatic teams and directed research and knowledge management efforts. Before joining C40, Amanda worked for New York City Mayor Bloomberg's Office of Long-Term Planning and Sustainability and served as Climate Protection Advisor to the Mayor of Seattle, Washington under the administrations of Mayor Greg Nickels and Mayor Mike McGinn. Prior to her work with city governments, Amanda held positions in the California State Assembly Speaker's Office and California State & Consumer Services Agency, where she led efforts to green state building investments, fleet management, and procurement.

Kerem Yilmaz has worked in a variety of capacities, from Fortune 500 companies and global philanthropic organizations, to small businesses, and start-up NGOs. Currently he is President of Sprout Solutions, a boutique consulting firm that specializes in helping organizations conceptualize and deliver on long-term, strategic sustainability initiatives. He also serves as Head of Strategy to The Resilience Shift, shaping the organization's direction to promote greater resilience through influencing policy, driving practice, and sharing key learnings. Previously, he served as Strategy and Operations Director for the Global Covenant of Mayors for Climate & Energy in Brussels, Belgium. Kerem received his Master of Public Policy from the University of Southern California and Bachelor of Arts from the University of California, Berkeley.

Bruce Katz and Luise Noring

The Governed City – Cities can tackle climate change if and only if they have institutions with the capacity, capital, and community standing necessary to get the job done. Capable governance and quality finance are essential, but often overlooked, elements of climate solutions.

Bruce Katz is the founding director of the Nowak Metro Finance Lab at Drexel University in Philadelphia. Since its inception in 2018, the Nowak Lab has strived to help cities and regions design, finance, and deliver transformative initiatives to drive innovative, inclusive, and sustainable growth. Previously, Bruce served for 21 years at the Brookings Institution, including as vice president and founding director of Brooking's Metropolitan Policy Program and as the Institution's inaugural Centennial Scholar. He is a Visiting Professor in Practice at the London School of Economics, and previously served as chief of staff to Henry Cisneros, Secretary of the United States Department of Housing and Urban Development during the first term of the Clinton Administration and staff director of the United States Senate Subcommittee on Housing and Urban Affairs. In 2008/2009, he co-led the Obama Administration's housing and urban transition team. Bruce is co-author of *The Metropolitan Revolution: How Cities and Metros Are Fixing our Broken Politics and Fragile Economies* (Brookings Institution Press, 2013) and *The New Localism: How Cities Can Thrive in the Age of Populism* (Brookings Institution Press, 2018). He is also the editor or co-editor of several books on urban and metropolitan issues, and a frequent media commentator.

Luise Noring is Research Director and Assistant Professor at Copenhagen Business School, where she also attained her PhD in supply chain partnerships. In 2016, she founded City Facilitators, which operates out of Copenhagen with clients across Europe and the US. City Facilitators offers niche consultancy specializing in urban governance and finance. Luise's work is captured in the City Solutions providing vehicles for deepening and accelerating urban problem-solving. The City Solutions offers a source of applied research on the most promising models of urban governance and finance that are emerging to tackle hard economic, social, and environmental challenges and fuel investments and value creation in cities. The City Solutions reveal new institutional models and finance mechanisms covering areas such as urban redevelopment of deindustrialized areas, infrastructure financing, affordable and social housing, devolved municipal power, pooling of municipal borrowing requirements, inclusive growth, climate investments, and pension funds, to name but a few. It aims to speed up the process by which solutions invented in one city are captured and codified and then adapted and adopted to other cities.

Leah Lazer and Nick Godfrey

The Decoupled City – Cities are a critical vehicle for delivering the emissions reductions needed to limit global warming. National governments can drive economic prosperity and address climate emergency by supporting sustainable, equitable cities.

Leah Lazer is passionate about just, sustainable cities. She serves as Research Analyst at the World Resources Institute, where she has authored numerous publications on urban planning, sustainable transportation, climate justice, and the circular economy. With the Coalition for Urban Transitions, she worked as researcher and project manager for a major global initiative supporting national governments to secure economic prosperity and tackle the climate crisis by transforming cities, based on a partnership of more than 35 of the world's leading institutions and companies. Previously, Leah was part of Siemens' Urban Development team in London, and a food justice NGO in Philadelphia. She holds an MSc in Regional and Urban Planning Studies from the London School of Economics and Political Science, and a BA in Food System Studies from Tufts University, Massachusetts.

Nick Godfrey is a Senior Adviser for the Grantham Research Institute at LSE. He was formerly the co-founder and director of the Coalition for Urban Transitions, a special initiative of the New Climate Economy, and a major global initiative supporting national governments to secure economic prosperity and tackle the climate crisis by transforming cities based on a partnership of more than 35 of the world's leading institutions and companies.

The initiative is co-hosted by the World Resources Institute and C40 Climate Leadership Group. Before this, Nick was a member of the Executive Team and Head of Policy and Urban Development for the New Climate Economy, a major international initiative to examine how countries can achieve economic growth while dealing with climate risks led by a Global Commission of 26 former heads of state, finance ministers, CEOs, and thought leaders.

Justin Keeble and Molly Blatchly-Lewis

The Responsible City – A responsible city is guided by a compelling mission and purpose harnessing business to bring environmental and societal value.

Justin Keeble is Managing Director of Accenture's European Sustainability practice. He has spent 22 years working with companies to harness environmental and social pressures as drivers for business transformation, growth, and innovation. He has worked across consumer industries, financial services, high tech, energy and utilities sectors, and the public sector including municipal and city administrations. He recently built an eco-house in south Oxfordshire where he lives with his wife and three daughters and has a penchant for amateur pugilism.

Molly Blatchly-Lewis is a Strategist within Accenture's Global Cities, Transport and Infrastructure Practice. She has worked with a wide range of government and private sector clients in the UK and internationally, specializing in sustainability, urban mobility, and emerging technologies such as 5G and digital twins. Her focus is on the role of systems thinking in tackling urban challenges, developing innovative, practical policies and solutions for sustainable impact. Her experience includes working with the World Economic Forum to develop integrated solutions to decarbonize cities; shaping a clean mobility strategy with Transport for West Midlands; driving innovation across construction firms and infrastructure agencies to accelerate decarbonization and enhance communities; helping to grow an innovation ecosystem in East Asia and co-creating a leadership development framework for a global climate change NGO.

Pete Daw

The Energized City – The energized city thinks of energy as something that can be used more efficiently, optimized in the way we supply it, consume it, and plan for it.

Pete Daw is director of Global Urban Futures, advising public and private organizations on climate change and sustainability. He recently supported the London Waste and Recycling Board in developing the case for their five-year business plan focused on driving down consumption-based emissions and making the London circular economy. He is now on assignment with the Greater London Authority, heading the climate change mitigation and adaptation teams.

At Siemens he was director of Urban Development and Environment at the Global Centre for Cities, where he worked with cities globally to help them understand the role technology can play in tackling their challenges. He worked extensively on Siemens' smart city approach in China, India, Italy, and Saudi Arabia. He also headed the Siemens partnership with C40 Cities, where he produced thoughtful leadership pieces on topics ranging from connected and autonomous vehicles to climate financing. He developed Johannesburg's first-ever greenhouse gas inventory as part of Siemens' work with C40 Cities.

Previously Pete worked in London government for 12 years. He was Policy & Programmes Manager for Climate Change Mitigation & Energy for the Greater London Authority between 2008 and 2013, where he led the development of the city's Climate Change Mitigation & Energy Strategy and its Air Quality Strategy. Prior to that he was Waste Policy manager at the London Development Agency, where he designed the concept and secured £24 million of funding for a waste infrastructure fund.

Julia Thayne DeMordaunt

The Agile City (Part I) – The chapter first takes a look back and then defines a path forward for how cities, people, and technology come together to deliver transportation systems that offer what people need when they need it.

An expert at the leading edge of systemic change for transportation and infrastructure policy, **Julia Thayne DeMordaunt** helps governments, private companies, and NGOs translate ambitious visions into actionable plans that benefit the communities they impact. Julia is Principal of Urban Transformation at the Rocky Mountain Institute. Prior to this Julia developed mobility innovation programmes for the Los Angeles Mayor's Office, she worked as director of Urban Development at Siemens, consulting on initiatives for 35 cities across Asia, Europe, Latin America, and North America. She is also founder and board member of the public–private innovation hub Urban Movement Labs, and an educator at USC Sol Price School of Public Policy. Her work has been featured in publications including *Fast Company, The Washington Post, CityLab, Vox, Bloomberg, Governing Magazine Quartz, Tech Crunch*, and *Curbed*.

Jonathan Laski

The Agile City (Part II) – What is a city without the ability of citizens to move around safely, inexpensively, accessibly, and without fear of sickness from pollution?

Jonathan Laski is a sustainability professional and lawyer based in Toronto, Canada. His professional career began in the corporate/commercial practice group of a large independent law firm in Toronto, following which Jonathan transitioned to a career in sustainability. He has directed innovative city-level research and impact programmes through roles with the C40 Cities Climate Leadership Group, World Green Building Council, and Waterfront Toronto. Highlights include managing the first of C40's peer-to-peer city networks on private sector building energy efficiency in 2012–2013 and launching WorldGBC's Advancing Net-Zero global initiative.

Following postings and education abroad, including time in London, Sydney, and Lund (Sweden), Jonathan is now firmly based in Toronto with his partner and two young daughters. At the time of writing, Jonathan is director of Sustainable Finance Solutions with Sustainalytics, one of the world's leading providers of ESG research and ratings. In this role he leads the delivery of "second-party opinions" for corporate and bank clients in the EMEA and Americas regions, looking to issue green, social, and sustainable debt to finance ESG projects which are aligned with the Paris Agreement and science-based targets initiative.

Olivia Nielsen

The Habitable City (Part I) – The chapter explores how cities can address one of their biggest challenges: housing a growing urban population in an affordable, sustainable, and climate-resilient way.

Olivia Nielsen is an Associate Principal at Miyamoto International, a global structural engineering and disaster-risk reduction firm, where she focuses on making housing affordable, sustainable, and resilient for all. From post-disaster Haiti to Papua New Guinea, she has developed and worked on critical housing programmes in over 35 countries for the World Bank, USAID, and Habitat for Humanity, among others. She has over a decade of experience in housing policy, finance, housing public–private partnerships, post-disaster reconstruction, and green construction. Prior to joining Miyamoto, Olivia was a principal at the Affordable Housing Institute, where she developed housing policy and finance solutions in Haiti, Sub-Saharan Africa, and the South Pacific for the World Bank and USAID. Olivia also managed CEMEX's housing and infrastructure projects in Latin America and the Caribbean, where she focused on leading the cement company's reconstruction efforts after the devastating 2010 earthquake in Haiti.

Olivia is originally from Paris, France, has a Bachelor's degree in Anthropology from McGill University, a Master's in Sustainable Management from the United Nations Mandated University, and an Executive Master's in Management from the London School of Economics. Through her work, she seeks to ensure that all families around the world have access to affordable, sustainable, and resilient homes.

Nicky Gavron and Alex Denvir

The Habitable City (Part II) – We must design and build our future housing in a way that promotes density over sprawl and locks away carbon with greener, cleaner, and more circular methods.

The former Deputy Mayor of London, **Nicky Gavron AM,** has served on the assembly's Housing, Environment, and Planning committees since 2008. She is a member of the London Sustainable Development Commission. An elected politician since 1986, Nicky has been at the forefront of developing integrated land-use, housing, transport, and environmental policy at every level of government. Throughout the 1990s she led the Labour group on the London Planning Advisory Committee (LPAC), becoming the chair in 1994. In this role she commissioned research and formulated strategies to create a more sustainable London, including on congestion charging and affordable housing. In the late 1990s she held positions on national committees and commissions. In 2000, she became London's first statutory Deputy Mayor, working closely with Mayor Livingstone to set up the Greater London Authority's working processes and policy frameworks. She led on the first London Plan, which set out the vision and long-term policies to make London an exemplary sustainable world city.

Leading London's response to climate change, Nicky introduced policies and programmes to reduce CO_2 emissions across energy, water, waste, transport, and sustainable design and construction. Her initiatives include establishing the London Climate Change Agency and C40 Cities. She firmly believes that cities working collaboratively are pivotal in the battle against climate change. Nicky is internationally recognized for her work on urban planning and the environment and has and continues to advise cities and city networks. Her advisory roles have included Chief Project Advisor to the London School of Economics (LSE) Stern Cities Programme on the economics of green cities, a member of the Rotterdam International Advisory Board, and honorary adviser to the Joint US China Collaboration on Clean Energy (JUCCCE). Nicky has many passions including furthering the nature/climate nexus and its relationship to accelerating carbon-free construction – the subject of her chapter.

Alex Denvir is an experienced advisor and researcher who has worked with senior politicians at a national and local level in England and London primarily on housing, planning, and regeneration policy. He began working with developers and communities on large urban regeneration projects in London, before going on to work with a Shadow Minister in the House of Commons, advising on national planning policy and developing party positions. He has most recently worked with members of the London Assembly to shape affordable housing policy in the capital and to steer the cross-party response to the new draft London Plan through its many stages towards adoption.

Conor Riffle

The Resourceful City – This chapter looks at how to move our urban economies to circular economies that reduce reliance on landfill and prioritize conservation of resources. Old models of disposing of resources aren't compatible with Earth's urban future.

Conor Riffle is Senior Vice President of Smart Cities at Rubicon, a global technology company that provides waste and recycling solutions to businesses and government. In this role, he runs the company's software business for municipal governments, RUBICONSmartCityTM. RUBICONSmartCity has been deployed in more than 55 cities, including Atlanta, Baltimore, and Kansas City. In 2020, Conor was named a "40 Under 40" award winner by *Waste360* magazine.

Prior to Rubicon, Conor was based in London and served as the founding Director of Cities and Data Product Innovation at CDP, a global environmental organization. Under Conor's leadership, CDP's cities programme achieved global recognition as the de facto platform for city governments to report environmental data, growing to more than 500 global cities by 2016. More than 800 global cities now use CDP's platform annually. In 2013 and again in 2017, Bloomberg Philanthropies announced major investments in CDP's work with cities. Prior to his role at CDP, Conor served in various roles at the Clinton Foundation in New York. Conor graduated magna cum laude in History from Connecticut College and holds an MA in History of International Relations from the London School of Economics. Follow Conor on Twitter at @c_riffle. Jakob Geiger contributed essential research and support in preparation of this chapter.

Terry Tamminen and Peter Lobin

The Zero Waste City – Imagine a city without waste, where "trash" bins become sources of energy, fuels, and raw materials for products and buildings; and where we adopt exciting new technologies to cut our energy usage and bills in half, making the switch to renewables easier and faster.

From his youth in Australia to career experiences in Europe, Africa, China, and across the US, **Terry Tamminen** has developed expertise in business, farming, education, non-profit, the environment, the arts, and government. Governor Arnold Schwarzenegger appointed him Secretary of the California Environmental Protection Agency and later Cabinet Secretary, the Chief Policy Advisor to the Governor, where Terry was the architect of many ground-breaking sustainability policies, including California's landmark Global Warming Solutions Act of 2006, the Hydrogen Highway Network, and the Million Solar Roofs initiative. In 2010 Terry co-founded the R20 Regions of Climate Action, a new public–private partnership, bringing together subnational governments, businesses, financial markets, NGOs, and academia to implement measurable, large-scale, low-carbon, and climate-resilient economic development projects that can simultaneously solve the climate crisis and build a sustainable global economy.

Terry also provides advice through 7th Generation Advisors to Pegasus Capital Advisors, the Green Climate Fund, and numerous global businesses on sustainability and "green" investing, as well as assisting governments and philanthropists with climate solutions, including Fiji, India, Rockefeller Brothers Fund, and the Leonardo DiCaprio Foundation. An accomplished author, Terry's books include *Cracking the Carbon Code: The Keys to Sustainable Profits in the New Economy"* (Palgrave Macmillan, 2011). In 2011, Terry was one of six finalists for the Zayed Future Energy Prize, and *The Guardian* ranked Terry no. 1 in its "Top 50 People Who Can Save the Planet". For more information see https://en.wikipedia.org/wiki/terry_tamminen.

Peter Lobin is a globally recognized expert in the waste and recycling sector, with a deep knowledge of the efficiencies, technologies, and human behaviour that drive sustainable economic growth. With a 30-year track record serving multiple waste and recycling firms, private equity investors, foundations, NGOs, and advocacy groups, he has developed innovative programmes that reduce cost, expand markets, create new opportunities, and increase revenues throughout North America, the Middle East, North Africa, Sub-Saharan Africa, South America, and Asia. Peter is currently Managing Director of ZeroWaste Global LLC (ZWG), an international management-consulting firm focused on zero waste solutions; a partner at Scarab Technology, LLC, a disabled-veteran-owned waste service and recycling management consultancy focused on federal and state governments; and Managing Partner at Fiber Innovation Technologies, the leader in residual management for pulp and paper mills. He holds a BS and MA in International Relations from the University of Southern California and is proficient in Spanish, having lived in South America.

Sarah Wray and Richard Forster

The Resilient City – A resilient city is one that embraces a holistic strategy that puts resilience at the heart of investment.

Sarah Wray leads the editorial team at *Cities Today* and specializes in writing about the impact of technology on cities, particularly with regard to the use of data, digitalization, and transport innovation, with a focus on climate action, citizen engagement, and the delivery of equitable municipal services. She was previously part of the TM Forum team where she was the editor of *Inform*, a research and content hub for the telecom industry. She was editor of *Smart Cities World* before joining *Cities Today* and has written for publications including *Smart Cities Dive, Mobile World Live, Mobile Europe*, and *Computer Weekly*, covering topics such as the Internet of Things, smart cities, 5 G, and blockchain.

Richard Forster has been an editor and journalist for over 20 years, having trained at Euromoney Institutional Investor PLC. He has written for the *Financial Times*, *Euromoney*, International Financial Law review (IFLR), and Project Finance Institute, and has launched publications for the Inter-American Development Bank, Asian Development Bank, and UN-Habitat. He is Editor-in-Chief at PFD Publications, which launched *Cities Today* in 2010 as the first global magazine for decision-makers in urban development. He has edited publications for UN-Habitat, United Cities and Local Governments Asia-Pacific, and the Latin American Federation of Cities, Municipalities and Associations of Local Governments. He is CEO of the Cities Today Institute, which provides training, forums, and research for a network of city leaders, focusing on digital transformation, transport, and sustainability.

John de Boer

The Fragile City – A fragile city recognizes growing inequalities in the face of extreme events and how it can build resilient systems to function and thrive.

John de Boer is a thought leader who combines experience in business, government, academia, and international organizations to develop solutions to some of the world's most pressing challenges. John is currently Senior Director at BlackBerry, a global leader in intelligent security solutions, where he leads Government Affairs and Public Policy in Canada. Prior to joining BlackBerry, John was Principal at the SecDev Group, a digital risk consulting firm, where he advised large corporations, federal and municipal governments, and organizations including the UN and the World Bank on how to navigate digital risks. John has also served at the United Nations, where, as a Senior Policy Advisor, he helped establish the United Nations University, Centre for Policy Research. His work at the Canadian Government included responsibilities as Team Leader for Governance at the Afghanistan Task Force. He also served as programme leader at Canada's International Development Research Centre, where he spearheaded the institution's work on governance, justice, and security, and directed innovative research programmes on safe and inclusive cities.

John has published extensively on issues related to urban fragility, violence, and resilience. This includes co-editing volumes on *Reducing Urban Violence in the Global South: Towards Safe and Inclusive Cities* (Routledge, 2020) and *Social Theories of Urban Violence in the Global South* (Routledge, 2019), as well as two volumes on security-sector reform and citizen security in Latin America with Ubiquity Press and Siglo XXI. His work has been published in peer-reviewed journals including *Environment and Urbanization* and *Stability: International*

Journal of Security & Development, and by the World Bank, International Federation of the Red Cross, United Nations University, World Economic Forum, Reuters, iPolitics, and the *Guardian*, to name a few. John has taught at, and received fellowships from, Stanford University and the University of California at Berkeley. He holds a PhD from the University of Tokyo.

Seth Schultz and Eric Ast

The Data City – The chapter provides a vision for how cities can leverage the power of procurement to break the cycle of data dependence and lead the fight for a just, equitable, and safe future in a role that they play best: convener.

Seth Schultz is CEO of Resilience Rising, a new global non-profit consortium working together to accelerate a safe, resilient, and sustainable future for all. He has a long track record of building consensus and initiating change in the field of sustainable development, and of raising international awareness on the role of cities in tackling climate change. He is a passionate advocate for a safe, resilient and sustainable future and the need for transformative decarbonisation and long term resilience.

Over the past two decades, Seth has worked with many of the most leading and innovative organisations in this space to turn theory into practice, including the Louis Berger Group, the US Green Building Council, the Clinton Foundation, C40 Cities Climate Leadership Group, the Global Covenant of Mayors, the Intergovernmental Panel on Climate Change (IPCC) and the Resilience Shift.

Seth shares his expertise through involvement with various boards and advisory councils, is a sought-after speaker and guest lecturer, and has authored numerous articles, reports, blogs and thought leadership pieces around the world.

Eric Ast is Chief Data Officer of East Data, where he works with mission-driven organizations to increase the strategic impact of data. He previously led the data and analytics practice at C40 Cities, where he oversaw enterprise intelligence work and co-authored research, including *Climate Action in Megacities* and *Powering Climate Action: Cities as Global Changemakers*. Eric previously served as Managing Energy Analyst for Bright Power, where he advised clients including HUD's Office of Affordable Housing Preservation (OAHP) and owners of affordable multifamily housing on portfolio energy and water efficiency strategies, and at Capital One where he developed credit pricing strategies during the Great

Recession. Eric holds a BS in Systems Engineering and Economics from the University of Virginia and an MS in Sustainable Technology from the KTH Royal Institute of Technology, Stockholm, Sweden. He currently lives in Portland, Oregon.

Patricia McCarney

The Measured City – The chapter advances the need for globally standardized measurement in cities and examines what global standards exist for city data that propel city sound leadership on the global stage and enable local success.

Patricia McCarney is President and CEO of the World Council on City Data (WCCD) and is Professor of Political Science at the University of Toronto, Canada. She has published widely in the fields of city governance, data governance, and the role of global cities in sustainable development planning.

Patricia received her PhD from MIT in 1987. Before joining the University of Toronto, between 1983 and 1994, she worked as a professional staff member in a number of international agencies, including the World Bank in Washington and UN-Habitat in Nairobi. She is Convenor of the Working Group on City Indicators in the ISO Technical Committee 268 and was integral to the development of the ISO 37120 Series, including ISO 37120, the first International Standard on Indicators for Sustainable Cities; ISO 37122, Indicators for Smart Cities; and ISO 37123, Indicators for Resilient Cities.

Having founded the WCCD in 2014, Patricia is building a globally standardized data platform for cities worldwide, where cities report data in conformity with the ISO 37120 Series for WCCD ISO Certification. As host of this knowledge platform, the WCCD is the leading global city database with ISO-certified and globally comparable city data for a growing network of smart, resilient, and prosperous cities.

Noorie Rajvanshi

The Smart City – The chapter explores how technologies can drive climate action in cities based on learnings from over 40 cities worldwide.

Noorie Rajvanshi is the Director of Sustainability and Climate Strategy for Siemens USA with more than a decade of experience in the field of environmental sustainability, energy, and urban development. In her current role, Noorie is responsible for supporting the strategy and e xecution of the Siemens US region's decarbonization plan to achieve net-zero operations by 2030 and works across the Siemens ecosystem of business and corporate units to develop and execute strategies based on data-driven insights. Noorie's previous work in urban development focused on evaluating environmental and economic impacts of growing cities and collaborating with more than 15 cities across North America to identify technology and infrastructure solutions that would enable cities reach their economic and environmental targets.

Noorie served as a Research Fellow for Project Drawdown where she provided technical analysis that served as the foundation for three chapters in The New York Times bestselling book "Drawdown: The most comprehensive plan ever proposed to reverse global warming."

Noorie graduated from the University of Florida with a PhD in Mechanical Engineering and a minor in Environmental Engineering. She is an active member of several organizations including the Corporate Eco Forum (CEF) where she has been inducted into the CEFNext **Community.**

Hayley Moller

The Just City (Part I) – The chapter is an investigation into the environmental, human, and economic costs of urban air pollution, and what we can do about it.

Hayley Moller is a communicator, strategist and entrepreneur with more than a decade of experience tackling the complex issues of climate change, clean energy, smart cities, and a just transition for all. She has crafted sustainability strategy for organizations of all sizes, from lean start-ups to the United Nations to some of the world's most well-known brands. A veteran advisor to the C40 Cities Climate Leadership Group, the Coalition for Urban Transitions, and the Global Covenant of Mayors for Climate & Energy, Hayley has delivered climate action campaigns on all seven continents.

An advocate for women and underrepresented groups, Hayley is passionate about making inclusivity the norm. She served several years on the board for the Women's Energy Network DC Chapter and co-founded a local reproductive rights advocacy group in Washington DC. Early in her career, Hayley researched global environmental issues at the Earth Policy Institute.

Hayley holds an MBA with distinction from INSEAD, where she led the school's Environment & Business Club, and graduated summa cum laude with a Bachelor's Degree in Environmental Science from UCLA. Always a Californian at heart, she currently lives in Paris.

Jane Burston and Matt Whitney

The Just City (Part II) – Air pollution is a hidden killer, causing millions of deaths each year and affecting almost every city on Earth. Creating a compelling case for change starts with data. This chapter reveals how emerging technologies promise a revolution in how we understand – and ultimately solve – the air pollution problem.

Jane Burston runs the Clean Air Fund, a global philanthropic initiative that supports organizations around the world working to combat outdoor air pollution, improve human health, accelerate decarbonization, and address climate change. The organization finds, funds, and scales projects that provide clean air for all. It shares expertise, data, and best practice from across sectors and geographies to ensure that clean air can become a reality for everyone. Previously Jane worked as Head of Climate and Energy Science in the UK Government, responsible for the UK greenhouse gas inventory and a £45 million science programme. As Head of Energy and Environment at the National Physical Laboratory she managed a team of 150 scientists working in air quality, greenhouse gas measurement, and renewable energy. She has been named as a "Young Global Leader" of the World Economic Forum and as one of the "40 under 40 European Young Leaders" by Friends of Europe, and is a previous UK Social Entrepreneur of the year.

Matt Whitney is Portfolio Manager at the Clean Air Fund. Matt leads the strategy and programme development for the Fund's work on air quality data. Previously he was at the UK National Physical Laboratory, working with 150 scientists to increase the impact of its environmental science programme, which included air quality, greenhouse gas measurement, and renewable energy. Matt holds a Master's degree in Environmental Dynamics and Climate Change and an undergraduate degree in Physical Geography.

Jenny Bates

The Just City (Part III) – Will air pollution on a death certificate for the first time lead to a better London? London's air pollution problem has now risen up the agenda, and the solutions are clear – they just need implementing.

Jenny Bates is a campaigner with Friends of the Earth England, Wales and Northern Ireland, covering transport climate emissions and related air pollution issues. Jenny began her campaigning as co-ordinator of Greenwich Friends of the Earth voluntary group in East London while a professional photographer and photographic artist, and became a Friends of the Earth staff member in 2003, covering campaigning in London. Fighting proposed new road river crossings in East

London, Jenny realised how bad air pollution was in the capital, and this issue became a key area of concern and activity. Jenny got Friends of the Earth signed up as a founding member of the Heathy Air Campaign in 2011, which had started after the UK missed its legal targets for nitrogen dioxide in 2010. She liaised with Friends of the Earth's regional staff and local groups around the country on air pollution while working to persuade Friends of the Earth to run a major campaign on it. The Clean Air campaign started in 2016 and the issue is now incorporated within wider climate campaigning.

Colin le Duc

The Invested City – The invested city is the incubator for innovative and sustainable models to be tested and perfected into mainstream solutions.

Colin le Duc is a Founding Partner of Generation Investment Management. Generation is a dedicated sustainable investing firm with a mission to demonstrate the investment case for sustainability and advocate for sustainable capitalism. He is also a founding member of the Growth Equity strategy and a member of the Firm's Management & Investment Committees. He is head of the firm's San Francisco office. Prior to joining Generation, Colin worked for Sustainable Asset Management in Zurich, Arthur D. Little in London, and Total in Paris. Colin sits on the boards of the NGOs NatureBridge and Ocean Conservancy. He lives in California with his family and has lived in nine countries across the world throughout his life.

James Close

The Financed City – Redirecting and scaling up investment to make cities fit for the future is a global priority. Mobilizing finance for low carbon, economically, environmentally, and socially viable cities will create better cities, reduce risk, and support livelihoods.

James Close has spent his career working at the interface of the public and private sectors, with a focus on mobilizing finance. He is now committed to working on some of the most challenging global issues focusing on sustainable development and climate change. He is currently the head of climate change at NatWest Group, supporting the implementation of the bank's purpose-led strategy, which includes the role of principal sponsor of COP26. He is also a member of the Energy and Natural Resources Advisory Group for the ICAEW and trustee for the Trust for Sustainable Living. Prior to this role, he was head of the circular economy programme for London, building an ecosystem for financing entrepreneurship and infrastructure.

Before returning to the UK, James spent five years in Washington, DC as Director for Climate Change at the World Bank. Prior to joining the World Bank, he was a partner in the Corporate Finance Practice of the accounting firm EY, and also worked for HM Treasury. James has a degree in Chemistry from the University of Durham and is a member of the Institute of Chartered Accountants of England and Wales.

Adam Freed

The Adapted City – The chapter offers a global tour of the devastating impacts climate change is having on cities today, how poor urban planning and design has increasing climate risks, the actions cities are taking to adapt, and the changes needed to accelerate this work.

Adam Freed has more than 20 years of experience working on local and global urban issues. Adam leads the Sustainability Practice at Bloomberg Associates, a non-profit consultancy, where he works with cities around the world to craft and implement sustainability strategies covering a wide range of issues, including energy, greenhouse gas (GHG) mitigation, climate resilience, housing affordability, green infrastructure, air quality, solid waste management, and neighbourhood revitalization.

Prior to joining Bloomberg Associates, Adam was the Deputy Managing Director of the Nature Conservancy's Global Water Programme, where he worked with cities in 33 countries to have safe, sustainable, and reliable water supplies and developed innovative financing strategies for natural infrastructure solutions. From 2008 to 2012, he served as Deputy and Acting Director of the New York City (NYC) Mayor's Office of Long-Term Planning and Sustainability, overseeing the implementation of PlaNYC and related sustainability initiatives and developing the city's first climate resilience programme. As part of PlaNYC, the city planted 1 million trees, created more than 240 new community playgrounds, enacted the nation's most aggressive green buildings legislation, achieved the cleanest air quality in over 50 years, launched a US$2 billion green infrastructure programme, and lowered its GHG emissions by 12%.

Adam is also a Lecturer at Columbia University and a member of the NYC Water Board, which is responsible for setting NYC's water rates to fund the city's water and sewer system's operating and capital needs (approximately US$3.8 billion annually). In addition, he serves on the Board of ioby, a national crowd-resourcing platform to support community-led improvement projects; is an External Advisor to Fannie Mae's Sustainable Communities Initiative; and is part of the Investor Advisory Group for the UN's Joint SDG Fund. He received his Master's in Urban Planning from New York University and was a Mel King Community Fellow at MIT.

Peter Bishop

The Open City – The open city celebrates its public spaces – its parks, squares, and streets. This is where the health of the city is judged. It needs to be protected, managed, and cared for. It is an essential ingredient of the richness and messiness of the twenty-first-century city.

Peter Bishop is a Professor of Urban Design at the Bartlett School of Architecture, University College London, and a partner of Bishop & Williams consultants. For 25 years he was a planning director at four different central London boroughs, and has worked on major projects at large and complex sites in the UK, including Canary Wharf, the BBC, and King's Cross.

In 2006 he was appointed as the first Director of Design for London, the Mayor's architecture and design studio, and in 2008 served as Deputy Chief Executive at the London Development Agency. In 2011 he carried out a policy review on behalf of the government, "the Bishop Review", on ways in which the quality of design in the built environment might be improved. From 2011 to 2018, he was director at the architecture firm Allies and Morrison. Recent projects include master planning frameworks for Old Oak Common (High Speed 2 interchange), the Palace of Westminster, and Ansan City Centre (Korea). In 2018 he was commissioned by the Government Architect of New South Wales to carry out a comprehensive review of its policies and programmes.

Peter lectures and teaches extensively, and has been a design advisor to the mayors of London, Bucharest, Ansan, and Zhuhai. He was on the jury for the Sochi Winter Olympics Legacy, Jabal Omar development in Mecca, and central Dallas regeneration project. He is an honorary fellow of University College London and honorary fellow of the RIBA, holds an honorary doctorate from the University of Kingston, and in 2017 was Distinguished Visiting Scholar at UTS Sydney. He is currently leading a significant research project on the ways to foster strong communities in housing regeneration.

His book *The Temporary City* (Routledge, 2012) explores the origins of current thinking on temporary urbanism. He also examined the political processes behind major developments in *Planning, Politics and City Making – A Case Study of King's Cross* (RIBA Publishing, 2016).

Carlo Laurenzi

The Natural City – Imagine for a moment that cities were havens for wildlife, the interaction with humans was entirely positive, and the dynamics between the two were mutually beneficial. It is not a big step to make, but a brave one.

Carlo Laurenzi has 34 years' experience of the UK voluntary sector, twice as a CEO. He now works as a consultant, mostly in the not-for-profit sector, and for the past several years has been an adviser to DEFRA on civil society matters. One of the founding trustees of Rewilding Britain and former CEO of the Wildlife Trust in London, he helped win the contract to create Europe's largest urban wetland scheme in 2015. He also volunteers in a small park near his home in north London.

Carlo has had short spells in government, as a Whitehall secondee and working for a local authority in London. He has sat on over 30 boards, quangos, and steering groups. A tutor and examiner for the Open University, he has co-authored one of the course books, *The Manufacture of Disadvantage* (Open University, 1990). He has edited two further short books on mental health. In 2000 he was awarded an OBE, and in the same year was runner-up in the Charity Times Awards for Director of the Year. He founded Hostage UK (now Hostage International) with Terry Waite in 2002. Currently he is its vice-president. He was awarded a Churchill scholarship for a six-nation study in northern Europe in 1989. He has enjoyed a lifetime of walking and cycling around this beautiful planet.

Mauricio Rodas

The Climate Resilient City – Cities are key to addressing the world's biggest challenges, but they lack proper access to finance. Reforms to the international financial system are needed, and the COVID-19 stimulus may yield that opportunity.

Mauricio Rodas is a Juris Doctor from Universidad Católica (Quito) and holds two Master's degrees in Government Administration and Political Science from the University of Pennsylvania (UPenn). He lived in Mexico, where he worked for the UN's ECLAC and as a policy consultant for the Mexican government. Later he founded and served as Executive Director of Ethos Public Policy Lab. In 2011, Mauricio returned to Ecuador and founded the SUMA political party. In 2013, he ran for President of Ecuador; the following year he was elected Mayor of Quito (2014–2019). During his term, he was the hosting mayor of the UN's Habitat III Conference. He played a leadership role in city networks: two terms as world Co-President of UCLG, and as a member of the boards of C40, ICLEI, and the Global Covenant of Mayors. He was a Young Global Leader and member of the Global Future Council on Cities of the World Economic Forum. In 2019, he was named one of the 100 World's Most Influential People on Climate Action by Apolitical; he also received UPenn's World Urban Leadership Award.

He is a Visiting Scholar at UPenn, working on the Cities Climate-Resilient Infrastructure Financing Initiative. He is also a Distinguished Fellow on Global Cities at the Chicago Council on Global Affairs and a Senior Fellow of the Adrienne Arsht-Rockefeller Foundation Resilience Center at the Atlantic Council.

Sophie Hæstorp Andersen

The Green City – The chapter describes Copenhagen's journey to a green and sustainable city with cleaner air, less noise, smart buildings, and green transport. Working with stakeholders from idea to implementation ensures a fair and feasible transition.

Sophie Hæstorp Andersen has been the Lord Mayor of the City of Copenhagen since January 2022. She has the overall responsibility of the city's plan to half CO_2 emissions from citizens' consumption towards 2035, half CO_2 emissions from the municipality's own procurement in 2030 and be the first carbon neutral capital after 2025 and climate positive with phasing out biomass before 2035. She is currently working on identifying the next steps of action to ensure the city reaches its goals.

By virtue of her role as Lord Mayor, Sophie Hæstorp Andersen is the chairwoman of Greater Copenhagen in 2022. Greater Copenhagen is a collaborative organisation promoting sustainable growth and development in the largest Nordic metropolitan area with 4,4 million citizens across 85 municipalities and 4 regions in Southern Sweden and Eastern Denmark. During her chairwomanship she aims to drive forward green growth, development, and collaboration across the entire region.

Before her time as Lord Mayor, she was chairwoman of the Region Council in the Capital Region in Denmark from 2014–2021 and member of the Parliament for the Danish Social Democratic Party from 2001–2005 and 2007–2014. Sophie Hæstorp Andersen holds a master's degree in Political Science from University of Copenhagen.

Mark Watts and Sarah Lewis

The Powerful City – The chapter takes a look at how entrepreneurial big-city mayors drive progress on climate change through bold and innovative action. Cities are where the future happens first.

Mark Watts is Executive Director of C40 Cities, a network of the mayors of the world's 100 most powerful cities. C40's 250 + international staff support mayors to deliver the most ambitious science-based climate action, focused on halving global emissions within a decade while reducing poverty and inequality. Prior to joining C40 in 2013, Mark was Director at pioneering engineering and design firm Arup, and before that was a senior adviser to the Mayor of London, in which role the *London Evening Standard* described him as "the intellectual force behind Ken Livingstone's drive to make London a leading light of the battle against global warming". In addition to being a climate activist, Mark's other passions are music, mountain running, narrowboats, and exploring places you can get to by bicycle.

Sarah Lewis is a Research Assistant with C40 Cities. Prior to this, Sarah worked as a medical secretary and hospital care assistant. In her spare time, Sarah blogs about sustainable lifestyles and has been interviewed for the BBC, *The Guardian*, *The Times*, and ABC's Future Tense Podcast on circular economy and low-consumption lifestyles. She was also a theatre critic for the *Hackney Citizen* newspaper for a number of years. Sarah holds an MSc in Sustainable Resource Economics, Transitions and Policy from University College London. Her dissertation focused on municipal energy companies in the UK. She also holds a BA in Middle Eastern Languages from the University of Manchester, and has lived in Alexandria, Jerusalem, Tel Aviv, the West Bank, and Amsterdam, as well as London. Sarah enjoys cycling, yoga, hiking, and beaches.

Introduction

Martin Powell

If you walk along London's embankment, you can see a gleaming river, lots of trees lining the streets, clean streets, some expansive green spaces and open areas, and Peregrine falcons nesting and flying between the buildings, and you breathe clean air.

Actually, none of that is entirely true. The river is regularly filled with sewage after heavy rain when the Victorian sewers merge storm water with sewage and the overflow goes into the Thames. This is being resolved with a super sewer being bored under the river as you read this. There are not enough street trees, but they are planted and added every year across the city. The green spaces are generally eroded over time. The wildlife is there to be seen, including the Peregrine falcons, but it is not exactly teeming with life, and the air regularly exceeds the World Health Organization (WHO) limits, as do all big cities, with some streets in constant breach of air quality limits.

It's not good enough, but the reason London is considered a successful city is that it's better than most. It's moving faster than most to rectify these problems, which are constantly exacerbated by a rising population, a rising demand for goods, changing demographics, changing habits, and, above all, the need to mitigate and adapt to the existential threat of climate change.

An Expanding Problem

A staggering two billion people live in extreme poverty and 789 million people do not have access to electricity.[1] Many of these people will seek a new life in the city, they will seek employment, and they will seek a home. They will join the rest of the population as consumers of energy and water, and makers of waste, and demand will go up. We have no right to deny anyone this opportunity, but it makes the challenge of reducing CO_2 emissions even greater.

If you look back in history, all of the great cities have progressed through economic focus and trade and always to the detriment of the environment. This has, in nearly all cases, been a problem that has reached a critical level, public outcry has ensued, and the

---

1 IEA, World Energy Outlook – 2019, based on WHO Household energy database and IEA World Energy Balances 2019.

The Climate City, First Edition. Edited by Martin Powell.
© 2022 John Wiley & Sons Ltd. Published 2022 by John Wiley & Sons Ltd.

error rectified. If I were a true cynic, I would say this is how we solve all of our problems. This constant iteration of growth and advancement followed by a "fix" is what has enabled this walk along the London embankment to be all the more remarkable. Cities through time have developed governance structures that are so in tune with the life of the city that the problems can be identified, captured, analysed, debated, and "fixed" without upsetting all of the other critical elements that make city life so enchanting.

The myriad of governing entities developed over time are able to respond and enact new policy, with the benefits to the wider city being the overriding decision. Perhaps this is what was so troubling about the rapid rise of ride-sharing. They just came. They used public advocacy for cheaper and more convenient travel but at a cost to the overall balance of city life. I don't blame the ride-sharing companies who see this service as invaluable to the citizen, and there is clearly demand for their service, but they can't possibly understand or calculate the overall impact to the whole city population.

In his book *If Mayors Ruled the World: Dysfunctional Nations, Rising Cities*,[2] Benjamin Barber makes the case that modern cities are best poised to meet the challenges posed by the global economy. Ben's book reflects on urban manifestoes with the theme that local governments are uniquely positioned to save the planet. Bruce Katz and Jennifer Bradley made this case in *The Metropolitan Revolution*.[3] Ben advocated for a parliament of cities, which would ratify a shift in power and political reality that, he argues, has already taken place.

"Because [cities] are inclined naturally to collaboration and interdependence, cities harbor hope," Barber writes. "If mayors ruled the world," he says, "the more than 3.5 billion people who are urban dwellers and the many more in the exurban neighborhoods beyond could participate locally and cooperate globally at the same time – a miracle of civic 'glocality' promising pragmatism instead of politics, innovation rather than ideology and solutions in place of sovereignty."

Barber claims that national sovereignty is a handicap. Cities, by contrast, can capitalise on diversity, share intelligence on security and the environment, and face up to inequality in housing, jobs, transportation, and education.

"My proposal for a parliament of mayors is no grandiose scheme," Ben writes, "no mandate for top-down suzerainty by omnipotent megacities exercising executive authority over a supine world. It is rather a brief for cities to lend impetus to informal practices they already have in place." And he notes that, "in changing the subject to cities, we allow imagination to cut through the historical and cultural impediments to interdependent thinking in the same way a maverick Broadway cuts through Manhattan's traditional grid".

So if cities are so effective at rectifying problems, then why did the air get so bad? The water so polluted? The traffic so congested? The answer to this is simple. Firstly, nobody fully assessed the external factors impacting these basic elements of air, water, and nature; secondly, nobody assessed the impact on our health and wellbeing; and thirdly, our guesses of future growth and future expectation have been underestimated.

---

2 Barber, B.R., 2014. *If Mayors Ruled the World: Dysfunctional Nations, Rising Cities*. Yale University Press, New Haven, CT.

3 Katz, B., Bradley, J., 2014. *The Metropolitan Revolution*. Brookings Institute, Washington, DC.

My journey since advising the Mayor of London on the environment has enabled me to look at technology and how it can tackle climate change and how it can meet city targets for climate change mitigation. This has enabled me to visit some amazing cities, such as Ephesus in Turkey; Madain Saleh in Saudi Arabia, which is the first settlement after Petra; Matera in Italy (Figure I.1), where you can stay in a hotel room where a cave-dwelling troglodyte once slept; and cities like Venice that help challenge the conventional wisdom of how cities can thrive. These trips have given me insight into why people live in cities and what is important to them, something that every mayor understands all too well.

Madain Saleh, for example, now a UNESCO World Heritage site, can tell this story. While not as well known as Petra, the Nabatean's second-largest city was once a thriving metropolis along the ancient spice route and played a crucial role in building a trade empire. Today, its monumental stone-hewn tombs are some of the last, and best-preserved, remains of a lost kingdom. The Nabateans achieved wealth and prosperity from their ability to source and store water in harsh desert environments. They also held a monopoly on desert trade routes as far southwest as Madain Saleh and north to the Mediterranean port of Gaza. They extracted taxes from camel caravans – laden with frankincense, myrrh, and spices – that stopped at their garrisoned outposts for water and rest. The tomb inscriptions, written in Aramaic, give insight into names, relationships, occupations, laws, and the Gods of the people who lie there. Walking around Madain Saleh (Figure I.2) you can understand how people came together in cities to live in proximity as a community, to do business together, to be safe, to honour those that pass, and to live by a set of rules that make sense to that community.

Figure I.1 Matera, Italy. Many consider this to be the oldest inhabited city on Earth. (*Source:* Siempreverde22/Getty Images.)

Figure I.2 Left: Madain Saleh, Saudi Arabia. Right: The Editor in Madain Saleh. (*Source of left photograph*: amheruko/Adobe Stock.)

In today's cities nothing has changed, but these basic requirements can be overshadowed by events, obscured by temporary nuances of daily life, which upset the fine balances that keep this ecosystem functioning. We see parts of the city that have been segregated, vulnerable subcommunities that have been distanced, cut off from the collective attributes of the city and so no longer receiving the benefits of city living.

Cities like London, New York, Copenhagen, and Singapore are far from perfect, but they have ensured a governance setup that responds to the needs of their city, the principles and policies of urban living. The rationale for investment is anchored to these principles that ensured a thriving Mesopotamia, Ancient Rome, and Madain Saleh.

The climate crisis will not change these basic requirements. We don't have to give things up. We will innovate without compromise, but the longer we leave it the more difficult the challenge becomes.

I learned something very powerful in my time in public office – "good ideas are hard to kill". If we can align the interests of the wide set of stakeholders that need to come together to enact change, with programmes that strike at the core of the problem and offer co-benefits that improve people's lives, then they will be successful. We can deliver on serious commitments in the timescales required. The mayor's office is designed to speak and transact with its citizens. People are generally motivated to seek information that validates their world view, and so the city must inform citizens of their place in a status quo in the most highly diverse communities – and in some places highly segregated, with the greatest inequality within the same spaces – that exist anywhere on Earth and in a world where we stream information and news that polarizes our views by the choices we make. If cities can overcome this to keep this delicate ecosystem in balance through communication that helps people have understanding of "why" they must compromise in order to co-exist, then they can communicate the need to tackle climate change even if it means citizens will have to compromise more than they ever had in history.

Acceptance and Reframing

When I was at school, I learned that "air" is 78% nitrogen and 21% oxygen, and the balance of 1% comprises noble gases and 0.03% CO_2. My youngest daughter recently asked me if I knew this and I proudly recalled these percentages. She told me I was close – but CO_2 was 0.04%. In the 1980s, CO_2 molecules were 335 ppm; today that figure is 412 ppm. To me, 0.03% could have been a fundamental constant, and while I recognize that our natural cycles can fluctuate, this is simply climbing.

When I first read about orbital mechanics, it made sense to me why we can expect an ice age every 10,000 years or so. Every book I have read since validates why we have climate change and a balance of global warming, and how this will lead to short-term climate variability that will target vulnerable coastal cities – where 75% of the world's cities reside. I have been immersed in this topic and no longer need to question the reality of the situation. Well, I will draw a line under this whole discussion, because whether you don't believe in climate change or you do believe in climate change, this book is for you! Renewable energy is clean, nobody is breathing coal dust into their lungs, electric vehicles are clean, and nobody is putting carcinogenic particulates into their lungs. The ideas we propose here make sense in an awakened world that wants good short-term living and a future for our children. The decoupling of economic progress and environmental disbenefit is being driven in cities with ideas and innovations that benefit society and mitigate the carbon emissions we can no longer live with. Cities have learned to innovate when we need to and to solve the problems in front of us without the need to compromise and without telling people how to live their lives.

I sat in the Andlinger Centre for Energy and the Environment a couple of years ago waiting to make my presentation and Professor Alain Kornhauser started to talk about how we count passengers on a journey. I was stunned by how he managed to bring to focus a way of looking at some of the biggest problems in urban society in a new way that will ensure we start making better decisions about urban mobility.[4] I shall not attempt to mimic his storytelling, but I will give you my own version.

"There are an average of 4 spaces available in the car for passengers, I drop my daughter off in town to meet her friends, leaving 2 vacant spaces on the outbound journey. When she is finished, I come and run her home, leaving 2 vacant spaces on the inbound journey. This would ordinarily equate to 50% occupancy but I have not yet counted my journey back from dropping my daughter off which is 3 vacant spaces and my journey back to collect her which is another 3 vacant spaces. These two trips equate to 25% occupancy. So overall occupancy is 37.5%. Now, my daughter's journey has a purpose – but I am just a driver, I have no purpose in this scenario! I am not participating in my daughter's evening entertainment with her friends. So, of the 16 spaces available in this car, the number of purpose-filled seats is just 2. Occupancy is 12.5%."

If we begin to consider how people move, where they go, and why they are going, we can provide a transportation network that more appropriately meets our purpose to travel.

---

4 https://www.princeton.edu/news/2019/03/06/autonomous-vehicles-could-be-environmental-boon-or-disaster-depending-public-policy.

The late David McKay's wonderful book *Without the Hot Air* gave me such inspiration in how to look at a problem from a perspective of factual numbers before looking at practical implementation. He made a lovely statement, "if everybody did a little, we would achieve ... just a little!" We often hear, if everyone picked up just one piece of litter, well we would still be creating litter faster than we are getting rid of it. If everyone planted a tree? Well, let's explore that one ....[5]

Thirty percent of the planet is covered in trees, and since human civilisation began we have cut down half of those trees. Now, if 7.7 billion people planted a tree, noting there are three trillion trees on the planet, we would add just one-quarter of 1% to the total. Now if we planted 1.2 trillion trees, we would cancel out 10 years of CO_2 emissions by the time the forest matured, but now each of us needs to plant 160 trees ... all said, this would capture 100 $GtCO_2$ on top of the 400 Gt captured by existing trees.

So, let's look at the problem from a different perspective: 15% of all greenhouse gas emissions are a result of deforestation, because, of course, they emit carbon instead of absorbing it, and we cut down 15 billion trees per year, that's 500 per second. My conclusion is of course that we should all plant a tree! But this is not the problem to solve. We should find mechanisms that encourage nations to profit from protecting their forests instead of cutting them down. You will read more about this in Chapter 1.

In the same way, if we come up with ideological solutions based on technological logic, we might try to build giant solar parks across the royal parks to power London, but these would never get approved ... so this is less effective than picking up a piece of litter or planting a tree.

We must understand scale, practicality, and cause and effect while forming ideas, even if this means driving policy change or changes to laws and regulations, or we risk trying to solve unsolvable problems because we are constrained by things we consider immovable. Cities have a lot of tame problems to solve, but it is the wicked problems[6] that we have to approach differently, and it is the wicked problems that define a mayoral term of success or failure. A wicked problem is almost unsolvable from the point of looking at it; it is almost too hard to grasp exactly what is being solved and how to solve it, with, in some cases, too many possible approaches, any of which could be right or wrong. A wicked problem is solved by taking a path you think will work but accepting it may not. This allows time and space to learn and progress. Inevitably we have to change direction or emphasis, even start on a new path that was not even available or visible when we began.

How do you solve traffic congestion in New York City? It is only through tackling this as a wicked problem and applying system thinking across city infrastructures, understanding societal needs, and achieving cultural alignment that you realize that the answer lies, at least partly, in solving homelessness. There are a significant number of people who avoid the New York Subway because of perceived dangers and high numbers of homeless

---

5 BBC ideas; if everyone in the world planted a tree. https://www.bbc.co.uk/ideas/videos/what-if-everyone-in-the-world-planted-a-tree/p084ttpq

6 Rittel, H.W.J., Webber, M.M., 1973. Dilemmas in a general theory of planning. *Policy Sciences* 4, 155–169.

Figure I.3 Homeless people in the Subway, New York City. (*Source*: Eric Kitayama/Getty Images.)

people using the train for shelter (Figure I.3). These people take a taxi or ride-share and continue to take this surface transport. High-occupancy metros/subways/underground networks around the world do not house the homeless. In fact, some have used increased revenues to support homeless shelters and food kitchens and even programmes to support the homeless with finding a job. In the case of New York, I have simplified the situation, but reducing congestion through congestion pricing or by providing more roadways would only solve part of the problem and only for a limited time.

In June 1991, Mount Pinatubo, in the Philippines, erupted, spewing vast amounts of white ash and sulphates as high as the stratosphere. Around 15–17 million tonnes of this volcanic material spread into a lazy haze covering much of the globe. During the following months, scientists discovered a second surprise: this particle cloud had formed a protective sun-shield, reflecting a significant proportion of the sun's rays back into space. As a result, the average global temperature that year dropped by 0.6°C. And for some researchers that raised an interesting possibility. Could we do this on purpose? Deliberately producing artificially engineered reflective clouds to reduce global warming? There have been many suggestions for artificial reflective surfaces: launching mirrors into space to orbit around the Earth; building wind-powered ice machines over the Arctic, or scattering it with trillions of silica beads; other geo-engineering solutions such as using phyto-plankton to sequester CO_2 and store it deep in the oceans.

I do feel that even if these solutions worked, and they are not strewn with unintended consequences that saddle us with a new set of disastrous outcomes to get our heads around, we would miss the opportunity that this book sets out to achieve. If we can create city habitats that fully net-off the outputs from their growth, that produce no waste that cannot be returned to the system, that consume energy from renewable resources, that minimize our energy use and make our lives more affordable, then we will have solved the climate crisis but also delivered co-benefits that enhance our health and wellbeing, that bring more people out of poverty and improve society as a whole.

The Logic of the Book

Everybody is an environmentalist, be it bringing a plant back from the brink, recycling responsibly, or volunteering to clean up a beach. The authors of this book have discovered the environmentalist inside them that they have put to good use, and each one has made their own tremendous impact in their field.

The chapter authors of this book have been appointed advisors to Lord Stern, Boris Johnson, Ken Livingstone, Al Gore, Arnold Schwarzenegger, Leonardo Di Caprio, Ban Ki Moon, Michael Bloomberg, Richard Branson, and Bill Clinton. The authors are or have been mayors, deputy mayors, mayoral advisors, leaders of institutions critical to cities, academics or professionals who have led significant bodies of work related to city progress, or communication experts observing and reporting on leading practice. The chapters present some leading practices that can be emulated and financed in another city. The book will inform national governments who can promote, regulate, scale, and signal to the markets where the capital should flow. Furthermore, it will inform businesses looking to invest in our urban and climate future.

The book identifies components of the city in 30 chapters, each with an expert author and in some case co-author. The sum of those parts make up the complete ecosystem of *The Climate City*. It does not address music, art, theatre, or wider cultural elements, but it does not ignore their role or importance. It builds the story of how cities are using innovative techniques to advance progress, which makes them the ideal leader in tackling climate change.

Each chapter addresses key ideas and how they can be applied in both the developed and developing world and formal and informal settlements, how practices can be replicated and scaled, and what impact this would have on climate targets. The authors provide the reasons behind why cities are leading and describe "systems thinking" as a means of creating true impact and real progress as our cities continue to grow.

The book highlights the steps that cities are taking today and follows the trends we see today, and shows how this translates into the reality of tomorrow. It identifies the contribution cities will make to combating climate change and their role in shaping policy at every level of government. It offers a fact-based economic, technological, and policy-driven journey that every individual can relate to in modern society.

Society has a role to play in both our professional and personal activities, and this book aims to point out the near urban future that is nearly upon us and highlight the opportunities to engage and develop the ideas for keeping us in work, keeping us moving, keeping us warm, and keeping us safe, and ensuring we progress as a society. It will break down the challenge into "numbers" and "stories", giving us the input we need to drive the right technologies into our urban centres and promoting a progressive society for a better future.

The need to decouple growth from the environmental dis-benefits it brings in cities is wrapped up in our need to solve other challenges such as poor air quality and the inefficient use of resources. Cities will endeavour to show the world how to reach the future needed to avoid the impacts of climate change we are currently hurtling towards. The book outlines the fundamental journey we need to embark on to maximize the contribution of cities to the climate crisis and provide insight into how these practices can be adopted by national governments to promote the economic, social, green, and just outcomes.

The scale of this problem is increasing. In Figure I.4, we can see the number of people and the number of cities on Earth that will be at risk by 2050 due to extreme heat, extreme heat and poverty, water availability, food security, sea level rise, and sea level rise plus power plant. Our growth needs to be governed by new rules that can mitigate against these risks.

The risks and complexities as we transition into this future city will be outlined. The radical shifts in mindset and urban policy to make this a reality will be discussed. How we solve the climate crisis will inevitably lie in solving these problems at a scale and at a pace that is more aggressive than the national governments the cities reside within. This book offers data to show our future path and our role in achieving the truly progressive and liveable city. In the end, we can all play a more valuable role today if we can get a glimpse of the urban world we want to inhabit tomorrow and what it will take to achieve it.

This is a book about meaningful action at a local level. A city or municipality can do more than a little; it can show a pathway for other cities and their national governments to follow and turn something more than a little into something significant.

The image of Fleet Street in London transformed into a place for people (Figure I.5) is actually a depiction of what is required to cut our carbon emissions and adapt the city to

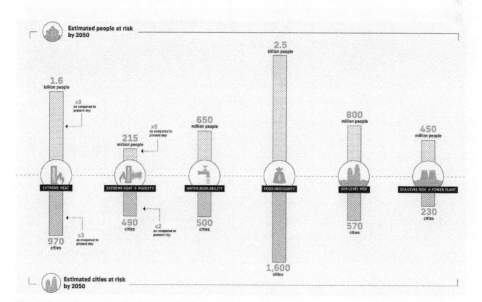

Figure I.4 Summary of critical risks that cities and their residents will encounter as a result of climate change. (*Source*: https://www.arup.com/-/media/arup/files/publications/c/arup-c40-the-future-of-urban-consumption-in-a-1-5c-world.pdf.)

Figure I.5 Left: Fleet Street today. Right: A reimagined Fleet Street. (*Source*: © WATG (Wimberly, Allison, Tong & Goo) and Pixelflakes.)

the inevitable consequences of climate change. That is the opportunity we have, to deliver a far more liveable city. However, we have to cut our energy use and we have to switch to renewable energy – fast – to solve the climate crisis; everything else in this book is to help protect us further into the future.

I always loved the expression "When you are hanging on by your fingernails, you don't go waving your arms about". This is the delicate balance of the city ecosystem. The time has come for cities to find a way of drawing attention to climate action – while still hanging on!

The clock is ticking. On the Mercator Research Institute on Global Commons and Climate Change[7] carbon clock, at time of publication, the carbon budget will be depleted

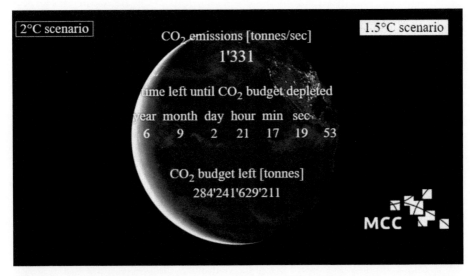

Figure I.6 Mercator Research Institute's countdown to limit the global temperature increase to 1.5 °C. (*Source*: https://www.mcc-berlin.net/en/research/co2-budget.html.)

---

7 https://www.mcc-berlin.net/en/research/co2-budget.html.

in 6 years, 9 months, 2 days, 21 hours, 17 minutes, and 19 seconds, after which limiting climate change to a 1.5°C increase will no longer be possible (Figure I.6).

Flow of the Chapters

To position this book, I wanted to include a little bit about the history of cities, a short journey through time, the historical make-up of life in our cities, and what has driven society through time. I wanted to position the trust we should place in cities to deal with the existential threat of climate change because of their ability to innovate and create working outcomes. The book will show the basic premise that urban life is more carbon efficient, and so the role of cities in retaining their growing populations will be critical to reaching our global climate goals and to continue the shift from their traditionally wasteful existence to a highly efficient one. How can the ideas they generate be applied across other cities and what will it take to make it happen?

In Chapter 1, *The Ambitious City*, Peter Boyd first reminds us that the city is where the ambitious in society have congregated for centuries; and that this ambitious leadership is what is needed more than ever to solve society's environmental challenges. It sets out a case for high ambition to be twinned with high clarity, proposing a bold and clear definition of what "Net-Zero" could mean to lead cities in a just transition to sustainability. It covers a definition that could include considerations of scope, of emissions reduction trajectory, of Paris-Agreement compliance, and even a cumulative approach to emissions that could help restore equity between the Global North and Global South. It concludes with considering the need for urban leaders to appreciate and embrace systems thinking if we are to manage this transition successfully – realizing how all the complex issues and opportunities outlined in subsequent chapters can and do connect.

In Chapter 2, *The Civilized City*, I take us through a historic journey to how we arrived at the modern city. The chapter looks at how cities have evolved, and how they have coped with the economic, social, and environmental challenges through time and how this shaped local governance. It seeks to explore and explain why this has positioned the modern city to be able to respond to the climate crisis while protecting the key liveability factors that govern our daily lives. It identifies some of the mechanisms needed to manage this transition and restore the balance.

In Chapter 3, *The Emerging City*, Professor Austin Williams examines the extraordinary rise of China, from a peasant economy a generation ago to one of the world's leading economies today. China's development has been unprecedented in world history, with around 850 million people lifted out of poverty in just 40 years, according to the World Bank. By contrast, the chapter compares the successes of China with the continuing plight of Malawi in central Africa – a country that remains in penury, with the limited chance that it will rise from its position near the bottom of the World Economic League Table.

While China can now take advantage of its wealth to reform its productive activities to provide better environmental conditions for its citizens, Malawi has the opposite experience. Its economic conditions are being mandated by others outside its control, with environmental conditions set by unaccountable global institutions, which ensures that Malawi will remain in a state of underdevelopment.

While China is the world's largest producer of renewable power, leading in solar, wind, batteries, and hydropower, it still cannot shake off the image of an environmental pariah. Malawi, on the other hand, is an ally. It is feted as a member of the Commonwealth of Nations and yet its lack of development – prioritizing the protection of its sacrosanct environment at the expense of modernization – is enforced by a neocolonial relationship to its supranational paymasters.

In Chapter 4, *The Sustainable City*, Patricia Holly Purcell sets out the path of the United Nations (UN) Sustainable Development Goals (SDGs) and looks at the universal challenges and the irrefutable truths that numbers bring to all cities. In 2015, the UN adopted the 2030 Agenda for Sustainable Development, which includes the 17 SDGs accompanying the Paris Agreement, which aims to limit global warming to 1.5°C. The SDGs are unique in what Patricia refers to as their "universal applicability", enforcing that all countries, whether they be north or south, rich or poor, have a responsibility to implement and achieve these goals. But the SDGs are also unique in the way they emphasize the important role cities play.

Numbers literally make a city and are what separates it from any other inhabited area. Patricia also examines urban agglomeration and how cities have a special responsibility to be sustainable and the steps taken to improve linkages between local and international governance levels.

In Chapter 5, *The Vocal City*, Amanda Eichel and Kerem Yilmaz identify that the 2015 Paris Agreement cemented climate change as one of the most important existential threats of our time and thus reinforced the importance of collective action "to address and respond to climate change, including those of civil society, the private sector, financial institutions, cities and other subnational authorities". Recognizing cities and the voice of cities in diplomacy was the culmination of a nearly 30-year effort from advocates, city networks, and cooperative initiatives.

There are, however, limits to what cities and the community that supports them can do alone. Even where cities have political will and available political, financial, and human resources, they face fundamental limits to their ambitions. And, in areas where cities can address climate change, they cannot achieve the economies of scale and transformative outcomes obtainable by national governments or through provincial/regional action. The voice of cities must better connect with the capabilities, skills, and learnings from other levels of government, as well as outside perspectives, to deliver action that both is locally appropriate and ensures the most impactful outcomes. This must include investing in the city/metro area and the city/region as part of an all-encompassing system, which simultaneously respects the need for equitable development while protecting and preserving natural resources and systems.[8]

In Chapter 6, *The Governed City*, Bruce Katz and Luise Noring expand on the idea that in a world of global climate summits, urban governance and finance is a critical but often-overlooked element of transformational change. This chapter uses Copenhagen (and Denmark more broadly) to distill the lessons not just around the design ("the what") of

---

8 https://unfccc.int/process-and-meetings/the-paris-agreement/the-paris-agreement/
key-aspects-of-the-paris-agreement.

policy initiatives but also around the plan ("the how") for delivering the climate commitments that an increasing number of cities and countries are making. Copenhagen and Denmark are models for how ambitious plans get implemented through capable municipal governments, empowered consortia of local governments, creative special-purpose institutions, and leading public pension funds. The creation of highly professionalized municipal governments and public institutions enables cities to use expert knowledge and sophisticated mechanisms to translate policy into action, leverage publicly owned assets, capture value appreciation for investment in public infrastructure, and deploy capital at scale. The adaptation of these models to cities and countries across the world is a necessary part of climate action. Our conclusion is simple: cities can tackle climate change if and only if they have institutions with the capacity, capital, and community standing necessary to get the job done.

In Chapter 7, *The Decoupled City*, Leah Lazer and Nick Godfrey base their ideas on the Coalition for Urban Transitions' report "Climate Emergency, Urban Opportunity", which quantifies the opportunities that urban climate action presents to fuel economic growth, create a more equitable society, and mitigate climate change. By taking an active role in supporting sustainable cities, national governments around the world can both drive economic prosperity and address the climate emergency.

Five case studies show that a rapid urban transition is possible with leadership from national governments, and quantitative analysis finds that investing in low-carbon measures in cities could be worth almost US$24 trillion by 2050 and that low-carbon measures in urban areas can support 87 million jobs by 2030 in sectors such as clean energy and public transport. Cutting 90% of emissions from cities would require an investment of US$1.8 trillion but would generate annual returns worth US$2.8 trillion in 2030 and US$7 trillion by 2050.

National governments can drive this transition by (1) developing a strategy to deliver shared prosperity while reaching net-zero emissions, with cities at its heart, (2) aligning national policies behind compact, connected, clean cities, (3) funding and financing sustainable urban infrastructure, (4) coordinating and supporting local climate action in cities, (5) building a multilateral system that fosters inclusive, zero-carbon cities, and (6) proactively planning for a just transition to zero-carbon cities.

In Chapter 8, *The Responsible City*, Justin Keeble and Molly Blatchly-Lewis write how a responsible city is an inclusive city, harnessing the catalytic potential of business for environmental and societal value. The city is guided by a clear and compelling mission and purpose. Its leaders know that responsible use of technology and innovation can shape sustainable outcomes. The authors draw on digitalization, the circular economy, and new modes of collaboration to accelerate efforts in combatting climate change through collective action. This chapter explores the critical role of businesses in reaching city goals.

In Chapter 9, *The Energized City*, Pete Daw makes it clear – the way cities use energy needs a total rethink if we are to make them sustainable. Energy is integral to our lives and our economies, but our cities rely heavily on fossil fuels to provide electrical power, to heat homes and offices, and to fuel our road transport. This drives many consequences for cities from polluting the air we breathe, causing premature deaths and diseases and contributing to the climate crisis. As more and more people choose to live in our cities the demand for energy is increasing. Increasingly we need to think of city infrastructure as one interconnected system, understanding that interdependence is critical to enabling a shift to clean power driving more and more of our systems. This requires thinking about transport, buildings, energy grids, water management, and waste management as integrated parts of the city. City governments can play an important role in driving that system change. They must have a clear and actionable plan, they can drive new zero-carbon development through their planning powers, and they can drive retrofitting programmes and renewable energy programmes, engaging their communities while creating jobs. And cities continue to innovate, blazing a trail for others to follow. But there is potential to do much more. National and state governments can and must unlock the full potential of cities through a new deal with city governments if we are to make the progress we need to see in tackling climate change.

In Chapter 10, *The Agile City (Part I)*, Julia Thayne writes how the onset of the COVID-19 pandemic in early 2020, and the racial justice protests that followed this, marked a tidal-wave change in how city governments view their roles in managing urban transportation networks. Both events exposed what was already becoming clear through heightened awareness of climate change: cities must leverage agile decision-making to create and operate agile transportation systems that respond to what people need when they need it. The "Agile City" is about how to build those new systems. It starts by exploring the impacts of how the design of cities' transportation networks – looking at how not only people but also things move across cities – has affected urban economies, environmental footprint, social inclusion, and even their physical shape. It then looks at new transport modes and technologies and how they might be used to supplement or supplant existing urban transportation networks. It acknowledges the current disconnect between the vision and the reality of urban transportation and discusses the multiple reasons why this might be the case. It ends with a note of pragmatic optimism on a way forward, a path based on one fundamental assertion: How people move, and how easily people move, in cities is directly linked to cities' wellbeing.

In Chapter 11, *The Agile City (Part II)*, Jonathan Laski asks the reader to contemplate if a city can truly be sustainable without allowing its citizens and visitors alike to move around safely, inexpensively, accessibly, and without fear of sickness from air pollution. Prioritizing walking, cycling, and mass transit offers governments a cumulative saving opportunity of almost 10% of global GDP. The chapter summarizes the environmental, social, and economic co-benefits of designing cities to prioritize active mobility, covering everything from reduced healthcare costs to increased support for local businesses. The chapter concludes by referencing two current events – the COVID-19 global pandemic and mass protests for racial equality – both of which have nudged cities further towards recentring people and their free movement at the heart of cities.

In Chapter 12, *The Habitable City (Part I)*, Olivia Nielsen walks us through some of the challenges of housing the world's population when financial and environmental resources are scarce. The world's current housing deficit is estimated at 1 billion units, and the UN predicts that by 2030 this deficit will affect close to 40% of the global population. By then, 60% of the world will live in urban areas, and the burden of housing these growing populations will fall primarily on cities. Cities must find ways to house 200,000 newcomers pouring in from the countryside every day. Addressing the global housing deficit will require us to build millions of new units while continuously upgrading and investing in our current housing stock. *The Habitable City* explores how cash-strapped cities can utilize new ideas and solutions to address this challenge, while minimizing our impact on the environment and integrating disaster and climate resilience as critical components of housing policies.

In Chapter 13, *The Habitable City (Part II)*, Nicky Gavron and Alex Denvir move the housing discussion to a new level. The construction sector is responsible for more than 23% of global greenhouse gas emissions, but cities will, and must, continue to grow. To do so while meeting global climate change targets and providing affordable higher quality homes, there needs to be a technological revolution in the way cities are built and designed. Leaders need to be bold and ambitious in the way they use that most precious resource: land. Density must be preferred to sprawl, to build faster and smarter, optimizing land. Cities must spearhead a new industrial sector in precision manufactured housing (PMH). PMH offers new jobs; reduced build times and costs; greater quality control; cleaner and safer construction processes; better air quality; and low construction waste and energy use.

Leaders must be bolder still. Rather than building homes with steel and concrete, by combining PMH with low- or zero-carbon materials such as engineered timber, we can dramatically reduce the carbon footprint of new buildings. Green belts can contribute to growing these homes of the future with trees and forests, while protecting and enhancing the crucial interrelationship between urban areas and their rural hinterland. These are homes that will lock away carbon, in greener, denser cities and be part of a circular, sustainable, and environmental economy.

In Chapter 14, *The Resourceful City*, Conor Riffle takes on one of the oldest challenges facing cities: how urban areas manage and dispose of their resources. Since the earliest Roman civilizations, cities have disposed of waste in landfills. As the world hurtles to 9 billion people and beyond, as our cities absorb more than half of the Earth's residents, and as climate change threatens our species' survival, the landfill model is unsustainable. Instead, leading cities are moving towards solutions that keep resources circulating for as long as possible – replacing the old, one-way paths to landfill with circular economies. This chapter highlights urban policies and technologies that are helping cities finally kick their landfill habits – and creating healthier, wealthier cities in the process.

In Chapter 15, *The Zero Waste City*, Terry Tamminen and Peter Lobin ask us to imagine a city without waste. "Trash" bins become sources of energy, fuels, and raw materials for products and buildings. Forests remain standing, because we no longer cut them down, only to throw away the resulting paper, cardboard, and wood in landfills or incinerators, nor do we discard half of the food grown on cleared land, necessitating the clearing of even more land to grow more crops. We no longer send armies around the globe to secure barrels of oil, only to throw away that valuable resource in the form of plastic, much of which we used only for a few minutes. We cut our energy usage and bills in half, making the switch to renewables easier and faster. *The Zero Waste City* shows how ending the concept of waste saves money, protects ecosystems, and creates new jobs using resources that are literally under our feet, with examples that can be rapidly scaled up to address climate change and ecosystem destruction. The chapter highlights the exciting new policies, technologies, and finance that now allow us to convert up to 95% of today's "waste" into tomorrow's valuable resources.

In Chapter 16, *The Resilient City*, Sarah Wray and Richard Forster show us that while COVID-19 has knocked the world sideways, the past year has also witnessed a series of parallel and compounding crises, including social unrest, intense weather events, and the economic fallout from the pandemic. This perfect storm looks set to be a pivotal moment in changing attitudes to resilience, with a clear focus on learning lessons to strengthen communities against future shocks and stresses – whether economic, social, or environmental.

The Chief Resilience Officer (CRO) role in cities is a relatively new one, but the advent of COVID-19 has seen CROs receive greater support in terms of funding, staff resources, and proximity to the mayor due to a recognition of the urgent need for a holistic strategy which can help cities face up to the various threats they face. Smart city initiatives are playing a key role, too, with cities using digital tools and data innovation to support resilience efforts.

While COVID has shown that cities must adopt a holistic approach to digitalization and resilience, it is equally important that they have the means to implement this. Resilience is becoming a growing consideration for investors, and cities are also increasingly incorporating resilience considerations into their own spending decisions. This chapter highlights examples from the cities of Rotterdam, New Orleans, Edinburgh, and more.

In Chapter 17, *The Fragile City*, John de Boer points out there is growing recognition that the cumulative impact of converging environmental, social, political, and economic risks is straining the ability of many cities to deliver essential services to residents in times of shocks and stresses. The COVID-19 pandemic brought cities around the world to a halt, causing massive disruption and suffering for hundreds of millions of people. The pandemic exposed the fault lines in our cities that make them fragile. This includes growing inequalities in income, gender, race, and opportunity, as well as structural factors linked to exposure to violence, poverty, extreme pollution, and natural disasters. This chapter assesses the sources of fragility rooted in our cities and explores approaches that could help cities develop more resilient urban systems, enabling them to function, and even thrive, in times of crisis.

In Chapter 18, *The Data City*, Seth Schultz and Eric Ast propose a bold and pragmatic vision for the role that cities can play in ensuring a just, equitable, and safe future for humanity. There's good reason to be optimistic about the long-term ascent of data-centric techniques within the political sphere and the potential for collaboration between public and private sectors around the globe. However, due to the immediate nature of the climate crisis, a clear-eyed view of our current trajectory dictates that intrepid action towards accelerating action is necessary. By leveraging their power over procurement processes and budgets to dictate the conditions for how data are collected and accessed, and tapping into rich local technological and research ecosystems, cities can embrace a new and more effective role within the data ecosystem. This approach, the Procurement + Platform Pivot, pulls cities around the world out of a cycle of dependence and data poverty and into the driver's seat in a role that they play best: convener.

In Chapter 19, *The Measured City*, Patricia McCarney positions cities in a highly connected world and advances the need for globally standardized data to empower sound city leadership on the global stage. Until recently, this interconnected world was traditionally reserved for national governments, connecting through trade, security, and global monetary policy, all supported by sound, standardized measurements – Gross Domestic Product

(GDP), Gross National Product (GNP), and other national income and monetary measures. Cities have been rising in stature as critical sites in this highly connected world. Cities are critical sites where investment, invention, prosperity, climate mitigation, security, health, and social wellbeing can either succeed or fail. However, globally standardized, comparable measurement, so valued at national level to drive data-informed global relations, has lagged at city level. The chapter advances the need for globally standardized measurement in cities and examines what global standards exist for city data to drive and enable the "Measured City". The chapter provides a look into the International Organization for Standardization (ISO) and the recent emergence of the *ISO 37120 Series* that has created a global standard for city data that enables comparative apples-to-apples data for the first time. The chapter answers the core question "Why is 'The Measured City' so important for cities today, and how are cities embracing global standards to propel their success?"

In Chapter 20, *The Smart City*, Noorie Rajvanshi tells a story, through data, of how technologies can drive climate action in cities. In this chapter we dive into learnings from over 40 cities worldwide, highlighting results of technology modeling with the Siemens City Performance Tool. Cities, irrespective of where they are located or their climate or socioeconomic standing, share a common understanding that these three actions will produce deep carbon reductions and lead the way to zero-carbon cities – decarbonization of the electric grid, reducing energy usage in buildings and transport, and electrification of everything. Many of the technologies that will enable the implementation of these actions already exist and have been proven to work on a large scale, but there is always room for innovation!

In Chapter 21, *The Just City (Part I)*, Hayley Moller explores the interlocking challenges of public health, climate change, and economic inequality through the air quality of our cities. The chapter reviews the key drivers and impacts of air pollution in cities around the world and investigates how poor air quality often visits the greatest harm upon poor communities and communities of colour. It argues that addressing air quality can have massive climate and equity co-benefits, and it explores how to maximize these benefits using examples of cities in the Global North and Global South that have successfully tackled this invisible adversary. It concludes with an in-depth look at an innovative solution in the city of Seoul and extrapolates lessons relevant to all cities.

In the same chapter, *The Just City (Part II)*, Jane Burston and Matt Whitney show us how London's fight for clean air is based on data. How do we solve an issue like air pollution? It is an almost universal issue: billions of people are breathing dirty air and millions are dying prematurely each year as a result. The solutions to air pollution are within grasp. A shift to clean energy and sustainable transport can improve air quality and bring real improvements to people's health, almost overnight. But adoption of these solutions must accelerate. Improving air quality will not only improve health, but also drive down the carbon emissions that can help to avoid the climate crisis. Air pollution is often invisible – it is highly damaging to health long before it forms smogs thick enough to be visible to the naked eye. But data can make it visible, and in doing so illuminate the sources and solutions needed, as well as the consequences of inaction. This chapter reveals how measuring air quality is on the verge of a revolution. Technological innovation is promising a shift in how cities measure air quality, enabling a new understanding of the issue and helping policymakers to design effective solutions.

In *The Just City (Part III)*, Jenny Bates asks the question "Will air pollution on a death certificate for the first time mean nine-year-old Ella's tragic death leads to cleaner air and better health for others?" London has a serious air pollution problem, as I became aware of as I worked for Friends of the Earth covering London. For too long, despite the great work of some, there wasn't enough public awareness or action. But with Sahara dust, Dieselgate, legal actions, campaigning, and more, it has risen up the agenda, alongside climate change. The solutions are clear, including the need for cleaner and also fewer vehicles, not adding to the problem such as with road-building or airport expansion, and updating our standards to align with WHO guidelines – they just need implementing. In a post COVID world this is all the more important and will also benefit the economy. Ella's death could help lead to a better London.

In Chapter 22, *The Invested City*, Colin le Duc puts cities at the forefront of the transition to a more sustainable form of capitalism. Capital allocation is increasingly a function of risk and return, as well as explicit impact considerations. Cities act as hubs for the financial system itself, but also for how new technologies and innovations are tested and implemented. Cities play a crucial role in mainstreaming sustainable investing and enabling environmental, social, and governance (ESG) factors to be fully integrated into capital allocation decisions. Additionally, in areas of critical societal needs such as building, transport, food, and energy, cities are incubators of new, innovative, sustainable models that can be tested and perfected to become mainstream solutions. Generation Investment Management's Chairman Al Gore often says: "The 'Sustainability Revolution' is upon us, it has the magnitude of the Industrial Revolution and the speed of the Digital Revolution."

In Chapter 23, *The Financed City*, James Close reminds us of the challenge that 70% of emissions come from cities and over 50% of the world's population live in cities. Cities are the foundations of our modern society and economy. As a result, they are central to managing the transition to a low-carbon, resilient future.

Cities will need to transition from their historic trajectory of high-carbon development to address climate change. Cities are well equipped to make this transition because they are dense, homogenous, and concentrated in terms of both population and infrastructure. Their long-term plans need to be informed by a compelling vision of the future and the mobilization of capital at scale for investment in businesses, communities, and infrastructure.

Net-zero carbon cities are an important aspiration. Net-zero cities will also need to reduce consumption-based emissions by adopting circular economy principles so they can eliminate their contribution to climate change. A clear vision and systemic approach reduces risk and decreases the cost of capital, supporting climate-smart investment, sustainable development, and a people-centred approach.

In Chapter 24, *The Adapted City*, Adam Freed puts cities on the front lines of the climate crisis, dealing with the catastrophic impacts of climate change-driven heat waves, coastal storms, droughts, and wildfires. But it's not just climate change that is putting people at risk; it is also decades of poor urban planning and design. The shape and form of our cities will help decide how well we can withstand today's extreme weather and what lies ahead. The chapter looks at the practical actions cities around the globe are taking to adapt to rising temperatures, too much water, and too little water, and the changes needed to scale up these actions to address the urgent reality of the risks we face. It outlines six key principles for mayors and city leaders to embrace to protect their residents from climate change, and it highlights several case studies that provide a roadmap for urban leaders on how to accelerate the breadth and scale of their work.

In Chapter 25, *The Open City*, Professor Peter Bishop shows us that "an open city" is spatially diverse, is generous, and celebrates its public spaces, parks, squares, and streets. They are places where citizens meet, exchange goods and ideas, debate, linger, play, and celebrate. This is where the civic life of a democratic society takes place. You can judge the health of a city by its open spaces. Public space is not a commodity, and the market will not provide it (except under very limited conditions). It is public – that is, communally owned and maintained for the use and enjoyment of all. It needs to be protected, managed, and cared for. Where it is lacking it needs to be provided, not as a luxury but as a necessity for urban living. At the time of writing, a global pandemic is causing many individuals to relearn the value of public services and community spirit and value clean air, parks, open spaces, and gardens. This chapter traces the theory and practice of providing public spaces in the city as an essential ingredient of the richness and messiness of the twenty-first-century city.

In Chapter 26, *The Natural City*, Carlo Laurenzi considers a range of disparate issues from asking how an artificial phenomenon, like increasing urbanization, on a planetary scale, can ever be compatible with the natural world. Architectural trends, natural geomorphological forms, and planning issues are seen under the microscope of whether they, in reality, help or hinder cities becoming more natural. Parallels are drawn between human migrations and the associated social diversity this brings to our cities, and how these compare with recent biodiversity winners and losers, as well as how terms about unwelcome visitors enter our language, discussed alongside questions about land-use and food growing. Controversial subjects like children's education and the individual's right to keep pets are not avoided, along with issues around health, and mental health in particular, as seen under the prism of achieving a natural city. London is used as a canvas to paint strategies and to examine what works and why; and hopefully some of these ideas will have relevance beyond the UK capital. The barriers to achieving a natural city are not centred, for once, around money or technology, but the political and social will to make it happen.

In Chapter 27, *The Climate-Resilient City*, Mauricio Rodas places cities as first responders to the world's most pressing issues, such as climate change, migration, and pandemics like COVID-19. The Paris Agreement and other agendas will not be met if cities don't take effective action, but the obstacles for cities' direct access to international finance are hampering the required investment. Structural reforms to the global financial architecture are urgently needed to make it cities-friendly. While there is a growing supply of financing mechanisms available for cities' climate-resilient projects, most of them are chained to national guarantees and are highly politicized. On the other hand, cities often lack regulatory certainty, project preparation capacity, and creditworthiness. There is a clear dissonance between financial supply and demand, which impedes capital flows from coming. In spite of these shortcomings, particularly in the developing world, cities find ways to finance infrastructure projects. It was thanks to innovation that during my mayoral term in Quito, the municipality managed to successfully build the first metro line in Ecuador. Innovation is key, but more remains to be done. There is a great opportunity to disburse resources directly into cities through COVID-19 stimulus packages. If this unfolds properly, it can become a historic milestone, recognizing cities' need to improve their access to finance as the only way to develop the infrastructure transformation required to foster a more climate-resilient future.

In Chapter 28, *The Green City*, The Lord Mayor of Copenhagen gives us the story of green leadership in Copenhagen through more than a decade. The chapter outlines the city's journey from 2009 when it adopted the goal of becoming the first carbon neutral capital in the world to where it is today. In 2025, Copenhagen will be a city with cleaner air, less noise, energy-efficient buildings, and green transportation. Already, the city has reduced its carbon emissions by close to 50%, while at the same time experiencing an increase in the social economic index. Copenhagen has shown and continues to show that through new solutions, green investments, and new habits, cities and citizens can enjoy green growth and green jobs in a green city. The city's approach has been refined over the years, but the principle is the same: Copenhagen incorporates data, research, analyses, and stakeholders in its initiatives. It strives to continuously develop its cooperation with businesses, universities, and research institutes as it implements its plans revolving around energy consumption, energy production, mobility, and administration initiatives. The green city aims to inspire others in the green transition, just as it always seeks to gain inspiration from around the globe.

In Chapter 29, *The Powerful City*, Mark Watts and Sarah Lewis show us "a shared responsibility". This chapter outlines how entrepreneurial big-city mayors are driving global progress on climate change. It shows that overcoming climate breakdown requires governmental leadership to set a clear policy direction, and create and shape markets necessary to meet science-based climate targets, working with a dynamic, mission-driven private sector that responds to the opportunity to create a new economy based on sustainability and fairness. Using examples from the C40 group of the world's 100 most influential mayors, the chapter considers how mayors are often working beyond their formal powers and working together across geographic and political boundaries to demonstrate how to deliver the future we want, rather than the one we are hurtling towards, creating opportunity for green investors, entrepreneurs, and communities to thrive along the way.

The conclusion to *The Climate City* (Chapter 30, *Epilogue*) is a "manifesto of actions" from the text. We are fast approaching tipping points that will destroy life on our land, in our rivers, and in our oceans. The impact will be on all of the inhabitants of the planet, and the role of cities in doing their part but also paving the way for wider improvements will make a fundamental contribution to the future we end up with. The keys to our future are in the hands of the mayors of our great cities. Enjoy the book.

1

The Ambitious City – Introduction

Cities run on a unique momentum. They combine a vision of the future with constant innovation. It is no mystery as to why our cities come to symbolize the best we can offer as nations. Ambition is what defines a city, the ability to look forward and the constant pushing of the boundaries of what is possible. It is clear that we need ambition, but ambition alone is problematic. All too often we are bound to empty leadership, PR stunts, and meaningless words that lead to unsubstantial policy. Short-term goals are ineffective and long-term outcomes are lost. Simply citing "a net-zero city" as a strategic goal is not enough.

So ... how do we retain ambition but keep clarity?

Throughout this chapter, Peter Boyd will discuss this challenge and outline his plan for ambitious but accountable cities. He argues that the "clarity" we seek for the ambitious city is achieved through "connected" leadership both to the environment and the city it serves. Peter writes, "As faith in national governments and global institutions falters, it is in the city level of democracy and leadership that we can locate a stable point of leadership to embrace this level of responsibility."

This need for connected leadership is not to be underestimated, especially as we move towards our ambitious (but possible) "Net-Zero" targets. Peter looks at what earns a city its "Net-Zero" status and considers how four powerful descriptors – Fully Scoped, Science-Based, Paris-Agreement-Compliant, Cumulative – bring clarity to the concept itself. Peter then goes beyond this, quite literally, in what he refers to as "Net Zero and Beyond", as he outlines further steps we can take beyond the reduction of greenhouse gas (GHG) emissions so that cities become "climate positive" rather than solely aiming for "Net-Zero" status.

All leaders act and all leaders are defined by their actions. But in the case of the ambitious city, action alone is not enough in regards to any future leader. They must go "beyond" this. Meaningful, clear, clean action is the only option in the ambitious city.

The Climate City, First Edition. Edited by Martin Powell.
© 2022 John Wiley & Sons Ltd. Published 2022 by John Wiley & Sons Ltd.

1

The Ambitious City

Peter Boyd

For cities to truly lead on climate change – a just transition to "Net-Zero" and beyond – high ambition needs to be coupled with high clarity of vision and a multidimensional, connected-leadership approach with an appreciation and embrace of systems thinking.

Ambitious Citizens in Ambitious Cities

Cities have been magnets to ambition for centuries. From fairy tales to ancient Rome (Figure 1.1), to contemporary migration trends, and everything in between, people have come to the city to seek their fortune – to "be somebody", to "make it" and make things happen.

If cities are to truly take the lead in combatting climate change, we need an ambitious city with purpose-driven city leaders to accelerate a just transition to a sustainable future. We need them to help define the city's goals and for *individual* ambition to inform and feed into *collective* ambition for a city we all want to live in, and where we think our children and grandchildren will in turn want to stay or return to.

But alongside ambition we need *clarity*. "If you don't know where you are going, any road will take you there," said Lewis Carroll's Cheshire cat in *Alice in Wonderland* (Figure 1.2). There are cities all over the world with a number of sustainability goals, but to truly take the lead on climate change we should reach and then surpass "Net-Zero" (capital letters to be explained) into carbon negative, eco-restorative territory – as soon as each city is able, and well before 2050.

Figure 1.1 A depiction of Ancient Rome, a concept shaping many cities that followed. (*Source*: ZU_09/Getty Images.)

The Climate City, First Edition. Edited by Martin Powell.
© 2022 John Wiley & Sons Ltd. Published 2022 by John Wiley & Sons Ltd.

Figure 1.2 If you don't know where you are going any road will take you there. (*Source*: EllerslieArt/Adobe Stock.)

To achieve this necessary climate goal while managing all the other priorities of the complex modern city, we need leaders that are truly connected: to themselves; to the team around them; to their communities and stakeholders; and to an awareness and appreciation of the entire system they are trying to manage and change.

What could this look like? A city would have a bold, clear, co-created goal that the city leadership, community leaders, and citizens readily understand and support. A city's climate action plan would target a fully scoped, science-based, Paris-Agreement-compliant, cumulative "Net-Zero" well before 2050; and it would do so in a way that also makes the city greener, more equitable, more resilient, smarter, friendlier, healthier, and all the other qualities to be examined in later chapters.

And just as Maya Angelou told us as individuals to "Be yourself, only better", cities will have achieved this in a variety of complex, adaptive, ingenious ways – peculiar to their own unique assets and their own unique, ambitious citizens.

Cities: The Need for Ambition and Clarity

Cities have been hives of ambition since they were first created. Birthed by the ambitious, they have attracted like minds in search of their fortune, of significance, of life's meaning, and much more. Today, we need these hives of ambition to assume leadership of the just transition to a sustainable world.

Thomas Berry, in *The Great Work*, describes these decisive decades in the context of a long multicentury arc as one when we have to work out how to become a benign force on the planet – with ourselves as fellow humans and with all creation. This is, in essence, the "Great Work" that lies before us. And as faith in national governments and global institutions falters, it is in the city level of democracy and leadership that we locate a stable point of leadership to embrace this level of responsibility.

Reaching the goals of the 2015 Paris Agreement will require ambitious actions from all sectors and levels of our society, but especially in our cities. More than 55% of people already live in urban areas, and this is forecast to rise to 68% by mid-century.[1] Urban areas account for more than 60–70% of the world's greenhouse gas (GHG) emissions, consuming 66–78% of the world's energy, while occupying less than 2% of the land.[2] Even as we decouple economic growth from emissions – already achieved by over twenty countries before – the sheer scale of the amount of people living in cities means that climate leadership by cities requires ambition and talent at city-scale as much as ever before in history.

But ambition on its own will not be enough. A mayor will always fall short when simply citing "a net-zero city" as a strategic goal. It may sound good, but without clarity in describing that destination it may even come to be seen as just a PR stunt that sounds good but lacks substance (Figure 1.3). At the other extreme, a crystal-clear set of cautious environmental objectives is simply not going to be bold enough in this time of climate emergency. So, how do we retain the ambition but increase the clarity on what cities can and should be aiming for to lead on climate?

As I've argued with co-author Casey R. Pickett in a recent paper[3] – from which this paper draws – we need a consistent definition of "Net-Zero" that cities (and organizations, companies, and countries) can use and measure progress against. If we are to maximize the probability of a just transition to a sustainable society, all actors have to explain what they mean by "net-zero" in addition to their intended *deadlines* and *paths*. We suggest four measurable criteria for any undertaking of "net-zero" to be worthy of capitalizing to "Net-Zero".

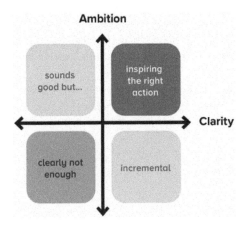

Figure 1.3 The need for both ambition and clarity will drive success. (*Source*: Boyd, P. and Pickett, C., 2020. *Climate Ambition: A Case for Net-Zero Clarity*. Yale Center for Business and the Environment. https://cbey.yale.edu/research/defining-net-zero.)

1 UN DESA: 2018 Revision of World Urbanization Prospects. https://www.un.org/development/desa/en/news/population/2018-revision-of-world-urbanization-prospects.html.

2 UN estimates > 60% of GHGs and consuming 78% of global energy; C40 estimates > 70% of GHGs, consuming two-thirds of global energy. https://www.un.org/en/climatechange/climate-solutions/cities-pollution. https://www.c40.org.

3 Boyd, P., Pickett, C.R., 2020. Defining Net-Zero: Addressing climate change requires a clear, bold explanation of our shared global goal. Yale Center for Business and the Environment. https://cbey.yale.edu/research/defining-net-zero.

"Net-Zero" and Beyond for Cities

In a refreshed and robust definition, a strategy for "Net-Zero" GHG emissions earns its capital letters if it is: *Fully Scoped, Science-Based, Paris-Agreement-Compliant*, and *Cumulative*. Each descriptive term imparts a dimension of clarity. "Net-Zero" can be a powerful and useful goal at the city level if the city embraces a concept of "Net-Zero" that is:

1) *Fully Scoped*: articulating the city's defined scope of responsibility. This should include all GHG emissions from scope 1 (GHG emissions from sources located within the city boundary); scope 2 (GHG emissions occurring as a consequence of the use of grid-supplied electricity, heat, steam, and/or cooling within the city boundary); and scope 3 (all other GHG emissions that occur outside the city boundary as a result of activities taking place within the city boundary).[4]
2) *Science-Based*: incorporating a destination-based[5] target for "Net-Zero" that demonstrates the city is assuming bold, appropriate responsibility for emissions reductions consistent with the Paris Agreement and *at least* proportional to its contribution to climate change.[6]
3) *Paris-Agreement-Compliant*: specifying if and to what extent carbon credits and external investments in carbon removal factor into the strategy. Any offsetting investments should be linked to the global carbon budget as defined in the Paris Agreement.
4) *Cumulative*: acknowledging the city's historical emissions of GHGs, not just their current level.

And Beyond ...

We will unpack each of the terms above and what it means for cities, but we should first explain the "and Beyond" of the title, as its meaning is designed to be taken in at least two ways. For cities to set climate action goals solely through a lens of GHG emissions would be too narrow and not recognize the complex set of interconnected goals and stakeholders. All climate goals in a municipal environment are often phrased within the Brundtland Commission definition of sustainability,[7] encompassing sustainability for social, economic, and environmental benefit. The UN Sustainable Development Goals (SDGs) have

---

4 Global Protocol for Community-Scale Greenhouse Gas Emission Inventories. https://ghgprotocol.org/sites/default/files/standards_supporting/gpc_executive_summary_1.pdf.

5 By "destination-based" we mean to distinguish an absolute target from a relative one. Many climate targets are relative, for example, "80% less emissions than our 1990 baseline". Relative targets do not tell a reader if their achievement is consistent with a world in carbon balance. Destination-based targets do.

6 Science-based targets have historically focused on corporate action, but the Science-Based Targets Network has more recently started to "translate the science that defines the limits of nature ... into actionable targets for cities. https://sciencebasedtargetsnetwork.org/earth-systems.

7 Our Common Future. https://sustainabledevelopment.un.org/content/documents/5987our-common-future.pdf. In Commissioner Brundtland's own foreword: "the 'environment' is where we all live; and 'development' is what we all do in attempting to improve our lot within that abode. The two are inseparable".

built on this important work over decades, and we have since heard most sustainability targets referring to not only the destination but also the nature of the transition, and the ambition to make the transition a just one. Beyond climate and human equity, biodiversity, water, and land stewardship are also critical considerations for any city.

The second intended orientation of setting "Net-Zero and Beyond" goals is leaving the "Net-Zero" market behind in the past. Cities hopefully become "Climate Positive" rather than just aiming for "Net-Zero". "Climate positive" is an increasingly used alternative to "carbon negative" that emphasizes the benefits of aggressive climate action. "Net-Zero" hopefully becomes a well-defined marker that many cities leave behind – and ideally well before 2050.[8]

Ambition and Clarity at the Global Level

Providing net-zero context at the global level, the Paris Agreement, agreed in 2015 and ratified the following year, created a vital destination-based target for the world. It set the objective of "Holding the increase in the global average temperature to well below 2°C above pre-industrial levels and to pursue efforts to limit the temperature increase to 1.5°C ..." by achieving "a balance between anthropogenic emissions by sources and removals by sinks of greenhouse gases in the second half of the century, on the basis of equity, and in the context of sustainable development and efforts to eradicate poverty".[9]

While the first phrase, from Article 2, outlines the temperature target, Article 4 effectively defines net-zero GHG emissions as reducing human-caused emissions to the level that natural climate solutions and other methods of CO_2 storage and removal can effectively absorb. It succinctly describes a global state of ecological balance, even if the results of *past* emissions have not been fully absorbed. If this state is achieved by mid-century, and if emissions decline further to net-negativity in the back half of the century, maintaining a 1.5°C world becomes likely.[10]

Paris-Agreement goals build on strong conclusions from the Intergovernmental Panel on Climate Change's recent assessment[11] and the US government's own 4th National Climate Assessment.[12] A recent report released in *BioScience*[13] listing 11,000 scientists as

---

8 In the lead-up to the next global climate talks in Glasgow, the UN champions of the "Race To Zero" campaign are calling for net-zero targets "in the 2040s".

9 UNFCCC, 2015. Paris Agreement. 3–4. https://unfccc.int/sites/default/files/english_paris_agreement.pdf (accessed 3 June 2020).

10 Rogelj, J., Luderer, G., Pietzcker, R., et al., 2015. Energy system transformations for limiting end-of-century warming to below 1.5°C. *Nature Climate Change* 5, 519–527. https://www.nature.com/articles/nclimate2572.

11 Masson-Delmotte, V., Zhai, P., Pörtner, H.O., et al. (eds), 2018. Global warming of 1.5°C. An IPCC Special Report on the impacts of global warming of 1.5°C above pre-industrial levels and related global greenhouse gas emission pathways, in the context of strengthening the global response to the threat of climate change, sustainable development, and efforts to eradicate poverty. https://www.ipcc.ch/sr15.

12 US Global Change Research Program, 2018. *Fourth National Climate Assessment.* Vol. II. https://nca2018.globalchange.gov/downloads.

13 Ripple, W.J., Wolf, C., Newsome, T.M., Barnard, P., Moomaw, W.R., 2020. World scientists' warning of a climate emergency. *BioScience* 70(1) January, 8–12. https://academic.oup.com/bioscience/advance-article/doi/10.1093/biosci/biz088/5610806.

contributing authors was equally unequivocal. We must adjust course dramatically and achieve net-zero emissions by 2050 if we are to avert the worst impacts of climate change and remain on target for 1.5°C of warming. These global goals transmit across language, culture, and ideology. They are vital to the transition to a sustainable world, but their applicability and measurability tend to burn off on impact as they enter a single enterprise.

Ambition at the City Level

City mayors have been stepping up and leading through a variety of fora, most notably the Global Covenant of Mayors for Climate & Energy, C40, and the International Council for Environmental Initiatives (ICLEI). In the former, 9,000 municipalities could potentially achieve savings of 1.4–2.8 $GtCO_2e$ versus "business as usual" if their combined pledges are achieved. C40's "Deadline 2020" is a commitment from the world's leading cities to urgently pursue high-ambition climate and has attracted 119 committed cities at time of writing.

Where and how does "net-zero" appear in these commitments? Some of these commitments clearly articulate goals, techniques, and timing. Many do not. Some of the commitments include room for flexibility, often by design, to create a wide net of inclusivity and attract more signatories. Well and good; but with four UN climate conferences behind us since Paris and COP26 in Glasgow this year (2021), it is time to tighten definitions – especially concerning the mid-century destination of net-zero. For the "Race to Zero", 449 cities have already joined this campaign.

The risk is that these campaigns are intentionally broad and open-sided tents. Only 12 of the 119 committed cities to C40's Deadline 2020 have Paris-Agreement-compatible climate action plans.[14]

Race to Zero lists a "Minimum Criteria for participation in Race to Zero"[15] but does not display as obviously a vanguard of bold, best-in-class targets and a "lead peloton" willing to show others how this is done.

Ultimately, the hope is that a clear, bold, ubiquitous definition will empower the world's most progressive cities to lead the world's just transition past "Net-Zero" to true sustainability and a climate-positive, restorative back-half of the century. So how can we describe "net-zero" better to be both clear and bold?

"Fully Scoped": What Does This Mean?

There is a flexibility and power in the net-zero concept. It can apply at all levels, from the globe down through nations and states to cities, communities, households, and individuals. For the world to achieve global net-zero emissions, all relevant entities should assume

---

14 C40: Deadline 2020. https://www.c40.org.

15 UNFCCC, 2021. Who's in Race to Zero? https://unfccc.int/climate-action/race-to-zero/who-s-in-race-to-zero#eq-4.

responsibility for direct emissions caused by sources they directly own and control (scope 1). For cities, this has been translated as emissions created within their boundaries.

Scope 2 includes targeting GHG emissions occurring as a consequence of the use of grid-supplied electricity, heat, steam, and/or cooling within the city boundary. This is typically the electricity and heat generation from utilities that may supply the city but not be located within the boundary.

Scope 3 emissions for a city are all other GHG emissions that occur *outside* the city boundary as a result of activities taking place *within* the city boundary. This may be upstream activities such as emissions going into the production of materials consumed in the city, as well as downstream activities – emissions caused by the consumption of goods/services produced inside the city but consumed elsewhere.

If every city in the world achieves net-zero emissions for scopes 1 and 2, the urban world should expect to realize an equilibrium among sources and sinks of emissions. But this leaves no margin for error and also relies on other communities to target net-zero too. A more airtight approach would acknowledge that some entities are not targeting boldly enough and would task those that can act more assertively to take responsibility for scope 3.

C40 analysis conducted with the University of Leeds, the University of New South Wales, and Arup adopted a consumption-based approach to investigate GHG emissions of 79 cities, capturing both direct and lifecycle GHG emissions (Figure 1.4).[16] This usefully captures the upstream effect of emissions arising in supply chains outside the city to meet the demand and consumption for goods and services inside the city; but from a "high-ambition" point of view it is unfortunate that it subtracts the downstream effect. It gives the city a (double) break on the emissions – both by netting out emissions created by producing goods and services inside the city that are consumed elsewhere and the emissions associated with the actual consumption (Figure 1.5).

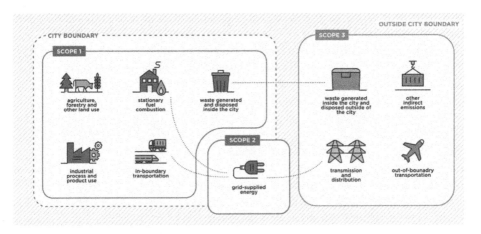

Figure 1.4 C40 analysis: consumption-based GHG emissions of C40 cities.
(*Source*: Consumption-based GHG emissions of C40 cities, March 2018. C40 Cities Climate Leadership Group, Inc. https://www.c40.org/researches/consumption-based-emissions.)

---

16 C40: Consumption-Based GHG Emissions of C40 cities. https://www.c40.org/researches/consumption-based-emissions.

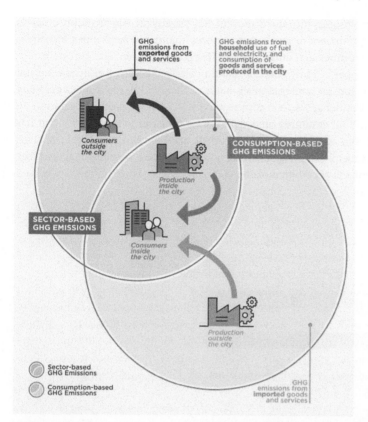

Figure 1.5 C40 methodology showing overlap and distinction between sector-based and consumption-based emissions. (*Source*: Consumption-based GHG emissions of C40 cities, March 2018. C40 Cities Climate Leadership Group, Inc. https://www.c40.org/researches/consumption-based-emissions.)

In the corporate realm, for example, Unilever's "Wash at 30" campaign arose from internal analysis that showed most of their "full-scope" emissions were caused by consumers *using* their products – in this case washing with hotter water than needed to clean clothes given the detergent technology Unilever now provides. Volkswagen, bouncing back from their diesel-emissions scandal, presented at the World Climate Summit 2019 a fully comprehensive value chain analysis showing that only 15% of their targeted emissions reductions to comply with the Paris Agreement are from their vehicle production and supply chain. Meanwhile, emissions from fuel supply and customers driving vehicles accounted for 79%. For Volkswagen, assuming responsibility for scopes 1–3 is propelling a more ambitious corporate strategy (with cleaner, electric-powered cars at its heart) and in turn a beneficial impact on world CO_2 emissions.

If cities are to truly lead on climate action, ideally we would define "Fully Scoped" as assuming responsibility for scopes 1, 2, and 3, upstream and downstream – sector-based *and* consumption-based.[17] The C40 analysis shows how the consumption-based method

17 US EPA, n.d. Greenhouse gases at EPA. https://www.epa.gov/greeningepa/greenhouse-gases-epa (accessed 3 June 2020).

"offers complementary insights into the drivers of GHG emission – recognizing cities as both consumers and producers of goods and services – and can help cities identify a broader range of opportunities to reduce global GHG emissions".[18]

In addition to "all scopes", "Fully Scoped" should clearly include all GHGs[19] and not just carbon dioxide. We recommend that this is made explicit within the concept of "Fully Scoped": i.e. all activities, all gases. A recent report in *McKinsey Quarterly*[20] shows – in Figure 1.6 – how important assuming responsibility for all gases is for any city that has significant industrial and agricultural sectors; but it also demonstrates that *all* cities should consider targeting all GHGs, considering waste has such a strong global warming potential outside the effects of carbon dioxide.

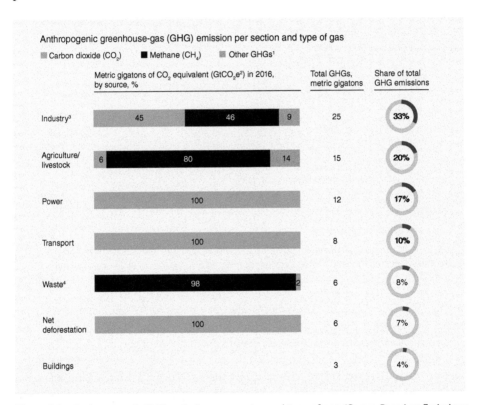

Figure 1.6 Anthropogenic GHG emissions per sector and type of gas. (*Source*: Based on Emissions databases for Global Atmospheric Research (EDGAR), 2015; FAOSTAT, 2015; IEA, 2015; Mc Kinsey Global Energy Perspective 2019; Reference case 1.5c Scenario Analysis.)

---

18 C40. Op. cit., p. 15.

19 Greenhouse gases are "any gas that has the property of absorbing infrared radiation (net heat energy) emitted from Earth's surface and reradiating it back to Earth's surface, thus contributing to the greenhouse effect. Carbon dioxide, methane and water vapor are the most important greenhouse gases. (To a lesser extent, surface-level ozone, nitrous oxides, and fluorinated gases also trap infrared radiation.)" Encyclopaedia Britannica, n.d. Greenhouse gas. https://www.britannica.com/science/greenhouse-gas (accessed 3 June 2020).

20 McKinsey Quarterly, 2020. Climate math: What a 1.5-degree pathway would take. https://www.mckinsey.com/business-functions/sustainability/our-insights/climate-math-what-a-1-point-5-degree-pathway-would-take.

"Science-Based": What Does This Mean?

The Science-Based Targets Initiative (SBTi) has focused on providing the corporate sector "with a clearly defined pathway to future-proof growth by specifying how much and how quickly they need to reduce their greenhouse gas emissions".[21] The targeting process involves a widely accepted methodology from the World Resources Institute, CDP, UN Global Compact, and World Wildlife Fund to calculate reduction pathways for different sectors. That methodology apportions the Paris Agreement's emissions-reduction goals to each entity according to their capacity to reduce. The methodology helps organizations calculate the emissions reductions to target in accordance with any of three aligned approaches: a sector-based division of the global carbon budget; an absolute approach based on emissions reductions required; and an approach that uses an entity's relative economic contribution to determine its emissions reduction target. As of December 2019, over 740 companies had committed to science-based targets.

The SBTi has not yet created similar pathways and guidance for cities but (1) points to the C-FACT methodology as a useful corollary of the method[22] and (2) has inspired a science-based approach.

Importantly for our summary here, it provides a principle for cities when targeting "Net-Zero": looking to the best-available science to help inform the appropriate share of emissions reductions for the city and defining how quickly the reduction should occur to be in line with global climate stabilization targets.

The contraction and convergence approach developed by the Global Commons Institute back in the 1990s, but championed more recently by C40 in its Climate Action Plan guide,[23] helps cities measure and "converge" on an emissions per capita that is becoming equal to its relevant peers and also "contracting" enough for global net-zero emissions to be achieved by 2050.

Keeping global temperature rise under 1.5–2°C requires leaders to focus their strategies on both the *total* amount of global emissions reductions and detailed accountability for each city's appropriate *share* of their relevant national total. To say a city's net-zero target or pathway is science-based communicates that the city has considered its appropriate obligations under the Paris Agreement and has set is emissions-reduction target accordingly. Of the many strong examples of "science-based" leadership to net-zero emissions, the Mayor of London's "Zero carbon London: A 1.5°C compatible plan"[24] explicitly links the plan to the global temperature goal in the Paris Agreement.

---

21 Science Based Targets, n.d. What are "science based targets"? https://sciencebasedtargets.org/what-is-a-science-based-target (accessed 3 June 2020).

22 City Finance Approach to Climate-Stabilising Targets, 2018. https://www.autodesk.co.uk/sustainability/business-operations/cfact-cities.

23 https://resourcecentre.c40.org/climate-action-planning-framework-home.

24 https://www.london.gov.uk/sites/default/files/1.5_action_plan_amended.pdf.

"Paris-Agreement-Compliant": What Does This Mean?

Many cities may approach the idea of reducing to zero emissions and conclude that with best efforts and all available technology they will not be able to get there on their own. There will be an emissions gap. This gap may be a hard-to-abate industrial sector within the city's boundary or it may just be the need to be connected to a grid that is not on a zero-carbon trajectory and lies outside the city's sphere of control.

As with leaders in the corporate world, these municipal leaders may conclude the need to purchase carbon credits, or invest in negative emissions technologies, in order to offset the emissions that they cannot reduce on their own account. These outside-entity investments *can* help us all along the journey to a net-zero world *if* they are tied to the global carbon budget. To achieve this, the unit of carbon credit supply needs to be counted in the appropriate nation's single, unified GHG inventory. Only then will the credit be netted against the nation's other carbon sources (e.g. logging and carbon-emitting factories) and sinks (e.g. forest preservation, afforestation); and only then will there be a unified picture of reality at the global level. Most organizations fund these sinks, though, by purchasing carbon credits in a flawed marketplace. Subnational, project-based carbon credits may often have a variety of real-world benefits but fall short from the point of view of atmospheric integrity because they do not form part of the relevant nation's GHG inventory.

However, encouraging directions announced at COP25 – for example, by REDD.plus – point to some fixes on the horizon.[25] Powered by the Coalition for Rainforest Nations, this platform will be making United Nations Framework Convention on Climate Change (UNFCCC)-registered REDD+ forest carbon credits issued by nations under the Paris Agreement, and these will be made available to subnational and private actors such as cities to purchase for the first time. This could trigger a significant, voluntary transfer of capital from wealthier, industrial cities to developing countries to reduce deforestation and encourage forest preservation. Under any global scenario that gets the world to 1.5°C, McKinsey point out in their "Climate math" report that a minimum 77% of deforestation's current emissions need to be abated by 2030.[26] This level of reduction will need to be funded by those who can afford it and who cannot necessarily get to "Net-Zero" without purchasing offsets.

Different approaches to financing the emissions gap will affect countries' nationally determined contributions in different ways. Instead of purchasing carbon credits, a city could create quantifiable projects themselves (think urban tree plantings within the city boundary) or investments in negative emissions technologies, to remove carbon dioxide from the air. While these emissions-reduction activities could appeal to that city more than purchasing carbon credits – especially as the former has many spillover community benefits – they might not be captured in the relevant nation's nationally determined contribution as part of the Paris Agreement, and therefore would not contribute to global

---

25 Disclosure – The author is a Senior Advisor to, and supporter of, REDD.plus.
26 McKinsey Quarterly, 2020. Op. cit., pp. 9–11.

carbon budget accounting. For a city to be liveable and also truly "Net-Zero" under the definition of the Paris Agreement, I would encourage a blend of the two approaches: creating and funding high-quality, verifiable in-boundary ("on-site") credits and blending that with UNFCCC-registered credits that are part of the global carbon budget and transfer funds to the countries and communities to abate the vast majority of deforestation emissions within the decade, while avoiding the purchase of non-Paris-Agreement-compliant outside-boundary carbon credits. Most importantly, the city should say explicitly to what extent it is relying on carbon credits and external investments, to what standards these credits are verified, and to what extent its actions affect a relevant nationally determined contribution.

"Cumulative": What Does This Mean?

Now for a suggestion for cities to truly lead on climate. City leaders could, and one might argue should, account for all the carbon their city has sent into the atmosphere up to the point they reach "net-zero" emissions and then beyond. First-principles logic supports this. If a customer kept going to their favourite store and leaving with goods without paying, then started paying for goods after a certain number of visits, the shopkeeper would reasonably expect the customer to settle their old tab at some point. Only then would the shopkeeper be paid in full. On Earth, scientists track the response of the world's natural systems to historic emissions. In the couple of centuries since the industrial revolution, cities have been emitting GHGs at vastly different scales and increasing at different rates. Even fully scoped, science-based, Paris-Agreement-compliant emissions reductions that start today would not eliminate that historical debt.

A fourth descriptor – "Cumulative" – would strengthen the definition of "Net-Zero". It would also encourage cities in developed nations to fund the just transition to a sustainable future; not with unlimited liability but instead surrounded by some rational boundaries. These cities could calculate their historic emissions "debt" and fund carbon credits at a significant scale to help the less wealthy preserve the natural resources on which we all depend. Entities could also invest in carbon removal (just at a significantly higher cost per ton than preserving developing-countries' forests). The key is to calculate how many tonnes of emissions long-term emitters have produced *before* they achieve net-zero emissions for the first time, and to purchase carbon credits to settle the full tab for the city up to that point.

The issue is not just theoretical. Carbon dioxide and some other GHGs, once emitted, remain in the atmosphere for over 100 years. The vast majority of global GHG emissions occurred within the last 80 years. For GHGs like carbon dioxide, nitrous oxide, HFC-23, and sulphur hexafluoride, most of the molecules ever emitted by human activity are still up there, trapping the sun's energy and warming the planet. Like a debt owed to a shopkeeper, nothing substitutes for repayment. Box 1.1 highlights the activity at Microsoft as perhaps a model a city could adopt.

Box 1.1 Case study: If Microsoft was a city – a best-practice model?

Microsoft's definition and target of "carbon negativity by 2030"[27] harmonizes with the approach just outlined. Planners and evaluators can use the four descriptors to discuss, applaud, and challenge their statement.

Fully Scoped: Microsoft has defined its responsibility across scopes 1, 2 and 3 (complete with a jargon-light explainer video). The inclusion of scope 3 emissions sets Microsoft apart. Typical of companies with a complex product range and large reach, its direct emissions are dwarfed by those from its supply chain and its products in consumer use and disposal.

Science-Based: Microsoft's historic commitment to reducing what it calls "operational carbon emissions" runs across several years with reference to peer-reviewed science.

Paris-Agreement-Compliant: Microsoft has correctly spotted that historical carbon credits available for purchase have not been issued by nations themselves, nor tied to independent assessments of any nation's carbon budget as part of the Paris Agreement. Microsoft seems to view sparing trees as secondary to planting new ones and appears to conclude that investing in expensive technologies for reduction and removal is preferable to transferring wealth to developing countries to prevent deforestation. In fact, preventing deforestation and preserving existing forest cover is crucial to achieving Paris Agreement goals.[28] Despite Microsoft's current view on forest carbon with which I would disagree, the company has stated its case clearly and left its options open for future changes in direction.

Cumulative: Most resonant and forward-thinking, Microsoft is pledging to account for all previous emissions by 2050. This is a significant challenge from a 45-year-old technology company to older industrials and younger tech companies alike. Putting it into practice can create methodologies, develop staff experience, and potentially define a new space for intellectual and investment growth.

Microsoft's approach is one of the first corporate examples to have thoughtfully addressed all four criteria for a comprehensive "Net-Zero" strategy. It sets a new bar for companies of any age. Which *city* could be first to do the same?

From Bold Leadership Goals to System Change

To achieve bold, clear climate goals while managing all the other priorities of the complex modern city, we need leaders that are truly connected: to themselves; to the team around them; to their communities and stakeholders; and to an awareness and appreciation of the entire complex system they are trying to manage and change.

---

27 Microsoft, 2020. Microsoft will be carbon negative by 2030. https://blogs.microsoft.com/blog/2020/01/16/microsoft-will-be-carbon-negative-by-2030 (accessed 3 June 2020).

28 Sourcing analysis for scalability of abatement solutions (McKinsey Global Institute, Global Carbon Budget 2015, Coalition for Rainforest Nations). The top three sectoral abatement opportunities are Forests & Agriculture (estimated 10–12 Gt pa). Power at 10 Gt pa and Industry at 8 Gt pa.

There are many cautionary tales of unintended consequences of trying to do the right thing but in the wrong way – parachuting cats into Borneo in the 1950s is still one of the easiest and quickest to illustrate the point (Figure 1.7).[29] Our leaders need to be able to have a *curiosity for the potential consequences* – both intended and potentially unintended – of various actions by actors within a system and be able to map causal loops before they act.

Figure 1.7 Cats being parachuted into Borneo in the 1950s. (*Source*: U.S. Air Force / Staff Sgt. Manuel J. Martinez.)

Too often, leaders also try to change a system in a harmful or less-effective way because of their own perspective. Being aware of the "ladder of inference" is powerful (Figure 1.8).[30] The flow of our thoughts may build from the bottom of the ladder upwards, but it is vital to realize how our inherent subjectivity filters and creates a unique view as compared to others, even if we all "saw" the same "data" or "had the same experience". Even more important, the reflexive loop noted in the diagram shows how as we form our beliefs about the world, it further influences and filters what we even choose to "notice", what data we actually observe and retain from the objective experience that was in front of us.

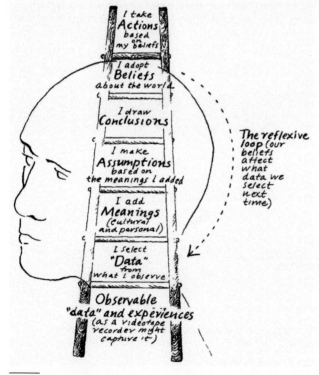

Figure 1.8 Chris Argyris's ladder of inference. (*Source*: Chris Argyris, Overcoming Organizational Defenses: Facilitating Organizational Learning, Allyn and Bacon, 1990. https://books.google.co.in/books?redir_esc=y&id=z7i3AAAAI AAJ&focus=searchwithinvolume& q=%22Ladder+of+Inference%22.)

---

29 Sustainability – Operation Cat Drop. https://www.youtube.com/watch?v=17BP9n6g1F0v=.

30 Argyris, C., 1990. *Overcoming Organizational Defenses: Facilitating Organizational Learning*. Pearson, New York.

In cities, leading an ambitious programme like climate action from our own narrow ladders, without taking the time to "get back to the base of the ladder" and truly understand others' point of view, would not only be unfair; the narrow view will likely be counterproductive to solving the problem at hand. Diversity of thought in the team and in the process not only can keep the individual leader's ladder more "true" and open to new data and experiences, but also can lead to better outcomes. CityStudio Vancouver, which brings city staff together with students, faculty, and community to design experimental projects that improve the city, has long championed a systems approach that builds time into the process to *understand individual perspectives and embrace others' points of view*.

The leaders of the ambitious city need to be *depth-finding* from a systems perspective. The "iceberg" tool is one of the more famous tools in systems thinking and illustrates effectively how little of the system is observable "above the water-line" (Figure 1.9). As the curious leader continually asks "why", the learning is increased while the ease of visibility decreases: from readily observable events to patterns of behaviour and results, to underlying structures of a systems, to the mental models and paradigms that created and sustain the system. The iceberg model can be readily applied to a newcomer arriving in a neighbourhood and taking time to understand how "everything works".

For the ambitious climate leader in the city, architecting a transition to "Net-Zero" would benefit tremendously from a completed "iceberg" diagram, as shown in Figure 1.9 from the Waters Center for Systems Thinking.[31]

The fourth dimension of systems thinking talent useful for leaders of the ambitious city is connected to action: the ability to analyse and *pull the levers* to change the system in ways that further the goal of addressing climate change while also increasing equity and

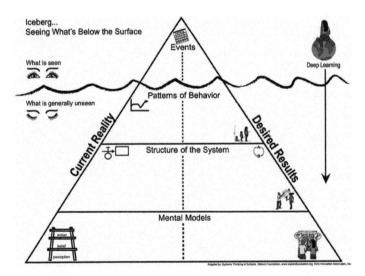

Figure 1.9 Waters Center's iceberg model. (*Source*: Adapted from Aki Fukutani (2015), Iceberg.. Seeing What's Below the surface. Retrieved from: https://yourlearningresource.wordpress. com/2015/02/12/iceberg-model/.)

---

31 https://waterscenterst.org.

social cohesion. Jay Forester defined leverage points "where a small shift in one thing can produce big changes in everything" and are often counterintuitive. Donella Meadows is arguably the definitive source on defining the levers of system change, and the league table reproduced here is drawn from her book:[32]

Places to Intervene (in increasing order of effectiveness):

12. Constants, Parameters, Numbers (subsidies, taxes, standards)
11. Sizes of buffers and other stabilizing stocks, relative to their flows
10. Structure of material stocks and flows
 9. Lengths of delays, relative to the rate of system change
 8. Strength of negative feedback loops relative to the impacts they are trying to balance
 7. Gain from driving positive feedback loops
 6. Structure of Information flows
 5. The rules of the system (incentives, punishments, constraints)
 4. The power to add, change, evolve, or self-organize system structure
 3. The goals of the system
 2. The mindset or paradigm out of which the system (its goals, structure, rules, culture) arises
 1. The power to transcend paradigms

Interestingly, much better researched and most-often tried are the shallower, less-effective methods, while the more-effective levers are less researched and less obvious. The applicability of "lever-pulling" to managing a city to address climate change is energizing, and would support efforts to increase listening, communication, and genuine understanding as deep paradigm-shifting system change efforts, and not just a box-ticking exercise on the way to "Net-Zero". "Net-Zero" is likely not to remain elusive unless these "deep" efforts to change beliefs and social structures are part of the plan.

Conclusion

For the ambitious city to succeed in leading the world on climate action, we have considered how four powerful descriptors could bring new power and clarity to the concept of "Net-Zero" and defined the crucial marker we have to pass to get to a regenerative, sustainable world:

Fully Scoped, Science-Based, Paris-Agreement-Compliant, Cumulative "Net-Zero".

We then considered how the ambitious city benefits from connected leadership that appreciates, embraces, and is trained in complex systems, creating:

Consequence-curious, Wide-perspective-taking, Depth-finding, Lever-pulling systems thinkers.

As Martin has challenged us all in his Introduction, "How we as leading cities solve the climate crisis will inevitably lie in solving these problems at scale and at a pace that is more aggressive than the national governments the cities reside within". The city has always been a magnet to ambition. Imagine your city pointing to a clear goal of "Net-Zero" by 2040; led by purpose-driven, connected leaders; aware of, embracing, and trained in the complex system thinking required to accelerate the just transition.

---

[32] Meadows, D., 1999. *Leverage Points: Places to Intervene in a System.* Sustainability Institute, Hartland, VT.

2

The Civilized City – Introduction

Peter's conclusion in Chapter 1 was for connected leadership within the modern city. In Chapter 2 I am going to take us into the past to get a stronger sense of why we need to be "systems thinkers" and how we can deliver Peter's clarity and ambition in today's city.

Cities themselves are stories. They are not constructed with pen and paper but are built with bricks and mortar. The one thing that remains constant, however, is the human imagination behind them, but just like the way we tell stories changes over time, the way we build and design cities must also change. In Chapter 1, *The Ambitious City*, Peter discusses how city leadership can aim "beyond" its goals; I too look at some of the ways that cities can move beyond their historical infrastructure to rise up and meet the environmental challenges that face the modern city, and, most importantly, how we can apply those learnings for a better future.

I look at the lessons learnt through the history of cities, from Uruk, the world's first city (Figure 2.1), to Silicon Valley. I explore what we have learnt to do and not do, along with the history of innovation, problem solving, and governance, and discover qualities that prove to us cities are going to solve the climate crisis, making the case for why local governments are more capable than national governments of addressing this challenge.

Figure 2.1 Uruk. (*Source*: SAC Andy Holmes / Wikimedia Commons / Public Domain.)

The Climate City, First Edition. Edited by Martin Powell.
© 2022 John Wiley & Sons Ltd. Published 2022 by John Wiley & Sons Ltd.

Many cities are increasing in size and power, leading to environmental degradation, social dis-cohesion, and cultural barbarism. This may seem to predict a dire outcome, but we must remember that the city, as a product of human design, attracts innovative people, absorbing them into its bloodstream, who proceed to invent, inspire, and attract other innovators who synthesize art and technology and move society forwards. From the cultural advancements of Athens and Florence, to the cotton mills of Manchester and the Industrial Revolution, and, most recently, the technological innovations of Silicon Valley (Figure 2.2), human invention has made all the difference in urban development.

The question is which story do we tell next, and, more importantly, how? As the great American writer Mark Twain said, "there is no such thing as a new idea", just as there is no reason as to why modern cities shouldn't benefit from lessons learnt in the past to find new ways of operating together in solutions that help clean up their delicate but improving urban ecosystems. Innovation relies upon the building blocks of the past as much as it does on imagination.

Figure 2.2   The "Spaceship" Campus, called the ring in Cupertino, California.
(*Source*: Dronandy/Getty Images)

2

The Civilized City

Martin Powell

From the dawn of time, well since 4500 BCE, people have been telling stories from great cities – from Uruk and Mesopotamia, to Athens and Rome, to London, Hong Kong, and New York – each city having formed an identity based on events through time attributed to their name. Cities are places where people go to, to live, to work, and to raise families. Cities are unique; they have a singular identity and an individual story. Jerusalem, the only city to exist twice, on Earth and in heaven; Venice, a road surface made of water; New York, the city that never sleeps. These identities were forged from inhabitants and visitors, reinforced through time, and ultimately immortalized in their name. The significance of this will become apparent as we navigate through the city archetypes, the governance structures, and the megatrends that are driving an unprecedented change in the natural course of city development – a change that only cities are truly equipped to handle.

In Italy, the "piazza" is the "city", its symbolic centre where people "descend" to celebrate or protest. It is the soul of local pride, a concept that is close to *"campanilisimo"* – the affection for one's own bell tower – and depending on the level of desire to contribute to local society is how much of a *"campanilista"* I am – how attached I am to my local bell tower. The significance of this place in all cities has defined the great cities across the world. It is a safe place to have a voice, to be heard, to be told, and most importantly for the citizens to transact with the government over the direction of the city.

The magnitude of the problems over which opinions are voiced in the "Piazza" has not changed over time. In Florence in 1784, the local archives[1] talk of a council gathering to discuss the construction of 50 new homes in the grand plan of the 500 needed. The population of Florence then, according to local records, was 78,537. A similar discussion held by the London Assembly, in 2020, spoke of the need to build 50,000 homes in a city of 8 million people – exactly the same magnitude of problem, proportionate to the scale of the city: the same constraints; the same concerns over affordability, access to transport, and access to green spaces; and the same people with divergent interests inhibiting progress. To say we have not learned how to do it would be wrong because cities today are underachieving against these targets to the exact same scale as they did in 1784. It's never fast enough, good enough, or just plain "enough", but it does get done to a point. The risk of

---

1 La Citta di Firenze, Archivio Storico. https://cultura.comune.fi.it.

The Climate City, First Edition. Edited by Martin Powell.
© 2022 John Wiley & Sons Ltd. Published 2022 by John Wiley & Sons Ltd.

trying to meet the demand required when we need it is that we get it so wrong we end up failing completely, and in an urban environment this is unforgivable.

We can show our greatest innovation when the need exists. In *Cities in Civilization*[2], Sir Peter Hall gives an optimistic account of urban history since Athens. Peter was a planner and distinguished academic who refutes predecessors, such as Lewis Mumford, with their dire prophecies that the modern metropolis is doomed. Mumford was clear that cities were "organic" entities that could not exceed their natural limits without terrible consequences. Cities were on a self-defeating quest for increasing size and power, leading to environmental degradation, social dis-cohesion, and cultural barbarism. He cites cities such as Rome and New York. Peter acknowledges that these cities were lawless, overcrowded, and unhealthy, but argues that cities are central to civilization because their sheer size and complexity make them the natural place for "the innovative milieu". Bringing together the critical mass of creative people takes a "great city", because the city enables this network of innovators that ultimately enables society to progress. It is clear the cultural centrality of cities will continue to intensify even in a connected world where virtual networks seem to nullify the need for this city co-location of talent.

Peter observes that past histories of success were attributed to the city being a cultural incubator – Athens and Florence – or a technological innovator – Detroit or Silicon Valley – but now that civilization has embarked on "a marriage of art and technology", a synthesis of these two forms of innovation will create a new culture. The synthesis will take place in large, diverse cities and attract a huge range of skills. It is because of this he believes the city is on the verge of a "golden age".

In the post-war era, the capitalist urban economy showed real resilience, strength, and agility as the entrepreneurial, individualistic, unplanned, marginal, and chaotic nature of cities shone through to define the great cities.

The artistic networks that grew Elizabethan theatre are analogous to the innovative networks that have created Silicon Valley. These networks are the lifeblood of creative urbanism. The specialism of cities is not just the collection of stories told about them but an act of leadership to be known for a specialism driven by the need to employ citizens and keep the ecosystem in balance. Cotton spinners in Manchester were the origins of the Industrial Revolution, ship builders in Glasgow were the centre of global shipbuilding, electrical workers in Berlin, and car makers in Detroit – all defined these cities. None of these cities were perfect, and the average citizen complained about crime, pollution, and a myriad of other displeasures, but they were thriving, on balance, for most citizens were happy to accept the dis-benefits for the wider opportunities the city offered.

It is at this juncture that the power of urban planning and reform would allow the innovative use of resources to improve the lives of those enjoying the booming city opportunities – perhaps it was simply cleaning up the mess of what historically cities have been for centuries – wasteful. The diversity of which cities have chosen to embrace different forms of urbanism is a direct result of the very people these cities attracted to live and work there.

The synthesis of art and technology needs to overcome the issues created by contemporary urbanism; traffic congestion, toxic air pollution, unprecedented rates of

---

2 Hall, Sir P., 1998. *Cities in Civilization*. Pantheon, New York.

urbanization, and the widening wealth gap and inequalities that exist are the wicked problems of our time, perceivably unsolvable issues that we have come to simply accept, or at least try to manage and accept. Cities will find innovative solutions to these problems because they contain the talent and governance to do so, and they are agile enough to make changes that deliver incremental improvement to the lives of all citizens. Cities are embracing the business community through private–public partnerships which now entwine urban planning with the uncertainty of politics, the economy, and business cycles. They are embracing these new ways of operating in order to maintain control over the cities' finely balanced ecosystem as new businesses come and disrupt traditional operations, offering new benefits to citizens but worsening other aspects of urban life in the process. How does the city account for these negative externalities? It builds a new ecosystem of partners and welcomes them to their city!

Why does this give us cause for hope in the fight against climate change? The answer is simple. The impacts are real and are making changes today that are upsetting the finely balanced ecosystem. Cities are having to react through the same innovative and creative approaches they have achieved for centuries – that bring the partners, finance, technology, and new ways of operating together into solutions that work. We must scale these solutions and provide them with context to allow dissemination, replication, and implementation across the planet.

As Jane Jacobs[3] said, the success of cities is security and safety; it is identifiable specialism and it is continual reinvention. With those in mind, I have selected a collection of cities and events through time – a set of cities, each with a lesson to share. It is these lessons that have shaped the modern city. The examples are Uruk, Mesopotamia, Memphis (Ancient Egypt), Rome, Venice, 1665 London, Jerusalem, modern-day London, the "eco-cities", and Christchurch, New Zealand.

Uruk

The Epic of Gilgamesh (c. 2150–1400 BCE) stands as the oldest piece of epic world literature.[4] This great Sumerian poetic work pre-dates Homer's writing by 1,500 years and tells the story of Gilgamesh the semi-mythic King of Uruk in Mesopotamia and his quest for immortality. The hero, King Gilgamesh, leaves his Kingdom following the death of his best friend to find the mystical figure Utknapishtim and gain eternal life. It becomes clear throughout the tale that Gilgamesh's fear of death is actually a fear of meaninglessness, and, although he fails to win immortality by the end of the tale, it is the quest itself that ultimately gives meaning to his life.

From antiquity to the present day, we have come to know this seemingly immortal theme well, its influences refracted across literature and drama whether it be Homer, Dante, John Milton, Ovid, James Joyce, or the Marvel universe! There is a reason this theme has had such wide-reaching consequences. There is something profoundly hopeful about it. Gilgamesh is defined by the pursuit of the unknown, and his bravery to face that

---

3 Jacobs, J., 1969. *The Economy of Cities*. Random House, New York.

4 https://www.ancient.eu/gilgamesh.

which he did not understand and to believe in a better outcome. All of this is reflected in the extraordinary city he ruled over.

Uruk accounts for a number of the world's firsts in the development of civilization. Among these are the origins of writing, the first example of architectural work in stone, the building of great stone structures such as ziggurat, and the development of cylinder seal.[5] Most importantly, however, it is immortal in our minds because it is considered to be the first true city in the world. According to the Sumerian King List, it was founded by King Enmerkar at around 4500 BCE and was located in the southern region of Suler (modern-day Iraq).[6] It began as two separate settlements, Kullaba and Eanna, which merged together to form a town covering 80 hectares; at the height of its development in the Early Dynastic period, the city walls were 9.5 km long, enclosing a massive 450 hectares, and may have housed up to 50,000 people.[7]

This population grouping was a major step forward for civilization and, technically, the very first instance of "urban development" we have on record. From two distinct settlements, construction, pooling resources, city planning, merging different peoples and doubling its population size, creating clear boundaries, and a lot of imagination, a city was born. It is no accident that Uruk was also the first city to develop cylinder seal, which the ancient Mesopotamians used to designate personal property or a signature on documents, clearly meeting the need for the importance of the individual in the collective community. We can all imagine the problems they faced that would call for such an innovation.

Uruk was a successful city also in that it was continuously inhabited from its founding until c. 300 CE when people began to desert the area. It clearly met the demands of its citizens for thousands of years, and although it took many hundreds of years after this for it to be excavated, Uruk has remained an immortal city in our minds. Uruk is a lesson in innovation and imagination.

Mesopotamia

By 3500 BCE, big cities were on the rise in Mesopotamia, and following on from Uruk we can look at some of the other specific challenges cities in this region faced and the innovation and urban technology they inspired.

The growth in population and city size created the same problems then that they do today, with excess garbage, human waste, and its accumulation. In smaller villages people would simply carry their waste to the edge of the village and leave it there. But with ever-expanding city walls and greater population density this became difficult, and by 2900 BCE this collective need led to the invention of individual deep-pit toilets, and by 2500 BCE there is even evidence of a bathroom at the site of Tello.[8]

Unfortunately, we are still struggling to transfer urban technology to underdeveloped rural areas today.

---

5 https://www.ancient.eu/uruk.

6 Ibid.

7 https://staff.cdms.westernsydney.edu.au/~anton/Research/Uruk_Project/history.htm.

8 https://en.wikipedia.org/wiki/Mesopotamia.

Figure 2.3 The remains of Mesopotamia provide key insights into the benefits of urban living. (*Source*: Fat Jackey/Shutterstock)

With a city and the increase in population there are a lot of people, mainly women and children, needing food and work. Mesopotamia recognized this and created a job market. This included standardization of the industries of pottery and thread-making, which up until then had been made individually with fine detail. By 3500 BCE, they were mass produced to a poor quality in what some records show as industrial-weaving factories that paid its workers in food and clothing. As is still true today, women were paid less than men.

Nonetheless, these advancements show a real ability to meet the new urban demands and ingenious industry development (Figure 2.3). Another fascinating example is found in Uruk, at around 3200 BCE, when a large religious complex was designed and took 100 years to be built, but once it was built it was immediately flattened and rebuilt.[9] There is no reason for this at all, so it seems that this unorthodox project was inspired by the government in an attempt to keep the city at full employment by providing 200 years of work. The government's dedication to its people's wellbeing is admirable in this sense. Mesopotamia is a lesson in providing its people with both convenience and opportunity.

Memphis – Ancient Egypt

The Greek historian Herodotus once said, "Egypt is the gift of the Nile", a saying that describes just how much Egypt relies, and still relies, on the Nile. Since the days of the Pharaohs, the Nile has been the main source of Egypt's water for agriculture. The yearly

---

9 Ibid.

flooding of the Nile was an important factor in ensuring a good harvest. The river was also an important trade route and mode of transport.

It can therefore be seen that Egyptian civilization developed along the river. The majority of Egyptian cities were located on the east side of the river, whilst the tombs and pyramids were built on the west side, mirroring the life and death cycle of the sun's daily route. All cities and settlements were built on the edge of the desert, with relative distance to the river so they would remain dry during the yearly floods. The land close to the Nile was considered too precious to build upon and remained the place for valuable crops to be grown and harvested.

Egypt's first Pharaoh, Menses, the source of many legends, unified Upper and Lower Egypt and established his capital just a few kilometres to the southwest of modern Cairo. Not wanting to favour Upper or Lower Egypt, Menses decided to build the new capital on the border between the two. The city was called Men-nefer, or, as the Greeks later called it, Memphis.

Memphis was one of the largest and most important cities of its day, with archaeologists predicting that as many as 100,000 people may have lived in it at the height of its power.[10] As the capital, it was also the seat in which Menses ordered Egypt's first irrigation system to be built in 3100 BCE.[11] The unpredictability of the Nile caused problems ranging from excessive flooding to droughts, but it did, however, also lead to the invention of water dams – one of ancient Egypt's greatest archaeological feats, with the foundations of this technology still in practice to this day. This allowed water from the Nile to be diverted into canals and lakes, considerably reducing the chance for water-related disasters.

The invention of water dams provided an infrastructure within the cities, such as Memphis, that lined the river's edges that faced a high environmental disaster risk and made them more resilient, whilst also ensuring the safety and welfare of its citizens. It is an excellent lesson of cities investing in resilience and an act of quick and decisive leadership and governance.

Rome

There is a reason why Rome remains so present in our minds when it comes to talking about cities, infrastructure, equality, subjugation and slavery, television dramas, movies, and even Shakespeare. Rome is dramatic, mysterious, and unique, and it is also like any other city today. It expanded, conquered, innovated, succeeded, and failed, and like all civilizations it eventually withered away and died. In many ways, Rome is inspiring, but there are also many lessons to be learnt from it.

Rome is the perfect example of how well positioned cities are to govern on a global level. Unlike the other cities we have looked at that fold back into the region in which they were established, Rome quite literally took over the world. I say "Rome" because it was a dominating identity that remains to be our prevalent descriptor in the course of history

---

10 https://www.laits.utexas.edu/cairo/history/ancient/ancient.html.

11 https://www.history.co.uk/shows/ancient-top-10/articles/10-things-the-ancient-egyptians-gave-us.

– whether you lived in the capital or not you were a Roman; there was no country but an empire – the Roman Empire.

For sustainable development and the economic growth that goes along with it to be truly effective, it must go hand in hand with social inclusion and respect for the environment. The Roman Empire lasted 500 years in part due to its ability to respect these unspoken bonds. It created social equality by giving the plebeians equal rights to the patricians, creating a democracy. It prioritized public service and the common good. It imported grain, servicing and feeding its population. It outsourced its material needs to foreign slaves (not necessarily something to be admired but in line with the times). This is governance we still see today with modern companies outsourcing the vast majority of things they buy abroad. China now produces 50% of the world's clothes and 70% of its mobile phones.[12]

Inevitably, the fault lines erupted, and it all went downhill with territorial expansion, corruption with the new wealth market not shared, wars and uprisings within conquered lands, national debt, and inflation.

Another, less well-known factor was climate change. As documented by Kyle Harper in *The Fate of Rome: Climate, Disease, and the End of an Empire*, the period between roughly 200 BCE and 150 CE is now known as the Rome Climate Optimum.[13] This was a period of a warm, wet, and predicable climate that helped harness the empire's agricultural crops. The climate became cooler and dryer in the third century, resulting in droughts and crop failures, and by the fifth century the Late Antique Little Ice Age arrived. The changing climate reduced the empire's resilience to a variety of shocks, such as a smallpox pandemic and the outbreak of the Plague of Justinian. The decade before the outbreak of the plague in the sixth century saw some of the coldest temperatures in millennia brought about by a series of massive volcanic eruptions in central Asia, which likely forced gerbils and marmots out of their natural habitats, causing the bacteria-bearing fleas they carried to infect the black rats whose population had exploded along Rome's network of trade routes.

A weakened population, mass deaths, a dying democracy, war, inflation, national debt, and corrupt governance, helped along by a climate crisis, saw the fall of Rome. But this does not take away from the lesson that Rome is perhaps the greatest example we have in recorded history of a city with effective, far-reaching leadership. They recognized that sustainable development and economic growth must go hand in hand with social inclusion and respect for the environment.

Venice

Venetians have lived on water ever since the first Venetians chose to settle in a mosquito-filled marsh in the northern Adriatic Sea. These fifth-century settlers were fleeing German and Hun invaders and probably picked the area, which consisted of dozens of disparate

---

12 https://www.imf.org/external/pubs/ft/fandd/2019/03/ancient-rome-and-sustainable-development-annett.htm.

13 Harper, K., 2017. *The Fate of Rome: Climate, Disease, and the End of an Empire*. Princeton University Press, Princeton, NJ.

islands surrounded by a 200-square-mile (517-km^2) shallow lagoon, for the protection it offered them.[14]

They harnessed the natural landscape to build a city by driving wooden poles into the ground, creating plank and marble foundations, and redirecting the rivers to the sea away from the lagoon[15] – problem solving at its finest. If the original Venetians hadn't done this, Venice as we know it today would be Italian coastline and not much else.

Venice is the city where the streets are made with water. There are no additional lanes or roads that have to be built, road traffic is purposefully kept to a minimum, and this has in turn made everyone walk. Venice is an inspiration for cities who want to reduce their car use and diversify their transport system. Venice was one of the first cities to hit its GHG emissions peak before the 2020 target.[16]

Venice faces many challenges today, including rising sea levels and shifting tectonic plates that technology and urban development are attempting to keep up with, but the original premise is still to be admired. For a city founded for the purpose of protection, it seems to still be in line with its original values (Figure 2.4).

Venice has endured, with a concept that is wholly unique. It has taught us a model for sustainable multimodal transport without car use. It is an enduring proof-of-concept.

Figure 2.4   Venice today, a city founded for the purpose of protection, now ironically under threat from rising sea levels. (*Source*: Miiisha/Shutterstock)

---

14 http://www.bu.edu/articles/2014/lessons-from-venice.

15 Ibid.

16 https://www.weforum.org/agenda/2019/10/carbon-emissions-climate-change-global-warming.

1665 London

History has shown us that London is a city that "applies" its learnings from its disasters.

The Great Fire of London

In 1665, the Great Plague swept through London, killing about 200,000 people, almost one-quarter of London's population.[17] The people of London were to face another disaster a year later when a fire started on 2 September in the King's bakery in Pudding Lane near London Bridge. It had been very hot that summer and there had been no rain for weeks. The wooden houses and buildings were tinder dry, so the fire spread rapidly across the city, its population powerless to stop it.[18]

As a result of the Great Fire, 80% of the city was destroyed: 13,200 houses, 87 churches, the Royal Exchange, Newgate Prison, and Bridewell Palace. It left 80,000 Londoners homeless, a fifth of the city's population at the time and equivalent to almost two million people in today's capital being made homeless.[19]

The Great Fire was not the first occurrence of disastrous fires to afflict London. St Paul's Cathedral was first built in 605 and 70 years later was destroyed by a fire that swept through the city and once again in another fire in 1087. Alongside these disasters were two medieval fires that caused massive damage to London in 1135 and 1212, which resulted in the first Building Act of 1189 that legislated standards for building materials and footprints but clearly did not go far enough.

The Great Fire inspired a series of measures to prevent future fires. Each parish now had two fire "squirts" (an early attempt at the fire engine), which set the foundation for today's fire brigade. The London Fire Prevention Regulations of 1668 also established a new water supply pipework and infrastructure, the origins are which are seen in modern fire hydrant systems.[20] But it was the London Building Act of 1667 that takes its place as the single most influential piece of construction-related legislation in British history. It regulated storey-heights, banned timber facades in favour of brick and stone, and banned thatched roofs.[21]

Many of these structures are still enforced in building regulations today, with the roof of the new Globe Theatre being the first thatched roof permitted in London since the Great Fire. The real success of the 1667 Act, however, was that it was the first-time money was allocated by the state to employ surveyors to enforce building regulations.[22] Thus, a profession that still exists today and our entire modern building control system was born.

It is also worth mentioning that the Great Fire helped bring about the end of the Great Plague by killing all the rats – another lesson London learned from disaster.

---

17 The Museum of London, 2011. The Great Plague of 1665. https://simple.wikipedia.org/wiki/Great_Plague_of_London#cite_note-mol-1 (accessed 28 January 2016).

18 https://www.historic-uk.com/historyuk/historyofengland/the-great-fire-of-london.

19 https://www.building.co.uk/focus/how-the-great-fire-shaped-modern-london/5083502.article.

21 Ibid.

20 Ibid.

22 Ibid.

Cholera

In 1854, at the height of a new cholera epidemic in the Soho district around Broad Street, John Snow, a physician, plotted the course of the cholera outbreak on a map he drew himself and worked out from the map that all the victims had used the same Broad Street water pump. As an experiment, Snow removed the handle from the pump so no one else could drink from it and from then onwards no one else in the area became infected. Snow had stopped the spread by removing the source of the disease – the water pump. He discovered the connection between contaminated water and cholera.

Sewage

The discovery that cholera was a waterborne disease was pivotal in the city's urban transformation. The Thames was little more than an open sewer system with zero wildlife and represented a public health hazard as much of the city's waste ran freely through the streets and thoroughfares directly into the river. The city depended on a system of local waste disposal such as night-soil collectors to empty local cesspits and the city's rivers, which also served as a source of drinking water and washing.[23] The increasing use of the flush toilet made things worse as it allowed the wealthy to flush their excrement directly into the river. The summer of 1854 saw the "Great Stink" engulf London. The hot weather helped expose the pollution of the water of the Thames, and waterborne diseases like cholera and typhoid swept through the population. The smell was so overpowering that *The Times* newspaper said that MPs had been "forced by sheer stench" to rush a bill through Parliament in just 18 days in the effort to provide money to construct a new sewer system for London.[24]

The engineer, Sir Joseph Bazalgette, was commissioned to design a network of enclosed underground brick main sewers to intercept sewage outflows, and street sewers to intercept the raw sewage that ran freely along them. At a time when London's population totalled 1 million people, Bazalgette made the ingenious decision to build it for 4 million people, "ensuring it will never need to be expanded". The results were enormous, with the end of the cholera epidemic, improved public health, and innovative transformation of London's sewer system.

With an ever-expanding population of 8 million, London still faces sewage issues to this day, with sewer overflow emissions happening about fifty times per year. This has resulted in a brand-new super-sewer, the Thames Tideway, being dug under London to intercept sewage that would otherwise pollute the river. Seven metres wide, it will run for 25 km, impressing Bazalgette I'm sure.

London is a lesson in reinvention. Its ability to enact new policy and reform to create a better place, learn from its disasters, and cater to its citizens' most basic needs with progressive change is a challenge we are still facing today.

---

23 https://www.museumoflondon.org.uk/discover/how-bazalgette-built-londons-first-super-sewer.
24 Ibid.

Modern London

The Great Smog of 1952 began with a simple veil of fog that was not unusual for grey, cool, misty London. Within a few hours, however, the fog began to turn a yellowish-brown colour as it mixed with the soot produced by London's factories, chimneys, and diesel-fuelled cars and buses. This combination produced a poisonous smog the likes of which London had never seen before.

A high-pressure weather system had stalled over southern England and caused a temperature inversion that sent a layer of warm air high above the ground, trapping the poisonous cold air at ground level. It prevented London's sulphurous coal smoke from rising and prevented the wind from dispersing the smog. It was a noxious, reeking, 48 km-wide air mass, which was so dense that Londoners were unable to see their feet as they walked.[25]

The Great Smog paralyzed London, with all traffic and public transport coming to a halt. Abandoned cars lined the streets, parents were advised to keep their children home from school, looting and burglaries increased, movie theatres closed, and a greasy grime covered all exposed surfaces.

The Great Smog lasted for five days and finally lifted when a wind from the west swept the toxic cloud away from London and out to the North Sea, but the damage was already done and the effects were lingering. Initial reports suggested that approximately 4,000 died prematurely in the aftermath of the smog, but many experts now argue that the Great Smog claimed between 8,000 and 12,000 lives.[26] The elderly, young children, heavy smokers, and those with respiratory problems were particularly vulnerable.

Heavy fog was a common occurrence in London, and as a consequence there was no sense of urgency. The British government was slow to act, until the undertakers ran out of coffins and flower shops ran out of bouquets. Following a government investigation into the link between deaths and the Great Smog, parliament passed the Clean Air Act of 1956, which restricted the burning of coal in urban areas and implemented smoke-free zones.[27]

The transition from coal, as the city's primary heating source, to oil, electricity, and gas took years, and there were other deadly fogs during this period. Slow action remains to be the greatest problem when it comes to cities enacting clear air quality protection measures, but the Great Smog did teach us about the lingering effects of air pollution and the disastrous human cost.

Ken Livingstone served as the Mayor of London from 2000 to 2007 and implemented a series of innovative measures in the city, including improved public transport and priority bus corridors. His most recognized initiatives, however, were a congestion charging policy and a Low Emission Zone. High-polluting vehicles pay a fee to come into the city, and all drivers (with valid exceptions) pay a specific congestion fee to access the central part of

---

25 https://www.history.com/news/the-killer-fog-that-blanketed-london-60-years-ago.

26 Ibid.

27 Ibid.

the city, creating space on city streets, redirecting the demand to drive elsewhere in the city to other modes of transport, and reducing carbon emissions. It is a policy that has inspired cities around the world, with New York and Milan adopting similar measures.[28] It also serves to illustrate perfectly how cities, which hold 50% of the world's population, are uniquely positioned, as well as having a unique responsibility, to tackle the climate crisis head on.

The lesson of how London implemented a largely unpopular set of measures is testament to its ability to listen to the plight of its citizens, respect the data, and above all to recognize the greater good for the city and be tenacious in putting it in place.

Given the lessons from 1665 London through to modern London, it can be argued that the Grenfell fire should never have happened. However, I am confident that London will apply the lessons from this tragedy to avoid such an event happening again.

Jerusalem

The holiest place in Judaism is Temple Mount in Jerusalem where the last remnant of the Second Temple, the outer Western Wall, stands. Each year, thousands of people, of all faiths and levels of religious observance, flock to the Western Wall (also referred to as the Wailing Wall) to pray, sight-see, and hopefully learn for themselves the incredible story behind this site.

The Western Wall is the last remnant of the Second Temple, meaning that before the Second Temple there was also a First Temple. The First Temple was built in 1000 BCE by King Solomon in the years after King David conquered Jerusalem and made it his capital, but the city was captured by the Babylonians and the Temple was then destroyed in 586 BCE.[29] In 538 BCE, the Jews returned from their exile to Jerusalem, and by 515 BCE they had built the Second Temple. The Second Temple spanned three significant historical periods: the Persian period, the Hellenistic period, and, finally, the Roman period.[30] During the Roman period it was eventually destroyed by Titus, along with the city of Jerusalem. Jewish tradition tells us that both temples were destroyed on the Ninth of Av on the Jewish calendar (usually July or August on the Gregorian calendar). Every year the destructions are commemorated by a day of mourning called Tisha B'Av.

Nowadays, the Western Wall is all that remains, and it stands at Temple Mount alongside the Al-Aqsa Mosque, the third-holiest site in Islam, and the gold-topped Dome of the Rock, one of the most famous symbols of Jerusalem. It is also in close proximity to the Church of the Holy Sepulchre, synagogues, and mosques. The site has become the seat of

---

28 https://thecityfix.com/blog/ken-livingstone-interview-london-mayor-congestion-charge-city-leadership-caroline-donatti.

29 https://www.moon.com/travel/arts-culture/jerusalem-history-first-second-temples.

30 Ibid.

the three Abrahamic faiths – Judaism, Islam, and Christianity – and a symbol of a diverse, tolerant, and international city (Figure 2.5).

This architectural palimpsest tells the extraordinary story of Jerusalem through a single structure. The cycle of destruction and reconstruction, of survival and preservation, is inherently hopeful and inspirational to the cities of today. Jerusalem is a resilient city. The city employs strict security, and behavioural measures are taken to protect the integrity of Temple Mount. The site itself has also proved helpful in attracting tourism, with it often being listed on various "Top Ten" things to do in Jerusalem, which by extension has provided economic benefits to the city.

In recent years the Western Wall has also become the place of social action on the subject of women's equality. For many years, ultra-orthodox men opposed women praying at the Wall, sometimes going as far as violently dispersing dedicated women's prayer services. The Wall itself is split into two sections. The larger section is dedicated for men, and the second, much smaller, section is dedicated for women's prayer. An advocacy group called Women of the Wall was founded, and in 2016, the Israeli government approved the creation of an egalitarian prayer space where non-Orthodox men and women could pray at the Wall[31] – a clear victory in the fight towards equality.

Jerusalem is the template for multifaith living in a small urban space, a place that has learned to be tolerant but has had to move with the times and be inclusive to all.

Figure 2.5 Jerusalem, the only city to exist twice – in heaven and on Earth. (*Source*: Josef/Adobe Stock)

---

31 https://www.jewishvirtuallibrary.org/history-and-overiview-of-the-western-wall.

Eco-cities

Dongtan, Shanghai

In 2005, the Shanghai Industrial Investment Corporation (SIIC) hired Arup, a design, engineering, and business consultancy firm, to design and plan a city that could bring increased sustainability to a region suffering with overcrowding and pollution. The result was Dongtan eco-city. Dongtan was conceived as a futuristic model for low-rise suburbs, accommodating spillover from the big cities and China's growing middle class. It was to be a carbon neutral and zero waste city, with a myriad of renewable energy systems; it would ban cars, recycle water, and surround itself with organic farms and forests.[32]

It failed for a number of reasons:

- The project stalled in 2006 after key officials, most significantly Shanghai's top bureaucrat Chen Liangyu, were arrested for bribery and fraudulent real-estate transactions and given long prison sentences.
- Dongtan's planners failed to comply with government land-use policy.
- Built on Chongming Island in the Yangtze delta, many of the people originally living there were displaced, calling into question ethical standards.
- During this time, Chongming county officials were able to outpace the SIIC and proceeded to confiscate farmland, relocate peasants, and invest in infrastructure.

In her book *Fantasy Islands*[33] Julie Sze, whose father hails from Chongming Island, argues that Dongtan failed due to not just corrupt officials but also the failure to take into account how "technology, engineering and politics are intimately woven together".[34] According to Sze, ambitious green projects, engineered "ectopia", have become symbols of technological excess that have overshadowed smaller more effective steps that can sustain environmental improvements.[35]

To succeed, it is clear that these ambitious projects cannot be sustained by political structures alone but require individual citizens who must themselves be part of the plans. This may also teach us the desire for incremental change as a means of grasping new realities more easily rather than radical new approaches that make people who have to accept the change nervous.

Masdar, UAE

First conceived of in 2006, Masdar promised to be the world's first planned sustainable city and spearheaded the plan to diversify the UAE's economy from fossil fuels to green energy. Developers envisioned a city of 50,000 residents and 40,000 commuters from Abu

---

32 https://chinadialogue.net/en/cities/7934-why-eco-cities-fail.

33 Ibid.

34 Sze, J., 2015. *Fantasy Islands*. University of California Press, Berkeley.

35 Ibid.

Dhabi.[36] It would be powered by renewable energy with driverless electric vehicles, water recycling, argon-insulated buildings, and wind towers.[37]

Masdar suffered a major setback due to the 2008 economic crisis, and since then barely any of the city has been developed and it has far exceeded its 2016 completion deadline, with it not predicted to have finished construction by 2030.[38] Despite its claims of embodying the future city, Masdar now remains in the past; first conceptualized as the world's first zero-carbon city, it may in fact remain just that – a concept. Its GHG-free status has also been abandoned, with the design manager Chris Wan arguing, "We are not going to try to shoehorn renewable energy into the city just to justify a definition created within a boundary. As of today, it's not a net-zero future[,] it's about 50 per cent."[39]

Currently, the world's only sustainable ghost town, Masdar houses only students studying sustainability at the Masdar Institute of Science and Technology, and it is another example of a well-intentioned but overly ambitious project that did not account for the overlap between technology and the political and economic climate that surrounds that innovation.

Neom, Saudi Arabia

Neom is a planned, cross-border, smart city in the Tabuk province of northwestern Saudi Arabia. What can Neom learn from Dongtan and Masdar that can save it from partial construction and untimely abandonment by its government? How can it become a fully functioning smart city and a working model for a sustainable city?

First of all, it actually has to get people to live there. We saw with Dongtan how the individual citizen was not included in the city planning, and Masdar houses only the students at its university. Neom, on the other hand, plans not only to be a smart city but also to function as a desirable tourist location, a positive investment in regards to the city's future population and economic growth. These smart eco-cities cannot just rely on a novelty factor of being "the first" or "the only" sustainable city in the world. People are what make a city – without them it is just an ambitious project. People will make the difference between success and failure for Neom.

These eco-city experiments show us the need to intertwine the people and the political landscape in the project. We have seen examples of new housing being built for inhabitants of favelas in Rio and townships in South Africa only to discover that they did not want to move. It is dangerous to predict how people want to live their lives. Perhaps investment in new improved services in the existing setup would be more acceptable or provision of transport infrastructure, like the cable car in Medellin, to allow those communities to access the job market. This gives them choice.

---

36 https://grist.org/climate-energy/the-worlds-first-zero-carbon-city-is-a-big-failure.

37 Ibid.

38 Ibid.

39 Ibid.

Peter Hall[40] looks at the inconclusive story of London's strategy to find a new airport, which after 40 years is no nearer resolution. This is a lesson in adopting big ideas, and one that Jane Jacobs warned those who attempted drastic surgery on the city. Small, she believed, was the future. Cities should be nurtured not traumatized.

Christchurch, New Zealand

In 2011, a violent earthquake killed 181 people in New Zealand's second largest city – Christchurch. On top of that, thousands were made homeless, and an area that was four times bigger than London's Hyde Park was deemed unhabitable.[41] Christchurch is still recovering and rebuilding from this natural disaster and is aesthetically and spiritually a completely different city to what it once was. With new constructions, gravel quadrants replacing multistorey buildings, and others that are still even boarded up, many of the city's inhabitants can no longer remember what it once looked like. A new Christchurch has been born.

In the immediate aftermath of the earthquake, the central government triaged the most important parts of a functioning city – roads and bridges had been destroyed, silt clogged up the sewage system, and powerlines were down. The government's main response was to establish a single body, the Canterbury Earthquake Recovery Authority (CERA), which was to be solely responsible for managing the rebuild.[42] This single-purpose organization was tasked with the demolition of buildings and residential homes, with almost 8,000 of the area's homes "red zoned", meaning the land was so badly damaged it was unlikely it could ever be built on again.[43] Within the space of a year, the population of the greater Christchurch area declined by 9,200, 2% of its total population, due to home loss.[44] The city's mayor, Lianne Dalziel, ushered in a new era of governance that focused on empowering community organizations to do things for themselves in a sort of grassroots-based resilience project.

Thus began the construction of a new Christchurch. CERA came up with the city's first urban blueprint plan by taking more than 100,000 ideas from the local community, and a team of local and international architects and designers spearheaded a super-project which involves 70 projects being constructed over the next 20 years or so.[45] The project imagines a central business district of low-rise buildings, a green frame and corridor of parkland, environmentally sensitive transport including a light rail network, pedestrian broad walks, and cycle lanes. The local government also saw the opportunity to improve Christchurch's economic infrastructure, with pre-earthquake data telling them that the city's retail areas were not competing well against the rise of shopping malls.

---

40 Hall, P., 1982. *Great Planning Disasters*. University of California Press, Oakland.

41 https://www.theguardian.com/cities/2014/jan/27/christchurch-after-earthquake-rebuild-image-new-zealand.

42 Ibid.

43 Ibid.

44 Ibid.

45 Ibid.

Christchurch is an example of efficient and vision-led local government action. We hear all too often of the failure of local and national governments to deal effectively with these disasters, with many agencies not funded adequately and consequently destroying all hope of a quick response.

When the loss of life is apparent and a new path must be forged, local governments are very effective at embracing the community and moving forward.

Conclusion

The purpose of this chapter is to show there is a precedence for everything we need to achieve in solving the climate crisis. We have overcome catastrophic disasters throughout history, and we have emerged and rebuilt better each time.

The lessons from Uruk, Mesopotamia, Memphis (Ancient Egypt), Rome, Venice, 1665 London, Jerusalem, modern-day London, the eco-cities, and Christchurch are lessons learned at the time and carried forward into the new cities formed around them. The modern city is a product of all these lessons, but with pressures of unprecedented urbanization and the need for transformation to address the climate crisis they once again have to rebuild to a new and higher standard.

While the housing problem of 1784 Florence shows us that the problems of today have not changed, it is clear that the scale of the problems has, and, therefore, so too have the consequences. This is our challenge.

3

The Emerging City – Introduction

Following on from the environmental challenges that urban cities face in cleaning up their urban ecosystems discussed in Chapter 2, *The Civilized City*, Austin now moves us forward in time to discuss the problems facing developing urban cities.

Throughout the chapter, Austin investigates urban conditions and the battle between the economy and the environment, using Malawi in Africa and China in Asia as examples, in an attempt to understand how cities within these countries and continents reflect their historical socioeconomic national conditions.

We see the struggle of Malawi and other underdeveloped African nations to balance sustainable development with poor economic infrastructure along with the legacy of colonialism, continuing interference from the developed world, and current neocolonialism. Austin also looks at Malawi's attempts to cater for the increased immigration into the cities from its rural surroundings.

Alongside this is China's progression from communist pariah state to one of the world's leading economies, and, a less well-known fact, the global leader in wind energy production, and as one of the first countries in the developing world to introduce sustainable development at a national and regional policy level (Figure 3.1).

Austin explores China and Malawi as examples of fast-growing economies within nations united by the fact that neither is classed as a "developed" country. The emerging city is at a great disadvantage here, as Austin writes: "underdeveloped countries are seldom able to control their own destiny in the way that it is hopefully expressed in their aspirations for urban renewal".

So whilst the civilized city must learn how to build sustainably, the emerging city cannot always afford to do so. This chapter explores how poor environmental conditions are a consequence of growth, but more importantly it explains how they can be rectified.

The Climate City, First Edition. Edited by Martin Powell.
© 2022 John Wiley & Sons Ltd. Published 2022 by John Wiley & Sons Ltd.

Figure 3.1 National policy is driving the scale and pace of wind turbines in China.
(*Source:* chinaface/Getty Images.)

3

The Emerging City

Austin Williams

This chapter is a snapshot of the emergent urban conditions in China and Africa and an attempt to understanding how cities reflect their historical socioeconomic national conditions, using urban examples from the Sub-Saharan African continent and also from east Asia. It is an exploration, in microcosm, of wider concerns and constraints on urban discourse, generalized over two continents to try to understand how particular responses to environmental and developmental issues are generated. It also explores how those responses are often not the result of free choice, democratic engagement, or sovereign will.

Malawi is a landlocked country bordered by Mozambique, Zambia, and Tanzania, with a border alongside and across Lake Malawi, one of Africa's Great Lakes. The region was arguably first brought to global attention by David Livingstone's missionary adventures up the Zambesi river in the mid-nineteenth century, an expedition that was intended to bring trade and Scottish Presbyterianism to remote regions in central Africa. As a consequence of this imperial adventure,[1] large swathes of land became part of Britain's colonial spoils: the nascent country of Malawi initially fell under British Protectorate status in various guises (including Nyasaland, tied to northern Rhodesia), and finally gained independence in 1964 under the government of President Hastings Banda.

Approximately 9,000 km northeast of Malawi's national capital Lilongwe lies the vast land mass of China, the most populous nation on earth. China is second only to the USA in national net worth, is over 60% urbanized, and – until coronavirus – was the fastest growing major economy in the world. This industrial, commercial, economic, and urban "growth miracle"[2] has occurred within one's lifetime, transforming the perspectives, lifestyles, and life chances of Chinese people within a generation. Its economy grew at an average rate of 10% since Deng Xiaoping initiated market reforms in 1978. One of its territories – Hong Kong Special Administrative Region – has just 1% of Malawi's land area but creates a GDP that is 2,500% larger.

---

1 Blog editor, 2013. African Nationalist or Imperial Agent – David #Livingstone analysed. 23 April 2013. https://blogs.lse.ac.uk/africaatlse/2013/04/23/ african-nationalist-or-imperial-agent-david-livingstone-analysed.

2 Yang, L., 2013. China's growth miracle: Past, present, and future. United Nations Research Institute for Social Development, Geneva. https://www.unrisd.org/80256b3c005bd6ab%2f(httpauxpages)%2f2893f14f4 1998392c1257bc600385b21%2f$file%2fchina's%20growth%20miracle%200808.pdf.

The Climate City, First Edition. Edited by Martin Powell.
© 2022 John Wiley & Sons Ltd. Published 2022 by John Wiley & Sons Ltd.

These two wildly diverse locations, histories, and economies, one in east Africa, one in east Asia, are united by the "fact" that neither of them is officially classified as a "developed" country. This anomaly is primarily because their GDP per capita falls below the internationally recognized benchmark for developed economic status. Malawi is clearly one of the most underdeveloped nations on Earth, while the economic behemoth of China is classified as a "developing" nation even though China's GDP is likely to rival America in the coming decade. Its "developing" status is due to the fact that its GDP share per head of population is still woefully low at around 15% of that of the USA. Partly because of its vast population, China's GDP per head is US$9,700 (compared to US$62,970 per person in the USA). Malawi's GDP languishes at around US$390 per person.

When President Xi Jinping came to power in 2012, he promised that China would become a "moderately well-off society" by 2021. That date coincides with the 100th anniversary of the formation of the Chinese Communist Party, and so there is a lot of political credibility riding on its realization. For almost 40 years, China experienced double-digit growth year on year, while Malawi's growth potential in 2017 was estimated at around 3.7–4.4% and from a much weaker base.

Since 1980, China's urban population has risen from 11% to over 60% (with the number of Chinese cities rising from 193 to the current level of 653).[3] Malawi's urban population was 11% in 1990,[4] rising to 17% today.[5] The disparities are not always straightforward, and even though Malawi also has 17.5 million of its population living on less than US$5.50 a day, China has 373 million in the same situation (albeit Malawian poverty represents 97% of the population compared to 26% of the entire Chinese population[6]).

Neither China-watchers nor financial specialists predicted the destabilizing effects of COVID-19 and the ultimate impact of a global pandemic on economies large and small. The consequences are yet to be revealed (and hence outside the scope of this contribution), so this chapter is intended merely to explore the generalized relationship of development and sustainable development using historical precedent. Through the prism of urbanization, this chapter compares two significantly differing countries – Malawi and China – and their communalities and differences, ambitions and challenges.

We will explore how Malawi, among many other African states, is still in hock to supranational finance and how any escape from subjugation needs to overcome environmental constraints. We will cast an eye over the dynamic shift in the direction of trade to see whether it is changing the terms of the debate. And we will look at the various historic and contemporary forces holding back – or in China's case, liberating – development.

3 Williams, A., 2018. *China's Urban Revolution: Understanding Chinese Eco-Cities.* Bloomsbury, London, pp. 176–177.

4 UN Department of Economic and Social Affairs, 2019. *World Urbanization Prospects: The 2018 Revision.* United Nations, New York.

5 World Bank, n.d. *Urban Population (% of Total Population).* World Bank, The International Bank for Reconstruction and Development, and the International Development Association, Lilongwe, Malawi.

6 Anon, n.d. The World Bank in China. World Bank. https://www.worldbank.org/en/country/china/overview.

Emerging from History

Malawi in its former guise was a country clearly held back by conflicting colonial ambitions during the Scramble for Africa at the turn of the twentieth century. The political and economic rivalries played out primarily between Britain and Portugal over the financing of railway infrastructure, amongst other things, ended with the country having to foot the bill for a rail network primarily designed for and demanded by foreign powers. The loan repayments imposed on Nyasaland for this and other infrastructure investments were punitive. Writing about events prior to and after the First World War, Leroy Vail, a specialist in African studies, says that the British Treasury "helpfully suggested that Nyasaland should practise economies and increase her taxes so as to meet the guarantee charges she was facing ... the Treasury was not impressed by the entreaties from Nyasaland, and the Protectorate entered upon what a later governor called 'the Times of Starvation'. Nyasaland was henceforth prevented from devoting her revenues to such things as African education, medical services, road building, agricultural development and the establishment of veterinary service."[7]

From such penurious and exploitative beginnings, Malawi continued to be mired in debt. Sixty years on from independence, the country remains one of the world's poorest in the world. According to the World Bank, 51.5% of the population live in poverty, agriculture is the primary source of employment, and only 11.4% of the population have access to electricity.[8] Even now, the country's manufacturing industry is virtually non-existent.

Poverty, in terms of US$ per day, is a measure that doesn't adequately do justice to the grinding privation of a country like Malawi that remains one of the most densely populated countries in Africa and yet one of the least urbanized, with more than 80% of its population living in rural areas. In 2013, the World Bank ranked Malawi 171 out of 189 countries to do business in. By 2020, it had risen to 109th (Eritrea is 189th). But the fact that a supranational institution has flagged up that foreign investors should be alert to the opportunities to make a profit in Malawi is not necessarily a good thing. Malawi is in desperate need of investment and financial assistance, but such external intervention has long imposed an overbearing and patrician burden on society.

Rural poverty continues to increase and Malawi requires significant development to pull it out of poverty. While "sustainable" development is the new *préfixe du jour*, it is premised on restraint and the precautionary principle rather than on societal ambitions and transcending "basic needs".[9] As one (controversial) US policy analyst writes, "Poor countries need sustained development, not sustainable development, if they are ever to take ... their rightful place among the Earth's prosperous people."[10] Conversely, a recent

---

7 Vail, L., 1975. The making of an imperial slum: Nyasaland and its railways, 1895–1935. *Journal of African History* 16(1), 89–112.

8 World Bank, 2019. *Malawi Economic Monitor, June 2019: Charting a New Course.* World Bank, Lilongwe, Malawi. https://openknowledge.worldbank.org/handle/10986/31929.

9 Earth Charter Commission, 2000. The Earth Charter. https://earthcharter.org/wp-content/uploads/2020/03/echarter_english.pdf?x46713.

10 Driessen, P.K., 2005. Sustainable development equals sustained poverty: Keeping developing countries cute, indigenous, electricity-poor, impoverished, & disease-ridden. *The Progressive Conservative* VII(55).

submission to the UK's International Development Committee dealing with Malawi's priorities condemned its development and population growth and lauded the country's "direct dependence on natural resources". It resolved that Malawi "must conserve its valuable environmental resources", noting that the government "faces many challenges including ... satisfying foreign donors that fiscal discipline is being tightened".[11] Neocolonialism aside, a modernization, industrialization, and urbanization strategy surely cannot be built on those flimsy fawning foundations (Figure 3.2).

China's development by contrast is reasonably well documented. From a low point in the 1970s, when it was reviled as a Communist pariah state, to delivering the opening speech at the home of global capitalism at the World Economic Forum in Davos, China has turned its fortunes around and raised the living standards of its people to a remarkable degree. It has transformed its socioeconomic status out of all recognition to a generation ago.[12] It has built more shopping malls, hotels, office buildings, and housing estates (as well as golf courses and theme parks) than any other country in the world. According to some, since 2005 it is on track to build 20 cities a year for 20 years, having already urbanized the equivalent of the entire US population in the first decade of this millennium.

Figure 3.2   The stifled development of Lilongwe, Malawi. (*Source:* MarcPo/Getty Images.)

---

11 Shumba, H.T., 2012. The development situation in Malawi, International Development Committee, **Session 2012–13.** https://publications.parliament.uk/pa/cm201012/cmselect/cmintdev/writev/malawi/mal27.htm.

12 Anderlini, J., 2016. Mao Zedong's China unrecognisable 40 years on from his death. *The Financial Times*, 9 September 2016.

What are less well known are China's environmental credentials. Within a short time, China has become the global leader in wind energy and aims to provide 100 million homes with wind-generated electricity by 2021. One BBC report claims that China is installing one wind turbine every 30 minutes. In addition, China is now the biggest player in global cleantech: it has the world's largest hydrogen production capacity, it is the number-one lithium-ion investor, and it is the biggest solar panel manufacturer. Once known as the biggest polluter on the planet, the Chinese foreign minister now feels confident to criticize the former US President for leaving the Paris Agreement.

China has reinvented itself as an ecological champion. In this way, its environmental transition is an opportunity for growth and the next round of accumulation. China was one of the first countries in the developing world to strategically introduce sustainable development at a national and regional policy level and has been using it as a way to promote a new direction for its urban development. The government announced in 2010 that 300 new cities would be built by 2025, of which approximately 20 would be eco-cities. By late 2015, it announced that they already have at least 284 eco-cities dealing, in some way or another, with a list of ecological problems. These include smart cities, green cities, sustainable cities, low-energy cities, recycling cities, sponge cities, etc. These generic eco-cities – admittedly many of which are indistinguishable from non-eco-cities – are a test bed for some of the new industries, research facilities, innovation themes, quality assurance, social policies, business opportunities, and ways of working that will continue China's insatiable drive towards modernization (Figure 3.3).

Figure 3.3 Rapid urban development in China. (*Source:* FilippoBacci/Getty Images.)

For example, the coastal city of Rizhao in Shandong province was given central party approval to "seriously implement the scientific concept of development, based on the advantages of the sun".[13] Dutifully, all houses were mandated to incorporate solar panels and solar thermal water heaters. The traffic signals, street-lights, and park illuminations were converted to photovoltaic solar cells, thus reducing the need for coal-fired grid power by, some say, up to 30%. The fact that Rizhao is one of China's largest liquid petro-chemical ports and home to vast asbestos mines didn't distract from the city achieving the designation "National Model City of Environmental Protection".

While the West advocates low growth and sustainable development, China still needs and wants material progress and real development. Not for China the sacrosanct nature of restraint or limited growth. Here, sustainability and consumerism are not seen as con-tradictory: development is good, sustainable development is growth. It may be guided by the global discourse on the environment, but it clearly intends to be in control of its own destiny and use eco-labels to its own advantage. Beijing, says one commentator, "does not appreciate being criticized, lectured, or even mentored".[14]

Emerging from Nature

When Malawi gained independence in 1964, it formally withdrew from its capital city in Zomba and established a new capital at Lilongwe. Malawi's second city, Blantyre (named after the Scottish birthplace of David Livingstone), still remains the industrial and finan-cial centre of the country, but Lilongwe is the administrative, commercial, and political capital. The population in the national capital continues to increase and now stands at around 1.1 million, and it seems likely to continue to rise to a predicted 1.6 million by 2030.

In 1968, the city implemented the Lilongwe Master Plan to cater for increased immigra-tion into the city from its rural surroundings and to create zoned districts for key, man-aged development. As one report notes: "From the beginning there was a concern to create a high-quality environment with spacious living standards, as befits a capital city."[15] This follows the urban ambitions and constitutional settlements of many newly liberated nations: whether India's Chandighar or Abuja in Nigeria. The expression of modernity and change, of political rebirth through conscious masterplanning of new towns and cities has also been a hallmark of urban development and national pride in places such as Brasilia in Brazil (Figure 3.4) and Islamabad in Pakistan (Figure 3.5).

After the Malawian government had implemented a number of structure plans and zoning schemas, a draft masterplan was drawn up in 2013 (with the help of the Japan International Cooperation Agency). Its intention was to expand the city limits within the constraint of a new green belt, improve transportation, and upgrade infrastructure. It

13 Williams, 2018. Op. cit., p. 165.

14 Green, C., 2012. Green on Vogt, 'Europe and China: Strategic partners or rivals?' H-Diplo, H-Net Reviews. http://www.hnet.org/reviews/showrev.php?id=36590.

15 Malawi Government, 1986. Lilongwe Outline Zoning Scheme. Malawi Government, Office of the President and Cabinet, Lilongwe, Malawi.

Figure 3.4 Brasilia. (*Source:* tirc83/Getty Images.)

Figure 3.5 Islamabad. (*Source:* Danish Khan/Adobe Stock.)

sought to realize a vision of an "environment-friendly urban development ... by introducing land use control, gradual conversion of unplanned settlement to other land uses, and greenery policy for land use plan", the latter including forests, parks, wetlands, and natural sanctuaries.[16]

While this sounds perfectly reasonable, it has been acknowledged by the World Bank that the general trend in Malawi's planning policy proposals has been, in fact, in favour of rural development and has been somewhat anti-urban, using green belt restrictions as "a means to curtail and contain the process of urbanization ... (and) reduce rural–urban migration".[17] Even though there is this tendency by Malawians to maintain and safeguard their economy dominated by agriculture and low-scale rural industry, the resistance to urban growth is not a positive response. The World Bank might want urban development in order to oil the wheels of trade and business (in the same way that the British Protectorate had tried to do 120 years earlier), but Malawians need to set challenging targets and not revel in rurality: to urbanize is a reflection of a country's sociocultural ambition not just its economic advances. While the fifth edition of the *Malawi Economic Monitor* positively speaks of the need to speed up the country's urbanization programme, it implies that this can only happen if sufficient wealth is available – which requires the agricultural sector to rely on good weather.[18] In other words, while donors prevaricate, Malawi's urban development is in the lap of the weather gods. Remaining vulnerable to the quirks of nature, its inability to transcend as far as possible the vicissitudes of the environment is itself a guaranteed sign of Malawi's parlous status and the continuation of its anti-urban underdevelopment.[19]

Underdeveloped countries are seldom able to control their own destiny in the way that it is hopefully expressed in their aspirations for urban renewal. To build a city, to expand an existing metropolitan area, or merely to create a new urban district takes money, time, and skills – attributes often denied the poorest of countries through no fault of their own. Indeed, a report by the Centre for Economic Business Research forecasts Malawi rising up the World Economic League Table, "from 148th place in 2019 to 140th place in 2034".[20] This is not the economic dynamism required to create a metropolitan masterpiece, and so it seems that Malawi will remain in thrall to outside forces – foreign aid or weather patterns – and will find it difficult to escape. One example of this catch-22 will suffice.

A corruption scandal around the time of the election of Mrs Banda (no relation to Hastings Banda) for her second term in office led to a substantial drop in foreign aid

---

16 The Urban Structure Plan of Lilongwe City DRAFT 17th June 2013. http://www.lands.gov.mw/phocadownload/land_policies_plans/urban%20structure%20plan%20of%20lilongwe%20city%2017%20june%202013.pdf.

17 Government of Malawi, 2011. Malawi growth and development strategy II 2011–16. Ministry of Economic Planning and Development, Lilongwe, Malawi.

18 World Bank, 2017. *Malawi Economic Monitor, May 2017: Harnessing the Urban Economy.* World Bank, Lilongwe, Malawi. https://openknowledge.worldbank.org/handle/10986/26763.

19 Formetta, G., Feyen, L., 2019. Empirical evidence of declining global vulnerability to climate-related hazards. *Global Environmental Change* 57, 101920.

20 Centre for Business Economic Research, 2019. World Economic League Table 2020, pp. 145–146. https://cebr.com/reports/world-economic-league-table-2020/.

donors as they feared for the legitimacy of their aid and investment. *The Economist* magazine claims that an Australian mining giant invested US$500 million in a uranium mine
as a "rare example" of foreign capital investment in the country. Investors are driven by
returns, and Malawi, like many other parts of Africa, is simply buffeted by the vicissitudes
of the world market. The collapse in uranium prices closed the mine, with the loss of jobs,
wealth, and prospects.[21]

As this example shows, very often it is the stabilization of sociopolitical relations that is
more important for investors than the actual material conditions on the ground. As the
Chinese one-party state is fond of announcing to its own people, social stability remains
the central party's foremost concern: its one-party authority rests on the maintenance of a
harmonious society. Similarly, the authority of World Bank, International Monetary Fund
(IMF), European Union (EU), or other loan guarantees relies on creating satisfactorily
stable conditions for payback. As a commercial body, it claims reasonably that: "the institution cannot help unless it preserves its assets as a revolving stock of lendable resources".[22]
For example, while Malawi's national debt is US$5 billion (equal to 65% of Malawi's
GDP),[23] such a degraded credit rating will bring its own threat to the country's finances.
Investors will not be lining up unless there are other inducements. Coincidentally, there
have been no restrictions on the controversial rights of foreign nationals to purchase land
in Malawi.

Malawi's reliance on its agricultural sector is a millstone around its neck. As China has
shown, overcoming natural barriers is essential to the development of a thriving economy. To treat nature as sacrosanct, farming as indigenous, and the soil as sacred is a recipe
for the maintenance of its impoverished status quo. It is no surprise to find that development is often measured by the percentage urban to rural population. Malawi's urban
population is just 17%; China's current urban population is 62%, but it was 17% in 1976 at
the beginning of its upward trajectory towards modernization. China fully embraced
urbanization as an engine of growth, creating 465 new cities in a 40-year timeframe.

That said, a number of researchers have pointed to the poor status, the insubstantial
public space, and the crowded, noisy, or polluted feel of China's cities. Researchers Xiao
et al. say that "the government pays too much attention to the increase of the proportion
of urban population, but ignores the quality of urbanization",[24] which is a charge that
might have been levelled on Engels' treatise on Victorian Manchester, or Chicago before
the City Beautiful masterplan was presented. China's race to urbanize was no different to
other developing countries. Rapid masterplanning of an emerging urban economy often
resulted in beauty and aesthetic subtlety being down-played. They became values that
were hankered after in retrospect.

21 Anon, 2014. What a job. *The Economist*, 24 May 2014.

22 Boughton, J., 2001. 16 digging a hole, filling it in: Payments arrears to the fund. In: *Silent Revolution: The International Monetary Fund, 1979–89*. International Monetary Fund, Washington, DC.

23 Nkhoma, P., 2020. Malawi public debt worrisome, says Finance Minister Mwanamvekha. *Nyasa Times*, 15 June 2020. https://www.nyasatimes.com/malawi-public-debt-worrisome-says-finance-minister-mwanamvekha.

24 Xiao, Y., Song, Y., Wu, X., 2018. How far has China's urbanization gone? *Sustainability*, 20 August 2018. https://doi.org/10.3390/su10082953.

The need for rapid development often precludes the desire for quality and nuance. China had to urbanize fast, and it therefore had to build millions of homes and high-rise apartments rapidly, together with new services and infrastructure. Typically – especially in the early years – these were inadequately designed and badly detailed, using low-grade materials precisely because they were constructed by people without the necessary skills and training in order to fulfil an urgent need. Planning legislation and building codes were virtually non-existent. Build fast, repent at leisure was, for China, a completely understandable maxim. Once the urban population had been settled, once economic dynamism had been established, China had the breathing space to rectify problems and provide decent replacement housing and urban upgrades.

China's rapid rise was done in extraordinary circumstances and at a speed unknown in history, but the ambition to build a new world – to lift millions out of poverty with the promise of economic growth, employment prospects, and material wealth – trumped the social or environmental issues. Indeed, those were considered to be necessary casualties. The Kuznets curve hypothesis of urbanization[25] suggests that a rapid process of development is required in order to yield social benefit in the longer term, from which the authorities are better placed to repair the harms along the way. It is only after China had created sufficient wealth to improve the living conditions of its citizens that it was able to turn its attention to the quality of those living conditions – unlike Malawi which is stuck in the vicious circle of maintaining rather than transcending its relationship with nature.

Admittedly, in many instances, China's urbanization has a long way to go to repair some of the damage to communities, heritage, and the environment caused by its drive for social and economic improvement. Coronavirus and an authoritarian one-party state aside, as the Chinese economy has continued to grow, the country now seems to be in a position to make local, rural, and environmental reparations. If the hype is to be believed, it is constructing 283 eco-cities, it is making contextual improvements in the countryside, and it is investing in parks, gardens, clean-air corridors, public transport, etc. This can only be done from a position of economic confidence.

Plus ça Change

The infamous structural adjustment programmes of the IMF and World Bank are essentially loans with conditionalities that benefit the lendee institution and place onerous controls on the debtor nations. Fundamentally, the IMF offers financial inducements with strings attached, most commonly in the form of political interference. Countries are pressured to restructure their economy to suit the needs of the loan payers. In its heyday in the 1970s and 1980s, the conditionalities imposed on indebted countries were financially and socially oppressive – described as "economic circumcision"[26] by one writer.

---

25 Dongfeng, Y., Chengzhi, Y., Ying, L., 2013. Urbanization and sustainability in China: An analysis based on the urbanization Kuznets-curve. *Planning Theory* 12(4), 391–405.

26 Anon, 2001. Helping Africa. *The Economist*, 21 February 2001.

Just after her first election victory, UK Prime Minister Margaret Thatcher applauded Malawi for having "the courage to take difficult decisions on your economy when they have been required by the World Bank and the IMF because you knew that in the long-term it was the best thing for your people".[27] The measures insisted, *inter alia*, that Malawi liberalize its import processes, remove credit controls, and reform the Post Office, etc.[28] Interference in the governance of an erstwhile independent country was *de rigueur*. As public health academics Thomson et al. point out, even today, "recipient countries are required to reform various macroeconomic and fiscal policies according to a neoliberal rubric, typically cohering around economic stabilisation, trade and financial liberalisation, deregulation, and privatisation ... critics argue such adjustment comes at a high social cost, while the recidivist nature of program participation also suggests that gains to macroeconomic stability are underwhelming."[29] Once again, Malawi's hopes of urban renewal are reliant on outside forces.

At least, in the 1970s and 1980s, there was mounting consternation among left-wing, anti-capitalist groups and other concerned citizens in the West at the structural adjustment programmes being imposed on the Third World (as it was known). These loans came with demands for social reforms and were (rightly) seen as authoritarian. Countries would have their payments revoked, for example, if they didn't change certain domestic policies that the IMF and World Bank considered to be a hindrance to market liberalization. But a shift occurred at the turn of the twentieth century when intervention in the affairs of developing nations was re-posed in a more morally legitimate way. Now aid would be given in order to protect the environment, and "good governance" became the buzzword; it incorporated the notion of responsible or sustainable development. In this way, sustainability became a vehicle for western intervention in sovereign states in the guise of virtuous social justice. Millennium Development Goal 11, for example, insisted that "no effort must be spared to free all of humanity ... from the threat of living on a planet irredeemably spoilt by human activities".[30] In order to restrict development, the World Bank indicates that Malawian businesses and landowners may be "eligible for compensation if operations are restricted for reasons of wildlife protection".[31] No wonder Malawi is torn between developmentalism and protectionism, between urbanism and anti-urbanism.

Nowadays, criticism of environmental interventions tend not to rebuke the intervenor for meddling in the affairs of another country, but to chastise the dominant party for not

27 Thatcher, M., 1989. Speech at dinner given by the Malawi President (Dr Hastings Banda), March 31, 1989. https://www.margaretthatcher.org/document/107622.

28 Hicks, R., Brekk, O.P., 1991. African Department Assessing the Impact of Structural Adjustment on the Poor: The Case of Malawi. WP/91/112. International Monetary Fund, Washington, DC.

29 Thomson, M., Kentikelenis, A., Stubbs, T., 2017. Structural adjustment programmes adversely affect vulnerable populations: A systematic-narrative review of their effect on child and maternal health. *Public Health Review* 38, 13.

30 Millennium Development Goals, Environment and Sustainable Development. United Nations Development Programme. www.undp.org.

31 Christy, L.C., Di Leva, C.E., Lindsay, J.M., Takoukam, P.T., 2007. *Forest Law and Sustainable Development: Addressing Contemporary Challenges Through Legal Reform.* Law, Justice, and Development Series 40003. International Bank for Reconstruction and Development. World Bank, Washington, DC.

intervening sufficiently stringently. Instead of debt repayments, sustainable aid cam-paigners demand "transparency" and environmental accountability. As such, environ-mentalism provides incentives to interfere and to penalize non-compliant states. The New Economics Foundation calls these actions "sustainable adjustment programmes", whereas free marketeer Paul K. Driessen describes it as "eco-imperialism" or "a virulent kind of neo-colonialism". Same intervention, different language.

Meanwhile, China is now forging relationships with Africa, seen by some as an oppor-tunity and by others as a contemporary version of colonial enslavement. There is some credibility to such criticisms, with China taking substantial amounts of raw materials out of Africa while providing loans for infrastructure projects. In Lilongwe alone, in return for political allegiance, trade relations, and exports to China of timber and minerals, China has built a series of presidential villas, the Bingu International Conference Centre, the New Parliament Building, a five-star hotel, and Bingu National Stadium, and is set to invest heavily across the country. In other areas there are new roads, hospitals, highways, and the Malawi University of Science and Technology. In 2016, a US$1.79 billion finance agreement was signed to construct a power plant and international airport.

China's intention is to provide infrastructural spending as a way of emulating its own rise out of agrarianism. By creating more dynamic economies around the world, China wants to lift people out of poverty and coincidentally increase its share of international trade. Even though there is the threat of default and crippling debt from loans from any source, it is widely acknowledged that instead of a neoliberal economic shakedown of poor countries that was common under the World Bank and IMF interventions, China has implemented a more contractual business relationship and reiterates – somewhat ironically – that Beijing does not interfere in other countries' internal affairs.[32] However, as we have seen, both the structural adjustment policies and the interventionist environ-mental policies are direct political acts of interference in the sovereign rights of an inde-pendent nation state, so a straightforward transactional deal seems, on the surface, to be more appealing to the Malawians.

In her book *The Dragon's Gift*,[33] author Deborah Brautigam shows that China, in its early stages of development, recognized that it needed infrastructural investment in order to kick-start its own economic plans. With little or no foreign direct investment in the mid-1970s,[34] China had to utilize a trade arrangement with Japan, which was granted rights freely to extract quantities of raw materials from China. But instead of having to pay hard cash, Japan was required to build roads, bridges, and other infrastructure within China to literally lay the foundations for its rapid rise towards development. China's engagement in Africa follows a similar model. While clearly neocolonialist in some areas, it is also a facilitator for real, necessary development in others. In terms of China's

32 Mumuni, S.M., 2017. China's non-intervention policy in Africa: Principle versus pragmatism. *African Journal of Political Science and International Relations* 11(9), 258–273.

33 Brautigam, D., 2011. *The Dragon's Gift*. Oxford University Press, Oxford.

34 Ford, J.L., Sen, S., Wei, H.X., 2010. FDI and economic development in China 1970–2006: A co-integration study. Discussion Paper. University of Birmingham, Birmingham. https://www.researchgate.net/publication/276045974_fdi_and_employment_by_industry_a_co-integration_study.

business model, there tends not to be any social, political, or environmental conditions attached to its loans, and this is important for Africans who are concerned about the re-emergence of an imperial intervention. Even though the gloss is wearing thin on Chinese carpetbaggers taking advantage of Malawian opportunities on the ground, the international deals still form a straightforward contractual relationship that seems to benefit both sides.

Admittedly, debts are amassed for services rendered (like Djibouti's US\$3.5 billion loan to build its International Free Trade Zone, or the US\$100 million for Malawi to build its new coal-fired power station), but the past-President Mutharika says: "We have chosen to stop depending on aid. Much of the aid to Africa was spent on services and consumption, and not so much on production. As such we have chosen to move from aid to a trade. But you cannot trade if you don't produce goods. Therefore, we have chosen to become a producing and exporting nation."

Conclusion

The coda to this chapter examines a short documentary, "Buddha in Africa". Made in 2020 by director Nicole Schafer, it tells the tale of a Chinese orphanage in the heart of Malawi where the children take lessons, learn Chinese, study Buddhism, and are trained in martial arts. Ordinary Malawians, too poor to look after their children or grandchildren, hand them over to the monks who tour the region in minibuses to take them away. The youngsters, some as young as five years old, fear that they are going to be eaten by the mysterious Chinese child-catchers, but they are whisked away with the opportunity to get an education. For years they engage in a strict martial arts regime, learn a range of subjects, and seldom get to visit their families. After a young lifetime at the school, the students are offered the chance to go to university in Taiwan, an opportunity unheard of for their relatives and friends. In this documentary, the main protagonist, Enock Alu, a Malawian teenager, is uncertain whether he wants to go or stay in order to retain his cultural roots. The head monk tells him: "The world is changing. The poverty of Africa now can't be blamed on you because you are a victim. But in 20 years, if Africa still hasn't developed, you are not a victim anymore, and should be blamed to a great extent."[35]

---

35 Schafer, N. (dir.) 2019. Buddha in Africa. Documentary. https://www.imdb.com/title/tt10323416/

4

The Sustainable City – Introduction

In Chapter 3 Austin showed us how to make progress as a society and how to fix the environmental blunders with the spoils of our economic success. Patricia now sets out the path of the UN SDGs and looks at the universal challenges and the irrefutable truths that numbers bring to all cities.

It is estimated that nearly 10 billion people will inhabit the Earth by 2050. Seventy percent of these people will live in urban areas. In 2015, the UN adopted the 2030 Agenda for Sustainable Development, which includes 17 Sustainable Development Goals (SDGs), accompanying the Paris Agreement which aims at limiting global warming to 1.5°C. The SDGs are unique in what Patricia refers to as their "universal applicability", committing all countries, whether they be in the Global North or Global South, rich or poor, to have a responsibility to implement and achieve these goals. But the SDGs are also unique in the way they emphasize the important role cities play.

Numbers literally make a city and are what separate it from any other inhabited area. Patricia will also examine urban agglomeration and how cities have a special responsibility to be sustainable. The same way Peter Boyd discussed in Chapter 1 how the city needs connected leadership in order to reach sustainable goals, the "sustainable city" needs connected policy that encourages the growing role of city leaders to collect data to provide for their national government. Patricia will also go on to discuss steps taken to improve linkages between local and international governance levels.

In recent years there has been a palpable energy and excitement, both socially and in the media, behind global environmental activism and technology, led not only by leading industry and technology figures but most poignantly by our youth. In 2018 Greta Thunberg propelled the School Strike for Climate movement into global organization and since then has addressed the 2018 UN Climate Change Conference and infamously sailed across the Atlantic to speak at the 2019 UN Climate Action Summit. Her words "How dare you" seared across headlines and continents, angering sitting presidents, impressing many, but most importantly confronting us with our own failings (Figure 4.1).

Before the COVID-19 pandemic hit, we were behind the 2030 goal, with 2 billion people still suffering from extreme poverty and 789 million without access to electricity, resulting

The Climate City, First Edition. Edited by Martin Powell.
© 2022 John Wiley & Sons Ltd. Published 2022 by John Wiley & Sons Ltd.

Figure 4.1   A girl begins a mural at Trullo, a suburb of Rome, depicting Greta Thunberg. (*Source:* Fabio Amicucci/Getty Images.)

in uneven energy demand growth and fast-growing urbanization. Figure 4.2 shows those countries with the largest increases in extreme poverty and Figure 4.3 shows the countries that will be most impacted by COVID-19 on extreme poverty. The escalating nature of the climate crisis also means we are fast approaching irreversible "tipping points", with scientists warning we may already have surpassed two critical points in the retreat of the Greenland glaciers and the deforestation of the Amazon basin. In 2019, fossil fuel emissions reached an all-time high of 148% of preindustrial levels, and unsustainable

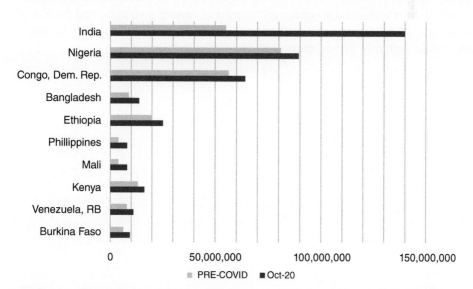

Figure 4.2   Countries with largest likely increases in extreme income poverty headcounts compared to baseline, 2020 (absolute numbers of people). (*Source:* Homi Kharas, 2020. The impact of COVID-19 on global extreme poverty. The Brookings Institution. Retrieved from: https://www.brookings. edu/blog/future-development/2020/10/21/the-impact-of-covid-19-on-global-extreme-poverty/.)

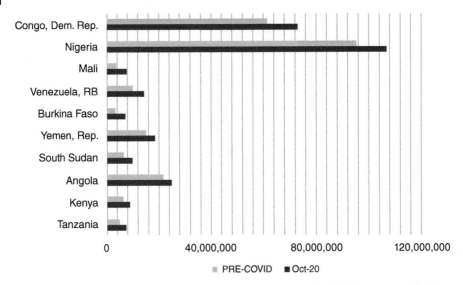

Figure 4.3 Long-term impact of COVID-19 on extreme poverty in 2030, by country (absolute numbers). (*Source:* Homi Kharas, 2020. The impact of COVID-19 on global extreme poverty. The Brookings Institution. Retrieved from: https://www.brookings.edu/blog/future-development/2020/10/21/the-impact-of-covid-19-on-global-extreme-poverty/.)

consumption and production has contributed to the depletion of the world's wildlife and natural assets, with an alarming 1 million species on track for extinction by mid-century.

COVID-19, the greatest post-war crisis we have known as a global community, magnified the pre-existing barriers to sustainable development and the current and projected impacts of the climate change emergency in terms of poverty, inequality, and the role of cities. These crises are interconnected and undeniable in proving that numbers do matter and, more importantly, that numbers don't lie.

4

The Sustainable City

Patricia Holly Purcell

This chapter is about putting cities in numbers: it sets out the United Nation's (UN) global framework for tackling climate change and the role of the UN in putting the world on a sustainable path through the Sustainable Development Goals (SDGs). It explores the role for cities in that process and identifies the levers city leaders need from their governments and the international community to tackle climate change and its wider implications, not only in cities, but also globally. It also indicates the growing need for city leaders to access data and other critical resources to support their asks of national governments. We will illustrate "the world in numbers", and how this breaks down into Global North and Global South and into formal and informal settlements, and the role of UN-Habitat and other international organizations in supporting cities' quest for a sustainable future.

Numbers matter. Everyone reading this book will remember the start of the 2020s as the year that numbers mattered because of the tragic number of deaths and illness unleashed by the COVID-19 pandemic. Many others will also remember it as the year that lost millions of jobs and industries in the ensuing economic collapse. The pandemic was moreover a reminder that numbers matter in terms of wealth and inequality, as the virus disproportionately impacted on the most vulnerable, most notably the estimated 1.5 billion informal and migrant workers[1] who could not work from home and had limited or zero access to any social protection, as well as the estimated 460 million children[2] who went without education because they – incredibly – didn't have digital access, or – more incredibly – lived without electricity. At the time of writing, the Human Development Index – basically the UN measure of wellbeing beyond GDP – was set to drop for the first time since the measurement began in 1990, as the impacts of the pandemic threatened to force more than 100 million people into extreme poverty.[3]

The pandemic also showed just how much numbers matter in terms of human rights, as global lockdowns were weaponized to oppress minority groups and political rivals, resulting in a steep decline in the quality of democracy across dozens of countries, leaving

---

1 ILO, 2020, https://www.ilo.org/wcmsp5/groups/public/---dgreports/---dcomm/---publ/documents/publication/wcms_734455.pdf.

2 UNICEF, 2020, https://www.unicef.org/press-releases/covid-19-least-third-worlds-schoolchildren-unable-access-remote-learning-during.

3 https://www.worldbank.org/en/topic/poverty/brief/projected-poverty-impacts-of-COVID-19.

The Climate City, First Edition. Edited by Martin Powell.
© 2022 John Wiley & Sons Ltd. Published 2022 by John Wiley & Sons Ltd.

the most vulnerable even more helpless.[4] The pandemic was also the ultimate indicator that numbers really do matter in terms of how human activity is negatively impacting on the natural environment, most notably measured by rising levels of climate-warming greenhouse gas (GHG) emissions, air pollutants, and ocean waste, and the mass eradication of species and destruction of ecosystems – conditions that are fast destabilizing the health of the planet and, with it, threatening the health and wellbeing of all 7.8 billion people who inhabit the Earth.

For cities, numbers matter because urban agglomeration – the concentration of people in a place – has a profound effect on all other numbers: that is, where and how people live, work, and interact within an urban space; the goods, services, food, water, and energy they produce and consume; the type of economic activity generated; and the less "countable" features such as the art, culture, knowledge, and innovation harvested in cities, all interact to create a "system" that has immediate, mid-term, and long-term implications, not only for people living in cities but for all of humanity.

As with any system, the city system is a complex, interdependent network that must consider not only its "form" in terms of the number of people living in an urban area and its physical infrastructure (buildings, parks, transport, etc.), but also its structure and function.[5] Urban planners have long approached the city as a system to solve current challenges and anticipate solutions to future ones – for example, how to limit congestion, accommodate new housing for growing urban populations, preserve or expand cultural activities, and so on. However, it was not really until 2015, with the adoption of 17 SDGs, which include a dedicated "cities" goal (SDG 11), that cities' role as contributors to both the causes and solutions to the wicked problems of our time, principally climate change and the interlinked challenges of poverty and inequality, assumed global recognition and significance. With the onset of COVID-19, the acknowledgement of cities' role as facilitators and catalysts of sustainable human development has never been more prevalent or urgent.

Adopted by the UN in 2015, the 2030 Agenda for Sustainable Development includes the 17 SDGs and the accompanying Paris Agreement, which aims at limiting global warming to 1.5°C. This crucial "tipping point" is, in a very real way, the difference between whether the planet can accommodate the nearly 10 billion people who will inhabit the Earth by 2050 – nearly 70% of whom will live in urban areas – or not. The broader 2030 Agenda also includes the Sendai Framework for Disaster Risk Reduction, the Addis Ababa Action Agenda on sustainable finance, and the New Urban Agenda, which is adopted every 20 years. Unlike its predecessor framework, the Millennium Development Goals, the SDGs are distinct in their universal applicability to all countries – Global North and Global South, rich and poor – meaning that all nations have an individual and collective responsibility for achieving the goals. The SDGs are also unique in their transversal nature – that is, the notion that no single goal can be achieved without positively impacting on all others (and vice versa). By their very design, then, the SDGs aim to connect all people in the common pursuit of a more peaceful and prosperous world that protects the most vulnerable, as well as the natural environment upon which all lives and livelihoods ultimately depend.

---

4 https://www.economist.com/international/2020/10/17/the-pandemic-has-eroded-democracy-and-respect-for-human-rights.

5 https://paginas.fe.up.pt/~tasso/pdf/sp/sp16-the%20city%20as%20a%20system.pdf.

Testing the Decade of Delivery

It did not go unnoticed, then, that the pandemic struck at precisely the time when nations' commitments to sustainable development made five years previously were to be put to the test. Before COVID-19, 2020 was heralded as the "Decade of Delivery" for the landmark UN 2030 Agenda framework aimed at charting a new, decisive course towards a greener, healthier, and more inclusive future by the close of the decade. Ambitious, certainly; impossible, no. The world was the richest it had ever been, with global wealth topping US$400 trillion (albeit unequally distributed, at least there was a greater share pouring into "green ventures" than ever before), and there was an almost euphoric energy that surrounded global climate activists, most notably the teenage-sensation Greta Thunberg. It was hard to find a day that the global media did not cover the climate crisis and the significance of the forthcoming UN conventions on climate change and the "Super Year for Nature". Indeed, what was not so long ago an almost esoteric scientific, academic, and diplomatic discourse had, at last, hit the mainstream. This momentum was propelled by "tech celebrities" like Elon Musk with his ambitious vision for sporty, all-electric transport, and futuristic instruments that could "scrub" carbon from our atmosphere. Microsoft splash-landed into 2020 with a promise to achieve carbon negative status for every year since the company was founded in 1970, with huge implications for its customers and supply chain. Even the "grand-old daddies" of the fossil fuel industry like BP entered 2020 with the promise of a clean energy revolution, while traditional credit card companies – needing to keep competitive with their Silicon Valley and Beijing rivals – transformed into modern-day "fintech" firms, and once "underground" crypto-currencies conspired to forever disrupt traditional finance, offering opportunities to the unbanked at a pace never seen. In January 2020, the most recognized captains of industry convened at the World Economic Forum in Davos to declare the value of "stakeholder capitalism" in line with the SDGs' overarching thesis to "Leave no one behind". It was, in many ways, the start of a social, economic, and environmental renaissance.

However, for all the fanfare, the reality was that, globally, we were way off course for meeting the 2030 Agenda well before "life in the time of COVID" had become the norm. Today, thanks to the economic walloping from the pandemic, the sustainable finance gap needed to fulfil the SDGs has nearly doubled from US$2.5 trillion per year pre-pandemic to over US$4 trillion annually by 2030. Meanwhile, nearly 2 billion people are still suffering from extreme poverty and some 800 million people do not have access to electricity, or access to reliable and affordable forms of energy. The places with some of the fastest pace of energy demand are also those countries experiencing the fastest rates of urbanization. Over 90% of new urbanization is occurring in developing countries in the Global South.[6] Sub-Saharan Africa (SSA), for instance, which is home to many of the world's most rapidly growing cities and megacities, is also the most energy impoverished region on the planet, with roughly 550 million people[7] living without access to electricity. Granted, the

---

6　UN, 2016, https://stg-wedocs.unep.org/bitstream/handle/20.500.11822/11125/unepswiosm1inf7sdg. pdf?sequence=1.

7　https://www.brookings.edu/blog/africa-in-focus/2019/03/29/figure-of-the-week-electricity-access-in-africa.

most affected live in harder-to-reach, "off-grid" rural areas, but their conditions cannot be disconnected from how we consider urbanization – that is, both in terms of the reasons that drive people to cities in the first place, and the interdependencies between urban and rural communities for food, water, labour, and so on.

At the current rate of progress, a projected 620 million people globally would *still* lack access to electricity in 2030 – and this estimate doesn't even take into account the full impacts of COVID-19. At the same time, since 1990, two-thirds of both developed and developing countries reported widening levels of income and wealth inequality, while other measures of wellbeing, such as nutrition, and access to health, housing, and education have also been unevenly distributed.

Against this backdrop, the world is also facing an escalating nature–climate crisis on the verge of breaching irreversible "tipping points" linked to rising GHG emissions and the widespread destruction of ecosystems and biodiversity loss. For example, in August 2021, the Intergovernmental Panel on Climate Change (IPCC), in its most devastating report on the current and projected impacts of climate change,[8] echoed previous scientific warnings that humanity may have already surpassed two critical, negatively reinforcing tipping points: the retreat of Greenland's glaciers – the largest contributor to global sea level rise;[9] and the massive deforestation in the Amazon basin resulting from logging, farming, mining, and other human activities (legal and illegal), which may have passed the threshold whereby tree loss would start to "feed on itself", triggering a catastrophic feedback loop that will accelerate global warming.[10]

Scientifically speaking, it is clearer than ever that traditional economic growth models are coming at a grave cost to the planet. According to the IPCC,[11] the evidence is now "unequivocal" that humans have caused climate change, with each of the last four decades successively warmer than any decade that preceded it since 1850. Some of the most notable temperature increases and associated climatic impacts such as lethal storms and hurricanes have coincided with bouts of profound economic growth, as seen in the 1980s and more recently in the boom years following the 2008 financial crisis. Worse, the economic disruption from global lockdowns provided only a temporary environmental reprieve. According to NASA,[12] 2020 was the hottest year on record. Meanwhile, unsustainable consumption and production patterns had already contributed to wiping out two-thirds of the world's wildlife in just 50 years and degrading 60% of the planet's natural assets, including a loss of 10 million hectares of forests every year. Moreover, by 2021, an additional one million species were on track to extinction by mid-century. To put this in perspective in economic terms, at the time of writing, COVID-19 had cost the global economy about US$26 trillion. But if we fail to do nothing about climate change, by mid-century, the economic damage induced by global warming will halve projected

---

8 IPCC, 2021, https://www. archive.ipcc.ch/report/ar6/wg1/downloads/report/ipcc_ar6_wg1_full_ report.pdf.

9 https://www.firstpost.com/tech/science/greenland-glacier-melting-crosses-tipping-point-wont-recover-if-global-warming-ended-today-study-8722631.html.

10 https://www.economist.com/briefing/2019/08/01/the-amazon-is-approaching-an-irreversible-tipping-point.

11 IPCC, 2021, https://www. archive.ipcc.ch/report/ar6/wg1/downloads/report/ipcc_ar6_wg1_full_ report.pdf.

12 NASA, 2021, https://www.nasa.gov/press-release/2020-tied-for-warmest-year-on-record-nasa-analysis-shows.

global wealth, amounting to about US$550 trillion. Ultimately, such crushing losses would be borne by those nations, cities, and communities least able to absorb their impacts.

Extraordinary Circumstances and Extraordinary Solidarity

By the end of 2020, the pandemic had unquestionably impacted every person, city, and business on the planet, while the environmental damage already "locked-in" to the atmosphere pursued a merciless path across the globe in the form of record flooding, heatwaves, wildfires, cyclones, and hurricanes. The much-heralded "Decade of Delivery" fast became a day-to-day sprint for survival.

The Heroes of Cities and Communities

Here, in the midst of the most profound post-war planetary crisis the UN had encountered since its foundation 75 years ago, something extraordinary happened. Those prevailing as the "heroes" of the crisis were in fact not sat in the chambers of the UN, or in the highest seats of national power, or in the boardrooms of global industry. They were the people gathered in city halls; in the makeshift community centres of gyms and basements; in the city hospitals and emergency rooms; in the backyards of friends and families; in the city squares and public spaces handing out food to those most devastated; and in the streets, lanes, and highways of every city centre across the world. They were the thousands of mayors and state and regional governments, from New York City to Port Vila, and the countless more community and civic leaders and small businesses who stood up and stood together. There is no one reading this book who will not recall the nightly applause to the frontline health workers across the globe who put their own lives at risk to save the lives of others.

Numbers matter. But not necessarily in the way we think – or have been trained to think – they do. COVID-19 showed that numbers matter not only in the tragedy it inflicted, but – more profoundly – in the extraordinary solidarity and unexpected places where this shared human tragedy was manifested. The pandemic pointed to an alternative reality that most people thought only happened in movies. This is not unlike the reality we are already facing and will continue to face from climate change. We have a choice to change the numbers in our favour. And not unlike the response to the pandemic, we are likely to find the extraordinary right in our own backyards and cities.

Where Do We Start?

Experiencing the Earth's orbit on the equator is a very special event. Without fail, the sun will rise and set on or about a few minutes before or after the six-o'clock mark. There is a narrow window when the sky will open to an incredible expanse where every beauty and every flaw is exposed, followed by a sudden transition from light to dark – as if borrowing time that will be paid back the next day. It was on such an evening in mid-2015 – that small weigh station between dusk and dark – that a UN mission was making its entry into

Jomo Kenyatta International Airport, Nairobi. Looking out over the immense, rust-coloured expanse, one colleague turned to the rest and asked, *"Where would you start?"*.

It was an important – even prolific – question. Nairobi, the capital city of Kenya, located in East Africa, is home to nearly 5 million people; by 2030, the city will accommodate an additional 2 million inhabitants. Nairobi is small by comparison to its continental neighbours like Lagos (Nigeria), whose population has swelled to over 14 million people and will add a further 6 million residents by 2030. Still, the challenges associated with its urbanization – poverty, lack of decent and affordable housing, access to energy, education, and jobs to meet the needs of growing youth populations, among others – are emblematic of all fast-growing urban centres, most of which are concentrated across the Global South. Anyone who has visited Nairobi knows that the flight path into the city will give you a bird's-eye view of the immense sprawling suburbs and slums, which are home to some 60% of the city's urban residents and represent among the largest slum communities in the world. This is a fantastic and surreal juxtaposition to a city that is also home to one of the continent's and the world's most-celebrated national parks that sits in spectacular view of that same flight path and is teeming with wildlife most people only read about or see on National Geographic. Abutting the National Park are some of the city's most underprivileged communities, who rely solely on the protection of their natural assets, which are being steadily eroded.

Nairobi is home to the United Nations Human Settlement Programme, better known as UN-Habitat, which is the effective custodian of the New Urban Agenda and, for that matter, the accompanying cities goal of the SDGs (SDG 11). Nairobi is also headquarters to the United Nations Environment Programme (UNEP), the global steward of the natural world (Figure 4.4).

Figure 4.4   Nairobi, home of UN-Habitat. (*Source:* millerpd/iStock/Getty Images.)

In many ways, as the only UN agencies based in the "Global South", these sister organizations hold a special responsibility and insight into the realities of their host city. The question *"Where would you start?"*, then, was not rhetorical. Rather, it was a recognition of the multifaceted challenges confronting a city bestowed with spectacular natural and human resources that are at once immense and impoverished. At the same time, the question recognized the opportunity to reconcile the very real, everyday concerns of poverty, unemployment, housing, education, healthcare, and so on with the seemingly intractable threat from climate change. One of the vantage points of being in the Global South is being able to avoid the same mistakes as one's richer neighbours to the north, and "leapfrog" – technologically, socially, economically, and environmentally – into a sustainable future.

In hindsight, of course, that question held more global significance than any of us on that flight could have possibly imagined – that is, the crucial part cities around the world would play in the response to the pandemic, which has only served to magnify their import in averting the even deadlier threat from climate change – now, for the coming decades, and probably beyond.

Sustainable Development Won't Happen if it Doesn't Happen in Cities

The New Urban Agenda (NUA) "is about people, the planet, prosperity, peace and partnerships in urban settings". Adopted at the United Nations Conference on Housing and Sustainable Urban Development (Habitat III) in Quito, Ecuador, in October 2016, it is effectively a local-level roadmap for global sustainable development. At its core, the NUA recognizes that cities – as home to roughly 4 billion people, over half of the world's population, generating two-thirds of global GDP – are essential to every aspect of human activity on this planet – social, economic, and environmental. Accordingly, it lays out practical standards and principles for the planning, construction, development, management, and improvement of urban areas, for example through urban legislation and regulations, urban planning and design, and municipal finance.

However, the true spirit – or *philosophy* – of the NUA is grounded in cities' historic role in human progress. As anyone who has ever lived in or visited New York, Nairobi, Paris, Istanbul, Barcelona, or any number of cities will attest, it's not just the number of people living in one place that makes a city; rather it is often the numerous features you can't "count" that give a city its character. The world's oldest recognized cities in the region of Mesopotamia dating back to 10,000 BCE (spanning modern-day Iraq, Syria, and Turkey) grew thanks to the advancements in agriculture (the Neolithic Revolution) that enabled people to stay in one place. As they did, these cities were synonymous with the first examples of civilization, united by common law, and shared forms of communication in the way of language and art, beliefs, and culture. Ultimately, people who moved to cities shared a common purpose, which, in turn, defined the fabric of the city. The ensuing first, second, third, and fourth industrial revolutions gave more people the ability to move to cities, mass-produce food and goods, and – fast forwarding to the present day – transform global communications, education, trade, politics, and economics. The greatest innovations and leaps stem in one way or another from urban agglomeration. And, despite some urban flight resulting from the pandemic, cities are still on track to become the dominant residence for the majority of humanity – projected to accommodate some 7 billion people – by mid-century, or an additional 2.5 billion inhabitants.

Of course, for all of the romance of the city, the reality is that for most people – past and present – the urban journey is propelled by a genuine need for social and economic betterment, principally the opportunities for decent jobs and education. For too many, and an increasing number, the migration to cities is forced out of necessity such as escaping conflict and climate-related impacts like drought, or both, and which will only worsen with unabated climate change. Today, the majority of the 25 million refugees and 40 million internally displaced persons in the world live in cities and urban settlements. Indeed, since the adoption of the SDGs, the picture in cities is getting gloomier. A recent progress report from UN-Habitat noted that growing inequality, social exclusion, and spatial segregation continue to have an impact on people's lives in most of the world's cities, producing a dramatic concentration of disadvantages. Urban areas are also increasingly epicentres of crises, insecurity, and violence. Housing remains largely unaffordable in both the developing and the developed world. As a result,, unsustainable models of urbanization persist, seen in the proliferation of informal and unplanned urban extensions, that gobble up dirty forms of energy, crowd people into slums that are often without access to water and sanitation, and facilitate unsafe living conditions – especially for women and girls.[13] Globally, 1.6 billion people live in inadequate housing, of which approximately 1 billion live in slums and informal settlements lacking basic services. Figure 4.5 shows "Rocinha" slum, one of the biggest in Brazil. Overall, cities account for 70% of global GHG emissions, 60% of global energy consumption, and about 50% of global waste. Figure 4.6 shows cities' share of the pie.

At the same time, all the failings of modern society – the stubborn poverty, the protracted inequality, the contribution to climate change, and, indeed, the onset of a global pandemic (and the conditions that will create the risk of future ones), which are most pronounced in concentrated urban areas – are exactly where the solutions to these seemingly intractable challenges lie.

Globally, it is estimated that at least 105 of the 169 targets of the SDGs cannot be achieved without the proper engagement and coordination of subnational and regional governments.[14] Figure 4.7 shows the aims of SDG 11 to make cities and human settlements inclusive, safe, resilient, and sustainable. Research across Organisation for Economic Co-operation and Development (OECD) countries found that well-functioning and well-managed cities are central to all the fundamental aspects of people's well-being and sustainable development, from water to housing, transport, infrastructure, land-use, and climate change, among others. As UN-Habitat observes, the inclusion of a dedicated cities goal (SDG 11) is a strong recognition by the international community

---

13 UN Secretary General, 2018. Progress on the implementation of the New Urban Agenda. https://digitallibrary.un.org/record/1628008?ln=en.

14 https://www.oecd.org/regional/regional-policy/Governing-the-City-Policy-Highlights.pdf.

Figure 4.5 Rocinha slum, one of the biggest in Brazil. (*Source:* Donatas Dabravolskas/Adobe Stock.)

Figure 4.6 Cities' share of the pie. (*Source:* Author derived from C40 city data. https://www.c40.org/.)

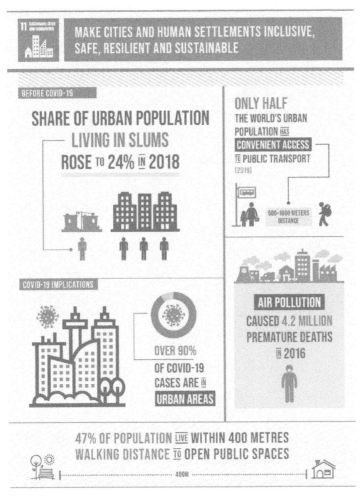

Figure 4.7 SDG 11 at a glance. (*Source:* Make cities and human settlements inclusive, safe, resilient and sustainable, 2021. United Nations.)

that urbanization is a transformative force for development that not only links to all other SDGs, but also underpins them. Figure 4.8 shows the indicators for progress against SDG 11.

But making the SDGs a reality in cities – and, correspondingly, leveraging cities' role in addressing the interlinked "wicked problems" of climate change, poverty, and inequality – will become even more complex as urban populations swell. For instance, energy demand is forecast to triple in line with urban growth over the next few decades, along with an equivalent demand for vital resources including water and food, as well as the need for housing, jobs, faster and better communications, and transport, placing ever more pressure on land and adding to already complex challenges and debates over land-use. Today, just 1.5% of the world's land is home to half of its production. If managed well, such concentration can be a positive force for development, as urban agglomeration boosts productivity, job creation, and economic growth, leading to more prosperous and

SDG 11 is the Theme of the 2018 Sustainability Report

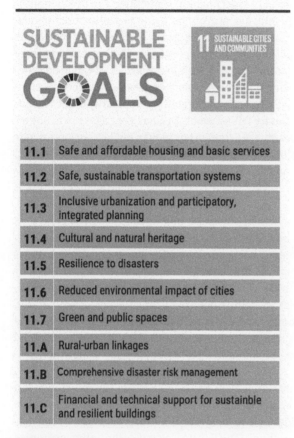

Figure 4.8   SDG 11 targets.

peaceful societies.[15] Consequently, the growth of urban settlements must be balanced with the effects on, and capacities of, rural and peri-urban settlements. As the World Bank observes, urban liveability and prosperity cannot be pursued effectively without distinguishing priorities for larger cities from those of smaller towns. For example, we must consider the environmental impacts – locally and globally – of cultivating land for farming to serve the growing number of urban populations, the additional roads that need building to bring supplies to markets and transport people from outer arteries to city centres, the associated energy and water demands for all this activity, and so on.

Put another way, it is the "circularity" of urban systems that will be crucial to the achievement of the SDGs and Paris Agreement. This does not just mean reducing waste or recycling more (though that's important), but also changing consumption patterns,

---

15 World Bank 2018, https://www.worldbank.org/en/topic/urbandevelopment/overview.

increasing the efficiency of urban "systems", and, crucially, ensuring that future energy demand is met by clean, renewable sources like solar.

For many developed cities, the concept of "sustainable consumption and production" is increasingly a reputational and competitive concern – be it due to self-imposed regulations, desire to comply with international "standards of excellence", evolving behaviours from citizens and businesses, or all of the above. Take, for example, the proliferation of urban farms growing on skyscrapers from New York to Tokyo aimed at leveraging nature-based solutions for heating and cooling and micro-scale food production (though cities like these have some way to go to meet tough climate targets, the popularity of such trends at least point in the right direction). But try telling people in Dhaka or Lagos not to buy air conditioners (if and when this is an affordable option) or to take public transport that is either non-existent or not safe. The backbone of the SDGs is to "Leave No One Behind", principally in terms of poverty and inequality – these goals do not elaborate on what individuals may consider equitable, like having an air conditioner when the temperature is topping 45°C.

Here, it is critical to understand where the most urban growth is occurring, and thus where the greatest investments will be needed, and where the greatest opportunities lie to maximize and sustain these investments. For example, smaller towns and municipalities naturally have fewer resources, but are likely to need greater support in the future as they serve as key arteries of expanding urban centres. The same holds true for rural agricultural settlements that will continue to help supply food for growing urban centres. Expanding the productive use of energy, and thereby enhancing income generation and productivity in these communities, will not only improve people's circumstances (and help prevent rural flight) but also provide more stability to supplies and bring other environmental benefits.

It is quite good news, then, that in March 2020, the UN Statistical Commission at last endorsed a new global method for classifying and comparing cities, urban areas, and rural areas in any part of the world.[16] Figure 4.9 illustrates the urban landscape. This new method, called "the degree of urbanization" (DEGURBA), classifies the entire territory of a country into three classes (cities; towns and semi-dense areas; rural areas) and it has two extensions: the first extension identifies cities, towns, suburban or peri-urban areas, villages, dispersed rural areas, and mostly uninhabited areas. The second extension adds a commuting zone around each city to create a functional urban area or metropolitan area. According to this calculation, in 2020, there were 1,934 metropolises with more than 300,000 inhabitants representing approximately 60% of the world's urban population.[17] At least 2.5 billion people lived in metropolises in 2020, which is equivalent to one-third of the global population.

But it's not just the numbers of people or the disbursement of those people that matter; equally – and perhaps even more important – it is understanding the conditions of where they are. As it happens, the places witnessing the fastest pace of urban growth are also those most vulnerable to the impacts of climate change. Indeed, over 90% of new urbanization is occurring in developing countries in the Global South, which is adding an

---

16 https://ec.europa.eu/eurostat/web/products-eurostat-news/-/ddn-20200316-1.

17 https://unhabitat.org/global-state-of-metropolis-2020-%E2%80%93-population-data-booklet.

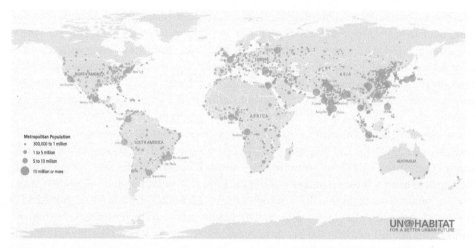

Figure 4.9 The growing urban landscape. (*Source:* https://urbanpolicyplatform.org/metropolitan-management/.)

estimated 70 million new residents to its urban population each year.[18] Of the world's 31 megacities (defined as those with at least 10 million inhabitants), 24 are located in the Global South. By 2030, the world will add 10 new megacities, all of which will be situated in developing countries. In countries in Sub-Saharan Africa and Southeast Asia, urban populations are set to double by mid-century. To put this in perspective, the percentage of people living in Europe's cities will double to just over 80% by 2050[19], where Africa's present rate of urbanization is nearly 11 times that of Europe's. Moreover, Africa's cities and surrounding peri-urban and rural areas will add 20 million people *per year* to the working-age population through 2030. By mid-century, Africa will be home to one-fifth of the global youth population (333 million people), many of whom are expected to migrate to ever-expanding metropolises in search of jobs and a better future.[19]

Against this backdrop, cities across Africa are experiencing some of the most severe disruptions and potential long-term "scarring" from the pandemic, threatening to erase all previous gains on sustainable development. To date, some 32 million *more* people have fallen into extreme poverty (earning below US$1.90 a day), adding to the pre-crisis levels of 571 million Africans who live in multidimensional poverty. If current trends persist, the United Nations Department of Economic and Social Affairs (UN DESA) estimates that by 2030, 88% of the world's poor (414 million people) will be in Africa, with most in urban centres.

Further, the pressures on the region's biodiversity and natural environment that are essential to the majority of the population's livelihoods (e.g. wildlife tourism) are only expected to rise in line with population growth and the accompanying demand for food

---

18 UN, 2016, https://stg-wedocs.unep.org/bitstream/handle/20.500.11822/11125/unepswiosm1inf7sdg.pdf?sequence=1.

19 UNDP Africa Promise, 2020.

and energy, potentially increasing the risks of another pandemic and impacts from climate change.[20] Since 2000, the region has lost two to five times more-intact landscapes compared with upper–middle and high-income countries, while only 18% of terrestrial areas and 12% of marine areas are protected across the region as a whole.[21] Moreover, the 2021 Dasgputa Review cautioned that the depletion of natural resources may prompt investors to reallocate current and planned investments or divest from existing ones.[22]

These conditions threaten to create a vicious circle. That is, the economic freefall caused by COVID-19 and the accompanying indebtedness of countries and cities means there are even fewer resources to invest in preventative and preparedness measures to avert a future pandemic and meet human development, conservation, and climate goals. For example, the United Nations Economic Commission for Africa (UNECA) estimates the Africa region will require an estimated US$92.8 billion in annual financing needs for 10 years to meet the SDGs as a result of an estimated US$29 billion loss due to COVID-19.[23]

Sadly, African cities are not alone. All cities face huge risks from climate change, but this is especially true for the 90% of urban areas located on, or in close proximity to, coastlines and low-lying areas that could be affected by myriad climate-related disasters.[24] For example, the 650 million urban dwellers expected to be living in delta and coastal areas by mid-century face serious risks from floods, water scarcity, and ecological and economic damage (UN, 2016). The number of urban residents exposed to cyclones and typhoons is expected to increase from 680 million in 2000 to more than 1.5 billion by 2050 (UN, 2016). All cities are also more vulnerable to the "heat island effect", since urban areas absorb more heat than suburban and rural areas.[25] One study of the effects of climate change on Asian megacities estimated that about one-quarter of Ho Chi Minh City residents are affected by extreme storm events, and this could reach more than 60% by 2050.[26] In Manila, a major flood under worst-case scenario climate change could cost the city nearly a quarter of its GDP, and that was before the pandemic.

Numbers matter.

20 OECD, 2020, https://www.oecd.org/env/resources/biodiversity/.

21 The Global Biodiversity Outlook 5 recognized that the world failed to achieve a single of the Aichi 2020 targets.

22 For example, some asset manager members of the Africa Private Creditor Working Group, such as Greylock Capital, have already pledged to incorporate climate and environment issues into investment decision-making.

23 UNECA African Green Recovery Report, March 2021.

24 UN, 2016, https://stg-wedocs.unep.org/bitstream/handle/20.500.11822/11125/unepswiosm1inf7sdg.pdf?sequence=1.

25 EPA, 2017, https://www.epa.gov/sites/default/files/2017-05/documents/reducing_urban_heat_islands_ch_1.pdf.

26 https://www.worldbank.org/en/news/press-release/2010/10/22/asian-megacities-threatened-climate-change-report.

From Ante-Theatre to Main Stage: Cities' Role in a Green, Sustainable Future

The centrality of the cities in achieving a global green recovery from the pandemic and reclaiming a path to realize the SDGs and Paris Agreement is undeniable. This is not only because of the volume of people living in cities, whose behaviour and demands for energy, goods, and services have a direct impact on the climate and natural environment, but also because cities are the engines of economic growth. In OECD countries, cities are responsible for some 60% of total public investments, particularly those related to climate action. In developing country cities, this engine is fuelled by the informal economy that also fills upwards of 60% of the national purse, comprised of countless local entrepreneurs and small and micro-sized enterprises.

Yet, till this day, cities and local governments are marginalized to the 'ante-theatre' of global discussions and decisions on the world stage. Official reporting on the state of the world's progress on both the SDGs and Paris Agreement is still done at national level (respectively, through the SDG High-Level Political Forum and Nationally Determined Contributions, or NDCs). In the most recent (February 2021) pulse check of countries' progress on climate change, undertaken by the United Nations Framework Convention on Climate Change (UNFCCC) that tracks such matters, the estimated emissions reductions "fall far short of what is required" to meet the Paris-Agreement goals of limiting temperature increases to 1.5°C or a maximum of 2°C this decade. In the entire synthesis report, there was just one reference to the role of "local governments" – an unkind reminder of the uphill battle cities face in accessing the necessary mandates and resources required to effect transformative change. Indeed, while local and regional governments' involvement in national-level reporting on SDGs has increased to 55% in 2020, up from 42% in 2016–2019, only about one-third of national governments bother to involve these actors in regular consultations.[27] Moreover, there remains a critical mismatch between the actual responsibilities of local and regional governments for critical public services and the revenues allocated to them.

Happily, local governments and networks themselves remain undeterred. In the face of the pandemic, thousands of committed mayors, city leadership groups, and regional authorities have doubled-down on their pursuit of a more inclusive and transparent delivery of the SDGs and Paris targets. For instance, at the height of the outbreak, the "Local 2030" initiative, comprised of several city and regional government representatives, issued an urgent call to ensure national climate action plans and corresponding budgets reflected the role of cities and local actors. The United Cities and Local Governments (UCLG), which represents more than 10,000 local and regional governments from 135 countries, the C40 Cities Climate Leadership Group, the Global Covenant of Mayors, ICLEI – Local Governments for Sustainability, and other groups all are mobilizing members to cut emissions in line with the targets of the Paris Agreement, and embed the SDGs into their local strategies, and insisting on building bridges between local and international governance levels – sometimes bypassing the national level. For example, the growing number of Voluntary Local Reviews (VLRs), which are the subnational version of sovereign SDG

---

27 UCLG, 2020, https://www.uclg.org/sites/default/files/report_localization_hlpf_2020.pdf.

progress reports, are demonstrating the concrete impacts of local governments' policies and investments in sustainable development and climate action. While the UN notes that VLRs "hold no official status", it nonetheless concedes that "sub-national reviews provide multiple benefits to the entities engaging in them and to SDG implementation at large".[28] "For now, what we see is that all these subnational, regional, civil society, and private actors are trying to exert influence in the context of the UN, without replacing the world body but rather having more of a say in its decisions."[29]

Perhaps one of the best moments of 2020 for cities was the solidarity demonstrated by the "We Are Still In" alliance that sprung up in reaction to former US President Donald Trump's decision to withdraw the USA from the Paris Agreement (a decision happily reversed by President Biden within hours of his taking office). Comprised of thousands of US mayors, governors, students, and businesses, the group sent a defiant message to all countries that cities count.

Numbers matter.

Conclusion

Thinking back to that flight coming into land in Nairobi, the question *"Where would you start?"* never seemed more relevant, or more challenging. This chapter has given only a glimpse of the many complexities and conditions confronting cities that we can "count". For many, the crisis is still too present to think about anything except *"When can we get back to normal?"*. This is understandable. The problem is that "getting back to normal" will not address the root causes that brought us here. We cannot continue to ignore the very real signs that we have already breached planetary boundaries that are, quite literally, killing us and the wonderful, extraordinary species with which we share the planet. We cannot continue to think it's ok, or look the other way, to the fact that hundreds of millions of people live in extreme poverty; that hundreds of millions of children don't have access to clean water or to an education. None of this was ever ok. In this way, the battle against COVID-19 has largely become a proxy fight against the nature–climate crisis and the poverty and structural inequalities that were already blocking progress on sustainable development well before the pandemic hit.

The power to correct course does lie with the local governments, communities, businesses, and networks. As the World Bank notes, "If urban areas are where COVID-19 impacts have been the most severe, it also means that interventions in cities and towns can have the biggest impact."[30]

So where do we start? If there is one upside to the tragedy of the pandemic it is that work towards a safer, healthier, and more sustainable future has already begun in the context of the global recovery, with trillions of dollars unlocked for new green energy and infrastructure, social protection, healthcare, and education. Cities must have a say in how

---

28 https://sdgs.un.org/topics/voluntary-local-reviews.

29 Browne, S., Weiss, T.G., 2020. *Routledge Handbook on the UN and Development*. Routledge, Abingdon.

30 https://www.worldbank.org/en/news/feature/2021/03/02/future-of-cities-will-shape-post-covid-19-world.

these recovery efforts take shape on the ground, and be able to take advantage of quick wins that show "proof of concept" that a sustainable world is possible and creates more benefits for more people. For example, many cities around the world see the crisis as an opportunity to combat long-standing public concerns like rising prices for housing and education, and poor air quality and pollution, and are investing in greener, low-carbon strategies and technologies, such as building efficiency retrofits, nature-based solutions, and replacing aging energy infrastructure with clean energy capacity. Such investments are helping put people back to work, while enabling cities to build "forward" better. Looking further ahead, we will need more initiatives like the World Economic Forum's BiodiverCities by 2030 programme that is supporting city governments to create an urban development model which is in harmony with nature. Regional and national governments must also do their part, for example, by directing an even greater share of stimulus to green investments and ensuring these resources reach and benefit cities and local communities. In the meantime, cities will use their own resources for the recovery where they can. These should be backed up with direct access to funding from intergovernmental grant-making facilities and funds; scaling up of support from development banks and investments by the private sector, including the provision of innovative financial instruments like municipal green bonds. Existing programmes like the European Bank for Reconstruction and Development's (EBRD) Green Cities programme to support climate action planning; the Cities Climate Finance Gap Fund to support the early-stage preparation of low-carbon and climate-resilient urban infrastructure projects in Global South cities; and the Cities Climate Finance Leadership Alliance (CCFLA) will no doubt continue to play a central role in supporting cities' path towards the SDGs, as will international sustainable cities initiatives supported by the World Bank, UN, OECD, and others. More programmes like these are needed to scale what works and accelerate progress.

Only time will provide us with the wisdom to know if we have made the right choices to reset a path towards a sustainable future for all people. What we do know is we cannot tackle the world's most pressing challenges – be they a global pandemic or climate change – at the international and national levels alone. Cities and local governments have a rightful and critical place in shaping our future – not only for people living in urban areas, but also for everyone. Perhaps the more important question, then, is not *"Where do we start?"* but *"Where do we go from here?"*.

> "A city is not just an accident, but the result of coherent vision and aims."
> *Leon Crier, pioneer architect, urban planner, and visionary, considered the godfather of the "New Urbanism" movement.*

5

The Vocal City – Introduction

Patricia gave us the "North Star" of how the world is coalescing around the SDGs in Chapter 4, *The Sustainable City*. The "circularity" of urban systems will be crucial to their achievement. In Chapters 14, *The Resourceful City*, Chapter 15, *The Zero Waste City*, and Chapter 23, *The Financed City*, this concept is expanded upon as we seek the utopian self-sustained city.

In this chapter Amanda and Kerem take us through the rise of the city networks, how they developed, and what it takes to drive a productive environment as centres of collaboration that connect, align, and produce high-value outcomes, which ultimately justify the collaboration. They explore just how important the voice of the city can be in effecting progress and argue that climate change is the "tragedy of the commons", a collective global problem that needs collective global solutions (Figure 5.1).

Connectivity is an ongoing theme throughout this book, and the ability of a city to communicate through the networks it establishes to fight against climate change is

Figure 5.1 Data and the connectivity of information, creating powerful networks.
(*Source*: Imaging L/Adobe Stock.)

The Climate City, First Edition. Edited by Martin Powell.
© 2022 John Wiley & Sons Ltd. Published 2022 by John Wiley & Sons Ltd.

crucial to this effort. Amanda and Kerem talk about how this effort began at the first World Congress of Local Governments for a Sustainable Future, which saw 200 local governments from 43 countries under the banner of the International Council for Environmental Initiatives (ICLEI), in collected effort to organize the city voice and to increase communication and data sharing. There has been a growing need to centralize the importance of city government and its investment in climate protection as we have seen through the work of city mayors and the steps taken to implement initiatives – steps that move cities past individual actions within communities to be part of a global movement, encouraging city networks to move beyond what Amanda and Kerem call the "coalition of the willing", and provide real measurable change.

Collecting, measuring, and sharing data between cities demands connectivity. Just as Peter encouraged us in Chapter 1 to go "beyond" our goals, while some city networks seek only to connect, some networks are developed to go beyond the point of connectivity and develop further alignment and production stages, which Amanda and Kerem discuss in more detail in this chapter. They also explore how the city network must engage with regional and national governments to standardize complex data and keep up with the ever-changing world of communication and technology that unites all the disparate city voices (Figure 5.2).

Just as Patricia argued in Chapter 4, climate cannot be divorced from other problems affecting our society, and in the same vein our collective climate problems are inevitably shared.

Figure 5.2 A network is in place to connect people, but it needs to be designed to align people and be productive. (*Source*: Orbon Alija/Getty Images.)

5

The Vocal City

Amanda Eichel and Kerem Yilmaz

The 2015 Paris Agreement signed at the 21st Conference of the Parties (COP) for the United Nations Framework Convention on Climate Change (UNFCCC) is an important diplomatic milestone in responding to the climate crisis. It not only cemented climate change as one of the most important existential threats of our time, but also reinforced the importance of collective action "to address and respond to climate change, including those of civil society, the private sector, financial institutions, cities and other subnational authorities".[1]

Recognizing cities and the voice of cities in diplomacy was an earth-shattering moment. So, how did the voice of cities rise to the top of global diplomacy and how did the efforts of many become one alliance? Even more importantly, now that cities are trusted partners in fighting climate change, how can we channel the collective capabilities of everyone – organizations, governments, citizens, and businesses – to build a more sustainable and climate-friendly world?

The Building Blocks

To put this into context, we must go back more than 30 years to the establishment of the UNFCCC (established in 1992) and trace both the concurrent establishment and evolution of city networks focused on marshalling the voice of cities to be a part of the solution.

The Start of a Movement

In 1988 when the Intergovernmental Panel on Climate Change (IPCC) was established by the World Meteorological Organization (WMO) and the United Nations Environment Programme (UNEP) there had been no internationally sanctioned effort to understand the changing levels of greenhouse gas (GHG) emissions in the atmosphere. It was a landmark moment and was soon followed by efforts to create an international framework to address climate change – what would become the UNFCCC. These bodies were tasked with understanding the impact of human activity on the climate and negotiating an associated human response, respectively.

---

1 https://unfccc.int/process-and-meetings/the-paris-agreement/the-paris-agreement/key-aspects-of-the-paris-agreement.

The Climate City, First Edition. Edited by Martin Powell.
© 2022 John Wiley & Sons Ltd. Published 2022 by John Wiley & Sons Ltd.

By no accident, cities were present at the release of the first IPCC global assessment report at the UN in 1990. The first World Congress of Local Governments for a Sustainable Future was organized at the UN in New York, bringing together more than 200 local governments from 43 countries under the banner of the newly established International Council for Environmental Initiatives (now known as ICLEI – Local Governments for Sustainability). The purpose – to organize the city voice to play a role in the future fate of the environment. And while it would still be several years before this constituency would put forward its proposal for measuring municipal impact, this stake in the ground established a clear intention to be not only named but also counted.

In 1997 the Kyoto Protocol – the first internationally binding treaty on climate change – was adopted.[2] This was a huge accomplishment for the world. And we all waited, wished, and worked fervently to ensure that all countries would sign on. Unfortunately, in 2005 when the treaty entered into force, not every country ratified the agreement. While not alone, perhaps the most significant outlier was the USA.[3]

While local governments had been organizing generally to date in support of various climate and environmental issues, the significant absence of a critical player in the international climate negotiations created a specific space and opportunity for local governments in the USA to respond.

Building a Big Tent: Brave Firsts at the Local Level

From initial brave commitments made at the individual city level, city networks worked to provide a platform to capture commitments and provide political cover for individual mayors so that their effort might be recognized within both the local and global community. These city networks initially focused on engaging and confirming commitment from local governments in various parts of the world to be part of the solution. Boxes 5.1–5.4 highlight organizations that have harnessed the power of city networks.

Box 5.1   The Mayor's Climate Protection Agreement

In 2005 when Mayor Greg Nickels of Seattle, WA called upon his colleagues to commit to addressing climate change in line with what would have been the US contribution, an initial group of 141 mayors joined the Mayors Climate Protection Agreement, a campaign supported by the US Conference of Mayors (USCM). This movement grew to boast over 1,000 signatories in the USA and was marked by a 2007 convening of mayors in Seattle when Mayor Nickels and the USCM also organized a US congressional hearing to hear directly from mayors on the commitment and opportunity they saw for partnership with federal law makers. While the US national government was still not ready for this level of partnership, other governments took note,[4] but what became clear at this moment was that local governments were ready to make a stand and were dedicated to exploring all options available to them to lead on behalf of their constituencies.

---

2 https://unfccc.int/sites/default/files/08_unfccc_kp_ref_manual.pdf.

3 https://unfccc.int/process/parties-non-party-stakeholders/
parties-convention-and-observer-states?field_partys_partyto_target_id%5B512%5D=512.

4 https://www.covenantofmayors.eu/about/covenant-initiative/origins-and-development.html.

Box 5.2   The C40 Cities Climate Leadership Group

In 2005 Mayor Ken Livingstone of London initiated a conversation among 20 of his fellow megacity mayors with the question of what might be possible if they were to align their voices. What would become the C40 Cities Climate Leadership Group (C40) was born, and through leadership that has now passed from London via Toronto, New York, Rio, Paris, and Los Angeles this group of nearly 100 leadership megacities is currently led by London Mayor Sadiq Khan and boasts a leadership steering committee of vice-chairs heralding from every region in the world. The majority of these cities have now committed to the most ambitious climate mitigation standards possible, in line with a net-zero contribution to global emissions by 2050.

Box 5.3   The EU Covenant of Mayors

In 2008 the European Union created the EU Covenant of Mayors, with the aim of creating a platform for EU cities to actively support the climate and energy commitments established at the EU level. Inspired by the US Mayors Climate Protection Agreement, the EU Covenant of Mayors went one step further – to establish a requirement for measurement and accountability – celebrating those cities that were able to advance from commitment to action via a dedicated platform to track progress and regular assessments of progress towards the local contribution to broader EU climate and energy objectives.

Box 5.4   Establishment of ICLEI

Following the initial convening of the first World Congress in 1990, ICLEI established a focus on both sustainability and climate action and became recognized by the UN-FCCC as the focal point for the Local Governments and Municipal Authorities (LGMA) constituency. In addition to this important UN-facing role, ICLEI has grown to establish more than 20 international offices, in addition to a global secretariat, which together support nearly 2,000 cities around the world.[5]
(*Source:* Kale Roberts, 2020. Three Decades of Sustainability: ICLEI at 30 Enters Next 'Decade of Local Action', ICLEI.)

The first challenge for all of these efforts was to understand the willingness of cities to move beyond their individual actions within their local communities to be part of a broader movement. The response was significant, but it soon became clear that for these local governments to be taken seriously it would be important to find a way of accounting for action and providing support to cities to move beyond their proclaimed ambition to concrete action.

---

5 https://icleiusa.org/iclei-at-30.

Securing a Seat at the Negotiating Table: Building the Evidence Base and Calling Out Inaction

Within the city climate community, there was a great deal of hope connected to the 15th session of national governments negotiating an international climate accord. By the time of the Copenhagen climate negotiations in 2009, sustainability and climate-oriented city networks were reasonably well established, but still largely focused on consolidating political commitments. When an international accord was not reached in Copenhagen, what became clear within the city community was a need to move beyond a coalition of the willing – and therefore to provide real and measurable evidence for potential change.

In response, city networks evolved to respond to the needs identified by their members:

- The C40 evolved towards a peer-to peer network model, building action-oriented sub-networks focused on specific sectors of opportunity, bolstered and informed by a solid research arm.
- ICLEI launched the Mexico City Pact and invested in the carbon*n* Climate Registry (cCR) to track commitments, while also doubling down to develop a solid technical assistance offering for members.

Additionally, these networks identified some shared areas of activity and action. Following COP15 in Copenhagen, Denmark, city networks, including C40 and ICLEI, as well as affiliated organizations like CDP[6] and the EU Covenant began to showcase the value and importance of cities in tackling climate changes by establishing city-specific reporting platforms. At the global level, the earliest efforts, starting in 2010 with the CDP Cities reporting platform and the cCR, focused on building measurable, reportable, consistent, and verifiable data to quantify the impact and to highlight the latent potential of cities as drivers for climate action.[7,8] When combined with regional efforts, including those by the EU Covenant of Mayors "My Covenant" platforms and the Canadian Partners for Climate Protection, what started as a few hundred cities reporting on their climate ambition, and directly accounting for their impact and progress in 2010, grew into well over 5,000 cities publicly reporting progress by COP21, and is now over 10,000 cities and local governments today (Figure 5.3).[9]

While more and more cities were choosing to invest resources to report on their climate action, there was still no established, global standard to allow for a true understanding of local impact at the national or global level, or to allow for comparison across jurisdictions. To ensure that reporting would actually influence policy, and associated investments, outside of city boundaries, a common standard was imperative. Such a standard needed to not only allow for consistent benchmarking and measurement, so that each community knew where to focus its resources to deliver the greatest benefit,

---

6 Previously known as the Carbon Disclosure Project.

7 https://climateinitiativesplatform.org/index.php/carbonn_cities_climate_registry_(cccr).

8 The CDP Cities reporting platform and cCR have now consolidated efforts for city reporting through the CDP–ICLEI Unified Reporting System.

9 GCoM internal research from 2018.

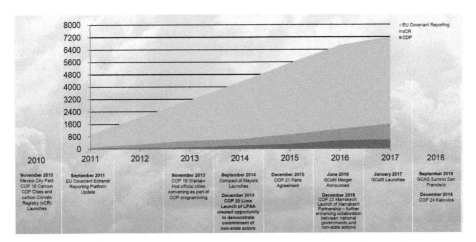

Figure 5.3 City-level commitment to transparency helps accelerate global climate diplomacy. (Source: https://www.globalcovenantofmayors.org/.)

but also provide confidence to policymakers and investors – whether they be state, national, or international governmental institutions, but also, with growing importance, the private sector.

Emissions accounting methodologies, such as the Global Protocol for Community-Scale Greenhouse Gas Emission Inventories (GPC), and follow-on reporting frameworks, specifically the Global Covenant of Mayors (GCoM) Common Reporting Framework, created an international, comparable standard for calculating emissions, and subsequently activity related to climate adaptation and access to energy, that could help cities and relevant stakeholders better understand their community's emissions drivers, localized impacts from climate change, and opportunities to increase local access to clean and affordable energy.[10] With these insights, there just was not a one-size-fits-all solution. Now, effort and resources could focus on the right sectors for each city and the right policy instruments to drive action.

In addition to facilitating more robust planning processes, consistent measurement and reporting ushered even more ground-breaking research around cities. Efforts like C40's Climate Action in Megacities series,[11] the Stockholm Environment Institute's global urban potential analysis, the Coalition for Urban Transitions global reports, and the GCoM annual aggregation of global urban emissions[12] proved to be seminal pieces of research that amplified the voice and leadership of cities. These early efforts spawned a much broader research agenda that has: (1) cemented cites' importance on the international stage; (2) advocated for and encouraged the investment opportunity that cities represent; and (3) highlighted the interconnected nature of solutions at the city level.

---

10 https://ghgprotocol.org/.

11 https://www.arup.com/perspectives/publications/research/section/climate-action-in-megacities-cam-30.

12 https://www.globalcovenantofmayors.org/impact2019.

With so many cities reporting progress and burgeoning research, there was also a need to create knowledge sharing and peer to peer groups – both for sharing ideas and for providing political support and cover for local leaders to take ambitious climate action. Both C40's and ICLEI's networks, for example, served as platforms for cities, towns, and local governments to share information, expertise, and success stories. These communities allowed cities that were a part of them to become critical policy innovators across their spheres of influence, demonstrate leading practices, and build support for climate action. The network effect, particularly in the early days after COP15, was tremendous. By 2013, cities in C40's network, for example, reported a 500% increase in cycle share programmes, and 80% of those cities had introduced dedicated cycling lanes.[13] And, 90% of respondents to the Climate Action in Megacities 2.0 Report had introduced LED street-lights.[14]

This was the decade of establishing the baseline of city evidence and also the requisite support mechanisms for city officials and mayors to take the political risk to be visible on the global stage. In some cases, mayors were at odds with their national counterparts. And, in most cases these interactions were provocative. In reflection, tension and pressure were necessary to ensure that the role of cities would be acknowledged at an international level. This tension resulted in a common mantra amongst the city community that "nations talk while cities act". However, as the global community coalesced around a common understanding of the science, and the associated imperative, to act, it ultimately became quite clear that the only path forward would require collaboration between levels of government and across actors – public, private, and including civil society – to advance a viable solution.

Evolution of the City Network

The city network community and associated climate action movement had been supremely successful in bringing recognition to the importance of city governments as critical actors in realizing climate protection outcomes, in part because it has continued to evolve to address the needs of both mayors and local actors in line with the demands of the broader global community.

To put this in context, research around climate networks and their mobilization identifies three key types of networks that overlay each other like concentric circles (Table 5.1). At the bottom, all networks are built on a foundation of *connectivity* that links people and organizations to each other. Some networks end there, having achieved their mission to build a connection. Other networks, however, develop *alignment*. This alignment occurs when network members strongly share a sense of identity and/or a value proposition. For some networks, alignment may be the end goal and for others they move into *production*. Production occurs when network members seek to accomplish something specific in collaboration that is above and beyond merely connecting or aligned on an identity.

---

13 https://www.c40.org/.
14 Ibid.

Table 5.1 Key types of networks.

	Connectivity	Alignment	Production
Definition	Connects people to allow easy flow of and access to information and transactions	Aligns people to develop and spread an identity and network value propositions	Fosters joint action for specialized outcomes by aligned people
Desired network effects	Rapid growth and diffusion, small-world reach, resilience	Adaptive capacity, small-world reach, rapid growth and diffusion	Rapid growth and diffusion, small-world reach, resilience, adaptive capacity
Key tasks of network builder	Weaving – helping people meet each other, increasing ease of sharing and searching for information	Facilitating – helping people to explore potential shared identity and value propositions	Managing – helping people plan and implement collaborative actions

(*Source*: *Guidebook for Building Regional Networks for Urban Sustainability 2.0*, 2012.)[15]

The city climate network community has navigated through these various stages and is now squarely focused on the last stage of evolution, *Production*. However, as part of this evolution, it has become clear that greater partnership, reaching both across the individual city network efforts and, importantly, outward towards other potential collaborators in the public and private spheres, will be essential if local contributions are to be fully realized as a critical and willing contribution to a global climate solution.

Evolution of Data Through Collective Action

The community of city networks/organizations since around 2008 has championed the need for precise data both to bring legitimacy to cities and local governments and to highlight the underlying opportunity that comes with investing in them. The primary objective of data collection, measurement, and reporting has been "policy-driven", focusing on the areas where cities have the greatest authority to pursue climate action and where investment can support the delivery of those policy proposals.[16] The truth is that this entire process *is* flawed – data acquisition and modelling are not perfect sciences and, in many cases, do lead to underreporting of key metrics, such as GHG emissions.[17] Models are only as good as the quality of data available and the capability of individuals employing them to make measurements.

---

15 Plastrik, P., Parzen, J., 2012. *Guidebook for Building Regional Networks for Urban Sustainability 2.0*. Urban Sustainability Directors Network and Innovation Network for Communities.

16 https://www.c40.org/blog_posts/cam2

17 https://www.reuters.com/article/us-usa-climate-cities-iduskbn2a226l

While the frameworks and guidance to report progress on climate action are now globally consistent,[18] the underlying information including its availability and the way it is reported is not. For some cities and local governments, empirical data are readily available through national or regional governments (and their affiliated agencies). Cities in the State of California in the USA, for example, have detailed, hyperlocal vehicle information by fuel type publicly available through the State's Department of Motor Vehicles website.[19] Access to this information is an important resource in supporting local governments to assess vehicle trends, develop emissions inventories, and assess policy options. For many cities and local governments, especially in areas of the world where emission growth is rapidly accelerating, this infrastructure does not yet exist. There are still significant data gaps to measure, plan, and track progress across all key sectors. The insights required to make critical policy decisions are therefore difficult to access, and they often lack the granular information necessary to make truly data-driven policy in the local context.[20]

To address these issues, the philanthropic community has invested in numerous technical assistance programmes since around 2005. While each programme does yield promising initial progress, the results are generally due to extremely resource-intensive, geographically focused, and time-limited engagements. As a result, each of these programmes encounters difficulties in scaling and replicating similar models to the tens of thousands of cities and local governments that require support. These programmes also cannot keep up with the pace of data and intelligence being generated. We need to find a better way to overcome this measurement hurdle, effectively scale support, and channel information to cities and local governments that need it the most.

Like the collective action efforts of networks to build awareness on the importance of cities, the space requires a similar solution to resolve data access and technical assistance challenges. The world is on the cusp of a significant transformation in how we process, harness, and incorporate data across all facets of society. Recognizing these trends, it is time to rethink the existing, bottom-up data collection paradigm and to begin considering areas and opportunities where economies of scale can create meaningful change and bring forth actionable data.

City networks must begin engaging with regional and national governments to standardize complex datasets. They must also work together and coordinate on empirical data collection and identify the channels through which it can be made available for decision-makers. For example, we can empower leading government innovation agencies and affiliated initiatives, such as Mission Innovation,[21] to galvanize this effort, unleash their collective capabilities, and expand the availability of information to data-poor regions of the world. They can harness their investment potential to incubate new technologies or innovative collection methods that can best support governments as they deal with addressing climate change.

---

18 https://ghgprotocol.org/greenhouse-gas-protocol-accounting-reporting-standard-cities.
19 https://data.ca.gov/dataset/vehicle-fuel-type-count-by-zip-code.
20 https://www.globalcovenantofmayors.org
21 http://mission-innovation.net.

It is important, moreover, to recognize that government alone cannot accomplish the scale and magnitude of change required. We must also seek ways to incentivize the private sector to bring forth their research and development efforts around data collection, artificial intelligence, and modelling for public benefit. Early efforts by leading technology companies, such as Google's Environmental Insights Explorer, to bring forward data to measure emissions and track progress are a helpful start. Practitioners and decision-makers need the ability to easily connect disparate datasets, mine that information, and assess decisions more rapidly.

Moving Together in Partnership as a Global Community

There are limits to fighting climate change alone. As data collection, measurement, scenario modelling, and climate planning have improved at the local level, it has become increasingly clear that we have reached the limit for what individual cities, or city networks, can do single-handedly. Even where cities have political will and available political, financial, and human resources, they face fundamental limits to their ambitions, especially if a significant number of cities are not similarly engaged and coordinated in pursuing GHG reductions, climate resilience, and energy access.[22] And, in areas where cities can address climate change, it will be difficult for the vast majority of them to achieve the economies of scale and transformative outcomes obtainable by national governments or even through coordinated provincial/regional action.[23] This is particularly the case for the cities and local governments whose names are not known outside their own national borders – these intermediary cities, some of whom are growing much faster than the current megacities that we all know by name, as well as the huge number of smaller local communities responsible for over 50% of current opportunity on climate. Over time this will only grow, and yet these are the cities that are least able to act on their own.[24]

Climate change is the most relevant and active example of the tragedy of the commons, and collective problems need collective solutions as well as broader coordination. The complexity that characterizes climate change at the global level is heightened at the city level to focus response on issues within local control, prioritize actions that are of greatest concern to a local constituency, and consider intercity flows (of resources, emissions, systems, and thus solutions).

As an initial response to address these issues, the city network community came together in 2014 to develop a new type of coordinating entity that would be able to channel the skills, resources, and abilities of global networks to assemble, coordinate, advocate, and mobilize resources on behalf of the city and local government constituency.

---

22 https://www.sei.org/publications/
what-cities-do-best-piecing-together-an-efficient-global-climate-governance.

23 https://mediamanager.sei.org/documents/Publications/Climate/SEI-WP-2015-15-Cities-vertical-climate-governance.pdf.

24 https://urbantransitions.global/urban-opportunity/

This effort, known initially as the Compact of Mayors,[25] which later evolved into the Global Covenant of Mayors for Climate & Energy[26] after a merger with the EU Covenant of Mayors, became the world's largest alliance of city leaders to have pledged reductions in local GHG emissions, enhance resilience to climate change, and track their progress openly and transparently.

In its short existence, the Compact and its successor, GCoM, has become more than a simple platform to centralize publication of climate commitments made through this partnership across networks and associate regional covenants of chapters that have now been established to support the action of individual cities around the globe. The alliance recognizes that no single organization, initiative, sector, or government, however innovative or impactful, can accomplish all the transformational, systemic changes required of the global community. GCoM has evolved into an organizing hub that seeks to move past *Production* and enable cities, local governments, and the networks that support them to bridge the gap from climate ambition to delivery. Its Secretariat operates at the intersection of knowledge, experience, and capabilities and works to channel the collective capabilities of all with the primary focus of mobilizing resources in support of city commitments, organized via a common platform which makes these commitments and associated reporting public and openly available. Perhaps most importantly, GCoM has established the connective tissue to secure political commitments from local leaders, secure buy-in at the global level, and translate that into a pathway for delivery.

The urgency of the climate crisis, coupled with ongoing economic, equity, and political crises at the global level, requires yet another evolution. In particular, climate can no longer be divorced from efforts to minimize and address human death and illness, climate migration, biodiversity collapse, and food insecurity, among others.[27] The execution of the Paris Agreement and the very recent and most welcome recommitment of the USA to this treaty have officially ushered in the era of implementation. While local governments will still bear the brunt of implementation, they should no longer have to shoulder the burden of climate leadership without support. The recognition of partnership is clear in the writing of the Paris Agreement and in many of the investments that national governments have made to date in their plans for implementation.

Many recent efforts have begun to recognize this need for partnership (see Boxes 5.5–5.7).

These are but a few examples of the important next phase of city network engagement and evolution that is required. It is time for city networks, local and national governments, and the broader community of stakeholders to be aligned in their investments and actions. We must enter more directly into a world of partnership and push the boundaries of *Production*. The city's voice through advocacy and accountability will still be critical,

---

25 The Compact of Mayors was launched in late 2014 by Michael R. Bloomberg, UN Secretary-General Ban Ki-moon, and global city networks ICLEI – Local Governments for Sustainability (ICLEI), C40 Cities Climate Leadership Group (C40), and United Cities and Local Governments (UCLG), with support from UN-Habitat.

26 https://ec.europa.eu/commission/presscorner/detail/it/ip_16_2247.

27 https://globalcovenantofmayors.org/press/newsummary-for-urban-policymakers-initiativeannounced).

Box 5.5   The Leadership for Urban Climate Investment Initiative

The Leadership for Urban Climate Investment (LUCI) initiative was launched at the UN Secretary General's Climate Summit in 2019, with the aim of aligning national government investments in financing climate action in cities with commitments for local action. While there are many projects that now fall under the banner of LUCI, one in particular recognizes the need for broader support for cities to develop investment-ready solutions for ultimate support by public and private financing alike. The City Climate Finance Gap Fund (Gap Fund) is an effort co-designed by the government of Germany and GCoM and now hosted by the World Bank and the European Investment Bank. Through the Gap Fund, cities will receive technical support to advance their climate action ideas to solid project proposals, ready for funding or finance.

Box 5.6   The International Ministerial Mission Innovation

In 2019 GCoM became one of only six collaborating organizations working with the international ministerial Mission Innovation, a collaborative of 22 member governments committed to increasing their investment in clean energy research, development, and demonstration (RD&D). As Mission Innovation moves into its next phase, GCoM is working with these member governments to confirm how committed resources can be directed to support city innovation, pilot technological solutions, and collaborate towards a broader vision of net-zero and resilient local communities.

Box 5.7   The International Coalition for Sustainable Infrastructure

The International Coalition for Sustainable Infrastructure (ICSI) was established in 2019, with the aim of bringing the international engineering community soundly into the global climate conversation – with a focus on action. As a founding member of ICSI, GCoM is working with the global engineering community to harness the significant expertise and resources of this critical stakeholder group to work collaboratively towards urban infrastructure solutions.

---

but maintaining this hard-fought seat at the table also requires a recognition of the limitations of city action. The ability of local governments to act alone is at its limit, and nor do cities wish to go it alone. Jurisdictional challenges and behaviour change require working across multiple geographic problems and stakeholders. New fora must be established to support better regional/city planning opportunities, especially around the provision of basic human rights to transportation, energy, water, natural resources, and housing, that is more inclusive of the patterns of a region. In other circumstances, the city level is not always the tipping point. More work needs to be done for greater national/local partnerships to ensure there is broader, more cohesive planning done.

Conclusion

Delivering on the Paris Agreement requires a change in how we discuss climate change, its impact, and the myriad ways it will change the fabric of society. We cannot and should not divorce climate from other critical investments including equity, community health, public safety, infrastructure, and economic opportunity, just to name a few. How we begin to frame climate change and its intersectionality with respect to so many societal challenges will be pivotal in galvanizing a much broader community of stakeholders.

We also need broader collaborations and holistic thinking to drive home big changes and to raise the connective tissue between cities, regions, national governments, the private sector, and civil society. We must also bring together the capabilities, skills, and learnings from other levels of government, as well as outside perspectives, to support successful policy planning, implementation, and monitoring that both is locally appropriate and ensures climate-friendly outcomes. And, most importantly, we must begin to think about and invest in the city/metro area and the city/region as part of an all-encompassing system, which equally respects the need for development while simultaneously protecting and preserving natural resources and systems.

The very good news is that city networks and the cities they work with stand ready and able to enter into this new phase of partnership. And it also appears that at least some, in the broader global community, stand ready to meet them.

6

The Governed City – Introduction

In Chapter 5 Amanda and Kerem showed how collaboration has amplified the voice of cities. They made reference to the Coalition for Urban Transitions, which you will read more about in Chapter 7, *The Decoupled City*. They also addressed the importance of mining and analysing the right data, which is the foundation of Seth and Eric's Chapter 18, *The Data City*, and if you have the right data what exactly should you measure to define success, which will be illustrated further by Patricia in Chapter 19, *The Measured City*.

We have walked you through the drivers of early and developing cities, and highlighted the "North Star" of the SDGs, and Amanda and Kerem have shown how city networks will allow more traction to be gained. That said, just understanding what steps to take does not guarantee a successful outcome. We therefore pivot to Bruce and Luise in this chapter, *The Governed City*. They use Copenhagen as a case study to highlight a model for achieving successful outcomes. Towards the end of the book, in Chapter 28, *The Green City*, you will hear from the Lord Mayor of Copenhagen, about what has been achieved.

In 2019, the Amager Resource Centre (ARC), a waste-to-energy power plant in Copenhagen, put an artificial ski slope on its roof. Jacob Simonsen, the chief executive of ARC, referred to it as an act of "hedonistic sustainability". In other words, they did it because they could, because they had spare capacity, because Copenhagen is far ahead of other global cities in that it is on track to become net carbon neutral by 2025.[1]

In Chapter 2 I said that every city is itself a story, but in the fight against climate change I would hasten to add that not every city is a "success" story. The Copenhagen climate story is one of the few examples of these success stories that we have to follow as a model for improving our cities and solving the problems they face in actively attaining their net-zero goals, so much so that they can become "hedonistic" in their sustainability (Figure 6.1).

Since the 1970s, Copenhagen has made a conscious transformation from a fossil fuel-reliant city to one of innovating green technology that strives to become the first global city to achieve zero carbon emissions by 2025. Famous for its obsession with bicycles, with the city's bike lines often more crowded that its roads, the city has also employed change

---

1 Robertson, D., 2019. Inside Copenhagen's Race to be the first carbon neutral city. *The Guardian*, 11 October. https://www.theguardian.com/cities/2019/oct/11/inside-copenhagens-race-to-be-the-first-carbon-neutral-city.

The Climate City, First Edition. Edited by Martin Powell.
© 2022 John Wiley & Sons Ltd. Published 2022 by John Wiley & Sons Ltd.

Figure 6.1 Cycle lanes in Copenhagen are busier than the roads, aided by an all-weather mindset. (*Source:* william87/Adobe Stock.)

through areas such as energy consumption, energy production, green mobility, and city administration. Copenhagen has managed to grow its population whilst (in direct contrast to the examples purported by Austin Williams in Chapter 3, *The Emerging City*) simultaneously decreasing its CO_2 emissions. As Bruce and Louise point out in the chapter, this "is not accidental but an example of good governance", specifically in regards to climate solutions. To explain how good governance has been achieved, Bruce and Luise have broken down the Copenhagen model into four institutional models:

1) a strong local government;
2) a municipal government that aggregates power across jurisdictions and gives local officials prominence in national policy in ways that are reflective of municipal realities;
3) investment in public–private "hybrid" institutions (such as City & Port, which has itself become symbolic of Copenhagen's regeneration and revival since the 1990s); and
4) active participation and investment of public pension funds.

By interrogating the four institutional models and the changes that could be made in cities across the world, it is clear that there are many lessons to be learned from the Copenhagen climate story. Bruce and Luise help to show there is a general disconnect, in short, between "the simplicity of climate goals and the complexity of climate finance", and that city governance must now focus on how policies can be implemented, make room for the growth of large-scale institutions with a holistic focus, and have access to financial sophistication. They advocate for a balance of power across the public and private sectors, the connection of different levels of government, and employing different institutional actions within cities.

Wouldn't it be wonderful if all cities could become "hedonistically sustainable"?

6

The Governed City

Bruce Katz and Luise Noring

With climate impact and climate advocacy on a meteoric rise, cities are at the vanguard of problem solving. This is because many cities, in collaboration with national governments, possess the powers and resources to reorient key sectors of the economy, particularly the energy, buildings, and transportation sectors that disproportionately drive carbon emissions. Cities also represent networks of public, private, and civic leaders and institutions that are pragmatic at the core and less likely to be hijacked by partisan rancour and ideological polarization, the curse of our times.

A small number of cities, disproportionately located in northern Europe, are leading the pack on the design, finance, and delivery of ambitious climate solutions. This chapter uses Copenhagen to distil the lessons not just around the design (the "what") of policy initiatives but around the plan (the "how") for delivering the climate commitments that an increasing number of cities are making. The Copenhagen example shows the continuum of good governance (e.g. capable municipal governments, consortia of multiple cities, special-purpose institutions, and leading pension funds) that makes transformational solutions and creative financing possible. Our conclusion is simple: cities can tackle climate change if and only if they have institutions with the capacity, capital, and community standing necessary to get the job done.

Case Study: Copenhagen

Copenhagen has emerged as the world's poster child for climate solutions, as it strives to become the first global city to achieve zero carbon emissions by 2025. In 2012, the city released the 2025 Climate Plan proposing that Copenhagen becomes the first carbon neutral capital city in the world. Copenhagen's plan is an intricate mix of concrete goals and initiatives that aim to drive change through four areas: energy consumption, energy production, green mobility, and city administration.

Given the complexity of climate change, it is not surprising that Copenhagen's strategy for zero carbon is data-driven and multilayered. On one level, the city (in close concert with the national government) is pursuing *policy innovations* around governmental investments, regulations, and commitments. Significantly, Copenhagen has established a series of subgoals that, if achieved together, enable the city to get to zero

The Climate City, First Edition. Edited by Martin Powell.
© 2022 John Wiley & Sons Ltd. Published 2022 by John Wiley & Sons Ltd.

carbon emissions. For example, the city has established the following energy production targets for 2025:

- District heating in Copenhagen is carbon neutral.
- Electricity production is based on wind and sustainable biomass and exceeds total electricity consumption in the city.
- Plastic waste from households and businesses is separated.
- The bio-gasification of organic waste is scaled.

These targets build on decades of smart national policy and local action. During the 1970s, for example, the Danish national government decided to make a strategic shift away from a fossil fuelled society towards green energy during the OPEC energy crisis. This led to a transition that today finds 100% of all Copenhagen households fuelled by district heating. More recently, government efforts and substantial public investments have resulted in green energy (based predominately on wind) becoming one of the largest industries in Denmark today, including exporting green technologies and services around the world (and even wind energy to Germany).[2]

Waste management in Copenhagen has also undergone large investments and changes. In 2010, 71% of waste was incinerated, 27% was recycled, and only 2% was sent to landfill. Importantly, the incineration of waste in Denmark does not contribute to air pollution. The 80-m-high CopenHill is the most recent incinerator built in Copenhagen that transforms waste to renewable energy for district heating (Figure 6.2). New rules of waste handling are currently being introduced that require citizens to separate waste in five different categories. Recycling is not a recent phenomenon in Copenhagen; the city recycles 96% of all plastic bottles and cans, with vending-style machines returning a deposit when you insert a plastic bottle or can.

A modernized sewer system has ensured that Copenhagen efficiently uses water. Consistently since 2010 the city has grown, while the amount of fresh drinking water used per capita has decreased. Thus, 108 litres were consumed per capita per day in 2010, which is 36% lower than in 1989. Additionally, only 8% of water is lost through the pipes, a measure that in many comparable cities is closer to 40–45%. This has largely been enabled through a close collaboration between Copenhagen and adjacent municipalities. Across Copenhagen and other Danish municipalities, local governments identify and pump up polluted water to preserve clean groundwater.

On another level, Copenhagen is either inventing or deploying *product innovations* that enable the deployment of concrete design and technological as well as financial advances. Using architectural and land-use norms to build sheltered and secure bike lanes and technology to monitor the energy use of buildings or accelerate traffic flow and reduce congestion are all examples of product solutions. At the same time, financing these and other actions is advanced by financial products (e.g. green bonds) and financial mechanisms (e.g. value capture) that can be standardized.

---

2 Danish Energy Agency, 2018. Department for Communities and Local Government. https://assets. publishing.service.gov.uk/government/uploads/system/uploads/attachment_data/file/590464/ Fixing_our_broken_housing_market_-_print_ready_version.pdf.

Figure 6.2 CopenHill – a new breed of waste-to-energy plant that is economically, environmentally, and socially profitable, combining industrial need with urban leisure. (*Source:* OliverFoerstner/Adobe Stock.)

Biking has enabled the city to grow its population and economy within the space available while reducing its CO_2 emissions. There are several product innovations underpinning this:

- Bike lanes are clearly demarcated and sheltered from pedestrians and car traffic.
- Green biking lanes cut diagonally through the city connecting nodes, such as university campuses, making it even faster to move by bike. These green lanes make the experience more pleasant and it is estimated that health costs are reduced by €0.77 per km cycled.
- Since 2010, two biking bridges have been built. These cut off substantial time and length of trips; importantly, those gains are only available to bikers.
- Green waves also ensure that cyclists are granted preference to cars when moving across the city. Only public buses can also ride the green waves.
- Public transit is integrated into the biking network and it is possible to take a bike on public transportation. Only one ticket is required for all modes of public transport, which is only available online and purchased with a mobile phone.

Finally, Copenhagen is perfecting *process innovations* by creating a series of public and public–private institutions with the capacity and capital to drive solutions at scale. As described below, Copenhagen's municipal government is strong and staffed with skilled individuals. The city benefits by being part of Local Government Denmark, which enables all cities to aggregate their political power and negotiate with national government on an equal footing. The city also is part owner (with the national government) of City & Port, a publicly owned and privately managed corporation that has been able to drive over 50% of the development of the core of the city since around 2010 and use the revenues generated by land sales and leases to finance the construction of a twenty-first-century subway system for the entire city. Finally, the city benefits from strong public and private pension funds that have the wherewithal to co-invest in projects that promote sustainability while providing adequate financial returns.

Copenhagen is the leading city in the world to marry economic growth with advances in environmental sustainability. In 2014, the World Bank counted Copenhagen amongst the top ten wealthiest cities in the world. At the same time, Copenhagen has one of the lowest carbon footprints of any capital city in the world: carbon emissions have dropped consistently between 1991 and 2012, from 7.9 to 3.2 tonnes per person.

Significantly, the benefits of the transition to a no-, low-carbon future are widely shared across the broad population. Cycling, for example, offers a healthier and less expensive means of transportation. It is estimated that cycling in Copenhagen alone results in €230 million annual healthcare savings. Cycling is also more cost-efficient; transportation costs only account for 3.4% of gross value, compared with 5.8% in Stockholm and 8.4% in London. Eighty-eight percent of Copenhageners say that they choose the bike over other means of transportation because it is faster and more convenient. Thus, Copenhagen has increased transport efficiency whereas most western cities experience increased car traffic and congestion as the economy grows.

Similarly, the cleansing of the harbour, which has largely resulted from major investments in a new sewer system, has been an important driver of the urban revitalization of Copenhagen.[3] The city and seven surrounding municipalities co-created and co-own the Greater Copenhagen Utility (HOFOR), which is responsible for building, maintaining, and operating the wastewater system, district heating and cooling, and gas supply for the city and metro region.[4] Around 2010, it would have been inconceivable to swim in Copenhagen harbour, as the water was highly polluted. Today, swimming in Copenhagen's harbour has become a symbol of Copenhagen's green agenda and the high quality of life for city residents.

A Continuum of Good Governance

Copenhagen's success is not accidental. It is the direct outcome of strong institutions that have the capital, capacity, and community standing to deliver climate solutions. We believe four institutional models help explain the Copenhagen model.

First, Copenhagen has a strong local government. Denmark has a highly decentralized governmental system, enabling municipalities to operate as strong partners with the

---

3 https://international.kk.dk/.

4 https://www.brookings.edu/research/why-copenhagen-works.

national government. Northern Europeans, in general, are devolutionists, using the nation state to provide a platform for market expansion, environmental sustainability, and social cohesion but allowing municipalities enormous latitude to innovate and align national direction to local realities. According to a 2009 OECD review, local governments in Denmark account for over 60% of government spending, the highest level among OECD peers. Copenhagen's ambition to be the first major global city to generate zero carbon emissions is thus enabled by a strong fiscal foundation.

Along with that local power and legitimacy comes more voter participation: while the USA suffers from voter turnout of around 20% in local elections, Copenhagen has consistently experienced turnout of 70%. Accordingly, local capacity is likewise strong. The knowledge and decision-making capacity of the public sector is robust, with a steady supply of highly educated public servants across technical, environmental, social, and business fields. The supply stems from the free tuition public sector educational system (which also greatly benefits the private sector).

The structural organization of the local government (and its strong partnership with other sectors) has a large impact. As the two of us wrote several years ago:

> Civic capacity in Copenhagen is reinforced by the city's "mini-mayor" system, which resembles a parliamentary arrangement. The lord mayor is appointed by the ruling party, but the city council elects, from varying parties, a cabinet of several "mayors per expertise," such as a technology and environmental mayor, an employment and integration mayor, a health and care mayor, and a children and youth mayor. While a mixed-party cabinet can lead to mayors with different priorities, it can also increase the incentive to collaborate and produce policies that enjoy widespread support and survive beyond the term of a single mayor. This governing arrangement results in a city government that is at once technically proficient, accountable, and incentivized to cooperate, enabling the city to be a stronger negotiator, partner, and investor.
>
> Public and private partnerships are supported by innovative civic entities. Some institutions are truly unique in structure. Realdania, an association that functions much like a U.S. foundation, is a member-based private philanthropic business capitalized by the sale of a home mortgage cooperative that dated back to 1851. Through smart investments in transformative local projects and strategic transportation and land-use planning, Realdania is at the center of new types of collaboration and innovation in Copenhagen's built environment. Through significant investments in the C40 Cities network, Realdania also ensures that Copenhagen remains a central player among global cities on climate change.[5]

Second, the Copenhagen municipal government operates within a system that aggregates political power across municipal jurisdictions and gives local officials the ability to shape national policy in ways that reflect municipal realities. Local Government Denmark (KL) sits at the core of that system.

---

5 Ibid.

KL is a cooperative that is owned and governed by all Danish municipalities. The municipalities pay a fee for their membership. It is through the universal membership and representation of all Danish municipalities emblematic of an "all for one and one of all" approach that KL is able to leverage political and fiscal power vis-à-vis national government, trade unions, and other key societal stakeholders. By spanning across party-politics, KL can focus on the actual substance of decision-making rather than the colour of politics. In short, KL is able to keep its eyes on the ball. Its political mandate stems from all different political parties represented across all city governments.

KL negotiates an annual municipal budget with the Ministry of Finance, representing the interests of its members. Once the annual budget it accepted, it is controlled through risk and reward sharing. Thus, if one municipality overspends in one year, all municipalities must repay this deficit in the subsequent financial year. If one municipality increases its taxes, all or some municipalities must reduce their taxes by equal measure. It is a carefully designed system of checks and balances that ensures that local spending does not go rampant and essentially ensures that politicians do not make unfunded promises. It fosters a politics of realism rather than wishful thinking.

KL plays an instrumental role in enabling and informing the climate plans of Danish cities. The organization has published concrete plans and roadmaps for areas of domestic policy that affect climate change, including public transit, road traffic, energy, and water. It provides guidance on how to finance and deliver distinct climate plans set by the national government and recommends changes in national law, where necessary. (To this end, KL recently proposed eliminating the ceiling on infrastructure expenditures for municipal investments in the mitigation of rising water levels.)[6] KL is effective because it applies the perspective of how to implement and finance climate plans within and across cities, recognizing that climate and its impacts are not confined within neatly set municipal boundaries.[7]

The Danish intergovernmental system has mechanisms to ensure that municipalities have the fiscal means needed to implement new political initiatives and laws. DUT, which in English translates to the Extended Total Balancing Process, is a formal process used to estimate the expenses of new national policies and when existing legislation or services are transferred between agencies. DUT sets the fiscal framework within which the *costs of implementing* new laws are determined. It is run by experts specialized in the area of a particular new law or initiative. These experts come from within national government, the national public administration, and outside of government.

The DUT process analyses 400–500 new proposals each year. Of these, approximately 100 proposals impact municipal budgets. Once it is clear that a new law and political initiative will impact municipal budgets, the national public administration and KL meet to negotiate a fiscal adjustment that will be included in the subsequent finance agreement. The fiscal settlement on the vast majority of new laws, perhaps 98 of the 100, is reached by specialized employees of the finance ministry and KL. The remaining two proposals of broader or more contentious new laws or governmental initiatives enter into the DUT

---

6 https://www.kl.dk/tema/klima/kl-udspil-klimatilpasning-for-fremtiden; https://www.kl.dk/media/22562/klimatilpasning-for-fremtiden-vand-fra-alle-sider.pdf.

7 https://www.kl.dk/politik/kkr/kkr-hovedstaden/trafik-infrastruktur-klima-og-miljoe.

process before reaching the annual finance agreement between the Ministry of Finance and KL. In this way, national government does not propose new laws and initiatives without having considered their fiscal implications, and local governments have the political and financial weight to claim that new initiatives and laws are appropriately funded before accepting to implement them. This ensures that problem solving is real and doable rather than "Alice in Wonderland" make-believe and that politicians are actually held accountable for their actions.

This framework for fiscal prudence applies to climate solutions. The national government might, for example, obligate municipalities to implement certain EU or national climate laws that impose additional costs that are not accounted for in the annual national finance agreement. A new EU law, for example, might require member countries to dedicate a certain percentage of land for nature preservation, preventing cities from realizing the economic gains from alternative use.[8] In such a case, the municipalities are granted the fiscal means for realizing the nature preservation.

The Danish governance system, in short, provides a formalization of institutions, procedures, and negotiators that allows the system to run in a predictable, uniform, and transparent manner. It leaves less room for unfunded political promises, excessive lobbyism, clandestine negotiations, flaring up of tensions and disputes, and many other ills we see across most western societies in today's world.

Third, Copenhagen and the national government have created public–private institutions that can both create value through smart and sustainable revitalization and then capture value for large-scale public benefit and impact. The most relevant for this discussion is City & Port, a public asset corporation that has been delegated the power to dispose of publicly owned lands and buildings in ways that spur large-scale urban transformation, particularly around Copenhagen's historic harbour and downtown, and use the revenue from such regeneration to fund infrastructure, affordable housing, and other societal benefits.

The origins of City & Port are worth exploring, because they explain, in part, Copenhagen's remarkable trajectory since the 1990s. In the mid-to-late 1980s, Copenhagen experienced unemployment as high as 17.5%. The tax base of the city had dried up, as there was extensive out-migration to the suburbs of Copenhagen. As resourceful citizens moved out, the city became overrepresented by pensioners and young people attending public universities, neither of whom contributed greatly to the city's tax revenue. By the late 1980s, the city was struggling with annual budget shortfalls of US$750 million.

The urgency of the situation heightened as the manufacturing base of the city, centred in Copenhagen's harbour, slowly withered. A study of the industrial activities in the harbour concluded that the harbour only utilized 5% of its existing land mass at any given time. In addition, with the construction of the bridge connecting Copenhagen to southern Sweden, the study estimated that activities in the harbour would decline a further 25% by 2000. The harbour also struggled with large pension liabilities. Harbour management began selling off waterfront land in a piecemeal manner to balance their annual budgets.[9]

---

8 https://www.klimatilpasning.dk/media/1680413/mufjo-praesentation-mst-klimatilpasningsmoeder.pdf.

9 Maskell, P., Füssel, L.R., 1989. *Den nødvendige havn.* Institut for Trafik-, Turist- og Regionaløkonomi, København, Handelshøjskolen, pp. 1–58.

In 1990, a historic alliance consisting of representatives from both national and local governments and across the political spectrum came together to re-envision the city and create a large-scale plan for its urban regeneration. In 1992, Ørestad Development Corporation was established by the municipality of Copenhagen and the national finance ministry (with a 55/45% split in ownership respectively). While, the national government allocated the land, local government rezoned it for residential and commercial use. Ørestad Development Corporation was a publicly owned privately managed organization tasked with regenerating 3.1 square kilometres of former military land coined Ørestad. The land of Ørestad is strategically located between the city of Copenhagen, the airport, and the bridge to Sweden.

In 2007, City & Port was established and tasked with regenerating the vast harbour and port areas of Copenhagen, along with the Ørestad area. In the process, City & Port absorbed the Ørestad Development Corporation and spun off separate purpose-driven organizations: Copenhagen Malmö Ports became tasked with managing the port operations and the Metro Construction Corporation managed metro construction activities.

City & Port is a hybrid organization, publicly owned and privately managed. City & Port uses land value capture to leverage the value of public assets through rezoning and infrastructure investments. It operates under a national statutory mandate to maximize revenue to fund large-scale urban regeneration along the harbour and Ørestad district and use the resulting value generated to finance city-wide infrastructure. This mandate shelters City & Port from political interference by obliging City & Port to always choose the investment proposition that yields the most revenue. The revenue yield is used to service the debt on a new city-wide metro system. In this way, City & Port has a holistic approach to regenerating the entire city that came about at a crucial moment in the city's history.

What has emerged in Copenhagen is a new kind of public–private institution that both creates value through smart revitalization and then captures value for large-scale impact. Over the past several decades, City & Port has driven one-half of the development in the city of Copenhagen in ways that maximize climate solutions. The subsequent regeneration has changed the city forever and provided a strong platform for not only the building of a world-class metro system but the myriad of local investments in sustainable investments.

Finally, Copenhagen's climate plan has benefitted from the active participation and investment of public pension funds. Danish pension funds are large, and privately owned and managed; they make up some of the largest pension funds in the EU, with the top two Danish pension funds counting among the twenty largest in Europe.[10]

For the most part, Danish pension funds operate as cooperatives that represent their members, the pensioners. They operate in a highly uniformly regulated market. Payment contributions are sheltered from politicians and public spending, which has enabled a fiscally solid and stable pension system to evolve. With a strong funding base, Danish pension funds are able to take the long view. They aim at a 3–4% return on investment on 80%

---

10 https://www.ipe.com/ipe-publishes-top-1000-european-pension-fund-list/5900.article.

of their invested assets and 5–6% on the remaining 20%. The large size of Danish pension funds makes them powerful players in the global market, leading the green climate transition both within Denmark and abroad.

Danish pension funds have pledged to support the national government's plans of reducing greenhouse gas emissions by 70% by 2030 (compared with 1990 levels) and investing US$50 billion in green assets.[11] These investments are mainly targeted to offshore wind farms and photovoltaic energy as well as climate-friendly properties and energy storage.

Danish pension funds have already invested more than US$18 billion in green assets, while another US$5 billion is already planned. To date, profitable green investments have been made in predominately green real-estate investments, windmills, and sustainable forest management.[12] In September 2019, the Danish prime minister, Mette Frederiksen, and some of the largest Danish pension funds visited New York to announce that they will invest US$52 billion by 2030 in the US green energy transition. These are investments in energy infrastructure and other green activities, such as green shares, green bonds, and investments in energy-efficient construction.[13]

Danish pension payments are directly linked to their accumulated savings and the yield generated through decades of wise long-term investments. At the same time, Danish interest rates have been suppressed since around 2012. In light of this, climate investments help Danish pension funds diversify and spread risk across their investments. Common to all the investments is that despite satisfactory profits, the investment horizon is substantially longer than most US investors generally operate with.

"We have long invested in projects in our part of the world. With our investments in the Danish SDG Fund in 2018 and Copenhagen Infrastructure Partners New Market Fund in 2019, we take our experiences with sustainable investments into the growth economies in Latin America, Africa and Asia," says Vice President of the Insurance & Pension Industry and CEO of PensionDanmark Torben Möger Pedersen.[14] To a large extent, despite Danish pension funds being privately funded and owned, they collaborate closely with the Danish national government, including supporting the government's mission for achieving climate goals and expanding the Danish climate industry.

Lessons Learned

The Copenhagen climate story presented in this chapter, and the particular institutional examples around governance and finance, yield seven lessons for other cities.

---

11 https://www.reuters.com/article/us-climate-change-un-denmark/danish-pension-funds-back-green-transition-with-50-billion-iduskbn1w80np.

12 https://groenforskel.dk/.

13 https://www.forsikringogpension.dk/nyheder/pensionsbranchen-offentliggoer-milliardinvesteringer-i-groen-omstilling.

14 Ibid.

City Governance Must Focus on How to Implement Policies Rather than What Policies to Implement

City leaders tend to focus on the "what" instead of the "how". It simply has more political appeal to explain *what* will be done rather than *how* it will be done. We believe that making the right decision (i.e. the "what") only gets you half-way to success. The other half depends on how you execute and deliver (i.e. the "how"). City leaders routinely announce policy agendas but often overlook the critical importance of the institutions and stakeholders driving the climate transition.

City Governance Must Make Room for the Growth of New Large-Scale Institutions

Setting a policy agenda that shifts from a solely economic to a climate economic focus requires cities to undergo an institutional overhaul. Yet city leaders often resist change, since they are woven into the complex fabric of stakeholders, agencies, authorities, and checks and balances. But change of the magnitude we are faced with requires an altered mindset, and new or consolidated institutions (e.g. City & Port) that have expanded powers, capacity, and capital. In most cities, we actually need fewer rather than more institutions.

City Governance Must Create Institutions with Holistic Focus

The institutions presented in this chapter were not born out of a narrow climate change perspective, as they fulfil multiple purposes. In fact, the power and impact of these institutions stem from their broad economic, societal, and environmental objectives. If these institutions only fulfilled, for example, narrow environmental purposes, we doubt that they would be as successful as they are, because economic, societal, and environmental outcomes are inextricably linked and should not be viewed in isolation.

City Governance Must Have Access to Financial Sophistication

The design and financing of climate solutions is complex, comprised of multiple investment sectors as diverse as renewable energies (e.g. on- and off-shore wind, solar farms), the transportation and building sectors (e.g. energy efficiency, transit-oriented development), water and waste management, and resilient development. Each of these investment sectors advances the common goals of mitigating (or adapting to) the impact of climate change. Yet each differs in how it is governed, regulated, owned, and operated. And, with regard to finance, each has a different mix of the source of capital (e.g. public, private, and civic), the nature of the capital stack (debt, equity, subsidy, and concessionary), and the interplay of project risk, revenue, and return expectations.

There is a disconnect, in short, between the simplicity of climate goals and the complexity of climate finance. In most cities and countries, this disconnect is exacerbated by the mismatch in knowledge and expertise between different stakeholders in the system, particularly public investors, cities, and public pensions on the one hand,

and asset managers and financial institutions on the other. By contrast, Copenhagen and other Danish cities have benefitted from institutions and intermediaries that have the capacity to understand the financing structure of complex climate solutions.

City Governance Should Balance Power Across the Public and Private Sectors

The global economy is dominated by large and powerful corporations and financial institutions. The creation of city institutions with power (e.g. municipal intermediaries, pension funds) enables cities to help set the investment agenda and shape our societies. We need to bolster the capacity, resources, and demand of large societal institutional actors. Cooperatives offer a way for states, regions, counties, municipalities, and pensioners to scale up their size to match that of large financial institutions. In fact, cooperatives erase the power imbalance between small localized askers and big globalized givers. Through universal membership, these institutions become powerful societal heavyweights that national governments and multinational corporations must reckon with.

City Governance Links Different Levels of Government and Sectors of Society

When there is no stable and solid institution acting as the connective tissue between national government and municipalities and pension funds, there is a disconnect between what goes on in national government and what is in fact implemented on the ground. Politicians are left free to set unfunded climate goals without anyone holding them accountable. Cooperative institutions fill the gap between fake politics and real policy execution on the ground. The Danish pension funds, in particular, are so large that few governments will ignore them. These are societal supertankers that shape and define northern European societies.

City Governance Links and Leverages Different Institutional Actions Within Cities

The institutional models described in this chapter have intricate relationships with each other. City & Port, for example, has attracted major pension fund investment for its regeneration activities. Two pension funds, Nordea Liv & Pension and PenSam, have invested almost US$430 million (2.7 billion DKR) in the South Harbour of Copenhagen, where they are financing 1,350 new affordable and social rental apartments.[15] Such developments enable City & Port to comply with the mandate to build 30% affordable and social housing in the areas of the former harbour. Both developments also advance sustainable objectives, since each has novel rainwater collection to avoid flooding, has natural vegetation to expand biodiversity, and is largely car-free to encourage street-level community activities and advance human health.

---

15 https://byensejendom.dk/article/nordea-og-pensam-bygger-ny-bydel-i-kobenhavn-for-27-milliarder--12540.

Conclusion

In a world of global climate summits, urban governance is a critical but often-overlooked element of transformational change. The creation of highly capable and professionalized municipal governments and institutions enables cities to use expert knowledge and sophisticated mechanisms to translate policy into action, drive creative financing, and enable effective implementation. Copenhagen (and Denmark more broadly) are models for how cities design and deliver ambitious plans through municipal governments, negotiate with national governments via aggregated political power (e.g. Local Government Denmark), leverage physical assets and capture value for broad public benefit (e.g. City & Port), and deploy capital at scale (e.g. Danish pension funds). The adaptation of these models to cities and countries across the world is a necessary part of climate action.

7

The Decoupled City – Introduction

In Chapter 6 Bruce and Luise outlined a model of governance that ensures financial commitment to meeting objectives before the agreement is struck. We could apply this principle to new growth by pricing the limit on CO_2 emissions to ensure all new development does not add to the contribution of greenhouse gas emissions.

In this chapter Leah and Nick introduce the energy problem, which Pete in Chapter 9, *The Energized City*, will build upon. They make the case for the efficiency of compact living and how cities have far outperformed their national counterparts. Figure 7.1 shows us how, despite growth, cities have decoupled this growth from environmental disbenefit. It has, however, spawned other local issues in terms of pollution and inequality, but Hayley, in Chapter 21, *The Just City*, will address some of those.

Leah and Nick highlight the importance of density and the benefits of the compact city. Figure 7.2 is an extract from the book *Shaping Cities in an Urban Age*[1] which wonderfully illustrates urban density in terms of population per kilometre^2. The significance of this will emerge in Chapters 10 and 11, *The Agile City*.

Cities inevitably lie at the epicentre of our climate, economic, and social problems. But more significantly, cities also hold the key to solutions. Science tells us to keep global temperatures from rising any higher than 1.5°C and global cities have a mid-century carbon net-zero target. The pandemic has brought us to a global inflection point and governments are not going to abandon continued economic growth anytime soon. This balance of goals leaves us with what Nick and Leah refer to as "the decoupled city" – a city with a growing economy and shrinking emissions.

Cities will be a crucial part of building back the economy post-COVID, but as Nick and Leah write it is not good enough to simply reconstruct what once was. We must once again go "beyond" and "build back better". Governments can, in fact, save money if they avoid measures that would boost the economy but exacerbate other crises. We must position cities to thrive in a world suffering the effects of climate change through low carbon investments and the creation of green jobs.

Nick and Leah show that all of these low carbon investments, even in the most conservative scenarios, can lead to greater prosperity, not only reducing emissions but in

---

1 Burdett, R., Rode, P. (eds), 2018. *Shaping Cities in an Urban Age*. LSE Cities. Phaidon Press, New York.

The Climate City, First Edition. Edited by Martin Powell.
© 2022 John Wiley & Sons Ltd. Published 2022 by John Wiley & Sons Ltd.

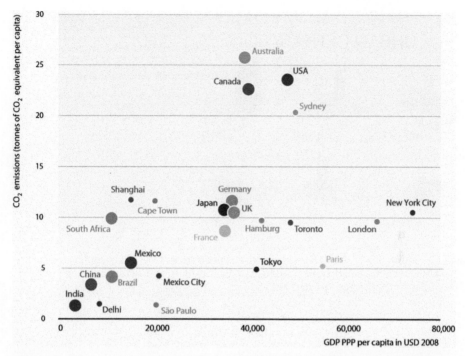

Figure 7.1 Decoupling of CO_2 emissions and GDP growth, highlighting the carbon efficiency of cities versus their national averages. (*Source:* Ricky Burdett, 2018. Shaping Cities in an Urban Age, Phaidon Press.)

terms of job creation and long-term economic growth both locally and nationally. The city and its people are one after all. Nick and Leah also show the need for city and national governance to work together through case studies of national and local coalition projects in South Korea, China, India, Namibia, and Chile. Much like Denmark's Copenhagen climate story, these too are "success" stories that prove to us what life can be like in a net-zero world. City action alone is not enough, and true success relies on connection and coordination on a national and global level. Nick and Leah analyse in detail the six priorities for national action laid out by the Coalition for Urban Transitions.

Most significantly, Nick and Leah show how the cost of inaction is far greater than the cost of action. Sea levels are predicted to rise by several metres over the next century, and 1.5–3°C of global heating is already "locked in", posing an overwhelming threat to cities. Once again, numbers don't lie. We have reached a short window of opportunity in which a real difference can be made. National governments now have an opportunity to place zero-carbon cities at the heart of national development and climate strategy. There are two options – prosperity and resilience – or decline and vulnerability.

A decoupled city is a resilient city, and that's the sort of city we need as we progress through this environmentally crucial century.

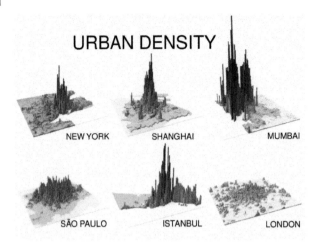

Figure 7.2 Illustration of population per square kilometre across cities – highlighting the variety of energy and transport solutions required to make the city function. (*Source:* Ricky Burdett, 2018. Shaping Cities in an Urban Age, Phaidon Press.)

7

The Decoupled City
Leah Lazer and Nick Godfrey

Cities at the Epicentre of Challenges and Solutions

The world is at an inflection point that will determine the future of the climate, economy, and socioeconomic development. The COVID-19 pandemic (the pandemic) has taken hundreds of thousands of lives,[1] and the lockdown needed to slow its spread triggered an economic depression that pushed hundreds of millions of people back into poverty.[2] As national governments scrambled to respond to the deadly virus and collapsing economy, the effects of climate change continue to wreak havoc and threaten those gains. Rising temperatures are already causing serious loss of life and threatening vital ecosystems, and further temperature increases pose an existential threat to entire cities and countries.[3]

Cities, with their concentration of people, economic activity, greenhouse gas emissions, and most recently COVID-19 cases, lie at the epicentre of the climate, health, and economic crises, but are also at the heart of their solutions. Over half the world's population lives in urban areas, which produce 80% of GDP and three-quarters of carbon emissions from final energy use.[4] These shares continue to grow rapidly, especially in Africa and Asia.[5] Meanwhile, science tells us that to keep global temperatures from rising by more than 1.5°C, cities have to achieve net-zero emissions by mid-century.[6] To enable these growing urban populations to enjoy safe, productive, dignified lives, governments around

---

1 WHO Coronavirus disease (COVID-19) dashboard. https://covid19.who.int.

2 World Economic Outlook Update, June 2020: A crisis like no other, an uncertain recovery. IMF, Washington, DC. https://www.imf.org/en/publications/weo/issues/2020/06/24/weoupdatejune2020.

3 Roy, J., Tschakert, P., Waisman, H., et al., 2018. Sustainable development, poverty eradication and reducing inequalities. In: V. Masson-Delmotte, P. Zhai, H.-O. Pörtner, et al. (eds), *Global Warming of 1.5°C. An IPCC Special Report on the Impacts of Global Warming of 1.5°C above Pre-industrial Levels and Related Global Greenhouse Gas Emission Pathways, in the Context of Strengthening the Global Response to the Threat of Climate Change, Sustainable Development, and Efforts to Eradicate Poverty*. IPCC, Geneva. https://www.ipcc.ch/sr15.

4 UN-DESA, 2018. *World Urbanization Prospects 2018*. United Nations Department of Economic and Social Affairs, New York. http://esa.un.org/unpd/wup.

5 Ibid.

6 IPCC, undated. *Global Warming of 1.5°C*. IPCC, Geneva. https://www.ipcc.ch/sr15/.

The Climate City, First Edition. Edited by Martin Powell.
© 2022 John Wiley & Sons Ltd. Published 2022 by John Wiley & Sons Ltd.

the world are likely to pursue continued economic growth. This means that the decoupled city – with a growing economy and shrinking emissions – is the way for countries to secure a climate-safe, prosperous, equitable future.

In the immediate future, pandemic measures are creating a global inflection point that may determine the future of the climate, economy, and socioeconomic development. This is the moment for governments around the world to pursue strategies and investments to deliver shared prosperity while reaching net-zero emissions. Recent research shows that investments in sustainable cities and urban infrastructure can generate green jobs and significant social, environmental, and health co-benefits in ways that are financially sustainable beyond the initial stimulus.[7,8,9] Governments can also save money if they avoid measures that would boost the economy by exacerbating other crises, such as slowing the phase-out of fossil fuel subsidies or bailing out fossil fuel-intensive corporations without conditions to improve environmental performance. National governments need to ensure that their fiscal stimulus packages will help them "build back better", creating green jobs and addressing basic needs today, while laying the foundations for more productive, inclusive, and sustainable prosperity tomorrow.

Cities will be a critical part of this strategy. National economic recoveries will depend on rapidly revitalizing cities while positioning them to thrive in a world suffering the effects of climate change. The leading countries of tomorrow will be those whose cities successfully make an equitable transition to a new urban economy.

Currently Available Measures Can Cut Urban Emissions by 90% by 2050

The IPCC 2019 special report showed that to keep global heating below 1.5°C, cities need to reach net-zero emissions by mid-century.[10] Without further action to tackle climate change, recent analysis shows that greenhouse gas emissions from urban buildings, transport, and waste could reach 17.3 billion tonnes of carbon dioxide equivalent (tCO_2e) in 2050 (Figure 7.3). This takes into account countries' 2015 nationally determined contributions (NDCs) under the Paris Agreement, yet is still 24% higher than in 2015, when the Paris Agreement was signed.

However, recent analysis shows that a bundle of currently feasible, widely available low carbon measures and investments can reduce greenhouse gas emissions from cities by almost 90% by 2050, compared to business-as-usual levels, bringing cities' global carbon footprint down to 1.8 billion tCO_2e by mid-century, close to net-zero.[11] The absolute value

---

7 Coalition for Urban Transitions, 2019. *Climate Emergency, Urban Opportunity*. Coalition for Urban Transitions, Washington, DC.

8 Gulati, M., Becqué, R., Godfrey, N., et al., 2020. The economic case for greening the global recovery through cities: 7 priorities for national governments. Coalition for Urban Transitions, Washington, DC.

9 Hepburn, C., O'Callaghan, B., Stern, N., Stiglitz, J., Zenghelis, D., 2020. Will COVID-19 fiscal recovery packages accelerate or retard progress on climate change? *Oxford Review of Economic Policy* 36(S1).

10 IPCC, undated. Op. cit.

11 Coalition for Urban Transitions, 2019. Op. cit.

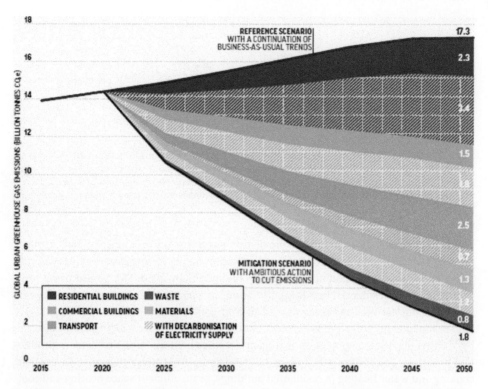

Figure 7.3 Technically feasible potential to reduce greenhouse gas emissions from cities by 2050, by sector. (*Source:* Coalition for Urban Transitions, Climate Emergency, Urban Opportunity: how national governments can secure economic prosperity and avert climate catastrophe by transforming cities, 2019. Available at: https://urbantransitions.global/en/publication/climate-emergency-urban-opportunity/.)

of this emissions reduction is equivalent to more than the 2014 energy-related emissions of China and the USA combined,[12] and is equivalent to 58% of the global energy-related emission reductions required for the International Energy Agency's 2°C pathway.[13] These measures also have the potential to tackle pressing political priorities, such as air pollution, traffic congestion, and access to basic services.

---

12 China and the USA had combined energy-related emissions of 15.1 $GtCO_2e$ in 2015, according to WRI, 2019. *Climate Watch Data Explorer.* World Resources Institute, Washington, DC. https://www. climatewatchdata.org.

13 Coalition for Urban Transitions, 2019. Op. cit. SEI's modelling draws heavily on the IEA's energy scenarios presented in Energy Technology Perspectives (2017 edition). The first of these is the baseline or reference scenario, which takes into account existing energy- and climate-related commitments by countries. The second of these is a decarbonization scenario consistent with holding the average global temperature increase to no more than 2°C. The third scenario is a more ambitious decarbonization scenario consistent with holding the average global temperature increase to "below two degrees", which is consistent with holding the average global temperature increase to no more than 1.75°C. This third scenario is based on the IEA's analysis of how far clean energy technologies could go if pushed to their practical limits. Urban sectors could deliver 44% of global energy-related GHG reductions needed for a 1.75°C pathway in 2050.

Figure 7.4 Technically feasible low carbon measures could cut emissions from urban areas by almost 90% by 2050. (*Source:* Coalition for Urban Transitions, Climate Emergency, Urban Opportunity: how national governments can secure economic prosperity and avert climate catastrophe by transforming cities, 2019. Retrieved from: https://urbantransitions.global/en/publication/climate-emergency-urban-opportunity/.)

The investments required in cities are distributed across the buildings, transport, materials, and waste sectors, with investments in decarbonizing energy systems critical across all sectors. Of the total potential emissions reductions by 2050, 58% would come from commercial and residential buildings, 21% from transport, 16% from materials efficiency (such as using less steel and cement), and 5% from solid waste management (Figure 7.4).[14] Half of the total potential emissions reductions come from decarbonizing urban electricity.[15] Other significant sources of abatement in cities include improved cement production processes, a shift from private cars to shared and active transportation, more efficient cooking and water heating in residential buildings, more efficient space heating and cooling in all buildings, and more efficient and electric vehicles.[16]

Investing in Zero-Carbon Cities Can Drive National Prosperity in the Short, Medium, and Long Term

This analysis shows that this bundle of low carbon urban investments would not only dramatically reduce emissions but also generate an economic return worth US$23.9 trillion in today's terms based on energy and material cost savings alone.[17] It could also create millions of good jobs: 87 million jobs in 2030 and 45 million jobs in 2050, in growth sectors like energy efficiency and mass transit, as illustrated in Figure 7.5. Whilst significant investment will be required globally – US$1.83 trillion (or about 2% of global GDP) per year – this would yield annual savings worth US$2.80 trillion in 2030 and US$6.98 trillion in 2050 (US$24 trillion in total), and many with short payback periods.[18] These figures are conservative, as they do not include wider, long-term benefits like improved public health and labour productivity – making this bundle of investments even more attractive.

---

14 Coalition for Urban Transitions, 2019. Op. cit.

15 Ibid.

16 Ibid.

17 Ibid.

18 Ibid.

Figure 7.5 Investments required to reduce urban emissions. (*Source:* Coalition for Urban Transitions, Climate Emergency, Urban Opportunity: how national governments can secure economic prosperity and avert climate catastrophe by transforming cities, 2019. Retrieved from: https://urbantransitions.global/en/publication/climate-emergency-urban-opportunity/.)

Since the net economic benefits of these measures vary with energy prices, interest rates, and technological learning rates, this analysis included a range of scenarios, as shown in Figure 7.6. Yet even in the most conservative scenario (an annual energy price increase of only 1% per year and a discount rate of 5.5%), the bundle of measures still has a present value of US$4.2 trillion.[19]

More widely, by fostering zero-carbon cities, national governments can also enjoy a range of long-term economic advantages that range far beyond these returns. When households and firms are located closer together in compact, connected cities, it becomes

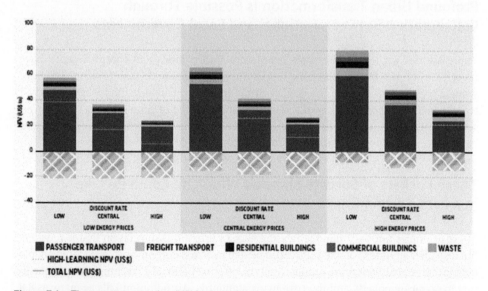

Figure 7.6 The net present value (NPV) of ambitious climate action in cities between 2020 and 2050 (US$ trillions). (*Source:* Coalition for Urban Transitions, Climate Emergency, Urban Opportunity: how national governments can secure economic prosperity and avert climate catastrophe by transforming cities, 2019. Available at: https://urbantransitions.global/en/publication/climate-emergency-urban-opportunity/.)

---

19 Ibid.

cheaper to provide infrastructure and services, because more people and buildings can be served by each length of water pipe and internet cable, as well as each bus stop and public park. This makes it more cost-effective for governments and private-sector companies to invest in infrastructure like metro systems or district heating. In addition, research shows that workers and businesses are more productive in larger, more densely populated cities, especially when combined with good transport networks. A review of over 300 studies on compactness found that, in higher-income countries, a 10% higher level of urban density was correlated with a US$182 higher annual gross value added per person.[20]

At that national scale, fostering low carbon, resilient, and inclusive cities can also help build the capacity of nations to produce and integrate innovations, which is a key element for enhancing national economic competitiveness. A 10% increase in urban population density is correlated with a 1.1% increase in patents per capita in Europe and a 1.9% increase in patents per capita in the USA.[21] Similarly, compact, connected, and clean cities can help countries in the global race to attract mobile talent and investment. Many countries hope to boost economic growth by expanding their output of tradeable goods and services, which often involves attracting high-value-added export-orientated firms and the skilled workforces they depend on. Both tend to be highly mobile, and are likely to be attracted to the benefits of zero-carbon cities: higher productivity, strong connectivity, direct cost savings, and better quality of life.[22]

Profound Urban Transformation Is Possible Through Collaboration Between National and Local Governments

There are not yet real-world examples of zero-carbon cities, but recent decades have revealed examples from around the world where national and local governments have worked together to improve the quality of life for urban residents, showing that urban transformation is possible within a relatively short period of time (Box 7.1). Moreover, several cities have already decoupled their economic growth from their (production-based) greenhouse gas emissions (see Figures 7.3 and 7.7).

These Pockets of Success Show Us What Life Could Be Like in a Zero-Carbon City

These examples not only demonstrate the pace of urban transformation that is possible, but also offer a glimpse of what life could be like in a zero-carbon city. They could enable people to live healthier, more productive lives, offering national governments an opportunity to eradicate poverty and improve living standards. Air pollution kills as many as 8.8

---

20 Ahlfeldt, G., Pietrostefani, E., 2017. *Demystifying Compact Urban Growth: Evidence from 300 Studies from across the World.* Coalition for Urban Transitions, Washington, DC. https://newclimateeconomy. report/workingpapers/demystifyingcompact-urban-growth/.

21 Coalition for Urban Transitions, 2019. Op. cit.

22 Coalition for Urban Transitions, 2019. Op. cit.

Box 7.1   Examples of urban transformation

Seoul, South Korea: Proactive planning for urban population growth

Proactive planning for urban population growth helped transform post-war Seoul into an ultra-modern megacity that powers the world's 11th largest economy in the space of only a few decades.[23] Seoul's population grew eight times over between 1950 and 1980.[24] National land reform policies enabled the rapid expansion of housing supply and critical infrastructure investment to meet the needs of this booming population.[25] Seoul's world-class metro, which now transports over 10 million people every day,[26] plays an especially pivotal role in keeping the city connected and affordable (Box Figure 7.1).

Box Figure 7.1   Inside view of the metropolitan subway in Seoul, one of the most heavily used underground systems in the world. (*Source:* ake1150sb/Getty Images.)

---

23 IMF, 2019. *IMF Datamapper: Datasets.* IMF, Washington, DC. https://www.imf.org/external/datamapper/datasets.

24 UN-DESA, 2018. Op. cit.

25 Kim, S.H., 2013. Changes in urban planning policies and urban morphologies in Seoul, 1960s to 2000s. *Architectural Research* 15(3), 133–141. Lee, S.K., You, H., Kwon, H.R., 2015. Korea's pursuit for sustainable cities through new town development: Implications for LAC. Inter-American Development Bank, Washington, DC. Seoul Metropolitan Government, Urban Planning Bureau, and Advisory Group for Urban Planning, 2016. Seoul, ready to share with the world! Seoul Urban Planning. https://www.metropolis.org/sites/default/files/seoul_urban_planningenglish.pdf.

26 Seoul metro. http://www.seoulmetro.co.kr/en/page.do?menuIdx=649.

China: Improving and scaling electric vehicles in cities

As of 2017, China was home to 40% of the world's electric passenger cars and 99% of electric buses and electric two-wheelers worldwide.[27] National R&D investments over the previous decade had dramatically improved electric vehicles; more recently, national policies had supported the deployment of electric vehicles and charging infrastructure in cities.[28,29] This national–local partnership led to innovations in business models and procurement policies to complement technological progress.[30] Today, China's cities and firms are leading the electric vehicle revolution: Shenzhen is the first city in the world to electrify its entire public bus fleet.[31]

Chile: Improving housing quality in cities

Over two decades, Chile reduced its shortfall of decent housing by two-thirds – despite rapid urban population growth.[32,33,34,35] Chile's housing policy was remarkable for successfully stimulating private-sector construction of housing for low- and middle-income urban families. This was accompanied by programmes to upgrade the quality of housing and services in *campamentos* (informal settlements) and provide better transport links to these communities. Together, these national policies enabled the massive expansion of good housing stock that was well connected to jobs and services across cities.

Indore, India: Sorting out solid waste management

Dramatic improvements in solid waste management moved Indore up the rankings from India's 149th cleanest city to the cleanest city in the country in just four years.[36] Thanks to the twice-daily, door-to-door waste collection for households and businesses (including those in informal settlements),[37] over 90% of Indore's waste

---

27 IEA, 2018. *Global EV Outlook 2018*. IEA, Paris.

28 Yu, P., Zhang, J., Yang, D., Lin, X., Xu, T., 2019. The evolution of China's new energy vehicle industry from the perspective of a technology–market–policy framework. Sustainability 11(6), 1711.

29 World Bank, 2011. The China new energy vehicles program: Challenges and opportunities. World Bank, Washington, DC. http://documents.worldbank.org/curated/en/333531468216944327/thechina-new-energy-vehicles-program-challenges-and-opportunities.

30 Chen, K., Hao, H., Liu, Z., 2018. Synergistic impacts of China's subsidy policy and new energy vehicle credit regulation on the technological development of battery electric vehicles. *Energies* 11(11), 1–19.

31 Poon, L., 2018. How China took charge of the electric bus revolution. CityLab. https://www.citylab.com/transportation/2018/05/how-china-charged-into-the-electric-busrevolution/559571.

32 OECD, 2016. Housing policy in Chile. OECD Social, Employment and Migration Working Papers No. 173. OECD, Paris.

33 Cociña Varas, C.L., 2017. Housing as urbanism: The role of housing policies in reducing urban inequalities: A study of post 2006 housing programmes in Puente Alto, Chile. PhD thesis, University College London. http://discovery.ucl.ac.uk/1571836/14/20170831_final%20thesis_viva%20corrections%20final_med.pdf.

34 Rojas, E., 2019. 'No time to waste' in applying the lessons from Latin America's 50 years of housing policies. *Environment and Urbanization* 31(1), 177–192.

35 Cociña Varas, 2017. Op cit.

36 Bansal, R., 2017. The curious case of a clean clean Indore. https://www.businesstoday.in/magazine/columns/story/the-curious-case-of-a-clean-clean-indore-76310-2017-06-10. Bhargava, A., 2017. How Indore became garbage-free and beat every other city to it. The Better India, September. https://www.thebetterindia.com/114040/indore-madhya-pradeshclean-garbage-free-india.

37 Ibid. Bapat, S., Bhatia, R.K., 2018. Comparative analysis of solid waste management in developing smart cities of India. *International Journal of Advanced Research* 6(10), 1330–1339.

Box Figure 7.2 Windhoek, Namibia, 2018. New property development around The Grove at the Mall of Namibia. (*Source:* Alexander Farnsworth/Getty Images.)

is now collected and sorted.[38] Through a public–private partnership, the city built a biogas plant that generates 800 kg of biogas every day, which fuels about 15 city buses.[39] The construction of over 12,500 household, community, and public toilets helped alleviate open defecation.[40] Improved waste collection not only makes the city cleaner and prevents disease, but also offers its residents a sense of dignity and civic pride.

Windhoek, Namibia: Providing affordable housing to city residents
Windhoek City Council pioneered an innovative solution to rapid population growth and deep urban poverty. The city government demarcated small plots of land and installed basic services (such as water points and toilet blocks) to accommodate new families.[41] This strategy allowed the city to shape land-use and avoided the health costs often associated with informal settlement. The national Build Together Programme then provided low-cost loans to urban residents, enabling them to incrementally upgrade their houses and services.[42] While it remains deeply unequal, Windhoek stands out among Sub-Saharan African cities for its low-cost shelter solutions and far-sighted land-use planning (Box Figure 7.2).[43,44]

38 Sambyal, S.S., Agarwal, R., 2018. Forum of cities that segregate: Assessment report 2017–2018. Centre for Science and Environment, New Delhi.

39 Smart City Indore, 2019. Solid Waste Management. Indore. www.smartcityindore.org/solid-waste.

40 Sinha, M., 2018. Swachh lessons for Noida, from Indore. Times of India. https://timesofindia.indiatimes.com/city/noida/swachh-lessons-for-noida-from-indore/articleshow/65801971.cms.

41 Chitekwe-Biti, B., 2018. Co-producing Windhoek: The contribution of the Shack Dwellers Federation of Namibia. *Environment and Urbanization* 30(2), 387–406. doi: 10.1177/0956247818785784.

42 Remmert, D., Ndhlovu, P., 2018. Housing in Namibia. Institute for Public Policy Research, Windhoek.

43 Weber, B., Mendelsohn, J., 2017. *Informal Settlements in Namibia: Their Nature and Growth.* Development Workshop Namibia, Windhoek.

44 Sweeny-Bindels, E., 2011. Housing policy and delivery in Namibia. Institute for Public Policy Research, Windhoek. https://ippr.org.na/wp-content/uploads/2011/10/Housing%20report%20IPPR.pdf.

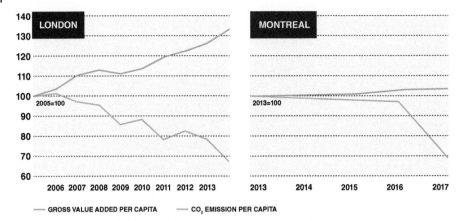

Figure 7.7 Examples of metropolitan areas that have achieved an absolute decoupling of per capita economic activity and per capita production-based greenhouse gas emissions. (*Source:* Coalition for Urban Transitions, Climate Emergency, Urban Opportunity: how national governments can secure economic prosperity and avert climate catastrophe by transforming cities, 2019. Available at: https://urbantransitions.global/en/publication/climate-emergency-urban-opportunity/.)

million people every year,[45] but by shifting to electric transportation, residents could enjoy cleaner air that reduces the health burden from asthma and cardiovascular disease.[46] Today, road crashes kill 1.35 million people each year,[47] costing most countries 3% of their GDP,[48] but in compact, connected cities, residents could enjoy the safety of protected pavements and cycle lanes, with greenery that absorbs stormwater and cools the city. Denser development means that residents could use shared and non-motorized transit to meet more of their travel needs, which would improve health, reduce noise pollution, and boost local economic activity, since foot traffic often benefits retailers and eateries. Energy-efficient building design would save money on heating and cooling bills. Figure 7.8 highlights the carbon efficiency of a compact city (Stockholm) compared to Pittsburgh – cities with similar populations.

Many of the measures to foster zero-carbon cities offer significant health benefits, mainly through improved air quality and increased physical activity. These benefits would accrue particularly to lower-income or marginalized communities, who are most likely to lack basic services like clean water and sanitation and to live in polluted areas.

---

45 Burnett, R., Chen, H., Szyszkowictc, H., et al., 2018. Global estimates of mortality associated with long-term exposure to outdoor fine particulate matter. *Proceedings of the National Academy of Sciences* 115(38), 9592–9597. doi: 10.1073/pnas.1803222115.

46 Requia, W.J., Moataz, M., Higgins, C.D., Arain, A., Ferguson, M., 2018. How clean are electric vehicles? Evidence-based review of the effects of electric mobility on air pollutants, greenhouse gas emissions and human health. *Atmospheric Environment* 185, 64–77.

47 Road Traffic Injuries. https://www.who.int/news-room/fact-sheets/detail/road-traffic-injuries.

48 Ibid.

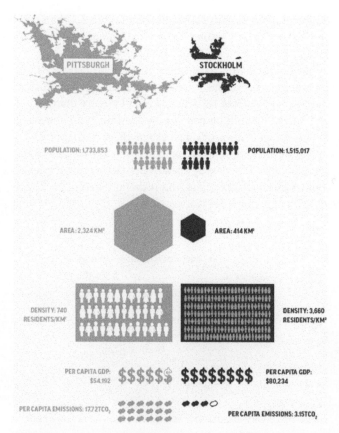

Figure 7.8   Urban extent of Pittsburgh and Stockholm, shown at the same scale. (*Source:* Coalition for Urban Transitions, Climate Emergency, Urban Opportunity: how national governments can secure economic prosperity and avert climate catastrophe by transforming cities, 2019. Available at: https://urbantransitions.global/en/publication/climate-emergency-urban-opportunity/.)

The Costs of Inaction Are Staggering

Without further action on climate change, sea levels may rise by several metres by the end of the century,[49] posing an existential threat to entire cities like Venice and Guangzhou.[50] No matter what happens next, 1.5–3°C of global heating is already "locked in",[51] meaning that the climate change will continue to be felt with increasing

---

49 Goodell, J., 2017. *The Water Will Come: Rising Seas, Sinking Cities, and the Remaking of the Civilized World*. Hachette, New York. Hansen, J. E., 2007. Scientific reticence and sea level rise. *Environmental Research Letters* 2, 024002.Vermeer, M., Rahmstorf, S., 2009. Global sea level linked to global temperature. *Proceedings of the National Academy of Sciences of the United States of America* 106(51), 21527–21532. Wallace-Wells, D., 2019. *The Uninhabitable Earth: Life After Warming*. Tim Duggan Books, New York.

50 C40: The future we don't want. https://www.c40.org/other/the-future-we-don-t-want-homepage.

51 AR5 synthesis report: Climate change 2014. https://www.ipcc.ch/report/ar5/syr.

severity,[52] threatening the hard-won development gains of recent decades. Urban policies and investments must therefore enhance resilience to climate impacts as well as reduce greenhouse gas emissions.

Climate risks are especially pronounced in coastal cities, including flooding, storm surge, salt-water intrusion, and ever-stronger storms.[53] Unfortunately, most coastal urban development proceeds without regard to climate hazards. As of 2015, more than 10% of the world's population (approximately 820 million people) lived within 10 m above sea level; 86% of those people lived in urban centres or quasi-urban clusters (Figure 7.9).[54] Low-elevation coastal zones are six times more densely populated than the world average (309 versus 56 people per square kilometre).[55] Looking ahead, population growth rates in low-elevation coastal urban centres are about 20% higher than other urban centres.[56] This means that sea-level rise overwhelmingly threatens cities.

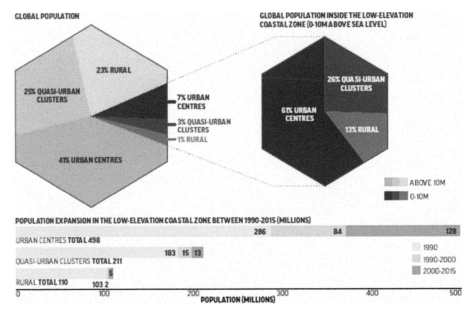

Figure 7.9 Share of global population outside and inside the low-elevation coastal zone, by settlement type, 2015. (*Source:* Coalition for Urban Transitions, Climate Emergency, Urban Opportunity: how national governments can secure economic prosperity and avert climate catastrophe by transforming cities, 2019. Available at: https://urbantransitions.global/en/publication/climate-emergency-urban-opportunity/.)

---

52 IPCC, undated. Op. cit.

53 McGranahan, G., Balk, D., Anderson, B., 2007. The rising tide: Assessing the risks of climate change and human settlements in low-elevation coastal zones. *Environment and Urbanization* 19(1), 17–37.

54 Coalition for Urban Transitions, 2019. Op. cit.

55 Ibid.

56 Ibid.

Coordinated Action Among Local and National Governments Is Needed to Drive the Zero-Carbon Urban Transition

Despite the enormous promise of a zero-carbon urban future and the staggering costs of inaction, questions remain about the remit and sequencing of urban climate policies. In recent decades, visionary local governments have led the way: nearly 10,000 cities and local governments have committed to set emission reduction targets and prepare strategic plans to deliver on them.[57] In addition to the climate and economic benefits, this action by city governments is critical to build political will and public appetite and test innovations.

However, city-level action cannot, on its own, achieve the pace and scale of the change needed to drive down urban emissions. Even the largest and most empowered city governments can achieve only a fraction of their mitigation potential on their own. National leadership and multilevel collaboration are essential to unlock the potential of zero-carbon cities. Over half of the urban abatement potential in cities comes from decarbonizing electricity grids, which are most often controlled by national or state/regional governments.[58] Notably, over half of global urban abatement potential is in cities that currently have fewer than 750,000 residents.[59] These smaller cities often lack the financial and technical resources of larger cities and are especially in need of support from higher tiers of government.

Worldwide, national and state governments have primary authority over 35% of urban mitigation potential (excluding decarbonization of electricity), including from energy efficiency standards for appliances and vehicles (Figure 7.10).[60] Local governments have

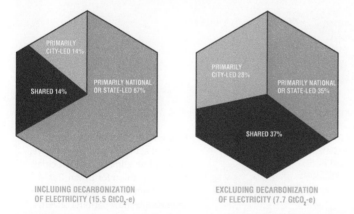

Figure 7.10 Proportion of 2050 urban abatement potential over which different levels of government have primary authority or influence. (*Source:* Coalition for Urban Transitions, Climate Emergency, Urban Opportunity: how national governments can secure economic prosperity and avert climate catastrophe by transforming cities, 2019. Available at: https://urbantransitions.global/en/publication/climate-emergency-urban-opportunity/.)

---

57 Global Covenant of Mayors for Climate & Energy, 2018. Implementing Climate Ambition. https://www.globalcovenantofmayors.org.

58 Coalition for Urban Transitions, 2019. Op. cit.

59 Ibid.

60 Ibid.

primary authority or influence over 28%, including urban form and waste management.[61] The remaining 37% depends on collaboration among national, regional, and local governments, including on building codes, renewable energy generation, and mass transit infrastructure.[62] With national governments involved in nearly two-thirds of urban mitigation potential (or 86% if including decarbonization of electricity),[63] proactive national leadership is needed to achieve these emission reductions and provide the enabling policies and investments that can spur the private and civic sectors to action.

There Is a Short Window of Opportunity Open Now for National Governments to Place Zero-Carbon Cities at the Heart of National Development and Climate Strategies

Decisions made about cities in the next decade will put countries on a path to prosperity and resilience – or decline and vulnerability. Pursuing zero-carbon, resilient cities in an inclusive way would simultaneously raise countries' living standards, tackle inequality, and address the climate crisis. For national leaders, this could yield short-term political dividends and secure long-term national prosperity. The Coalition for Urban Transitions' *Climate Emergency, Urban Opportunity* report[64] – a collaboration of more than 50 leading institutions and backed by the UN Secretary General – identified six priorities for national action that can foster zero-carbon, inclusive, resilient cities (Figure 7.11):

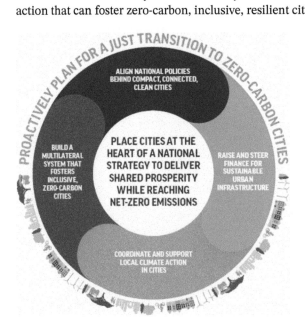

Figure 7.11 Six priorities for national action to achieve inclusive, zero-carbon, climate-resilient cities. (*Source*: Coalition for Urban Transitions, Climate Emergency, Urban Opportunity: how national governments can secure economic prosperity and avert climate catastrophe by transforming cities, 2019. Available at: https://urbantransitions.global/en/publication/climate-emergency-urban-opportunity/.)

---

61 Ibid.

62 Ibid.

63 Ibid.

64 Ibid.

1) *Develop a strategy to deliver shared prosperity while reaching net-zero emissions – and place cities at its heart.* National governments will need a robust, cross-cutting plan to drive economic and social progress in the context of climate change – but few today have such a strategy. Given the increasing urbanization of people, emissions, and economic activity, cities need to be at the heart of any national strategy to deliver shared prosperity while reaching net-zero emissions. This could align action across ministries and sectors, the private sector, and civil society – especially if developed in an inclusive way.

2) *Align national policies behind compact, connected, clean cities,* including:

Removing land-use and building regulations that limit higher, liveable density	Reforming energy markets to decarbonize the electricity grid by mid-century	Introducing net-zero building codes in all buildings with minimal use of offsetting by 2030
Banning the sale of fossil-fuel-powered vehicles from 2030	Adopting green alternatives to steel and cement by 2030	Shifting away from building detached housing in established cities

3) *Fund and finance sustainable urban infrastructure,* including:

Eliminating subsidies for fossil fuels	Strengthening land and property tax collection in cities to grow local tax bases for low carbon investment	Working with local governments to establish a pipeline of climate-safe, bankable projects
Scaling land-based financing instruments to fund sustainable urban infrastructure and limit sprawl	Shifting national transport budgets from road-building to public and active transport	

4) *Coordinate and support local climate action in cities,* including:

Creating integrated land-use and transport authorities for cities	Strengthening the capacities of built-environment professionals to pursue zero-carbon, climate-resilient development	Authorizing local governments to introduce climate policies and plans that are more ambitious than national policies
Establishing "regulatory sandboxes" for low-carbon innovations in cities	Allocating at least one-third of national R&D budgets to support cities' climate priorities by 2030	

5) *Build a multilateral system that fosters inclusive, zero-carbon cities*, including:

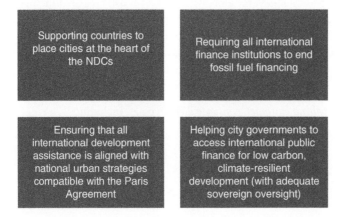

6) *Proactively plan for a just transition to zero-carbon cities*, including:

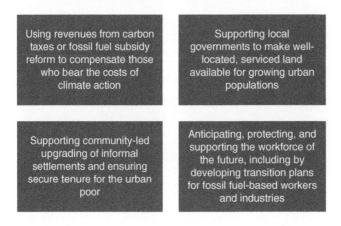

Conclusion

The leading countries of tomorrow will be those that invest in zero-carbon cities. This chapter has shown that pursuing zero-carbon, resilient cities in an inclusive way could simultaneously raise countries' living standards, tackle inequality, and address the climate crisis. It has shown that for national and local leaders, creating such cities would yield short-term political dividends and secure long-term national prosperity. And it has shown that this urban future is possible and within our grasp.

This is a chance not to be missed. We can and we must continue to encourage our leaders to seize the urban opportunity.

8

The Responsible City – Introduction

Nick and Leah in Chapter 7 showed us how GDP growth and reducing CO_2 emissions don't have to be mutually exclusive. They importantly gave us six priorities for national action to achieve inclusive, zero-carbon, and climate resilient cities.

It's time to shift to the key players within the city ecosystem that can affect change. Justin and Molly set out how businesses are leading the charge as well as highlighting some of the barriers that prevent businesses from acting. In order for cities to continue economic growth, carbon emissions must be curbed as a matter of urgency. A resilient city is also a responsible city – one cannot exist without the other.

Justin and Molly also write that the responsible city is an "inclusive city", adding yet another adjective to the growing list of qualities a net-zero carbon city must employ, suggesting openness, honesty, and innovative collaboration. Just as Peter spoke about the importance of collective and collaborative leadership in Chapter 1, Justin lays out in detail Accenture's Five Element Model of Responsible Leadership as relevant for cities, companies, and others committed to creating sustainable cities. For a city to succeed in its zero-carbon mission certain qualities need to be reflected across the board in national government, local government, large businesses, and small businesses, and on a global level the same rules apply, all aided by the UN Sustainable Development Goals (SDGs).

A responsible city naturally needs responsible businesses. Justin and Molly look in detail at the growing sense of responsibility among leadership companies and the desire to ensure businesses are positively contributing to the cities in which they operate, as well as how businesses can collaborate to further this aim through public and private sector hybrids, as we saw in the Copenhagen climate story in Chapter 6. Justin and Molly explore how BMW and Owens Corning put considerable scales of resources behind their efforts to build responsible cities. They discuss how the circular economy has captured the interest of business leaders interested in sustainability, in conjunction with the chapter in *The Circular Economy Handbook*.

Within this chapter, they highlight the importance of city leaders in reaching these zero-carbon goals, using Engie and Costa Rica as case studies, looking at the important role technology will play, and present a case study – "The West Midlands, A Reimagined, Responsible Region". Responsible cities need responsible leaders – that is for sure.

Justin and Molly write that there is "no turning back" from this collective responsibility we now face in the race against our 2030 net-zero carbon goals. But it is a shared problem, and therefore collective and responsible action is key and, more importantly, possible.

The Climate City, First Edition. Edited by Martin Powell.
© 2022 John Wiley & Sons Ltd. Published 2022 by John Wiley & Sons Ltd.

8

The Responsible City

Justin Keeble and Molly Blatchly-Lewis

A city is a powerhouse of change, innovation, and economic growth, a multifaceted organism that depends on its many constituent parts – policymakers, citizens, civil society leaders, and the private sector – to keep everything humming efficiently. In the best of circumstances, a city is also an engine that spurs change at the systemic level, taking an outsized responsibility to demonstrate leadership in creating a more sustainable world. System challenges require system responses, and, in that transition, cities rely on the private sector to be a responsible partner, helping to lead the charge. When city and business leaders join forces and apply their ingenuity to tackling our most wicked sustainability challenges, anything is possible.

The responsible city is an inclusive city. It is committed, creative, open, and honest. It is guided by a clear and compelling mission and purpose. Its leaders know that responsible use of technology and innovation can create new societal value. And it is constantly seeking out new knowledge. These characteristics distinguish not only thriving urban centres, but also the qualities that leadership teams will need to navigate the decade ahead. Shaped by input from more than 20,000 people around the world, Accenture has developed a Five Element Model of Responsible Leadership that is as relevant for cities in the twenty-first century as it is for companies and others passionate about creating sustainable cities and communities.[1] The Five Elements are shown in Figure 8.1.

1) *Stakeholder Inclusion.* Safeguarding trust and positive impact for all by standing in the shoes of diverse stakeholders when making decisions, and fostering an inclusive environment where diverse individuals have a voice and feel they belong.
2) *Emotion & Intuition.* Unlocking commitment and creativity by being truly human, showing compassion, humility, and openness.
3) *Mission & Purpose.* Advancing common goals by inspiring a shared vision of sustainable prosperity for the organization and its stakeholders.
4) *Technology & Innovation.* Creating new organizational and societal value by innovating responsibly with emerging technology.
5) *Intellect & Insight.* Finding ever-improving paths to success by embracing continuous learning and knowledge exchange.

---

1 Global Shapers Community, The Forum of Young Global Leaders, in collaboration with Accenture. 2020. Seeking New Leadership: Responsible Leadership for a Sustainable and Equitable World. https://www.accenture.com/_acnmedia/pdf-115/accenture-davos-responsible-leadership-report.pdf.

The Climate City, First Edition. Edited by Martin Powell.
© 2022 John Wiley & Sons Ltd. Published 2022 by John Wiley & Sons Ltd.

Figure 8.1 The Five Element Model of Responsible Leadership. (*Source*: "Seeking New Leadership: Responsible Leadership for a sustainable and equitable world," Global Shapers Community, The Forum of Young Global Leaders, in collaboration with Accenture, January 20, 2020, https://www.accenture.com/_acnmedia/PDF-115/Accenture-DAVOS-Responsible-Leadership-Report.pdf.)

For the responsible city to succeed with its mission – whether that is transitioning to low-carbon or zero-carbon or leveraging the benefits of digitalization in the smart city – these qualities need to be reflected across the board by everyone who has a stake in the future viability of our cities. And naturally, that includes business. Whether it's a multinational giant, a small to medium-sized enterprise (SME), or the smallest of disruptive start-ups, the private sector builds and operates much of what keeps cities running, from roads to buildings to housing to transport to waste management. As a major source of employment, capital, products, and solutions for more sustainable cities, leading companies see the inherent win-win in responsible leadership. Our analysis of more than 2,500 listed companies showed that companies that combine top-tier innovation with top-tier sustainability and trust outperform their industry peers on operational and market metrics. Their estimated operating profits are 3.1% higher on average, and they deliver a higher annual total return to shareholders.[2]

We've seen an upswing in the number of companies around the world that are redefining what responsible means, and this chapter is peppered with examples to illustrate that point. But a purpose or vision does not necessarily lead to action. There is still a long way to go to scale up the partnerships needed to bring about the transformational impact demanded by the UN Sustainable Development Goals (SDGs) – whether at the city, national, or global level.

In this chapter, we will look at how cities, which contribute about 60% of global GDP,[3] rely on businesses to play their part in realizing the vision of the responsible city, explore the potential of digitalization to unleash greater impact, the ability of the

---

2 Ibid.

3 Goal 11, UN Sustainable Development Goals. https://www.un.org/sustainabledevelopment/cities.

regulatory environment to help or hinder the private sector's role, and the viability of solutions for the developing and emerging economy as well as in the developed world. Not least, we return to our conviction that the real power behind transformation lies in collective action.

Shared Effort, Shared Gains

According to a recent UN Global Compact (UNGC)–Accenture study, a majority of chief executive officers (CEOs) acknowledge that action on climate change is critical to achieving the SDGs (also referred to as the Global Goals). Both the private and public sector recognize that they need each other if they are to succeed.[4] The same study found that 59% of UN leaders feel that business will be the most critical partner in the UN's ability to deliver the SDGs. And 83% of business executives believe governments need to step up their efforts to provide an enabling environment for business efforts on sustainability. Finding common ground will be key to executing the vision of the responsible city.

The good news is that both city and corporate leaders recognize that partnerships that connect business to local communities are increasingly critical for meeting the Global Goals. When sustainability and trust go hand in hand, as our Five Element Model of Responsible Leadership illustrates, it bodes well for a successful partnership.

There are myriad ways that cities rely on the innovation that business can bring to the table to achieve their sustainability goals. Public–private partnerships and collaborations have become essential in the face of the daunting global trends impacting cities. Cities are simultaneously grappling with a range of environment, social, and financial challenges, and public health crises like the COVID-19 pandemic. By 2030, 61% of the global population (5 billion people) will live in cities, passenger traffic is expected to double, and freight volume is forecast to grow by 70%. This will create new demands for infrastructure to safely accommodate the demand for housing and mobility while reducing greenhouse gas emissions and protecting air quality.[5] As Brian Owens, President and CEO of Owens Corning, a global producer of insulation material, roofing, and fibreglass, views it, "Buildings and homes contribute about 40% of the world's greenhouse gas emissions, and the best way to combat that is to design energy-efficient infrastructure."[6]

The ongoing shift from rural to urban life, along with the expanding middle class and growing population, are straining the ability to meet people's basic needs. Cities aptly illustrate the scale of the challenge, yet also provide spaces of opportunity for seismic shifts in the fight against climate change – which can be met only with leadership from the business community.

Overall, structural trends are driving business to take greater action on sustainability. Of the global issues prompting business to prioritize sustainability, 54% of CEOs that Accenture surveyed name environmental degradation, 39% cite the growing middle class,

---

4 UN Global Compact–Accenture Strategy 2019 CEO Study – The Decade to Deliver: A Call to Business Action, 2019. https://www.unglobalcompact.org/library/5715.

5 Ibid.

6 Ibid.

and 38% list urbanization among the top macro trends.[7] But alongside these macro trends is the growing sense of purpose among leadership companies – a desire to ensure that the business is positively contributing to the cities and communities in which it operates. That lays wide open a system change question: How can business collaborate to ensure improved quality of life in the communities where it operates?

For oil and gas company BP and ride-hailing firm Uber, the journey to answer that question led to a commitment to convert the entirety of Uber's fleet to electric cars within five years.[8] The partnership offers Uber drivers discounted rates across BP's network, with investment in dedicated charging infrastructure to close the loop between supply and demand to accelerate electrification. While large companies like automaker BMW and Owens Corning can put considerable scale and resources behind their efforts to help build responsible cities, there is a much-needed place for solutions from SMEs as well. Take the start-up Climeworks, which has developed technology to capture CO_2 directly from the air. The air-captured CO_2 can either be recycled and used as a raw material or completely removed from the air by safely storing it. Climeworks' direct air-capture machines are powered solely by renewable energy or energy-from-waste. Grey emissions are below 10%, which means that out of 100 tCO_2 that its machines capture from the air, at least 90 tCO_2 are permanently removed and only up to 10 tCO_2 are re-emitted.[9] In 2017, the Climeworks AG facility near Zurich was the first ever to capture CO_2 at industrial scale from air and sell it directly to a buyer. The plant captures about 900 tCO_2 annually – or the approximate level released from 200 cars – and pipes the gas to help grow vegetables. Climeworks aims to capture 1% of global CO_2 emissions by 2025.[10]

According to the UNGC–Accenture's 2019 CEO Study, CEOs see a growing need to work below the national and state levels to drive action and to move seamlessly between global and local implementation. Working with cities, in particular, CEOs believe, can help move the needle from high-level dialogue to greater action on the ground. As one CEO explained: "It doesn't succeed unless there is a strong willingness at the city, regional or governmental level, but the higher up you go, the larger the consensus that is needed to move forward."

Government-enforced regulation isn't always necessary to encourage businesses to act responsibly – the Five Elements of leadership as described earlier take on a momentum of their own in spurring action. As we noted, there is a clear competitive advantage for business in the pursuit of sustainability. This can make it unnecessary for governments to prod action through policy, particularly for the leading companies that recognize the opportunities. The circular economy illustrates this neatly. With the current take–make–waste economy, the world is already using approximately 1.5 planets' worth of resources every year. Based on the current pace, we'll consume three planets by 2050. Under this "linear system", cities consume over 75% of natural resources, produce over 50% of global waste, and emit between 60 and 80% of greenhouse gases.[11]

---

7 Ibid.

8 https://www.bp.com/en/global/corporate/news-and-insights/reimagining-energy/world-ev-day.html.

9 https://www.climeworks.com/page/co2-removal.

10 Climeworks makes history with world-first commercial CO_2 capture plant. Climeworks, 31 May 2017. https://climeworks.com/news/today-climeworks-is-unveiling-its-proudest-achievement.

11 Circular economy in cities. Ellen MacArthur Foundation. https://www.ellenmacarthurfoundation. org/our-work/activities/circular-economy-in-cities. https://www.ellenmacarthurfoundation.org/ our-work/activities/circular-economy-in-cities

As laid out in *The Circular Economy Handbook: Realizing the Circular Advantage*, when adopted strategically, the circular economy can create significant financial and economic value for business and society. This is what we call the circular advantage, based on analysis from over 1,500 company examples and 300 in-depth case studies conducted from 2015.[12] The US\$4.5 trillion value at stake in the circular economy is, for many companies, sufficient incentive to innovate and create new markets while also reducing harmful environmental impacts and improving socioeconomic outcomes.[13]

The opportunity of a circular economy has captured the imagination of business leaders, emerging innovators, governments and cities, designers and academics around the world. In its analysis of more than 120 case studies of companies that are generating resource productivity improvements in innovative ways, Accenture identified five underlying business models, as shown in Figure 8.2.

The path to circularity differs among industries. In general, consumer-facing industries have seen the largest volumes of circular activity, often driven by demands from consumers, governments, and employees. Other industries feel the pressure of regulatory drivers, like the household appliances sector, where increasing regulations on responsible treatment of products at end of use are pushing companies to focus on greater recovery of used machines. No matter the industry, there are common barriers and enablers. Companies across all industries need to scale innovation, build partnerships, focus on broader supply chain circularity, and support enabling policies and regulations.

To understand how the circular economy works on the ground in cities, consider Toronto, which aims to achieve zero waste, starting with diverting 70% of waste produced from landfill. New York City is committed to becoming a worldwide leader in solid-waste management by achieving the goal of zero waste by 2030.

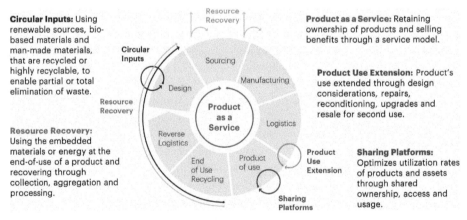

Circular Inputs: Using renewable sources, bio-based materials and man-made materials, that are recycled or highly recyclable, to enable partial or total elimination of waste.

Resource Recovery: Using the embedded materials or energy at the end-of-use of a product and recovering through collection, aggregation and processing.

Product as a Service: Retaining ownership of products and selling benefits through a service model.

Product Use Extension: Product's use extended through design considerations, repairs, reconditioning, upgrades and resale for second use.

Sharing Platforms: Optimizes utilization rates of products and assets through shared ownership, access and usage.

Figure 8.2 The five business models in resource productivity improvements. (*Source*: The Circular Economy Handbook, Accenture, January 13, 2020. Retrieved from: https://www.accenture.com/us-en/about/events/the-circular-economy-handbook.)

---

12 Lacy, P., Long, J., Spindler, W., 2020. *The Circular Economy Handbook: Realizing the Circular Advantage.* Palgrave Macmillan, London.

13 Ibid.

In the future, a number of emerging innovations could play a prominent role in helping companies to close the circular loop – and cities stand to benefit. For example, the city of Helsinki's fully owned energy company "Helen" provides solar technology for different properties that are then able to feed their excessive solar power to a grid. The company is also developing solutions to recover waste heat from properties, as well as new geothermal energy technologies.

In sync with cities like Paris that are working towards a mostly car-free inner city to curb the city's greenhouse gas emissions, the personal mobility industry has taken significant steps towards circularity. Jaguar Land Rover, Mercedes-Benz, Volvo, and other automakers have already committed to partial or full-fleet electrification by 2030 and even 2025. Vehicle sharing and pooling highlight another disruption, illustrated by the popularity of DriveNow and ZipCar. By 2030, one in ten cars is expected to be a shared vehicle.

Of course, as pointed out in *The Circular Economy Handbook*, you can't successfully tackle a systemic reshaping of the production and consumption model that has been dominant since the mid-eighteenth century without tight alignment of supply, demand, and policy. Governments must use their powers to share market conditions at the national and global level to create the right situation for change. Businesses can also accelerate the transition by adopting the circular economy in their own substantial organizations and supply chains through areas like public procurement.

The estimated US$300 billion renewable energy market[14] is an area where policy has been instrumental in driving greater progress. For example, where Renewable Energy Certificates (RECs), which represent the clean energy attributes of renewable electricity, and virtual Power Purchase Agreements (vPPAs), a specific type of a PPA contract, used to procure long-term renewable energy, are stimulating both public and private sector renewable energy demand. These mechanisms have helped spark the dramatic rise of renewable energy procurement by US local governments. Between 2015 and the first quarter of 2020, US cities signed 335 renewable energy deals totalling 8.28 gigawatts (GW), equivalent to nearly 1% of the total current electric generating capacity installed in the USA.[15]

In short, there needs to be a give-and-take on the role of government in the regulatory space – a too-much or too-little balancing act that can either help or hinder business from deploying the kinds of solutions needed. When it comes to responsible leadership, two examples illustrate both sides of the equation. The first is commercial energy provider Engie, which needed no government intervention to reinvent its transition from selling fossil-fuel-based energy to providing low-carbon energy and services. The second is the country of Costa Rica, where, under the leadership of President Carlos Alvarado Quesada (a member of the Forum of Young Global Leaders), the government has set out a holistic vision defining the country's commitment to become carbon neutral by 2050.[16]

---

14 Benefits of renewable power purchase agreements. Accenture, 22 May 2020. https://www.accenture.com/gb-en/insights/utilities/renewables-power-purchase-agreements.

15 Goncalves, T., Liu, Y., 2020. How US cities and counties are getting renewable energy. World Resources Institute, 24 June 2020. https://www.wri.org/blog/2020/06/renewable-energy-procurement-cities-counties.

16 Global Shapers Community et al., 2020. Op. cit.

Technology and innovation are at the centre of Costa Rica's plan, applying systems thinking and creating opportunities across a number of industrial sectors.

Both Engie and Costa Rica are among the case studies in Accenture's recent report on responsible leadership, in collaboration with the Forum of Young Global Leaders and Global Shapers Community.[17] Not only did Engie, formerly Gaz de France, return to profitability while making its transition, but also its radical transformation combined instinct and sensemaking to anticipate growing political and social momentum for action on climate change as well as tighter emissions regulation and rising carbon prices. In 2016, Engie set a goal to become the world leader in the zero-carbon transition. Since then, the company has been rapidly growing its portfolio of renewable energy assets, while shutting down or divesting coal-fired power plants. For example, in December 2019 it announced plans to close coal-fired power plants with a total capacity of almost 1 GW in Chile and Peru by 2024, following a commitment to build 1 GW of new solar and wind capacity in Chile, worth US$1 billion. The company's ongoing efforts to reduce emissions have intensified. Between 2012 and 2018, Engie cut its CO_2 emissions in half. Further, Engie shows the significance of a global business committing to carbon neutrality, where decisive action can help developing countries that are still financing coal-fired power plants to make strides in climate action. Similarly, Apple's pledge to become carbon neutral by 2030, which covers itself and suppliers, and Microsoft's promise to be carbon negative by 2030 (removing more carbon from the atmosphere than it emits[18]) have an outsized impact, due to the sheer size and scale of these tech giants – and their ability to positively influence others.

The need for business and government coordination is a linchpin of Costa Rica's decarbonization goal, because the success of its ambition depends on implementing an effective cross-industry, cross-sector plan. The government is engaging a number of industrial sectors: transport and sustainable mobility; energy, green building, and industry; integrated waste management; and agriculture, land-use change, and nature-based solutions. The plan articulates explicit strategies for each (including green tax reform, digitalization, open data and transparency, mobilizing public and private funds, and environmental institutional reform). Costa Rica is also evidence that for developing and emerging economies, the same solutions and approaches can apply, although the need for government intervention may be more critical in some developing countries to stimulate market forces.

As Zhao Kai, Vice President and Secretary General of the China Association of Circular Economy, puts it in *The Circular Economy Handbook*, "Cohesive national strategies are essential for accelerating the transition to circular economy, and we see this starting to have an impact in China. Best practice pilots or initiatives, such as creation of zero waste city pilots or circular industrial parks, are then needed locally to bring these policies to life and demonstrate the advantages of going circular."[19]

---

17 Global Shapers Community et al., 2020. Op. cit.

18 Kelion, L., 2020. Apple has announced a target of becoming carbon neutral across its entire business manufacturing supply chain by 2030. BBC, 21 July 2020. https://www.bbc.co.uk/news/technology-53485560.

19 Lacy et al., 2020. Op. cit.

There is no shortage of urban challenges where business can help make a difference, from contributing to reductions in greenhouse gas emissions, to spurring the transition to more sustainable transport solutions, to reducing waste. The questions are, do individual corporate actions make a dent in the overall problem, and how does one sort out the greenwashing from credible and meaningful impact – especially the kind of impact required to help achieve the SDGs? Recent Accenture research demonstrated the potential of business to move the dial on sustainability through ecommerce. A greener last-mile supply chain made possible by local fulfilment centres could lower last-mile emissions between 17 and 26% by 2025.[20] Using local fulfilment for even half of ecommerce orders between 2020 and 2025 could lead to significant impacts.

Accelerating Action: Digitalization Holds the Key

The advances in the circular economy that we've seen take off across multiple industries and in many cities and countries around the world often have digitalization as an accelerator. Digital technologies can help to shift the way we manage energy and reduce carbon footprint, provided we rethink the way we design buildings, industries, and, not least, cities.

In *The Circular Economy Handbook*, 27 key technologies with a central role in the circular economy are identified which fall into three realms: the digital, the physical, and the biological. In cities, we have seen how digital technologies, in particular, are spurring the smart city movement. In a smart city, digital–physical interfaces, sensors, smart software, and the Internet of things-centred technologies work together to enhance and streamline how the city runs. Cities globally are incorporating data and digital solutions into a more sustainable future, including artificial intelligence (AI), machine learning, blockchain, and big data analytics.

Breakthroughs in physical and biological technologies are also enabling much-needed innovation. In cities, for example, less than 2% of nutrients in food and organic wastes (excluding manure) are reinserted in nutrient cycles, losing potential value and adding to future environmental costs. Companies like Lystek, a Canadian waste treatment technology company, are taking organic waste from cities and converting it into regenerative soil enhancers.

Future 5G networks are enabling massive machine-type communication that can manage vehicle traffic and electrical grids, with the possibility for substantial savings through reductions in energy use, traffic congestion, and fuel, our research shows.[21] Smart public lighting concepts, for example, automatically dim public lighting when no pedestrians or vehicles are near, conserving power, while still keeping a neighbourhood safe. More broadly, smart cities have the potential to reduce traffic congestion overall, through smart traffic management systems.

---

20 The sustainable last mile. Accenture, 27 March 2021. https://www.accenture.com/_acnmedia/pdf-148/accenture-sustainable-mile-pov.pdf#zoom=40.

21 Smart cities: How 5G can help municipalities become vibrant smart cities. Accenture, 2017. https://api.ctia.org/docs/default-source/default-document-library/how-5g-can-help-municipalities-become-vibrant-smart-cities-accenture.pdf.

In Japan, Accenture and the University of Aizu are working on a joint research project to develop the first standardized marketplace for AI in public services. The research will focus on Aizuwakamatsu, a city in Fukushima with a population of 120,000, where Accenture is part of a broader smart city project to help the region recover economically after the devastating 2011 earthquake, with an emphasis on the use of data. Smart city projects in Aizuwakamatsu include programmes relating to mobility, fintech, education, healthcare, childcare, agriculture, Industry 4.0, and tourism. The application programming interface (API) marketplace site, where standardized API code for the creation and linkage of software can be shared, aims to help local governments and companies benefit from new data connections and integrations between smart city and industry initiatives.[22] Because it can be built once and deployed often, it's a prime example of an easily scalable solution. City applications can be shared across different smart cities. Leaders could share the cost of development or import applications developed elsewhere to their own city, providing the technological fabric for an open-source smart city of sorts. The Aizuwakamatsu model is founded on citizen trust, with an opt-in model underscoring the role of relations between city leaders and citizens to drive adoption and sustainable change.

Leaders Use Their Influence for Positive Impact

A range of opportunities emerge for cities when businesses use their influence for positive impact in the marketplace, their supply chains, as an employer, and as a corporate citizen. Business must play a leading role in transformative change, innovating for new forms of value creation. Again, technology is a key lever, and cities and their leaders occupy dual roles as enablers, clients, and partners in these digital ecosystems. Digital platforms allow companies and industries to more easily share resources. These platforms drive resource efficiency by helping to extend the use of products and by providing marketplaces for waste. The aim is that the excess capacity or waste of one business (or industry) can become the input for another business (or industry). That was the thinking behind Austin Materials Marketplace, an online platform for businesses and organizations in Texas to connect and find reuse and recycling solutions for waste and by-product materials. Since its launch in 2014, the platform has diverted over 55,000 cubic feet from landfill, generated over US$600,000 in cost savings and value creation, and avoided 900 tCO_2e.

Online platforms are also a powerful tool for enabling knowledge sharing and open innovation, as participants can easily exchange ideas, share insights, and co-create solutions for overcoming circular challenges. The open access innovation platform Circle Lab acts as a "Wikipedia for circularity": an online space for individuals, businesses, organizations, and cities to discover, discuss, and share circular business practices and strategies to tackle both local and global challenges.

22 Wray, S., 2020. Accenture to develop standardised API marketplace for Japan's smart cities. CitiesTodayInstitute,8July2020.https://cities-today.com/accenture-to-develop-standardised-api-marketplace-for-japans-smart-cities.

The West Midlands in the UK has been particularly ambitious in its climate and sustainability goals, and tapping into the knowledge of the community and the catalytic potential of business, using digital capabilities, has been part of its success. Our case study looks at what the West Midlands under the leadership of Mayor Andy Street has accomplished so far in a re-imagined region – and what remains to be done.

Case Study: The West Midlands – A Re-Imagined, Responsible Region

To support their journey to become a global hub for clean mobility, Transport for West Midlands (TfWM) partnered with technology services firm Accenture, design agency Fjord, and engineering consultancy Mott MacDonald. They co-created a vision and value proposition informed by user research, global benchmarking, and extensive qualitative research to help to guide TfWM trajectory in the future of mobility.

With an ambitious vision to reach net-zero within two decades, the West Midlands is building on its powerful manufacturing and automotive legacy to innovate and industrialize across the region. The region is a key growth engine in the UK economy and is home to Birmingham, the UK's second largest city, in addition to a constellation of towns and cities including Coventry and Wolverhampton. While continuing to serve the region's 2.8 million population, the region's transport agency TfWM also has a clear mandate to support sustainable regional growth and decarbonization.

Against the backdrop of rapid pre-pandemic disruption and accelerated technological change, environmental, social, economic, and public health imperatives are compounding pressure in the sector. Precisely because the movement of goods and people is so foundational to society, TfWM saw an opportunity for transport to become an enabler for place-wide change and more positive outcomes through a differentiated approach.

Accelerating Action in Practice

TfWM worked with a blended team from Accenture, Fjord, and Mott MacDonald to understand the future of mobility and the region's role in the evolving transport landscape.

The mission was to translate experiments into impact, providing accessible services for all, impactfully reducing emissions, and driving local economic growth. The team harnessed a holistic perspective to validate and challenge the West Midlands' existing projects and strategic intent through:

- Macro context
  - *Insights*: global forces of change and trends shaping transport and mobility
  - *Tools*: desk-based research, subject-matter expert interviews
- Market context
  - *Insights*: international challenges and opportunities, understanding other regions' approach to mobility and what "good" looks like
  - *Tools*: global benchmarking, capability assessment and gap analysis, ecosystem mapping, place archetypes

- Human context
  - *Insights*: citizen and business needs, customer archetypes development and challenges
  - *Tools*: user research and journey mapping, secondary research, quantitative analysis

Through this multidisciplinary approach, the team developed a vision and value proposition to set the course to become a *global hub for clean mobility*. Through local understanding and international insight, TfWM targeted four key areas to move the dial on sustainable and inclusive growth:

- *Attracting and growing businesses*: creating spaces for innovation and idea generation through a focused innovation ecosystem, with a targeted industry strategy and a constant eye on new entrants and opportunities.
- *Improving efficiency of freight*: expanding sustainable solutions to freight and supply chain through dedicated logistics infrastructure and last-mile initiatives.
- *Reducing car usage*: a multifaceted approach to transport informed by data and dialogue with citizens, embedding multimodality and active transport through Mobility-as-a-Service and access to micro- and e-mobility.
- *Expanding access to services and opportunities*: user-centred transport systems across the entire region, leveraging options such as demand responsive transport for underserved areas, ensuring accessible and affordable services to underpin inclusive growth.

In recent years, the region has begun to emerge as a mobility playground, home to a connected autonomous mobility lab, future transport zone, a digital data exchange (conVEx), and the UK's largest 5G mobility testbed.

Whilst these forward-thinking technologies are key, to really tackle the critical challenges surrounding transport, cities need to change the way people move; to change the way people move, cities need to understand *what moves people*. TfWM applied a human-centred approach and an inclusive design philosophy through purposeful citizen engagement to deliver benefit for people, business, and the planet amidst rapid change.

The depth of this holistic perspective has helped to set the region's trajectory towards a net-zero future through collaboration across the full spectrum of businesses, from start-ups and SMEs, to multinationals within and beyond transport, to the wider innovation ecosystem including academia and not-for-profits.

Responsible Vision and Leadership

TfWM recognizes that through partnership it can get the imaginative solutions it needs. This combined with leadership focused on positive impact were important factors in the success of this work. Collective action is pivotal to sustainability, and the West Midlands is tapping into this potential by building on the region's existing strengths such as:

- *Rich research and development*: access to a range of innovation assets from manufacturing hubs to R&D facilities with clean mobility potential, including the UK Battery Industrialisation Centre, UK Mobility Data Institute at Warwick Manufacturing Group, hydrogen research at the University of Birmingham, Tyseley Energy Park, and Energy Systems Catapult.

Figure 8.3 Cornwall Street, Birmingham. (*Source*: Tom W / Unsplash.)

- *Partnership and engagement opportunities*: a melting pot of industries, from transport technology to creative user-centred services – all open to innovation and collaboration. Sectoral analysis shows strong advances in automotive, logistics, and low carbon technology; clean mobility is at the centre of the collaboration agenda, whilst adjacent industries can accelerate innovation.
- *Technology and connectivity*: leveraging the catalytic potential of 5G roll out and real-world testing environments to accelerate practical applications of emerging and evolving technologies.

Transport is at the heart of the West Midlands' industrial strategy and economic development. Mike Waters, Director of Policy, Strategy, and Innovation for TfWM, states, "there are very few models of sustainability that don't put economic prosperity as an essential factor in having a sustainable system of any kind", and the West Midlands is a prime example of business in leading the charge in sustainability. Figures 8.3–8.5 show street view images of Birmingham and Coventry – a business-led sustainable transport infrastructure strategy for the region.

Conclusion

What leaders across the public and private spheres have consistently told us is that the real power is in collective action – and when walls come down, the potential for change can be significant. And where great leaps are not possible, incremental progress is still

Figure 8.4 Tramway leading up to the Museum and Art Gallery, Birmingham. (*Source*: Adam Jones / Unsplash.)

important for chipping away at the wicked challenges cities face. There is no more relevant platform for defining the scale of that collective action than the UN SDGs. When a global company aligns its strategy and core business with the delivery of the SDGs, significant scale is possible – especially with the right public–private partnerships.

Climate change is a collective problem that requires collective action. According to the UNGC–Accenture CEO Study, UN leaders feel that cross-sector alliances, networks, and partnerships are essential to accelerating progress on the SDGs. Some 59% of UN leaders feel that business will be the most critical partner in the UN's ability to deliver the SDGs. But what will it take to scale up partnerships to bring about transformational impact on the SDGs, especially in ever-growing, resource-strained cities?

In our 2019 study of UN leaders' and CEOs' views on what is required to achieve the SDGs by 2030, described as "the decade to deliver", three immediate areas of intervention emerge as necessary to pivot the organization towards deeper, strategic collaboration with the private sector:

1) the articulation of a clear value proposition for all partners;
2) effectively measuring the impact of the partnership;

Figure 8.5 Coventry. (*Source*: Georgi Kyurpanov / Unsplash.)

3) communicating the clear value of collaboration and why results could not be achieved in isolation.

While CEOs clearly see the role for global norms and standards, they also say local engagement is central to shaping solutions tailored to a specific place and situation. In 2016, 85% of CEOs said partnerships with governments, non-governmental organizations (NGOs), and international organizations that can connect business to local communities would be critical to enabling business to be a transformative force for achieving the Global Goals.[23] In 2019, our conversations with CEOs reinforced that solutions to many global issues only become relevant at the local level, such as collection of waste and recyclables.

---

23 UN Global Compact–Accenture Strategy, 2019. Op. cit.

However, lack of trust in the altruistic motives of business or anti-industry bias is a barrier to partnering with governments and development organizations, CEOs emphasized.

Collective action is what led to the Task Force on Climate-related Financial Disclosures (TCFD), created in 2015 by the Financial Stability Board to develop consistent climate-related financial risk disclosures for use by companies, banks, and investors in providing information to stakeholders.[24] The TCFD recommendations are just as relevant for cities, on the front line of climate change impacts. Yet another example of collective action was the business communities' reaction to the 2019 announcement by the former Trump administration to withdraw the US from the Paris Agreement. Within days, businesses contributing US$6.2 trillion to the US economy and states accounting for more than a third of the national GDP joined together to declare their intent to continue the path.[25] The withdrawal took effect on 4 November 2020, but on 29 January 2021, US President Joseph R. Biden recommitted the US to the Paris Agreement.[26]

There is clearly no turning back. For cities and metropolitan areas to continue to be powerhouses of economic growth, rising carbon emissions and unrestrained resource use will have to be curbed. SDG 11, to make cities inclusive, safe, resilient, and sustainable,[27] lies within reach – that is, provided business and local policymakers alike recognize their shared responsibility to create cities with social, environmental, and economic impacts in balance – for current and future generations.

24 https://www.fsb-tcfd.org.

25 Birkes, L., 2017. Our power is in collective action. The SustainAbility Institute, 9 June 2017. https://sustainability.com/our-work/insights/power-collective-action.

26 Davenport, C. and Friedman, L., 2021. Biden cancels keystone XL pipeline and rejoins Paris Climate Agreement. *The New York Times*, 20 January 2021.

27 Goal 11. Op. cit.

9

The Energized City – Introduction

Justin and Molly highlighted in Chapter 8 how governments must use their powers to share market conditions and adopt circular economy principles in their own supply chains. They highlighted the Renewable Energy Certificates (RECs) and virtual Power Purchase Agreements (vPPAs) as examples of stimulating public and private sector demand. US cities have signed renewable energy deals since 2015 totalling 8.28 gigawatts (GW), and big business is now signing vPPAs at a rate that will ensure fast growth of new renewable energy infrastructure (Figure 9.1).

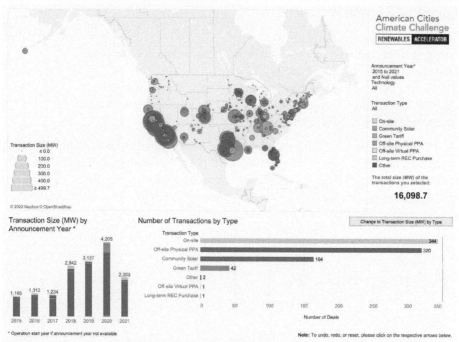

Figure 9.1 American cities signing vPPAs by transaction size and type. (*Source:* Abbott, S., Goncalves, T., House, H., Jungblut, W., Liu, Y., Roche, P., Rosas, J., Shaver, L., Tang, J., Vanover, A., and Walz, E. 2021. 'Local Government Renewables Action Tracker.' Washington, DC: Rocky Mountain Institute and World Resources Institute. Available online at: https://cityrenewables.org/local-government-renewables-action-tracker.)

The Climate City, First Edition. Edited by Martin Powell.
© 2022 John Wiley & Sons Ltd. Published 2022 by John Wiley & Sons Ltd.

Pete begins this chapter by stating: "This century belongs to cities." As previous chapters have stated over and over again, the majority of the world's population now lives in urban centres, and we are constantly met with figures and data that endorse Justin and Molly's requirement in Chapter 8, *The Responsible City*, that cities be just that – responsible.

Global energy demand is growing, and with electricity at the other end of a switch it is all too easy with large coal-, oil-, and gas-run power plants outside of cities that adjust to our demands for power. This unsustainable relationship needs to change, or to use Pete's phrase, it needs "a radical rethink", if we are to meet the energy demands of our growing cities and the need to tackle catastrophic climate change and air pollution crises in cities.

This "radical rethink" includes reducing our use of fossil fuels and better management of our investments in renewable energy, rethinking our infrastructure, and seeing it holistically as an integrated and interconnected system: in a word ... efficient. Our energy systems and their management must become more efficient.

Different cities face different challenges depending on their infrastructure, development, and climate. London's use of renewable energy has grown considerably, but it still struggles with controlling its road and housing emissions. Fossil-fuel-rich Saudi Arabian cities have seen the need to diversify to cleaner and more efficient power sources. The priority of action is different but the final result remains the same – our electricity must change. Pete lays out several clear universal actions all cities, irrespective of particular circumstance, can focus on:

1) having a clear, actionable plan;
2) using their planning powers to shift towards zero-carbon development;
3) the push for more efficient buildings;
4) increasing the use of renewables;
5) driving innovation and switching energy systems to electrical power; and
6) unlocking the potential of cities.

This final action of "unlocking the potential of cities" is particularly powerful for me and reminds me of Pete's previous statement – "this century belongs to cities". If we cannot decentralize our governmental power structures by acknowledging the importance of our cities and empower them to take meaningful action, then we have no hope of diversifying our energy industries.

9

The Energized City

Pete Daw

The Growing Demand for Energy and the Increasing Role of Electrification

The chances are you know that urban populations are rising around the globe. More than half of the world already lives in urban areas, and by 2050 two-thirds of the world's population will reside in them. This century belongs to cities. Cities drive global economic activity, through the services they provide and the demand for goods they generate. They also face challenges from tackling toxic air pollution, tackling inequalities, and climate change. Energy is an essential ingredient of the global economy, and cities are at the heart of both the challenges cities face and the solutions they need.

Energy is an integral ingredient of our lives and our economies, and our relationship with it has been a pretty straightforward one as users. We haven't as a society given a great deal of thought to the implications of providing it. We have typically generated energy at the point we need it. From switching on our car engines and combusting petrol, to combustion of gas or oil in boilers or stoves to heat homes and water, electricity has traditionally arrived at the flick of the switch or press of a button, supplied by large coal-, oil-, and gas-fired power plants many kilometres away from our cities, which adjust their power output dependent on our demands for power.

That relationship needs to change if we are to meet the growing demand for energy in cities and enable development and sustainable economic growth, whilst doubling down on efforts to tackle air pollution and climate change.

Driving Economic Growth, Tackling Air Pollution, and Climate Change – The Role for Electrification

Energy is needed to drive economic growth, support development, and improve quality of life. But our energy system needs a radical rethink. Put simply, continuing our profligate relationship with energy is not sustainable. It will result in catastrophic climate change, impacting on

The Climate City, First Edition. Edited by Martin Powell.
© 2022 John Wiley & Sons Ltd. Published 2022 by John Wiley & Sons Ltd.

all of us. In 2015, under the Paris Agreement, every country agreed to keep global temperatures below 2°C, yet current pledges will push global warming beyond 3°C by the end of this century, which risks natural tipping points such as the thawing of much of the world's permafrost – which could drive global temperatures uncontrollably higher. The impacts on humans and nature are vast: increased heatwaves and natural disasters, the loss of biodiversity, harming crop yields, the spread of disease, the displacement of people, and increasing levels of poverty and inequality. Typically, the impacts will hit the poorest and most deprived hardest.

Cities face an air pollution crisis, principally driven by ever-increasing numbers of internal combustion engines and increasingly grid-locked roads. Already globally the World Health Organization estimates that 4.2 million premature deaths are linked to ambient air pollution, mainly from heart disease, stroke, chronic obstructive pulmonary disease, lung cancer, and acute respiratory infections in children every year. Air pollution is responsible for 29% of all deaths and disease from lung cancer.[1] These impacts are most keenly felt where our populations are most dense in our towns and cities.

Global energy demand is growing; in 2018 we used 14,280,569 kilotonnes of oil equivalent (ktoe), 60% more than in 1990.[2] Fossil fuels continue to dominate meeting this demand. Coal, oil, and gas provide 81% of energy. But since 1990 there has been a sevenfold increase in solar and wind energy which provides 2% of demand.

Reducing our dependence on fossil fuels will require a shift to a greater share of our systems being powered by electricity (and from cleaner generation sources), replacing fossil fuels as the energy source in our heating and cooling systems and in our transport systems. A switch to powering more of our systems through electricity will place greater demand for electricity generation. The electrification of more of our systems and growing populations will lead to a 75% increase in electricity demand by 2050.[3] Cleaning up the energy mix for power generation is a key foundation for tackling climate change and air pollution.

But providing more of our power from renewable sources will create a more complex energy system. Wind farms need wind to generate power, solar panels need sunlight (Figure 9.2). These resources are not available 24 hours a day, seven days a week. We need to better manage the power system to ensure that we can store power in the system when it is being generated but not needed and make sure it is available when demand is higher, and ensure where we need fossil fuels in the transition to a zero-carbon world that they are used in the most efficient way and in the most efficient applications. Our energy systems will need to be more responsive too, so that high-energy users can power down use when necessary but also take advantage of clean electricity when there is a surplus available. Managing the energy system effectively through storage and managing demand, coupled with greater efficiency in buildings and appliances mean that we do not need to overprovide generation plants to meet demand, reducing costs and carbon impacts.

Advances in digitalization and automation mean the technology exists to realize these systems and manage them, whilst more accurately understanding demand. Increasingly we need to think of city infrastructure as one interconnected system. Understanding that

---

1 https://www.who.int/airpollution/ambient/health-impacts/en/#:~:text=an%20estimated%204.2%20million%20premature,and%20disease%20from%20lung%20cancer.

2 IEA, 2020. World Energy Balances. https://www.iea.org/reports/world-energy-balances-overview.

3 Schuster, W., de Miranda, P.P., Loos, J., Powell, M., 2018. *Better Cities, Better Life*. Booklink.

Figure 9.2 Off-shore wind farm off the coast of the UK providing renewable energy for vast urban centres. (*Source:* Professional Studio/Getty Images.)

interdependence is critical to enabling a shift to clean power driving more and more of our systems. This requires thinking about transport, buildings, energy grids, water management, and waste management as integrated parts of the city. For example, electric vehicles can be sent signals to charge when there is excess power available on networks, buildings can provide power to grids when they need it, they can store energy when there is too much available, and they can react to the needs of the power system by powering up or down their systems. Wastewater and waste systems provide opportunities to provide green power and heat, and effective management of those systems through smart technology means power demand can be reduced. As much as 40% of the costs of drinking water can be associated to energy. Better scheduling of pumping and more efficient use of water can reduce costs and power requirements. Waste heat from waste management and wastewater can be captured and utilized in providing heating or cooling to buildings, supported by large-scale heat pumps (powered by electricity) to increase or decrease the temperature as required.

It is pretty clear that electricity must play an increasing role in driving cleaner, more efficient systems. The starting point and the initial focus for cities are of course different across the world, but the end point must be the same.

The Context and Challenges for Cities Differ

The challenges facing cities around the global are not the same, and the drivers for action can be very different too.

The energy challenge facing cities can often be set against the stage of the city's development. For many European and North American cities they face replacing ageing infrastructure and the need to integrate new sources of renewable energy generation, while ensuring the grid can support greater electrical demand. In developing cities there is an urgent need for new generation capacity and to improve grid reliability to support fast growing populations, economic development, and the purchasing power of burgeoning middle classes.

Thinking about some of my own experiences working with different cities in different geographies, those differences in priorities and demands are clear. Here are some examples.

In London, most of the power demand for the city is met through the national grid. The power system has decarbonized significantly since the early 1990s. Coal, once the prevalent fuel for electricity generation, has been almost entirely removed from the national grid mix, and renewables have grown considerably, including offshore wind. Since 2010, the capacity of offshore wind has increased in the UK from 1 GW to around 10 GW, whilst construction costs have been driven down by around two-thirds. The UK currently has the largest share of offshore wind capacity globally. However, London faces significant challenges to drive down its emissions from heating and road transport, both of which also create significant challenges around local air pollution. Tackling these will be central to achieving ambitions for a net-zero London. The Mayor of London has introduced initiatives such as the Ultra-Low Emission Zone to tackle the most polluting vehicles, but the transition of road vehicles from combustion engines to low and zero carbon will be critical in reducing air pollution further and achieving net-zero. The UK's grid operator estimates that a 50% share of electric vehicles on the UK's roads would increase electricity demand by 12%.[4] London also needs to switch from its dependency on natural gas typically combusted in boilers for heating and cooking. Gas is responsible for just under 70% of emissions from domestic dwellings in London. Natural gas also causes around a third of greenhouse gas emissions from energy consumption in the commercial and industrial sectors. The large-scale deployment of heat pumps in buildings and supporting low- and zero-carbon heat networks are an essential to London's drive to tackle air pollution and delivery of its own net-zero plan but will increase the electricity demand.

By comparison, the drivers for change in Saudi Arabian cities are very different. Saudi Arabia has set itself a vision for 2030, aiming to transform its economy from its huge reliance on fossil fuels: 90% of the country's exports and 75% of government income are driven by fossil-fuel-related activities. Saudi Arabia is an urban nation, with 82% of people living in its cities, and energy consumption is growing quickly. Electricity consumption is also growing rapidly by about 7.5% a year, with 80% of electricity consumed by buildings, two-thirds of which is for air conditioning. Nearly all electricity is generated from fossil fuels, and the majority (55%) is from old, inefficient, polluting oil-based power plants. As oil has been abundant, there has been little effort to build efficient power plants. Recent oil prices have led to a rethink by the government and a realization that the economy needs to diversify, with greater focus on a more efficient power system and the addition of cleaner power sources. The government's vision sets a target of 9.5 GW of renewable power by 2030 from wind and solar.

---

4 Cities in the Driving Seat; Connected and Autonomous Vehicles in Urban Development, July 2018. Siemens, Munich.

By contrast, China is rapidly urbanizing, with an economy still growing at pace. Since 2000 the urban population has grown from 37% of the total population in 2000 to 59% now, and it will reach 71% by 2030. China is already home to three of the world's twelve cities with populations over 20 million and has 102 cities over a million; that figure will more than double by 2030. China's urban middle class is also expanding rapidly, with their increasing spending power. These trends have tripled energy use since 2000, with fossil fuels still providing the main source of energy (coal 60%, oil 19%). Two-thirds of electrical power is generated by coal. However, China is driving much of the growth in renewable investment globally, with solar and wind growing from 0 to 164 and 134 GW respectively (Figure 9.3).

This rapid economic and urban growth has led to well-documented urban challenges, including air pollution from industry, energy generation, and transport. Electricity consumption is set to double in China's households by 2040, and there are plans for 74 million electric vehicles by 2030. This will make electricity the leading source of final energy consumption by 2040. The vast majority of that electrification will be met by growth of renewable technologies, with 1.2 terawatts (TW) of wind and solar.

We can see from the above examples how cities in countries with good access to fairly clean grids should focus on replacing fossil fuels in transport and heating. This can be seen clearly in London's approach, where the focus is on improving the energy efficiency of buildings, replacing gas heating, and switching out combustion engines. Clearly the

Figure 9.3 Shanghai, solar array, providing highly effective power to its citizens. (*Source:* Aania/Adobe Stock.)

priority for cities with highly inefficient and polluting grids must be to improve the power mix and make buildings more energy efficient to reduce pressure on capacity from growing consumption.

With its heavy reliance still on fossil fuels for heat and a carbon-intensive grid, China must also prioritize the take-up of renewables to meet a fast-growing electrical demand and decarbonize heating systems. The massive growth in electric vehicles while improving local air quality needs to be matched by rapid decarbonization of the grid. The greenhouse emissions saved through switching to electric vehicles can vary greatly depending on the carbon intensity of electricity. If we imagine a city switching 40% of its passenger vehicles from combustion engines to electric, the greenhouse gas emission savings could vary from just 9% to 38% depending on whether the power grid is high carbon (like India) or low carbon (like France). The priority for action is clearly therefore very different in both of these illustrations.

The Role for Cities

The drivers and demands for electric power change from city to city. But equally the political and governance arrangements for cities vary too. Their ability to raise revenues, secure debt, or control taxation can also be very different. Additionally, their ability to legislate or their role in the electricity sector may vary.

While the different operating systems and the context for cities may be very different, cities are being creative and ensuring they are decarbonizing grids and making their power systems more reliant and efficient. There are some actions that all cities can focus on.

Have a Clear and Actionable Plan

Cities need to develop a plan for meeting their energy needs, taking account of their population and economic growth and climate change. This is fundamental to identifying where the challenges and opportunities lie and what should be prioritized. The plan must be supported by clear actions and progress must be reported. Many cities now report their environmental performance via the CDP Cities reporting platform. In 2020, 812 cities were ranked after reporting via the CDP and were awarded a core from A to D based on how they manage, measure, and tackle greenhouse gas emissions and adapt to climate-related risks; 43 cities[5] scored an "A" for their action, including Barcelona, Calgary, Canberra, Cape Town, Hong Kong, London, Paris, San Francisco, The Hague, and Taipei.

Using Their Planning Powers to Shift Towards Zero-Carbon Development

Cities have different functions in relation to urban planning, but many are responsible for developing citywide spatial plans. Essentially, citywide spatial plans allow cities to answer

---

5 https://www.cdp.net/en/articles/media/43-cities-score-an-a-grade-in-new-cities-climate-change-ranking.

two questions: (1) What infrastructure is needed to support the city and where should it go? and (2) How does it connect together and function at the city level?

An effective plan should drive social, economic, and environmental improvements. This has not always been the case, and we can see the effects of poorly planned cities driving unsustainable layouts leading to dependency on cars, congestion, social tensions, and a poor quality of life.

Comparing Barcelona and Atlanta illustrates the impacts of different approaches to planning. They are similar cities in terms of wealth and population (over 5 million people), yet their design is very different. Barcelona is a far more compact city, more than ten times smaller in land area than Atlanta (648 km^2 compared to 7,692 km^2). The results are a transport system in Atlanta that is car dependent and generates six times the greenhouse gas emissions of Barcelona.[6] Urban sprawl also increases the costs of deploying infrastructure, by increasing the distance that essential services such as transport, water, and electricity need to cover. The cost of sprawl to the US economy is estimated at US$1 trillion every year. In the USA, more compact cities spend US$500 on infrastructure per head of population compared to those with greater sprawl which spend US$750 every year. Sprawl also impacts on energy consumption. Equally, a study of 50 cities worldwide estimated that almost 60% of growth in energy consumption is directly related to urban sprawl.[7]

Cities through their spatial plans and development policies can design compact, connected places and drive zero-carbon development, supporting policies that minimize energy through design and support decentralized renewable energy. By driving new developments towards zero-carbon design, cities are driving developers to do more and reducing the scale of the retrofit task they face in the future.

C40 Cities Climate Leadership Group demonstrates the opportunity from the redevelopment of land in their cities to drive compact and zero-carbon planning outcomes. The Reinventing Cities programme is supporting 28 sites available for redevelopment, and winning projects have been announced in 20 cities.[8] The programme is working with developers, architects, engineers, public officials, and people to collaborate and drive new approaches. For each area identified by cities, bidder teams submit proposals for the development which include measures to drive energy efficiency and renewable energy, sustainable building materials, circular economy, water management, and other measures to drive zero-carbon, resilient developments. The Demain in Montreal, Canada will redevelop a 0.9-hectare site to a mixed-use development that will both reduce embodied carbon through its redevelopment and capture carbon during its operation. Proposals include 100% renewable energy and the use of renewable construction materials, increasing urban vegetation by 75%, on-site food production, and food waste recovery on site.

---

6 Bertaud, A., Richardson, A.W., 2004. Transit and density: Atlanta, the United States and Western Europe. Figure 17.2, p. 6. http://courses.washington.edu/gmforum/Readings/Bertaud.pdf.

7 Bourdic, L., Salat, S., Nowacki, C., 2012. Assessing cities: A new system of cross-scale spatial indicators. *Building Research & Information* 40(5), 592–605. doi: 10.1080/09613218.2012.703488.

8 https://www.c40reinventingcities.org.

The Push for More Efficient Buildings

Many cities have ageing, inefficient buildings, which pushes up energy costs and the energy needed to keep them warm or cool depending on the climate conditions. By making buildings more energy efficient, cities can take pressure off the demand for new energy generation, reducing capital costs, while also bringing other benefits such as reducing health impacts from poor indoor air quality and inefficient buildings as well as healthcare costs. Cold homes contribute to poor physical and mental health. Poor respiratory health, asthma, and common mental disorders have been associated with living in damp, cold housing.[9] Creating more efficient workspaces has also been shown to boost the productivity of workforces.[10]

A study into the energy efficiency opportunity of Saudi Arabia's five largest cities illustrated that even limited improvements in building energy-efficient homes and offices through upgrading building fabric, improved lighting, and basic levels of system automation could deliver a 14% reduction of air pollution emissions from buildings and 12% energy savings, equivalent to 121 million barrels of crude oil, while saving 13.7 million tonnes of greenhouse gas emissions. This would also deliver 11,000 jobs in 2030, helping to diversify the economy.[11]

Cities again have different powers in relation to building stock; while some are able to mandate energy efficiency levels in existing and new buildings through legislative powers or are able to establish trading schemes to drive improvements in the built environment, others without the powers to enforce improvements are driving improvements through incentives, technical support, or challenge programmes.

The Tokyo Metropolitan Government established its scheme in 2010, setting fixed reduction targets for greenhouse gases and energy consumption. Its Environmental Master Plan aims to reduce energy consumption by 38% by 2030. The Carbon Reduction Reporting (CRR) programme works with owners and tenants of all commercial and industrial buildings to monitor their greenhouse gas emissions and take action to reduce them. Building owners are encouraged to set individual emissions reductions targets which are publicly disclosed. The programme includes both mandatory reporting for larger buildings and high-energy users and voluntary reporting for smaller businesses. In total around 35,000 facilities report data, showing a reduction of 13.3% in greenhouse gas emissions over the first five years of the scheme, with nearly all of these improvements achieved through more efficient electricity usage.[12]

---

9 Public Health England, 2014. Minimum home temperature thresholds for health in winter: A systematic literature review. Public Health England, London.

10 World Green Building Council. 2014. Health, Wellbeing and Productivity in Offices. World Green Building Council, London.

11 The energy opportunity in Saudi Arabian cities, October 2017. Center of Competence Cities – Urban Development, Siemens, Munich.

12 Urban efficiency II: seven innovative city programmes for existing building energy efficiency, 17 February 2017. C40 Cities and Tokyo Metropolitan Government. https://issuu.com/c40cities/docs/urbanefficiencyii_final_hi_res__1_.

At the other end of the scale, the Retrofit Chicago Energy Challenge encourages, supports, and celebrates voluntary energy efficiency leadership among large commercial, municipal, and privately owned buildings. The challenge works in partnership with multiple stakeholders to drive voluntary action towards a 20% reduction in energy consumption over five years. The programme encourages best practice sharing and showcases ambitious leadership in the city. Currently the challenge includes 100 building participants spanning 5.2 million m^2.[13]

Increasing the Use of Renewables

More and more cities are setting targets for renewable energy. Four of the CDP A-list cities (Canberra, Paris, Minneapolis, and San Francisco) have already committed to a target for all the city energy coming from renewable sources, while Reykjavik is already entirely fuelled by renewable sources.

Cities authorities can use their own spending power and large energy footprints to support renewable energy generation, especially where they operate transport services. The Metropolitan Transportation Authority serves 12 counties in the State of New York, including New York City and two counties in the State of Connecticut, supporting nearly 3 billion rail and bus journeys each year. Its traction power demand for subway and rail services alone is 2.8 terawatt-hours (TWh) every year.

Power Purchase Agreements (PPAs) are becoming an attractive route for many cities looking to take advantage of reducing costs of renewable energy and to clean their energy supply. Essentially, a PPA is a long-term contract that utilizes the guarantee of a large power demand to purchase renewable energy directly from a generator. PPAs provide financial certainty to both city government and project developers, unlocking finance to allow the development of new renewable energy infrastructure. A number of variations of the PPA exist, and many cities in North America and increasingly Europe are looking to drive action through this route. Between 2015 and early 2020, 335 renewable energy deals were signed by US cities, totalling 8.28 GW, equivalent to 1% of the country's total generating capacity (Figure 9.4).[14] A spike can be seen following the federal government's announcement to withdraw from the Paris Agreement in 2017. This, combined with the declining costs of renewable electricity generation help us to understand the growing interest.

Driving Innovation and Switching Energy Systems to Electrical Power

Cities have always successfully driven innovation and new ways of meeting challenges. They play an important and growing role in transitioning energy systems and shifting thinking away from traditional approaches. Meeting the multiple challenges facing cities, including air pollution, climate change, fairness, and economic growth requires a system rethink.

---

13 https://www.chicago.gov/content/dam/city/sites/retrofitchicago/news/NRDC_Retrofit_report_productionREV_0717142.pdf.

14 https://www.wri.org/blog/2020/06/renewable-energy-procurement-cities-counties.

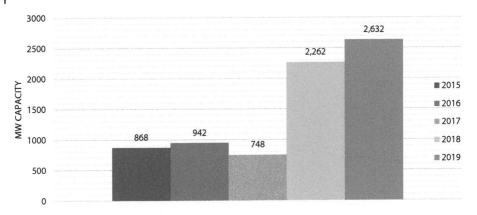

Figure 9.4 Renewables transactions by US local governments. (*Source:* Data from CITY RENEWABLES DEALS, 10 of the Most Noteworthy Local Government Renewables Deals of 2020. www.cityrenewables.org.)

In London, the power system has decarbonized quite rapidly since the 1990s. This has led to an increasing focus on reducing emissions from others sectors, in particular heating. To meet the city's heating demand, the current gas network needs to be decarbonized with strategic low- and zero-carbon heat networks and heat pumps. Secondary sources of heat including waste heat as a by-product of industrial and commercial activities, or heat that exists within the natural environment (in the air, ground, or water) could play an important role. Even under existing market and regulatory conditions in the UK, around 38% of London's heat demand could be met by these sources via heat networks.[15]

London has innovated in this space, looking to identify and heat buildings from secondary heat sources in its efforts to switch from traditional fossil fuel forms of heating. The Bunhill Heat and Power network, located in the London Borough of Islington, is a great example of this opportunity. The network was developed by a range of public and private stakeholders including the London Borough of Islington, Mayor of London, Greater London Authority, Transport for London, UK Power Networks (the city's power distribution network operator), and Ramboll (designers of the energy system). The project was also partly funded by the European Union through the CELSIUS project (Figure 9.5).[16]

The network utilizes the Northern Line, the oldest deep metro line in the world, as its heat source, capturing waste heat from the London Underground to provide heating to the network. It demonstrates the kinds of opportunities that are readily available in our cities but often overlooked. The project brings other benefits too; by displacing gas boilers there is a reduction in NO_2, produced by the combustion of gas; and by capturing trapped heat from the underground network, such systems can help cool the underground tunnel systems.

The project is built on an existing local heat network developed by Islington which provides heat to 850 homes through a combined heat and power (CHP) gas plant. The

15 Greater London Authority, 2013. Secondary Heat Study – London's Zero Carbon Energy Resource. Greater London Authority, London.

16 https://celsiuscity.eu.

Figure 9.5 Bunhill energy centre, illustrating well designed urban integration. (*Source:* Islington Council.)

London Underground's tube ventilation shaft, located nearby in a former underground station, vented air to the atmosphere at temperatures of 18–25°C. By capturing this heat via a heat pump, the temperature is upgraded to 70°C. Heat is captured in the ventilation shaft via a closed-loop water circuit which is put through a compressor to increase the heat of the liquid and the pipes in the circuit. This is then used to provide the hot water through well-insulated pipes connected to buildings in the network.

As part of the project's design, the existing fan extracting heat from the underground was upgraded, allowing it to be reversed. This will allow the district heating network to provide a supply of cooler air to the underground during the warmer summer months.

A new energy centre is located at the disused underground station that houses the ventilation shaft from the tube. The former station now houses a 500-kW ammonia heat pump, driven by the underground's waste heat. The system has been combined with two small gas CHP units to help with peak demand and is connected to the existing heat network, providing heat to an additional 1,300 homes, a primary school, and two leisure centres. The scheme also offers affordable energy to those who need it, cutting costs of existing communal heating systems by around 10% and by half for standalone heating systems in homes.[17]

---

17 https://www.colloide.com/colloide-launch-first-of-its-kind-renewable-energy-project-bunhill-heat-and-power-network.

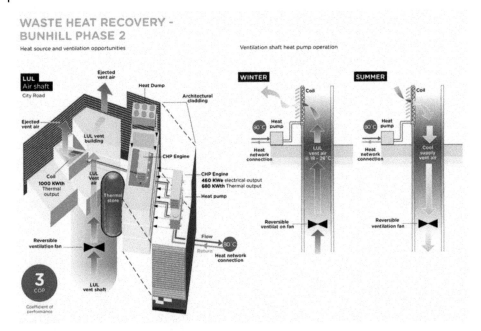

Figure 9.6 How Bunhill phase 2 works. Reversible ventilation fan. (*Source:* Copyright: Ramboll.)

Heat pumps can play an increasing role both in heat networks and in homes to decarbonize heating systems. As they are powered by electricity, they can significantly reduce carbon emissions where the grid mix is cleaner than natural gas. As heat pumps operate at lower temperatures than gas boilers, cities will need to ensure that homes are well insulated and draught proofed to minimize heat loss (Figure 9.6).

Conclusion

To unlock the potential of cities, city governments can play an active and important role in shaping the future of our energy systems. They are much closer to the populations they serve and have a greater understanding of the implications, challenges, and priorities in their areas.

Whilst there is growing action to tackle climate change, more and more institutions are recognizing the need to increase the pace and scale of activity between now and 2030, which the UN describes as the decade of action, before we lock ourselves into the horrendous consequences of runaway climate change.

Cities are demonstrating their leadership on climate change and in many nations driving national governments to do more. National governments must place city government firmly at the heart of their plans and commitments to do more to tackle climate change. That means providing them with devolved powers and funds so they can shape the economies of their regions and ensure a just transition.

The increased role for electricity across city infrastructure, together with its energy, water, transport, and buildings infrastructure are complex and cut across many regulated industries and actors. City governments will need to play an increasing role in shaping energy systems, convening stakeholders, and driving those system changes. National governments need to give them greater powers to ensure that happens.

The energy system needs to become more decentralized, flexible, and integrated within the city. That will require coordination at the city level, a city vision for energy, and the powers to deliver it. In short, decentralizing infrastructure needs to be coupled with decentralizing the political system. Cities need greater powers and funding to drive change. COP27 provides the perfect opportunity for national governments to shift from acknowledging the importance of cities to giving them meaningful powers to drive the changes at the local level that need to happen.

10

The Agile City (Part I) – Introduction

In Chapter 9 Pete highlighted further the uniqueness of cities, the UK, Saudi Arabia, and China all having to look at the problem differently based on the energy mix. Pete points out the importance of measurement and reporting of environmental performance, which Patricia will expand upon in Chapter 19, *The Measured City*. He also wants to see cities use their planning powers and turn exemplar projects into long-term policy. Nicky and Alex, in Chapter 13, *The Habitable City (Part II)*, will show pathways to zero-carbon development. Pete points out the importance of design, comparing Barcelona and Atlanta, in the same way that Nick and Leah showed for Pittsburgh and Stockholm in Chapter 7. The cost of sprawl is significant.

Pete also outlined the need to take extra steps in the energy efficiency of our buildings and utilization of waste heat in the Bunhill case study – we can see this circularity beginning to work.

As Pete talked about powering our cities, Julia now talks about powering our transportation networks and how transport choices will help reduce carbon emissions. As we can see in Figure 10.1, an example from the USA, the carbon abatement potential from renewable energy and from our buildings, and the way we build, makes up around 50% of the abatement potential. Public transit and smart growth, through alternative transit, make up the other 50%.

Julia also discusses the impacts of local air pollution and the inequalities this has led to. In Chapter 21, Hayley expands on this topic in *The Just City*. Using Los Angeles as a model, Julia will move us away from cars and look at how cities must leverage agile decision-making to create diverse transport systems.

Julia argues that there is a "disconnect" between the vision of how people could move and how they actually do move. This is significant because how people move and how easily they move is directly linked to any city's wellbeing. Transport technologies have effectively grown our cities. We have progressed from our feet, to horse-drawn buggies, to the railways, to cars, and to modern public transport, transforming the environment and making cities more accessible. Yet it has become clear that the most accessible cities are ones where people can use many modes of transport and do not have to travel long distances. This accessibility has a direct link to determining the carbon footprint of cities. Transport emissions account for one-third of CO_2 emissions. The more accessible a city is, the less carbon intensive its transport networks are.

The Climate City, First Edition. Edited by Martin Powell.
© 2022 John Wiley & Sons Ltd. Published 2022 by John Wiley & Sons Ltd.

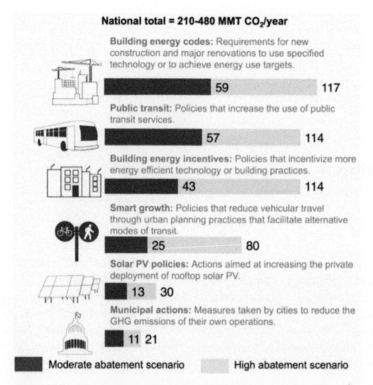

National total = 210–480 MMT CO₂/year

Building energy codes: Requirements for new construction and major renovations to use specified technology or to achieve energy use targets.

59   117

Public transit: Policies that increase the use of public transit services.

57   114

Building energy incentives: Policies that incentivize more energy efficient technology or building practices.

43   114

Smart growth: Policies that reduce vehicular travel through urban planning practices that facilitate alternative modes of transit.

25   80

Solar PV policies: Actions aimed at increasing the private deployment of rooftop solar PV.

13 30

Municipal actions: Measures taken by cities to reduce the GHG emissions of their own operations.

11 21

■ Moderate abatement scenario   ▢ High abatement scenario

Figure 10.1 Estimated national carbon abatement potential (million tonnes of CO_2 per year) of city policy areas. PV, Photovoltaics; GHG, greenhouse gas. (*Source:* Eric O'Shaughnessy, Jenny Heeter, David Keyser, Pieter Gagnon, and Alexandra Aznar, Estimating the National Carbon Abatement Potential of City Policies: A DataDriven Approach, National Renewable Energy Laboratory, 2017. Public Domain.)

Julia discusses ACES (autonomy, connectivity, electrification, and sharing) advancements and how they must respond to a city's current challenges rather than supplanting them if they are to have any hope of succeeding to reduce emissions. She looks at automation technology, hybrid electrical motors, ridesharing, and Mobility as a Service (MaaS), and explores the digitization of analogue infrastructure.

It is clear that the technology a city employs goes hand in hand with its commitment to sustainable values. A transport innovation "success" story, therefore, relies upon the character of the individual city as opposed to the technology itself.

10

The Agile City (Part I)

Julia Thayne DeMordaunt

On 8 April 2020, the air quality in the city of Los Angeles (LA) was cleaner than the top 97 cities around the world, resulting in stunning views of the LA basin ringed by snow-capped mountains and the Pacific Ocean. In fact, LA had just experienced the longest consecutive "good" air days since at least 1980. It is not climate science to figure out why this was: as with the rest of the world, LA was grappling with the novel coronavirus pandemic, or COVID-19, and had imposed a strict "Safer at Home Order", restricting people's movement to a short-list of "essential activities".

Effects of Restricting People's Movement

What were the effects of restricting people's movement? There was a precipitous drop in traffic on LA's notoriously congested streets, with vehicle kilometres travelled (VKT) falling by up to 70%, as the unemployment rate neared 20%. There was a less-precipitous decline in transit ridership, with buses and trains still moving between 34 and 37% of their typical weekly ridership to reveal a transit-dependent population of roughly 300,000 people in a 10 million-person region.[1] Angelinos enjoyed a steady rise in outdoor activity, with pedestrians, cyclists, and joggers jockeying for the newly freed space on pavements and streets (Figure 10.2). And there was an unprecedented level of coordination across city, county, and state governments to meet the new mobility – and immobility – needs of a region that, for decades, had seemingly relied on just one mode of transportation: the car.

The onset of the COVID-19 pandemic in early 2020, and the racial justice protests that followed, marked a tidal-wave change in how city governments view their roles in managing urban transportation networks. Both events exposed what was already becoming clear through heightened awareness of climate change: cities must leverage agile

---

1 Data from the LA Department of Transportation, the LA County Metropolitan Transportation Authority, Apple Maps, and the Transit App, March–June 2020.

The Climate City, First Edition. Edited by Martin Powell.
© 2022 John Wiley & Sons Ltd. Published 2022 by John Wiley & Sons Ltd.

Figure 10.2 Pedestrians on Hollywood Boulevard – space for more modes of active transport. (*Source:* SeanPavonePhoto/Getty Images.)

decision-making to create and operate agile transportation systems that respond to what people need when they need it. Creating such a system means meaningfully engaging with communities to understand underlying challenges and co-author solutions. It means setting clear goals and criteria for how innovation will be evaluated, which acknowledge historical inequities and disinvestment and which offer clear goals to the private sector for technology and partnership. And it means properly resourcing projects with people, money, and – importantly – the political will needed to sustain the projects long term.

This chapter is about how we build those new systems. Taking inspiration throughout from the city of Los Angeles, it starts by exploring the impacts of how the design of cities' transportation networks – looking at how not only people move across cities, but also things – has affected urban economies, the environmental footprint, social inclusion, and even cities' physical shape. It then looks at new transport modes and technologies and how they might be used to supplement or supplant existing urban transportation networks. This chapter acknowledges the current disconnect between the vision and the reality of urban transportation and discusses the multiple reasons why this might be the case. It ends with a note of pragmatic optimism on a way forward, a path based on one fundamental assertion: how people move, and how easily people move, in cities is directly linked to cities' wellbeing.

Impacts of Transportation Networks on Cities

Los Angeles at the beginning of the nineteenth century is practically unrecognizable from the city it has become today. New-to-town developers, taking stock of the immense natural resources of the region, wanted to differentiate LA from their hometowns. In their estimation, LA's "empty lands" should be big and clean, with flat and sprawling developments in stark opposition to the "dense and dirty" vertical neighbourhoods of US cities Chicago and New York City.[2]

With the advent of horse-drawn streetcar lines in LA in the 1860s, the bones of LA sprawl were set. The streetcar could cover 6.5 km in 30 minutes, which allowed residents to live in bucolic locations while still commuting daily to the denser, jobs-rich downtown. Once real-estate developers clued in to the fact that a new streetcar line and station increased residential land value in the suburbs, LA's fate was further sealed. The "City of Iron and Steel", as former *Los Angeles Times* architecture critic Christopher Hawthorne deemed nineteenth-century LA, was characterized by the hub and spoke of an extensive streetcar network, which would pave the way (quite literally) for the freeway networks, vast urban sprawl, and issues with air quality and transportation emissions of the twentieth and early twenty-first centuries.

This relationship between transportation technology, travel times, and sprawl can best be explained by an idea known as the Marchetti Constant, coined by the Italian physicist Cesare Marchetti in 1994.[3] That is, people have always been willing to commute for about a half-hour, one way, from their homes each day. Because of this, the physical size of cities is a function of the speed of the transportation technologies that are available. What transportation technologies are used, what their power source is, and how far they travel are then direct inputs into the overall emissions and air quality of the city.

One can see how the Marchetti Constant has played out over time by looking at the examples of cities developed during ancient versus modern times. Cities like Rome and Barcelona, which developed between 800 BCE and 1700 CE, generally did not grow beyond 3 km in diameter – a distance traversable by the transportation technology available (feet!) in 30 minutes. Between the 1840s and 1950s, the "city on rails", such as Paris, was redefined by above-ground and steam-powered rail lines. Trains could cover up to 16 km in 30 minutes, allowing affluent people to live in less-dense areas on the fringe of cities. When the 1900s saw those rail lines undergrounded, as in London, suddenly the suburbs were available to the working class as well. Even today, these cities maintain the built environment footprint of those transportation technology predecessors, with hyper-dense central areas and still dense suburbs reined in by the edge of the subway and train networks.

The transportation technology that forever changed the built environment of the city, however, was the car. Mass produced starting at the turn of the twentieth century, the motor vehicle met its infrastructure match in the 1950s: the "expressway" unlocked a kit

2 KCET, LA, https://www.kcet.org/shows/lost-la/iron-sprawl-how-trolleys-made-la-a-horizontal-city.

3 City Lab, https://www.bloomberg.com/news/features/2019-08-29/the-commuting-principle-that-shaped-urban-history.

of tools for land development in which travellers could suddenly cover 32 km in 30 minutes. This meant that homes and jobs in an expressway city could spread across more than 3,250 square kilometres – which is (not coincidentally) the current size of Los Angeles County.[4]

This relationship between transportation technologies, travel times, and land-use in cities has a direct impact on their accessibility, carbon footprints, and equity. Cities that developed predominantly during the age of "the walking city", "the streetcar city", or "the city on rails" retain the population, jobs, and dwelling density that those modes necessitate. At some point in their journey, a streetcar, train, subway, or bus rider becomes a pedestrian. Cities that have historically relied on these modes must therefore be dense, with services, work places, and residences clustered around transportation hubs.

Transport by car, meanwhile, does not have that pedestrian requirement built in. In an ideal world (where parking is not a pain!), cars can deliver passengers from point-to-point, from a person's place of work, to their home, to wherever else they may be travelling. Because cars are not mass transport options, nor are they used as frequently as buses or trains, they require a disproportionate amount of space. Cities built for cars must therefore be sprawled to accommodate the space needed for cars, and in order to traverse sprawled cities, cars are needed.

How easily urban dwellers can access work, services, and even leisure is a function of the type of transportation mode they must, or have access to, use, plus the distance they have to travel.[5] The most accessible cities are ones where people can use many modes of travel and do not have to travel far distances. The least accessible ones are those where transportation modes are limited and travel distances are long. By design, then, cities that are dense and leverage networks of shared transportation are more accessible than those that are sprawled and rely on individual transportation, like cars.

Accessibility is also a determining factor in the carbon footprint of cities. The more accessible a city is, the less carbon intensive its transport networks are. This is along a hockey-stick curve, with transport energy use (and therefore emissions) coming down significantly once the population density reaches 12,950 people per square mile (Figure 10.3). So, for LA, a region of 7,000 people per square mile, emissions per capita are much higher than in Mumbai at 76,790 people per square mile or even Bogota at 35,000.[6]

Because land-use and transportation in cities can take time to build or charge, cities can become "locked in" to certain combinations of urban form and transportation networks. Whether or not city decision-makers in LA were aware of it at the time, the choice to build vast road networks designed to facilitate vehicle speed was a choice to invest in higher carbon emissions and poorer air quality. In Figure 10.4, we can see how that decision was then perpetuated.

------

4 KCET, LA, https://www.kcet.org/shows/lost-la/concrete-fantasy-when-southern-californias-freeways-were-new-and-empty.

5 LSE Cities, Transport and Urban Form, http://newclimateeconomy.report/workingpapers/workingpaper/accessibility-in-cities-transport-urban-form.pdf.

6 Urban Observatory, https://www.urbanobservatory.org.

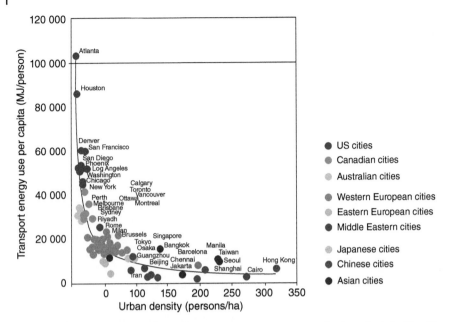

Figure 10.3 Population density and transport energy use per capita for selected cities. (*Source:* WHO (2011). Health in the Green Economy: Health co-benefits of climate change mitigation – Transport sector. Geneva, World Health Organization.)

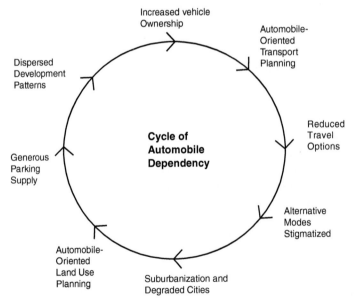

Figure 10.4 Cycle of automobile dependency and sprawl. (*Source:* Evaluating Transportation Land Use Impacts Considering the Impacts, Benefits and Costs of Different Land Use Development Patterns, Todd Litman, 23 April 2021, Victoria Transportation Planning Institute.)

The impacts of inaccessibility in cities are not just environmental. In LA, for example, just as highways were offering added convenience to some, their construction wreaked havoc for others. Despite entreaties from affected communities, engineers – proclaiming that highway construction was more "science" than art – bulldozed neighbourhoods and separated communities from each other. They also permanently impacted people who lived within 150 m of these highways with higher-than-average levels of noise and poor air quality, creating generational inequities that LA is still grappling with today.[7]

The choices that cities made in building urban transportation networks reflected not only the latest mobility technologies of the time but also a set of values. What city-makers learned from the transportation and land-use developments of the nineteenth and twentieth centuries is that there were trade-offs made in prioritizing one method of building over another. Those who invested in dense communities and mass transport technologies have minimized environmental impact and elevated connections to jobs and services. Those who did not – well, as Andersen and Thayne write, "Streets with singular purposes are detrimental to the creation of a coherent urban fabric, promoting inequality and disrupting communities instead."[8]

What Will Change? What Do We Have to Look Forward to?

The legacy of transportation investments and land-use choices leading up to the twenty-first century have paved the way for mobility innovation today. These innovations respond to cities' current challenges: rising populations, a disconnect between homes and jobs, and new ways of purchasing goods, not to mention outdated modes that either lack additional capacity to carry more passengers (think crowded minibuses in Jakarta or queues for the subway in Tokyo; Figure 10.5) or are themselves contributing to rising transportation emissions (like autos in New Delhi or sports utility vehicles (SUVs) in Atlanta; Figure 10.6).

Twenty-first-century transportation technology advancements are generally known by the moniker ACES: autonomy, connectivity, electrification, and sharing. Taken separately, they represent advancements in software and hardware that increase vehicle efficiency, improve safety, reduce emissions, and increase capacity and utilization. Taken together, they have the opportunity to achieve all of these ambitions, plus reduce the amount of space devoted to transport in cities, including shrinking parking places and allowing for flexible use of streets.

Automated vehicles are fitted with sensors and software that enable the vehicle itself to perform some or all of the driving tasks without assistance by a human. There are five levels of automation, ranging from level 0 (no automation) to level 5 (full automation), where the automated driving system can do all driving in all circumstances. Modelling has shown that automated vehicles are safer than their non-automated counterparts, because they can more easily detect and avoid obstacles. They are also more efficient, because they can moderate speed and drive more closely together.[9]

7 The Planning Report, https://www.planningreport.com/2019/04/29/future-streets-and-value-shared-multimodal-transit.

8 Gehl, https://gehlpeople.com/blog/streets-ahead/.

9 National Association of City Transportation Officials, Blueprint for Autonomous Urbanism, https://nacto.org/publication/bau2.

Figure 10.5 Tokyo, Japan in 2015: Commuters in the highly efficient subway system. (*Source:* olaser/Getty Images.)

Figure 10.6 Atlanta SUVs on the interstate heading towards Downtown. (*Source:* Sean Davis/Getty Images.)

If deployed with a people-first mentality, automated vehicles could also have a positive impact on street design. One best-case scenario visualizes a reduction in parking spaces made possible by shared automated vehicles, thereby giving way to new infill buildings. Shared bus and automated vehicle stops allow for wider pavements and active street fronts, and automated metro trains enable higher-density, mixed-use surroundings. Urban dwellers soon forget what a freeway ever was as neighbourhoods are stitched back together when unsafe-by-design roads are remedied.[10]

Automation technology can be applied to just about any type of vehicle with a motor. Since 2015, it has been applied to purpose-built passenger shuttles, heavy-duty trucks, public buses, unmanned aerial systems or drones, electric vertical take-off and landing aircraft, and even dockless electric scooters and pavement automated delivery robots. Whether automation becomes more prevalent will be largely dependent on whether the technology itself can advance to level 4 or level 5 – a level at which companies or mobility operators can operate shared fleets of vehicles that could more efficiently transport goods or people. If it does, there is also the question of whether this will create jobs – or reduce them, as workers are no longer required to perform the driving functions.

Automation technology is often coupled with a hybrid or all-electric motor. All-electric vehicles (EVs) include battery electric vehicles (BEVs) and fuel cell electric vehicles (FCEVs), which run on electricity and recharge from the electrical grid. They are therefore dependent on the electricity mix for exactly how much cleaner they are than internal combustion engine (ICE) vehicles. If, for example, they recharge from a grid powered by 100% renewable energy, then the vehicle has no operations-related greenhouse gas (GHG) emissions.[11]

Nonetheless, the greatest environmental benefit of EVs is that they have zero tailpipe emissions. Their only tailpipe emissions result from the particulate matter generated by friction between tyres and road surfaces and brake wear, amounting to approximately a 50% reduction in local particulate matter between an ICE and an EV. For cities that have consistently appeared at the top of poor air quality indexes, the immediate reduction in emissions and particulate matter offered by EVs has enticed massive public investment in their success. For example, in Shenzhen, China, the government invested in converting its full fleet of 16,000 public diesel-powered buses to all-electric.[12] Now, more than 26 of the largest cities around the world, from Rio de Janeiro, to Santiago, to Seoul, have committed to purchasing all-electric buses by 2025, setting the standard for clean skies in the future.[13]

One study estimates that in a city like LA, transitioning public buses and shared fleets to EVs could reduce citywide GHG emissions by 24% by 2050. The subsequent amount of money saved on gas per year could fund the purchase of more than 2,000 homes, increasing the affordable housing stock by 15%. Excluding chargers for private cars, the number of chargers needed for buses and shared fleets to reach this level of impact could be fewer

10 Andres Sevstuk, https://www.apta.com/wp-content/uploads/20190413_APA_SF.pdf.

11 US Department of Energy, https://www.energy.gov/eere/vehicles/batteries-charging-and-electric-vehicles.

12 Curbed, https://www.curbed.com/2018/5/4/17320838/china-bus-shenzhen-electric-bus-transportation.

13 C40, https://www.c40.org/other/green-and-healthy-streets.

than 3,000 across a city of more than 1,300 square kilometres.[14] Figure 10.7 highlights the transport modes by their average carbon emissions, illustrating the impact of transitioning car users to buses and shared fleets.

Technologies enabling shared fleets allow for mobility operators to offer urban transportation as a menu of options rather than a *prix fixe*. Ridehailing and ridesharing have emerged as powerful tools that could reduce private car ownership and usage and therefore reduce congestion and environmental footprint. With ridehailing, customers use an application to book an on-demand trip operated by a taxi-like system (driven by a human, or not). With ridesharing, customers are paired with other passengers going in the same, or similar, direction as they are, thus increasing the overall efficiency of the trips.

Although ridehailing and ridesharing are not entirely new (as anyone who has ever ridden in a taxi in Marrakech would know), the technologies underpinning ridehailing and ridesharing allow for greater scale and coordination of fleets. Taken to their extremes, the

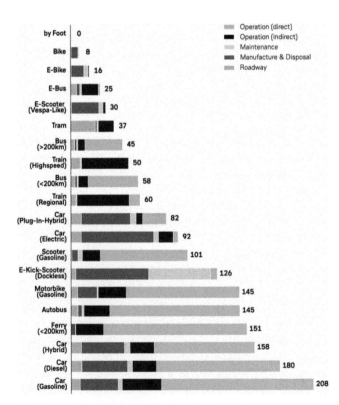

Figure 10.7 Average carbon emissions (in grammes per passenger kilometre) by transport mode. (*Source:* Lufthansa Innovation Hub, Mobitool, BMVI, UBA, Handelsblatt, Statista. travelandmobility.tech.)

---

14 Siemens, Powering the Future of Urban Mobility, https://assets.new.siemens.com/siemens/assets/public.1537788849.d3498f52-de30-4955-bb8e-f7bcccef80ea.powering-the-future-of-urban-mobility-final-092418-v2.pdf.

application interface partnered with algorithms for fleet management, routing, payment, and planning are the building blocks for Mobility as a Service, the Holy Grail of future passenger transportation. Under Mobility as a Service, customers would use a single application to plan, pay for, and schedule their trip across multiple modes – taking a shared bike ride to a transit station, then picking up a microtransit shared shuttle to finish the last leg of their trip to a doctor's appointment.

These new business and operating models that increase utilization per vehicle, whether passenger or goods, coupled with vehicle technologies that reduce vehicle emissions, not only offer ways to draw down transportation emissions, but also promise a seamless, frictionless way of travelling. Whether they do so is contingent on the policies levied by city policymakers to incentivize optimal behaviours. New infrastructure technologies, such as congestion pricing and curb management, are just two examples of policies expressed digitally, which could guide their deployment and lead to positive environmental outcomes.

The digitization of analogue infrastructure creates a digital map of a city's streets, pavements, and curb infrastructure. Then through ingesting multiple sources of data on how people and vehicles utilize infrastructure, systems can allocate physical space dynamically for different uses or set prices for their-use, based on decision-maker input. This enables decision-makers to leverage a "carrot and stick approach" for determining how the public right of way is used. For example, for activities deemed to be for the greater good (such as a neighbourhood block party), there might be no or low costs, whereas with activities harmful to the overall ecosystem (such as an increase in ICE delivery vehicles dropping off parcels) there might be costs determined by time of use and estimated negative impact. Similarly, for EVs driving at non-congested times there might be no toll, whereas heavy-duty, diesel-powered trucks might be restricted to off-peak hours at high toll rates.

These four advancements in transportation technology – autonomy, connectivity, electrification, and sharing – are best viewed in conjunction with the transportation tools policymakers already have, rather than as supplanting them: safe street design, privileging cyclists and pedestrians; reliable public transit, offering low-cost and high-frequency trips; and compact, connected communities, building services and jobs next to housing.

One vision articulates a perfect balance between old and new. Automated vehicles are required to drive slowly, and transit incorporates all of the new bells and whistles: real-time information, flex-route vans, limited ridehail services, and other new mobility technologies, such as dockless electric scooters or bikes. Curbs are reserved for parklets, green infrastructure, bike lanes, and small-scale vendors and kiosks. Despite the chaos that could have been imposed by overused, personally owned automated vehicles and EV chargers installed at every corner, this future vision organizes them.[15] Curb management, congestion pricing, and transit lane enforcement reinforce people- and passenger-first mentalities, in which the public realm is viewed less "like a static, permanent installation and more like a highly flexible puzzle".[16]

---

15 NACTO, https://nacto.org/publication/bau2.

16 Gehl. Op. cit.

The Disconnect between Vision and Reality

If implemented under a best case scenario, the combination of these new transportation technologies, plus the traditional policy and design tools, may right historical wrongs around the effects of transportation on the environment, on accessibility and job creation, and on communities, and set cities on a path for cleaner, safer, more equitable futures. If not, they could contribute to "a physical reinforcement of inequality" based on socioeconomic, racial, ethnic, and other lines.[17] As any city decision-maker reading this chapter knows, the success – or failure – of these transportation innovations lies entirely in how they are designed, deployed, and championed. Only with successive and principled efforts in cities will the dream of clean transportation come to light.

There are a multitude of paths via which a transportation technology project might fail – the most obvious of which is lack of political will. As an example, post-independence from Great Britain, Ghana established an Omnibus Service Authority to develop infrastructure and facilities for bus terminals, buses, and rolling stock, establishing a strong foundation for public transit. However, British colonial policy laid out the city development and road infrastructure for the capital city Accra such that there was a deficit of east–west corridors, limiting mobility between residential and commercial areas and resulting in transit having to compete for road and lane space with the increasing number of personally owned vehicles travelling into the central business district. The challenges with traffic led Accra to try, multiple times, to pilot a bus rapid transit (BRT) network, in which high-speed buses with limited stops would use dedicated lanes along populated east–west routes. The pilots were ultimately unsuccessful, however, because, as one interviewee said, "Whenever there is a change in government, the new government wants to do some 'editing' of previous projects." A winning idea, without the political champions, is ultimately not an implementable one.[18]

Sometimes, there might be a mismatch between public and private investment whereby the underlying economics of the novel transportation mode is unsustainable without public investment. For example, 2019 seemed to be the year of the dockless electric scooter, with hundreds of thousands of them showing up on city streets and pavements seemingly overnight. In short order, and with the onset of the COVID pandemic, scooters suddenly disappeared. Advocates of the mode, pointing to its role in connecting people to transit and replacing car trips, suggested that it was one mode ripe for public investment and intervention.[19]

Other times, the requisite infrastructure investment might not match up to the current technology demand. As another example, despite there being more than 1 million EVs on the road in the USA, in EVs' largest state market, California, they still only represent 1%

17 Ricky Burdett, https://www.ipcig.org/pub/eng/PIF37_A_new_urban_paradigm_pathways_to_sustainable_development.pdf.

18 Transforming Urban Transport – Sub-Saharan Africa.

19 BloombergCityLab,https://www.bloomberg.com/news/articles/2020-04-16/subsidize-e-scooters-cities-should-consider-it.

of all vehicles.[20] Because so few vehicle owners use EVs, private and public investors alike are reluctant to invest in widespread charging infrastructure; however, until there is widespread charging infrastructure, advocates believe that it will be more difficult to convince vehicle owners to purchase EVs: this is the classic "chicken and egg" problem so familiar to anyone trying to advance a cutting-edge solution.

The development of transportation technologies might outpace governments' ability to create mechanisms for oversight. Then the choice is either to shut down the technology's implementation entirely (such as some cities or states have done with pavement automated robots) or to quickly power up dynamic policy tools. One example of such a tool is the mobility data specification, which is a data standard for ingesting information on how transportation technologies are utilizing the public right-of-way.[21] This allows for cities not only to understand the operational implications of new vehicles on roads and pavements, but also to troubleshoot when things go awry – as the complex operational design domain of cities is wont to do.

Fortunately, there are examples of cities that have managed to combine new transportation technologies with the funding, infrastructure, and political will in order to achieve environmental, economic, and social benefits. Mexico's successful implementation of the BRT system, Metrobus, took more than 10 years to advance from idea to implementation. Through a series of mayoral-led efforts, the local government's leaders slowly but steadily used carrots, sticks, and ongoing negotiations to discipline private sector transport providers – who were providing uncoordinated and ultimately inefficient competing bus services – and create room for new transit services. Leaders linked the conversion of the proposed bus route to a larger urban policy vision: one that prioritized improved air quality, advanced environmental sustainability, and enhanced urban redevelopment of distressed areas of the city. They also linked the conversion to jobs. With the consolidation of bus services in the previously jitney-crowded corridors, jobs could have been lost. How leaders structured the consolidation, however, led to additional wealth not only for the private owners of the new bus lines, but also for the drivers, who were paid better, more reliable wages than while providing informal transport services. Ten years later, the Metrobus carries 300 million passengers per year on six lines with 125 km of exclusive bus lanes. Travel times have dropped by 40%, with 30% fewer accidents compared to jitney and independent bus services (Figure 10.8).[22]

The example of Mexico City shows that all is not lost when it comes to leveraging innovation to meet the challenges of cities today. Creating the transportation of the future, however, will require elevating voices not normally heard, pairing public and private sectors together with the public, and institutionalizing new processes that privilege the many over the few.

---

20 Edison Electric Institute, https://www.edisonfoundation.net/-/media/files/iei/publications/iei_eei-ev-forecast-report_nov2018.ashx.

21 GitHub, https://github.com/openmobilityfoundation/mobility-data-specification.

22 Transforming Urban Transport – Latin America, http://www.transformingurbantransport.com/downloads.

Figure 10.8 Thanks to Mexico City's BRT, the safer streets allowed more active transport to thrive. (*Source:* erlucho/Getty Images.)

Conclusion

For pedestrians navigating LA's super blocks, ill-timed intersections, and multilane crossings, it can be difficult not to judge the city. It has made decades of investments in car-based infrastructure and designed streets that prioritize efficiency of car travel over the experience and safety of the pedestrian, and the city is paying for those decisions now. While the temptation for decision-makers might be to wield transportation technologies as panacea for previous generations' infrastructure decisions, the reality is that the cure for poor air quality is not just EVs; it is also pedestrian- and cyclist-oriented street design. The cure for traffic is not autonomous vehicles; it is better land-use planning The cure for emissions is not shared private fleets; it is also investment in public transit.

The promise of the post-COVID period is that most transportation and environmental decisions will not be an "either/or" but a "both/and": cities will have to re-examine, and perhaps rebuild, the systems that were built prior to the pandemic, including our transportation networks. The goal is that they must do it by acknowledging and righting what was done before. That will require humility, empathy, and leadership in equal measure. It will also require innovation, and that includes technology. In a world that was remade in just a few months by a single virus, cities can absolutely remake it again – this time, by and for all of the people that live in them.

The reality of transportation post-COVID is that cities will also have to double down on their values in carrying out projects. This may mean that projects take longer to implement, because they need to be done with communities in a bottom-up approach, rather than to them in a top-down one. Projects may be more expensive, as the process of doing a project will be as important as the project outcome itself. There will be no way to sidestep quantifying impacts to jobs or to businesses: the economy itself needs to be rebuilt in the same measure as our physical environment.

As sociologist Richard Sennett says, "The physical environment should nurture the complexity of identity." The transportation network that emerges from this process should be as diverse as the people and things it carries. If it is, then city makers will have achieved what previous generations had tried to: vibrant and high-quality transportation networks that serve the cities and people that built them.

11

The Agile City (Part II) – Introduction

Julia warned us in Chapter 10 that cities must leverage agile decision-making to create diverse transport systems. The emphasis on diversity, meaning options or choices, is significant, as we can see in Figure 11.1 that despite GDP growth, car use is, in fact, declining. This is not just thanks to diverse forms of public transport, as Julia discussed, but is also due to a big shift towards active transport, which Jonathan discusses in this second part of *The Agile City*.

How would we explain our transport systems to an alien visitor? More importantly, how do our transport systems look from an interplanetary perspective? I'll give you a hint – there is room for improvement. Jonathan shows us that for a city to thrive it must place the individual at the heart of its urban system. In the case of transport systems, walking,

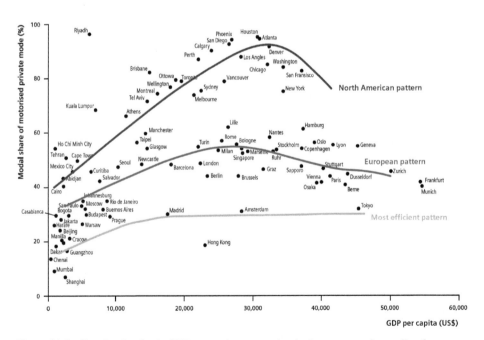

Figure 11.1 Despite the rise in GDP per capita, we are beginning to see a decoupling from car use. (*Source:* LSE Cities.)

The Climate City, First Edition. Edited by Martin Powell.
© 2022 John Wiley & Sons Ltd. Published 2022 by John Wiley & Sons Ltd.

cycling, micro-transit, and all active mobility in cities are the key to reducing urban carbon emissions, a prerequisite to a sustainable city.

We can see from Figure 11.2 that even a slight reduction in the use of vehicular transport can make a big difference in carbon emissions. Jonathan outlines the wide-reaching benefits of active mobility in cities; the environmental advantages seen through reduced air pollution (as Julia spoke about in Chapter 10); and the social benefits, with fewer pedestrian deaths, improved city parking, and increased green space, and the economic link between active transport and increased city retail prices. It is estimated that improving active transportation offers governments a saving opportunity of approximately US$14 trillion or 10% of global GDP.

Cities have a variety of tools at their disposal to make positive change both to disincentivize non-active transport and to incentivize active transport through promoting infrastructure and letting the results unfold. But as Jonathan says, it is important to remember that "active transport and mass transport fit together much better than either of them fit with personal transport". Once again, a collective mentality is required if we are to see real change.

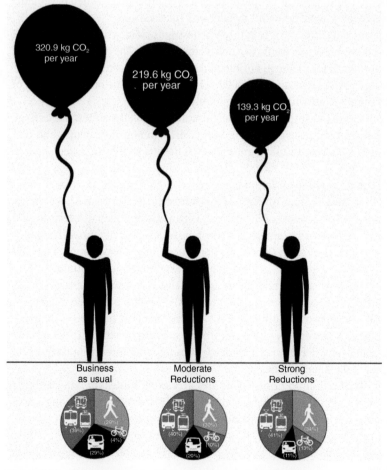

Figure 11.2 Through modal shifts away from car use we will see big carbon reduction gains. (*Source:* Transport Strategy Refresh, Transport, Greenhouse Gas Emissions and Air Quality, Prepared by the Institute for Sensible Transport, April 2018.)

11

The Agile City (Part II)

Jonathan Laski

An alien observing daily life in any major city around the world might be confused as to whether human beings or moving vehicles were in charge. In urban centres around the globe our alien friend would watch as people push past each other on crowded pavements on the margins of streets to avoid being hit by taxis and tuktuks.

In the busiest of our cities the alien might see pedestrians, weary from a long day of work, take a chance and run across traffic to catch a crowded bus. The penalty for missing that bus is a 20-minute wait for the next, during which time the pedestrian must stand on a 0.3-m (1-ft) elevated curb as cars and trucks whiz by, and try to tune out the revving of car and motorcycle engines, all while breathing highly toxic air.

How could we explain to our interplanetary visitor that walking the streets of our most famous cities has inspired centuries of philosophers, artists, and businesspeople and that increasing numbers of people continue to move to cities despite significant safety, health, and environmental risks?

We might try to convince our alien friend and ourselves that hazards to citizens in the form of road congestion, air pollution, and lack of safe infrastructure for people have always been around and are a necessary cost of living in an urban setting. Deep down we would know that this is neither true nor sustainable. Cities of the future can only thrive when they place the individual at the heart of every urban system, including transportation. The ability to move from "point a" to "point b" in a city safely, inexpensively, and without poisoning themselves or their fellow residents is, in fact, a prerequisite to a sustainable city.

Sustainable mobility in cities is key to reducing urban GHG emissions, to creating more equitable, safer, and healthier communities, and to driving support for local businesses. Many of the tools that are needed by city governments are largely within their existing powers. Recent global events have further nudged cities towards recentring people and their free movement at the heart of cities.

The continued costs of failing to provide safe and accessible routes for daily walking and cycling in cities are almost incalculable, but a fair estimate is that improving active transportation offers governments a saving opportunity of approximately US$14 trillion, equivalent to almost 10% of global GDP. In other words, transportation infrastructure which fails to prioritize active mobility carries up to a US$14 trillion annual cost. This is calculated as follows:

The Climate City, First Edition. Edited by Martin Powell.
© 2022 John Wiley & Sons Ltd. Published 2022 by John Wiley & Sons Ltd.

- Cost of deaths due to transportation-related pollution = US$6.3 trillion per year globally.[1]
- Cost of pedestrian and cyclist fatalities = $2 trillion per year globally.[2]
- Cost of congestion = $4.2 trillion per year (in the world's largest 75 cities only).[3]
- Cost of healthcare attributable to lack of physical activity which could be achieved through active transportation = $1.5 trillion per year.[4]

Environmental, Social, and Economic Co-Benefits Cannot Be Overstated

Active transportation and mobility will only become the norm in cities when all stakeholders grasp the environmental, social, and economic co-benefits offered by providing citizens with meaningful opportunities to walk and cycle and, where necessary, to integrate these activities with options such as mass transit and ridesharing.

Environmental Benefits

As of 2016, transportation-related emissions accounted for 24% of global CO_2 emissions. Of the approximately 7 gigatonnes (Gt) emitted annually in the transportation of people and goods, 5 Gt are road-related emissions and half of those are emitted in cities.[5]

Car and truck pollution is a primary driver of city-based smog and air pollution[6] and, as such, is cited as responsible for at least half of the 4.2 million deaths each year occurring as a result of exposure to outdoor air pollution.[7] As seen at the start of the COVID-19 pandemic, when people stop driving in significant numbers, smog and air pollution can improve dramatically in a short amount of time. For example, Delhi, which often has Air

---

1 WHO estimates 4.2 million people die each year as a result of exposure to outdoor air pollution; 50% of air pollution is estimated to be caused by transportation-related emissions. Economists have calculated the average human life to be worth US$3 million, so this equals US$6.3 trillion per year attributable to deaths due to transportation-related emissions.

2 WHO estimates 1.35 million traffic fatalities per year globally, of which 50% are pedestrians and cyclists, so 675,000 deaths per year. Economists have calculated the average human life to be worth US$3 million each, so this equals US$2.1 billion per year attributable to deaths due to traffic fatalities.

3 Global GDP is estimated to be US$142 trillion per year; C40 Cities account for 25% of global GDP (US$35 trillion per year); congestion is estimated to cost as much as 8% of a city's GDP; so that is US$2.8 trillion per year in C40 Cities × 1.5 for smaller and less-congested cities.

4 Daily physical activity, including walking and cycling, reduces the chance of chronic disease by 50% across entire populations. The USA spends US$117 billion annually fighting chronic disease, so there is a $58 billion saving opportunity. The USA represents 4% of the global population, so $58 billion × 25 = $1.5 trillion.

5 World Resources Institute, 2019. Everything you need to know about the fastest-growing source of global emissions: Transport, 16 October 2019. https://www.wri.org/blog/2019/10/everything-you-need-know-about-fastest-growing-source-globalemissions-transport.

6 Rojas-Rueda, D., de Nazelle, A., Andersen, Z.J., et al., 2016. Health impacts of active transportation in Europe. *PLoS ONE* 11(3): e0149990. doi: 10.1371/journal.pone.0149990.

7 WHO, 2021. Air pollution. https://www.who.int/health-topics/air-pollution#tab=tab_1.

Quality Index (AQI) scores above 150, saw 80% reductions in AQI scores in early April 2020 as strict quarantine and work from home measures were introduced in the Indian capital city.[8]

Social Benefits

Viewed from a social lens, active transportation offers significant benefits in the form of increased safety, reduced public healthcare costs, and improved equity.

The World Resources Institute (WRI) has estimated that 1.3 million people die in traffic accidents each year, of which 50% are pedestrians or cyclists.[9] Statistics show that pedestrians and cyclists have a significantly higher chance of surviving a collision with a car or truck if that vehicle is driving slower. So, simply put, fewer personal vehicles on the roads, driving slower will reduce pedestrian deaths.

In addition to the social costs of pedestrian and cyclist fatalities in collisions with cars are the physical healthcare and mental healthcare cost savings associated with active transportation for personal journeys as well as journeys to and from work. A 2008 study by the New Zealand Transport Agency estimated government savings of US$3.00 per kilometre walked and US$1.50 per kilometre cycled due to reduced morbidity and mortality, and avoided healthcare costs.[10] The city of Copenhagen's most recent "Bicycle Account 2018" confirms the economic cost savings to both city and national governments of approximately 10 DKK (US$1.50) for every kilometre cycled instead of driven, for a total annual cost–benefit gain of 467 million DKK (US$75 million) annually.[11]

From a social equity perspective, prioritizing roads for cars and their parking needs inherently favours certain demographics of citizens – namely, those wealthy enough to own or lease a car and to drive on a daily basis. Countless studies show that people who are racialized or economically marginalized, or who have special needs are disproportionately higher users of public transit. For those who are physically able, providing safe means of active mobility can dramatically improve social and economic opportunities. Even where public transit is reliable for such people, fares might be cost-prohibitive, while a combination of walking and cycling could save them a significant amount of money. Where public transit is less reliable, having a safe route to cycle could be the difference between people making it to a job on time or not. Even when city governments invest in public transit, if associated policies are not adopted to make such public transit fast, affordable, reliable, and safe, such investments do not provide marginalized communities with the service they need.

---

8 *New York Times*, 2020. India savors a rare upside to coronovirus: Clean air. https://www.nytimes.com/2020/04/08/world/asia/india-pollution-coronavirus.html.

9 World Resources Institute, 2014. Is your city safer by design? https://www.slideshare.net/embarqnetwork/is-your-city-safer-by-design-holger-dalkmann-embarq-transforming-transportation-2014-embarq-the-world-bank.

10 NZ Transport Agency, 2008. Valuing the health benefits of active transport modes.

11 City of Copenhagen, 2018. City of cyclists. https://urbandevelopmentcph.kk.dk.

We would also be remiss if we talked about active transit in cities and failed to mention parking. The average car is parked 95% of the time. For drivers who live or work in cities, this means their car is taking up valuable real-estate. It is estimated that there are 1 billion parking spaces in the USA alone,[12] with some cities having ratios of up to 27 parking spots for every citizen.[13] As the infographic in Figure 11.3 from the city of San Francisco shows, reimagining how limited road and parking space is used opens up the potential for increased social infrastructure such as affordable housing and green space.

The potential impact of reallocating both on-street and off-street parking for use of all citizens, instead of only those owning cars, might be the single most impactful action that city governments could take, at least in the developed world. Implementing such changes will require a combined policy effort, starting with reducing or eliminating mandatory parking minimums for new development and pricing parking based on its true cost, rather than a random amount of what drivers are used to paying.

Finally, there is a further equity component to the kind of active transportation options that are offered in cities. A 2008 New York City Department of City Planning Bicycle Survey found that women make up 45% of people who use cycling as their usual mode of transportation but only 15% of bicyclists in bike lanes that have no physical separation from traffic.[14] This means that when cities choose to install bike lanes but don't go the extra step and make them protected bike lanes, they are essentially investing in infrastructure that will be significantly better used by their male citizens as opposed to female citizens.

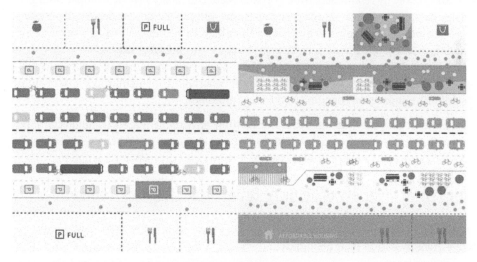

Figure 11.3 San Francisco has an audacious plan to reclaim land from cars. (*Source:* Plumer, B., 2016. Cars take up way too much space in cities. New technology could change that. https://www.vox.com/a/new-economy-future/cars-cities-technologies.)

---

12 Plumer, B., 2016. Cars take up way too much space in cities. New technology could change that. https://www.vox.com/a/new-economy-future/cars-cities-technologies.

13 Lindeman, T., 2018. In some US cities, there are over ten times more parking spaces than households. https://www.vice.com/en_us/article/d3epmm/parking-spots-outnumber-homes-us-cities.

14 Transportation Alternatives, 2012. East village shoppers study. https://www.transalt.org/reports.

Economic Benefits

While the economic co-benefits of increased active transportation are listed here after environmental and social, we recognize that for many policymakers and business owners, this is the consideration that will matter most.

In fact, while writing this chapter I happened to be having dinner with a friend currently serving as the chair of the business improvement association for one of Toronto's most established retail districts. While discussing the daunting circumstances faced by high-street retailers currently attempting to stay afloat during the COVID crisis, my friend remarked that the city "better not try to put cycling lanes" on the street. I proceeded to insist that study after study now shows that citizens who arrive to a shopping district by foot or bicycle support these retail establishments to a greater extent than drivers and those who travel by mass transit.

A brief summary of some US-based research may be helpful:

- A 2014 report from the New York City Department of Transportation showed streets receiving a protected bike lane saw a greater increase in retail sales compared to equivalent streets.[15]
- A 2012 survey of 420 pedestrians in New York City's East Village revealed that while 55% of respondents walk or bike as their usual means of transportation, this group accounts for 95% of retail dollars spent in the study areas.[16]
- A 2012 survey by researchers from Portland State University revealed similar findings. They summarized that "patrons who arrive by automobile do not necessarily convey greater monetary benefits to businesses than bicyclists, transit users, or pedestrians" and that "this finding is contrary to what business owners often believe".[17]
- Finally, a study released in 2020 studied 14 transportation corridors in six cities and found that street improvements for active transportation "had either positive or non-significant impacts on corridor employment and sales".[18]

Assuming the case has been made for the multiple environmental, social, and economic co-benefits of active transportation in cities, the next question is whether cities reasonably have the carrots and sticks to nudge citizens towards active transportation options.

Cities Have a Variety of Tools to Bring About Positive Change

Cities are fortunate to hold many of the levers of power required to significantly improve active mobility within their boundaries.

---

15 New York City Department of Transportation, 2014. Protected bicycle lanes in NYC. http://www.nyc.gov/html/dot/downloads/pdf/2014-09-03-bicycle-path-data-analysis.pdf.

16 Transportation Alternatives, 2012. East village shoppers study. https://www.transalt.org/reports.

17 Clifton, K., Morrissey, S., Ritter, C., 2012. Exploring the relationship between consumer behavior and mode choice. http://kellyjclifton.com/Research/EconImpactsofBicycling/TRN_280_CliftonMorrissey&Ritter_pp26-32.pdf.

18 National Street Improvement Study, 2020. Economic impacts of bicycle and pedestrian street improvements. https://wsd-pfb-sparkinfluence.s3.amazonaws.com/uploads/2020/03/Economic-Impacts-of-Street-Improvements-summary-report.pdf.

The first sets of important tools are those designed and implemented to disincentivize non-active transport. These include regulatory mechanisms such as congestion charges and low or zero emission zones. While never perfect in their design (some people are wealthy enough to pay the congestion charge daily without a second thought) or implementation (low emission zones promote electric cars, but such cars can still be unsafe for pedestrians and cyclists), there are now 250 cities in Europe with low and/or zero emission zones.[19]

Analyses of low emission zones show positive impacts, though they range from small improvements up to 32% reduction in NO_2 emissions in Madrid following the implementation of their hybrid low and zero emission zone.[20] Best practice implementation of low and zero emission zones is for cities to consider a combination of exemptions (generally for people with physical disabilities requiring a personal vehicle), subsidies (financial incentives to help businesses convert company vehicles and fleets), and other creative incentives. For example, in the Brussels capital region, the government offers a Bruxell'Air mobility package whereby citizens willing to deregister their vehicle licence plate and destroy their vehicle are eligible to receive free public transport for up to two years.[21] The challenge now is to scale up implementation on a variety of levels. Low emission zones must become zero emission zones in the next decade, and the rest of the world must quickly catch up to Europe in implementing these important disincentivization schemes.

A related "tool" is the driver experience. Studies show that of all the factors that may entice a driver to leave their car at home (or better yet, to not buy a car in the first place) it is not the financial cost of driving a car but rather the driving experience. Copenhagen unintentionally demonstrated this to some observers during the construction of its "*Cityringen*" (City Circle Line) between 2012 and 2018. As summarized in one article: "We don't like the Metro but ... right now, we love the Metro construction ... Copenhagen has 17 Metro stations under construction and this is having a massive effect on mobility patterns in the city. Driving is a pain in the ass ...."[22]

Cities around the world should bear in mind that active and mass transit infrastructure projects which inconvenience drivers not only are building infrastructure of the future but also will have the effect of nudging drivers to form new habits because their experience will be quite poor (Figure 11.4).

On the other side of the equation, city, regional, and national levels of government have many tools to incentivize active transport. These include financial incentives such as the UK's Cyclescheme,[23] which offers employees cycling to work discounts on bike and bike accessories and allows them to pay for such equipment through payroll deductions and not put any money up front. Since data show that perceptions of safety also impact a citizen's willingness to make cycling part of their daily routine, some

19 European Federation for Transport and Environment, 2019. Low emissions zones are a success – But they must now move to zero-emission mobility. https://www.transportenvironment.org.

20 Ecologistas en acción, 2019. Madrid: Air quality. https://www.ecologistasenaccion.org/114930/balance-delfuncionamiento-de-madrid-central.

21 Low Emission Zone Brussels. https://lez.brussels/mytax/en/alternatives?tab=primes#.

22 Copenhagenize.com, 2014. The greatest urban experiment right now. http://www.copenhagenize.com/2014/07/the-greatest-urban-experiment-right-now.html.http://www.copenhagenize.com/2014/07/the-greatest-urban-experiment-right-now.html

23 Cyclescheme. https://www.cyclescheme.co.uk/.

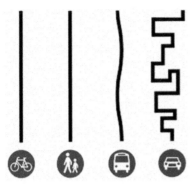

Figure 11.4 The surest way to reduce driving mode share is to make it significantly less convenient than the alternatives. (*Source:* Copenhagenize.com, 2014. The greatest urban experiment right now. http://www.copenhagenize.com/2014/07/the-greatest-urban-experiment-right-now.html.)

authorities are partnering with local cycling advocacy groups to deliver free coaching and training. In one example, the Brussels Capital Region offers all of the following: free training session for new riders; a personal coach to accompany cyclists on their route to help navigate and identify safety considerations; and the programme will even lend people a bike to try out the experience before investing in their own bicycle or in a cycle hire scheme.[24]

From an infrastructure perspective, active transportation is a clear example of the "build it and they will come" phenomenon. Data from Copenhagen reveal that as the number of total kilometres cycled by citizens doubled between 1995 and 2015, the number of deaths and injuries halved.[25] Thus, another "tool" to promote active transportation is piloting the infrastructure and letting the results speak for themselves. Cities around the world should recognize that building new car and even mass transit infrastructure should only be prioritized once all investments in active mobility have been made. For example, use of on- and off-peak pricing is currently underused as a means of flattening peaks in mass transit demand. Further aligning mass transit and active transit will mean that both become invaluable parts of citizen journeys, all at the expense of trips made with personal vehicles.

Finally, it is worthwhile acknowledging that even a focus on active transportation on its own will not serve every citizen, in every city, in all kinds of climatic and topographical scenarios. In this regard, it is vital to remember that active transport and mass transport fit together much better than either of them fit with personal transport. For example, bicycles can be loaded onto public transit buses and trams for first-mile and last-mile journeys. Even electrified autonomous car sharing, a service that is likely to be on our streets by 2030, should be designed to integrate with active transportation as opposed to cannibalizing it.

24 Bike Experience Brussels. https://bikeexperience.brussels/en/home.

25 World Economic Forum, 2018. What makes Copenhagen the world's most bike-friendly city? https://www.weforum.org/agenda/2018/10/what-makes-copenhagen-the-worlds-most-bike-friendly-city.

Recent Global Events Have Further Nudged Cities Towards Recentring People and Their Free Mobility at the Heart of Cities

Writing this chapter in 2019 or the first few months of 2020 would have carried significant urgency – as millions of people move into cities each year, the costs of the status quo prioritization of car-based transport over active transport continue to grow in the form of traffic fatalities and injuries, increased healthcare costs, lost productivity due to congestion, and more.

It is impossible to write this chapter in July 2020 without acknowledging the impact that recent global events have had on urban societies and the role that active transportation should play in the future of cities.

The first notable event is the COVID-19 pandemic and its impact on city life. Citizen usage of public transit in some cities has dropped by as much as 90%.[26] Transportation agencies are now faced with some impossible choices such as raising transit fares while simultaneously offering reduced service and the threat of health risks. At the same time there are many global cities with relatively high public transit usage where "opening the economy" without a fully functioning transit system simply won't work. Cities around the world have rightly focused on quickly expanding active transportation routes and encouraging residents to seek out their basic needs in their own neighbourhoods. In Paris, Mayor Anne Hidalgo was already championing the idea of the "15-minute city" in her re-election campaign – this is the concept that every citizen in a city should be able to meet their basic needs such as groceries, childcare, and other essential services within a 15-minute walk or bike ride. Following her re-election and due to COVID-19, Hidalgo quickly created an additional 50 km of bike lanes for Parisians afraid of taking public transit and she has since made this new infrastructure permanent.[27] In my hometown of Toronto, the city council approved the creation and extension of 25 km of cycle tracks in a matter of weeks – approvals that under normal circumstances would have taken years.[28]

As many cities started adjusting their mobility offerings for citizens impacted by COVID-19, a second notable social issue came to the forefront in the form of mass protests for racial equality. While protests for racial equality and transportation systems may not seem to have a lot in common, a deeper look reveals systemic bias in transportation planning and infrastructure. As already noted, mobility policies in cities currently disproportionately favour those wealthy enough to drive a car. The allocation of finite resources and space for cars therefore disproportionately impacts poorer communities, which for a variety of reasons generally include people of colour and other marginalized groups. As cities

---

26 *New York Times*, 2020. The worst-case scenario: New York's subway faces its biggest crisis. https://www.nytimes.com/2020/04/20/nyregion/nyc-mta-subway-coronavirus.html.

27 RFI, 2020. Paris's temporary bike lanes to become permanent after Hidalgo's re-election. https://www.rfi.fr/en/france/20200701-paris-temporary-bike-lanes-to-become-permanent-after-hidalgo-re-election-mayor-green-pollution-cars.

28 City of Toronto, 2020. COVID-19: ActiveTO – Expanding the cycling network. https://www.toronto.ca/home/covid-19/covid-19-protect-yourself-others/covid-19-reduce-virus-spread/covid-19-activeto/covid-19-activeto-expanding-the-cycling-network.

continue to prioritize parking over the provision of services such as affordable housing, poorer communities are pushed outside of downtown cores and, therefore, even further away from established cycling and public transit options.

City action to reprioritize active mobility is an attempt to level the playing field and to acknowledge that the right of all citizens to get to a place of employment, to an appointment, or to pick up groceries exists.

Conclusion

Transportation-related emissions are the fastest growing sector of GHG emissions globally. While there is reason to be optimistic about increased sales of electric cars and the possibility of "transportation-as-a-service" providing an autonomous electric vehicle service for middle-class people in many of the world's cities, there is, both literally and figuratively, nothing quite like a walk or bicycle ride.

Heeding the call to action laid out in this chapter can have the following impacts in thousands of the world's largest cities:

- Cities could reduce their emissions between 5 and 15%.
- Governments will save billions in healthcare costs.
- Hundreds of thousands of people will be saved from pollution-related death and disease.
- Billions will be saved annually in healthcare costs and billions more will be recouped in productivity from reduced congestion.
- Most importantly, lives will be improved through safety, accessibility, support for local business, and increased equity of opportunity

As I have demonstrated in this chapter, prioritizing walking and cycling provides related and significant co-benefits for individuals, local businesses, and governments at all levels. Even more importantly, active transportation is safer and more equitable. If the next hundred years of cities recentres the pedestrian, the benefits to people are almost incalculable.

12

The Habitable City (Part I) – Introduction

As we've read in Chapter 11, Jonathan is pushing for more active transport. If we can visualize a city centre that can sustain the same economic activity, without cars, we can begin to plan for truly people-centric places. In Copenhagen, the success of the high modal share of cycling is as much a mindset as it is providing the infrastructure for more sustainable forms of transport (Figure 12.1).

Anne Hildago, Mayor of Paris, has also set out a bold plan for Paris and the Champs-Élysées (Figures 12.2 and 12.3), and you will see in Hayley's case study of Seoul in Chapter 21, *The Just City*, something just as bold that has already been achieved.

Jonathan built a case for calculating the externalities associated with failing to consider active mobility and began to touch on the value of green spaces. As well as Hayley's case

Figure 12.1 The Copenhagen mindset to travel by bicycle is not impeded by winter weather. (*Source:* GBPerkins/Getty Images.)

The Climate City, First Edition. Edited by Martin Powell.
© 2022 John Wiley & Sons Ltd. Published 2022 by John Wiley & Sons Ltd.

Figure 12.2   The Mayor's vision for the greening of Paris and the Champs Élysées. (*Source:* Shamil/Adobe Stock.)

Figure 12.3   The Mayor wants to reduce car use and increase public spaces across the centre of Paris. (*Source:* PCA-STREAM.)

study in *The Just City*, Adam shows the value in Chapter 24, *The Adapted City*, Peter in Chapter 25, *The Open City*, and then Carlo, who reminds us that we are shared inhabitants of our cities and our planet, in Chapter 26, *The Natural City*.

Both Julia and Jonathan have tackled the topic of moving people around our cities. Figure 12.4 is a great example of photo-journalism from the Münster planning office, which has helped shift our thinking in how we use street space. Electric cars and

Figure 12.4 Poster in city of Münster Planning Office, August 2001. (*Source:* Press Office, City of Münster, Germany.)

autonomous cars have the potential to bring great benefits, but they both sit in the box on the left! The next topic that also defines a functioning city is the relationship between transport and housing. This topic has been handled in two chapters.

Olivia sets out housing in developing countries and informal settlements, and Nicky and Alex discuss housing in developed countries.

Affordable and safe housing is a global issue affecting nearly every city on the planet. The world's current housing deficit is approximately at 1 billion units, and by 2030 it is predicted that 60% of the world will live in urban areas. The burden of housing will fall primarily on cities, as we can see in the population density log in Figure 12.5.

Cities are already facing the major obstacle in doing so: lack of land. Olivia identifies and lays out a series of steps all cities must undertake in order to combat this challenge and the many other problems facing urban development, whether it be overcrowding, green upgrades, structural integrity, or disaster mitigation. She addresses:

- the quantitative housing deficit (building vertically, building green, building resilient);
- the qualitative housing deficit (vulnerabilities, green retrofits, disaster mitigating retrofits).

She provides detailed case studies addressing these issues such as: the Ethiopian Condominium programme; the green housing in Mongolia initiative; EDGE Green Building Certifications; the Global Program for Resilient Housing; the Alaska Housing Finance Corporation; and iBuild and Miyamoto.

The creation of affordable and sustainable housing poses a risk to urban sustainability in contributing to the significant increase in CO_2 emissions. Addressing the global housing deficit will require us to balance two important goals – building millions of new

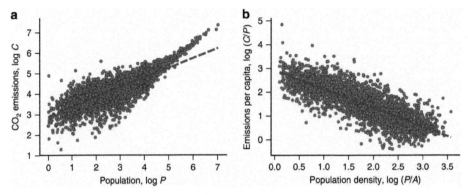

Figure 12.5 Conventional approaches for investigating urban emissions. **a** Urban scaling: scaling relationship between CO_2 emissions (C) and population size (P). Dashed line represents power-law fit with exponent $\beta = 0.48 \pm 0.01$. We observe that this model underestimates emissions for large population sizes. **b** Per capita density scaling: scaling law between CO_2 emissions per capita (C/P) and population density (P/A). Dashed line is a power-law fit with exponent $\alpha = -0.79 \pm 0.01$. In both plots, each dot is associated with an urban unit obtained from the city clustering algorithm and all quantities are expressed in base-10 logarithmic scale. Emissions are measured in tonnes of CO_2, population in raw counts, and area in square kilometres. (*Source:* Ribeiro, H.V., Rybski, D., Kropp, J.P. 2019. Effects of changing population or density on urban carbon dioxide emissions. *Nature Communications* 10, 3204. https://doi.org/10.1038/s41467-019-11184-y.)

housing units, and also, simultaneously, upgrading our current housing stock. As Olivia says, "one cannot be done without the other". We cannot afford – financially and environmentally – to build homes just to tear them down a few decades later. The homes we build today should last for generations to come.

12

The Habitable City (Part I)

Olivia Nielsen

Cities are places where people work, study, and travel, and, most importantly, where people live. Yet access to affordable and adequate housing is a global issue affecting almost every city on the planet. The world's current housing deficit is estimated at 1 billion units, and the UN predicts that by 2030 this deficit will affect close to 40% of the global population.[1] By then, 60% of the world will live in urban areas, and the burden of housing these growing populations will fall primarily on cities. Cities must find ways to house 200,000 newcomers pouring in from the countryside every day. Yet cities face the hardest obstacle in doing so: lack of land. Figure 12.6 highlights the trend towards increased inadequate housing conditions.

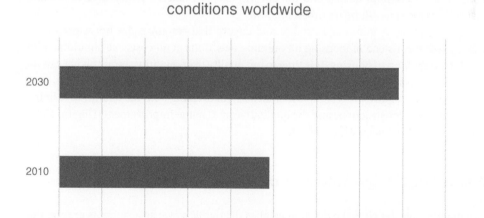

Figure 12.6   Number of households with inadequate housing conditions worldwide. (*Source:* United Nations, 2020. https://unhabitat.org/sites/default/files/2020/10/wcr_2020_report.pdf.

---

1 https://unhabitat.org/topic/housing.

The Climate City, First Edition. Edited by Martin Powell.
© 2022 John Wiley & Sons Ltd. Published 2022 by John Wiley & Sons Ltd.

The consequences of the global housing crisis have recently been exposed by the COVID-19 pandemic. Poor housing quality and overcrowded living conditions have increased mortality rates as billions of people have been asked to shelter-in-place. Beyond the pandemic, inadequate living conditions have been linked to a number of other issues. From disease, to stunted education, to mental health issues, poor-quality housing affects households and communities in a myriad of ways. Despite these disastrous consequences, the global housing crisis is expected to increase in coming decades.

In response to this crisis, many governments around the world have announced ambitious plans to develop millions of new housing units. In fact, 41 out of 54 countries in Africa have announced large projects of over 10,000 units. Nigeria alone has announced its intention to build 1 million affordable housing units every single year. "Housing for all" has become a motto cited by politicians worldwide, but with few successes. While India and Ethiopia have been relatively successful in rising to this challenge, most countries in Sub-Saharan Africa have failed to meet their self-declared objectives, causing frustration among their populations. There are many reasons for these failures. From institutional obstacles to lack of available land and financing – a multitude of barriers exist.

Those governments that do succeed in building at scale face additional problems, as new developments can quickly become slums without proper investments, maintenance, and connectivity to the larger urban framework. This has been the case in the USA, where public housing projects have been destroyed mere decades after their construction.[2]

Mexico, for example, succeeded in building millions of new units in its cities' peripheries only to see urban sprawl and traffic increase along with its programmes. Today, millions of houses remain unsold and the government has taken a policy U-turn, seeking instead to refocus its efforts on urban renewal.

It is time to learn from our mistakes and ensure that resources are not wasted elsewhere. Perhaps instead of focusing on building new units at all costs, we should seek to build the right kind of housing. It is time to put quality over quantity, as our planet can no longer afford to see its resources squandered on buildings that are not meant to last.

Doing so will require programmes that focus on addressing both the quantitative housing deficit (new construction) and the qualitative one (home improvement) (Figure 12.7). Cities will be at the heart of this effort.

Addressing the Quantitative Housing Deficit

The quantitative deficit is the one that catches the public's eye: governments around the world promise to fight housing inequality by building new homes at scale. Building a billion new units would require a considerable amount of resources in the construction and life of the home. In the USA, more than 50% of all buildings are residential and consume a large amount of resources. Building a billion new homes will need to be done

---

2 https://www.nytimes.com/2018/02/06/magazine/the-towers-came-down-and-with-them-the-promise-of-public-housing.html.

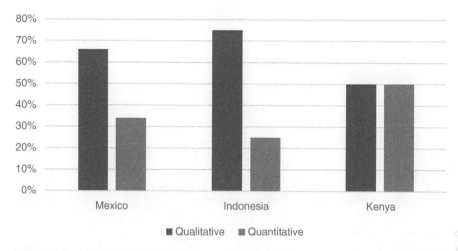

Figure 12.7 Housing deficit share by country. (*Source*: Created by Olivia Nielsen. World Bank, 2020. Mexico: https://documents1.worldbank.org/curated/en/605571574937088827/pdf/synthesis-report.pdf. Indonesia: https://documents1.worldbank.org/curated/en/299581479368540859/pdf/pid-prin t-p154948-11-17-2016-1479368535904.pdf.)

strategically if we wish to meet the needs of both our people and our planet. Yet when the cost of new homes is already a challenge, how can we hope to promote green building practices in emerging markets?

Build Vertically

Because land is scarce in cities, it has become the major bottleneck to housing affordability. In order to overcome it, we have no choice but to move upwards. From Shanghai to Manhattan, we have become accustomed to living in vertical structures. Yet many countries still lack the legal backbone that enables us to build vertically: a condominium law.

From Haiti to Guinea in West Africa, the condominium law remains either non-existent or difficult to implement, meaning that one entity would need to own all units of a building, as land cannot be subdivided vertically. Cultural preferences may be behind this reluctance to move upwards as people have long been attached to the land on which they live. Yet urbanization is forcing us to leave our preferences behind and adopt means of living that can seem uncomfortable at first.

As more and more households move to the cities and seek shelter, overcrowded informal settlements or "slums" tend to grow. Overcrowding[3] has been linked to public health hazards during the COVID-19 pandemic and can lead to major mental health

---

3 https://www.insidehousing.co.uk/insight/the-housing-pandemic-four-graphs-showing-the-link-between-covid-19-deaths-and-the-housing-crisis-66562.

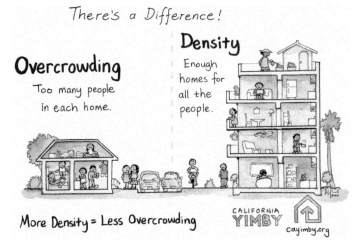

Figure 12.8 The balance between overcrowding and density. (*Source:* Alfred Twu / California YIMBY, www.cayimby.org.)

concerns as well. Vertical density can decrease overcrowding while minimizing land-use. Figure 12.8 highlights the difference between overcrowding and density.

Yet density needs to be done carefully in order to avoid creating mere dormitories, which have been referred to as "warehouses for the poor".[4] Building vertically also comes with a set of problems linked to the need to install elevators – often a wasted investment in cities where maintenance is not prioritized.

Gentle density should be considered instead, with walk-up buildings, shops, and restaurants on the ground level and investments in infrastructure, to ensure that a community is formed around the houses.

Box 12.1 describes the Ethiopian condominium programme.

Box 12.1 Case study: Ethiopian condominium programme

Since 2010, Ethiopia has engaged in one of the largest housing construction schemes in Sub-Saharan Africa. The government has led an ambitious drive to build hundreds of thousands of affordable condominium units in the heart of Addis Ababa, the capital city. By going vertical, the programme was able to work with scarce land and drive prices down. Homes sell for as low as US$8,000 – an astonishingly low price for the market. This has had the effect of changing cultural norms away from single-family units and has changed the city's skyline forever (Box Figure 12.1). The units are in such high demand that families need to sign up for a lottery system before they are given the opportunity to buy one.

---

4 https://www.theguardian.com/commentisfree/2019/aug/22/we-need-social-housing-not-warehouses-for-homeless-children.

Box Figure 12.1 Ethiopian condominiums have changed the urban skyline. (*Source*: Dereje/Shutterstock.)

The government has achieved this success by making well-located and transport-accessible land available for the project at close to no cost. By paying for construction, the government was able to bypass developer costs. Contractors were hired directly, and traditional masonry was used in order to create a great number of job opportunities. This is an important element, as it is often assumed that new construction technologies that promote faster and more efficient construction bring down housing costs. While this might be true in developed economies where labour costs are high, Sub-Saharan Africa has high unemployment rates and low wages, making traditional construction techniques more important for governments seeking to both boost employment and address the housing crisis.

Despite its numerous successes, the programme has encountered a number of issues linked to quality and sustainability. Large-scale high-rise projects eventually run into maintenance issues. If properties are not properly maintained, they quickly look rundown. The drive to bring down construction costs may have affected quality issues and accelerated decline. Yet if homes are to last several generations, they need to be not only built to last but also maintained accordingly. Addressing the high global housing deficit requires that each unit be built with sustainability in mind.

Finally, the programme has relied too heavily on government funding and, by bypassing the private sector, has not found a long-term strategy to fund its continuation. As government funding dries up, finding financial sustainability by working with private banks, developers, and investors becomes critical. Public–private partnerships are being considered in order to best leverage the government's main resource (land) with the private sector's (funding).

Build Green

Box 12.2 describes the financing of green housing in Mongolia.

Box 12.2   Case study: Green housing in Mongolia[5]

For half of the year, Ulaanbaatar, the capital of Mongolia, is one of the most polluted cities in the world. Winter months are met with extreme cold conditions, forcing the city's informal settlements to burn coal to heat their homes. These homes are called "gers" and consist of portable, circular structures made out of wood and canvas and insulated with felt (Box Figure 12.2). Poor insulation and ventilation means that households need to burn a lot of coal, facing health risks in the process. The coal used by households is a great source of air pollution in the winter. The situation is so bad that children are sometimes sent to the countryside in the winter in order to avoid the negative health effects of permanent fog. Today, close to 60% of the city lives in these informal settlements. (*Source:* Based on Ulaanbaatar Green Affordable Housing and Resilient Urban Renewal Sector Project: Report and Recommendation of the President, https://www.adb.org/projects/49169-002/main.)

Box Figure 12.2   Slum districts with gers, yurts, and poorly built houses in Ulaanbaatar, Mongolia. (*Source:* Baiterek Media/Shutterstock.)

---

5 https://www.adb.org/projects/49169-002/main.

To address these issues, the city, together with the Asian Development Bank, has put together an ambitious plan to upgrade these settlements and work with the private sector to build sustainable homes at scale. Properly insulated structures connected to the city's heating supply would significantly reduce pollution while increasing the population's health.

However, the challenges are significant and on many various fronts. On the financial side, making the economics of a large-scale redevelopment programme work can be very challenging. Indeed, most "ger" households have low incomes and may not be able to afford even a subsidized new home. Nevertheless, if the government truly wishes to leverage investments from the private sector, making the economics work is a necessity! Other complications are cultural – these ambitious plans would not only remap neighbourhoods but also change the people's way of life.

With billions of new units needed over the next decades, the resources consumed in the construction and lifecycle of each home are expected to be tremendous. Each home needs construction materials, water, energy, furniture, appliances, and more. Each step we can take to reduce upfront and ongoing consumption is critical.

Yet building green has often come at odds with building affordably. When it comes to affordable housing, every penny saved is important. However, slightly higher upfront costs can lead to cost savings in the medium to long term. Unfortunately, these factors are often overlooked or difficult to calculate.

To further add to this issue, the housing construction industry is ripe with "greenwashing", where technologies claim to be sustainable without much evidence. In fact, few materials are "green" in absolute terms, as their sustainability is context dependent. Climate variations and whether households use cooling or heating devices have a strong impact on the best type of material required, while distance from the manufacturing location will affect transportation emissions. A material can thus be green in one location but not in another.

Fortunately, new tools are being developed to support developers, governments, and financial institutions in evaluating the sustainable characteristics of new homes in emerging markets. These tools are context appropriate and provide support in choosing technologies that reduce both costs and environmental footprints. Beyond investing in technologies, these tools enable decision-makers to adopt passive designs and understand the importance of small architectural changes with regard to energy consumption. A small difference in design can go a long way.

Box 12.3 describes a green building certification programme for developing countries.

Build Resilient

Earthquakes and storms don't kill people, buildings do. Though this is a common saying, year after year, natural disasters continue to destroy both lives and homes.

Residential buildings represent more than 60% of the building stock, and it is not surprising that most of the structural damage from earthquakes is concentrated in the housing sector. In the 2010 earthquake in Haiti, 400,000 houses were destroyed within a matter of minutes. More than 10 years later, reconstruction efforts are still ongoing.

Box 12.3   Case study: EDGE Green Building Certification[6]

EDGE is a software and certification platform developed by the International Finance Corporation (IFC), the private sector arm of the World Bank. EDGE seeks to offer an alternative to LEED in developing countries (LEED is "Leadership in Energy and Environmental Design" and is the most widely used green building rating system). The software enables developers to make informed decisions about the design of the houses being built and see the environmental footprint of the buildings they plan on developing. In the affordable housing space, reducing upfront building costs while reducing lifecycle costs of the home is critical. EDGE offers an affordable alternative which can help in reducing the energy and water consumption of households, while decreasing upfront energy embodied in the materials. To be certified, developers must meet a 20% reduction on all three fronts. EDGE operates in over 100 countries in the world.

The certification is in fact so easy to use that it has been successfully implemented in some of the most difficult housing markets, for example, in Haiti. Through the HOME[7] programme, funded by the US Agency for International Development (USAID) and implemented by the World Council of Credit Unions (WOCCU), three developers did what no NGOs had done in the country before: prove that homes can be both affordable and green! Though hundreds of NGOs sought to rebuild the housing stock after the 2010 earthquake, few sought to incorporate green building practices. Part of the issue was in knowing how to do so while working within limited budgets. EDGE enabled developers to do so for free in a matter of minutes. The HOME programme incentivized them and trained them how to use the software in order to ensure that green practices continue after USAID funding ends. (*Source:* Based on EDGE Green Building Certification, https://www.edgebuildings.com/.)

Box Figure 12.3   EDGE-certified affordable housing in Haiti. (*Source:* Olivia Nielsen.)

---

6 https://www.edgebuildings.com.

7 https://www.woccu.org/programs/current_programs/haiti/haiti_home.

Strong storms are the world's costliest natural weather disaster.[8] Since 1970, the global population exposed to tropical cyclone hazards has increased three-fold[9] – a figure expected to continue to increase as climate change intensifies their frequency and strength. Since 1971, tropical cyclones have claimed about 470,000 lives, or roughly 10,000 lives per year, and caused US$700 billion in damages globally. The costs associated with cyclone damages, adjusted for inflation, are rising by approximately 6% per year.[10]

Research[11] has pinpointed five failure modes of buildings exposed to high winds: separation of components of the building envelope, overturning, sliding, excessive drift, and structural collapse.

In the USA alone, about 7 million houses are at risk of destruction from strong winds and storm surge. These numbers are even higher in emerging countries, as hurricanes disproportionally impact informal housing units, which are often self-built and more vulnerable to strong winds and structural collapse.

We need to build with resilience in mind. To help developers and investors do this, the IFC is piloting a new programme, the Building Resilience Index. The Index standardizes and quantifies disaster risk, gives guidance on risk management, and creates a reporting system on adaptation and resilience for the construction sector. The new tool will enable construction developers to identify ways of improving building resilience while minimizing costs. For investors and households, the Index will provide reassurance that the building can withstand significant hazards and protect both lives and properties. As climate change intensifies, resilient building practices are key to ensure that our housing stock lasts and does not need to be repeatedly rebuilt. Mitigation and prevention strategies need to be incorporated from the beginning to protect homes from earthquakes and strong winds. We simply cannot afford to waste resources on homes that will crumble within a generation (Figure 12.9).

Addressing the Qualitative Housing Deficit

Addressing the global housing deficit requires us to upgrade and maintain our existing housing stock – otherwise we will be locked in a continuous wild-goose chase to build new units as our old ones deteriorate.

Most of the world today does not need a new home, but a better one. Indeed, most families are not homeless, but rather live in substandard housing without access to basic infrastructure. They often lack security of tenure. The house itself can be dangerous and have

---

8 https://www.yaleclimateconnections.org/2019/07/how-climate-change-is-making-hurricanes-more-dangerous.

9 Peduzzi, P., Chatenoux, B., Dao, H., et al., 2012. Global trends in tropical cyclone risk. *Nature Climate Change* 2, 289–294.

10 Emanuel, K., 2017. Will global warming make hurricane forecasting more difficult? *AMS* 98(3), 495–501.

11 Merritt, F.S., Ricketts, J.T., Hinklin, A.D., et al., 2001. *Building Design and Construction Handbook*, 6th edn. McGraw-Hill, New York.

Figure 12.9 New homes in Indonesia destroyed before being habituated. (*Source:* Miyamoto International.)

negative health (and mental health) impacts. But despite this reality, most governments continue to push policies for new housing construction, and few home-improvement programmes have been implemented at scale.

This is a mistake. We will need both types of programmes if we wish to meet our global challenge while preserving our Earth's resources. Tackling the global housing shortage requires working on both fronts simultaneously: building new, sustainable, and resilient homes and ensuring that our existing housing stock is up to standard.

Identifying Vulnerabilities

Large cities contain millions of housing units, and it can be a daunting, time-consuming, and expensive process to assess the structural vulnerability of each one. Yet these risk maps are important to undertake, especially in cities at risk of earthquakes and/or hurricanes – structural retrofits can mean the difference between life or death. Figure 12.10 shows houses in Port-au-Prince, Haiti at risk from earthquakes and other natural disasters.

Fortunately, new technologies, such as machine learning and drone imagery, can help reduce the costs of the process and provide municipalities with detailed risk maps that will support decision-making and prioritizing investments.

Box 12.4 outlines the World Bank's Global Program for Resilient Housing.

Green Retrofits

The majority of the existing housing stock does not meet energy-efficiency standards, as older homes have been built without modern insulation techniques and units in emerging

Figure 12.10 Housing stacked up a hillside in Port-Au-Prince, Haiti. (*Source:* Sylvie Corriveau/ Shutterstock.)

Box 12.4   Case study: Global Program for Resilient Housing[12]

The Global Program for Resilient Housing was developed by the World Bank to protect lives, homes, and economies from the dangers of disasters. Disasters and climate-related events leave 14 million people homeless every year – but as the saying goes, "earthquakes and hurricanes don't kill people, buildings do".

Despite our knowledge of the issue, identifying buildings at risk is time consuming and expensive, as it has required individual assessments by trained engineers ... until today. The World Bank has leveraged new technologies in order to substantially bring down the costs of identifying buildings at risk so retrofit interventions can be targeted before the next disaster strikes.

The World Bank is using drones and street-view cameras to capture images of homes and entire neighbourhoods and cities, which are then processed through machine-learning algorithms to pick up on buildings presenting key risks. Buildings at risk could include soft-storey buildings where a first floor provides a wide opening for a shop and a garage and may not be structurally sound enough to carry the floors above in the event of an earthquake.

---

12 https://www.worldbank.org/en/topic/disasterriskmanagement/brief/global-program-for-resilient-housing#:~:text=benefits%20of%20resilient%20housing%20can,policy%20informed%20by%20best%20 practices.

The software can accurately assess thousands of buildings in a matter of hours and issues an ID for each one, providing occupants and policymakers with the key characteristics required to plan and implement investments to upgrade homes and neighbourhoods.

Programmes like this can reduce both time and costs substantially and enable policymakers to strengthen the housing stock before the next disaster strikes. After all, US$1 invested in prevention can save between US$6 and US$14 in repairs! (*Source:* Based on Global Program for Resilient Housing, JUNE 21, 2019, https://www.worldbank.org/en/topic/disasterriskmanagement/brief/global-program-for-resilient-housing#:~:text=Benefits%20of%20resilient%20housing%20can,policy%20informed%20by%20best%20practices.)

Box Figure 12.4 The World Bank is using drones and street-view cameras to capture images of homes and entire neighbourhoods and cities, which are then processed through machine-learning algorithms to pick up on buildings presenting key risks. (*Source:* World Bank, 2020.)

Box 12.5   Case study: Alaska Housing Finance Corporation[13]

Alaska is well known for its cold climate and rough winters. Yet more than 50% of houses in Alaska were built over 25 years ago and do not meet energy-efficiency standards. Energy costs per household are between two and four times the national average.

Small changes to houses may go a long way in reducing heating needs and thus energy costs. These changes include making sure that the home is well sealed and doesn't let in cold air, adding extra insulation, and window upgrades. In order to support low-income households in undertaking these retrofits, the Alaska Housing Finance Corporation (AHFC), a public entity, offers subsidies under its appropriately named "Weatherization" programme. To apply, households need to have their home evaluated by a state-certified energy rater who will make appropriate recommendations.

For households with higher incomes, the AHFC offers home energy loans for up to US$30,000, as well as an interest rate reduction while purchasing a new energy-efficient home. (*Source:* Based on Alaska housing Finance Corporation, https://www.ahfc.us/.)

---

markets have only recently started adopting green building practices. As a result, households consume much more energy and have higher utility bills. From Alaska (Box 12.5) to Mongolia, programmes are being put in place to incentivize homeowners to retrofit their houses by installing insulation, solar panels, low-flow faucets, better ventilation, and more. At the household level, these modifications can lower costs, but at the city level they can have a compounded effect and increase the quality of life for all.

Disaster Mitigating Retrofits

Every year, natural disasters destroy and damage hundreds of thousands of houses around the globe. These disasters disproportionally affect poor households who have built their homes informally. Retrofitting these houses before a disaster strikes can save countless lives. However, a retrofit is only effective if it is properly executed, and structural engineers are expensive. Figure 12.11 highlights the vulnerability of houses in Papua New Guinea.

Government retrofit programmes often focus on government buildings, as they are relatively easy to perform with a set type of design and direct control over the buildings. Housing, on the other hand, presents a whole new set of complexities, because tens of thousands of units may need to be retrofitted and each unit may be unique. Furthermore, units may be in geographically dispersed locations, sometimes hard to access by road. Under these circumstances, it can be very time consuming and expensive to send engineers to verify quality throughout the construction process. In fact, this cost can be so expensive it may be more than the cost of the works themselves. Figure 12.12 illustrates the difficulty in ensuring quality control in the housing process.

---

13 https://www.ahfc.us.

Figure 12.11 Traditional housing in Papua New Guinea. (*Source:* Olivia Nielsen.)

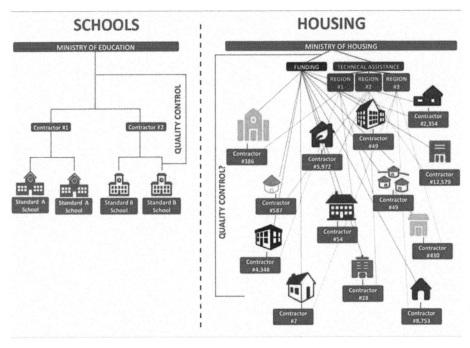

Figure 12.12 The difficulty in ensuring quality control in the housing process. (*Source*: Miyamoto International, 2020.)

Box 12.6   Case study: iBUILD + Miyamoto

The iBUILD + Miyamoto app is a tool that provides an end-to-end home construction platform. The iBUILD platform incorporates every stage of the construction process, allowing efficiency, determinism, and, perhaps most importantly, transparency for all stakeholders. The app provides households, financial institutions, and contractors suppliers with the ability to access data, connect, perform secure transactions, and share ratings. Bringing this model to home construction helps to formalize and professionalize some of the processes, such as documenting contracts and promoting higher building standards by making contractors more accountable to their clients.

The iBUILD + Miyamoto app also provides remote real-time assessment of damages incurred after a disaster. With the app, trained workers can instantly develop a budget for repairs and submit a subsidy request through the system. Quality of works performed can be tracked, and the app can release payments when works are performed to construction standards. The app has an embedded wallet feature that easily connects with leading mobile payment providers and banking systems. The wallet allows payment and transaction tracking throughout project life through verifiable and geo-tagged milestones. Box Figure 12.6 depicts critical steps in the iBUILD application, including the retrofit checklist, evaluation, and budget.

New technologies like this empower consumers to make more informed decisions, while reducing the costs of quality control. The technology encourages safer building methods and promotes affordability by simply collecting information in a single place that can be accessed by anyone with a phone.

Box Figure 12.6   Example of a retrofit checklist (iBUILD).

In order to address this major issue, new technologies are being developed that will guide homeowners through the retrofit process and enable remote verification as well as mobile disbursement of funding. These mobile apps are especially important in a time of pandemic, when social distancing and remote work have become critical (Box 12.6).

Conclusion

Addressing the global housing deficit will require us to build millions of new units while continuously upgrading and investing in our current housing stock. One cannot be done without the other.

When building new homes, we need to embrace gentle verticality and make the best out of the little urban land left. New construction should always be built with resilience and sustainability in mind – no matter how affordable the house. In the meantime, just like the Golden Gate Bridge in San Francisco, our existing housing stock will need continuous upgrades, maintenance, and retouches.

New tools and programmes are being developed in order to enable housing to be safe and energy efficient. For now, these programmes are still in the minority, but they are growing in popularity as technologies get cheaper and city dwellers demand better homes. As the COVID-19 pandemic has put housing inequalities back into the spotlight, governments from Mexico to Indonesia are taking up measures to ensure more inclusive housing for all. The homes we build today should last for generations to come, as we simply cannot afford – financially and environmentally – to build homes just to tear them down a few decades later.

13

The Habitable City (Part II) – Introduction

In Chapter 12 Olivia showed us the pathway to resilient and sustainable homes, ensuring we upgrade and maintain them to avoid unnecessary replacement.

In this chapter Nicky and Alex show us what can be done to make substantial reductions in GHG emissions and what can be achieved when we embrace sustainable thinking – from the materials we use, to how we construct, to how we operate our housing.

They also introduce the importance of the green belt in limiting urban sprawl and driving sustainable urban development within a boundary, not to mention the wider benefits of this ecological haven. They push us to think about circular economy principles within this sector, something that Connor builds on in Chapter 14, *The Resourceful City*, and Terry and Peter in Chapter 15, *The Zero Waste City*.

Half the world's population live currently in cities, and at this rate cities are only going to keep growing. As discussed by Olivia, this leaves us not only with the challenge of growing our cities to meet the population demand, but also with the urgency of significantly reducing their carbon emissions. We need dense but sustainable cities. Therefore, the question is not *if* cities will grow but *how* they will grow.

Nicky and Alex discuss this question. Following on from Olivia's conclusion that a business-as-usual approach to urban development poses great environmental risk, they further this discussion and argue that it is impossible for us to achieve the scale of production and carbon-free construction without innovation in the way we build our homes.

They examine the fight against urban sprawl by looking at London's green belts, increasingly the envy of planning professionals around the world, as an example of how cities must not only protect but also enhance their green belts. They offer a solution through precision manufactured housing (PMH) and the creation of modular homes. Although PMH is seen predominantly in countries like Japan and Sweden, it is exemplified as a cost-effective way of addressing the shortage of temporary homes through public housing building projects across the world, such as PLACE/Ladywell in South East London. They promote the standardization of PMH in conjunction with the PRiSM app which harnesses the latest digital technology, enabling architects, planners, developers, and landowners to quickly assess the viability of a site for PMH.

Within the prism of PMH technology, Nicky and Alex call for a materials revolution and advocate for the movement away from concrete, steel, and iron (these materials

The Climate City, First Edition. Edited by Martin Powell.
© 2022 John Wiley & Sons Ltd. Published 2022 by John Wiley & Sons Ltd.

contribute 15% of all global GHG emissions) to engineered timber products like cross-laminated timber (CLT), as Nicky and Alex discuss in detail through the Dalston Works case study. By combining engineered timber with PMH we will begin to unlock urban development in a way that is even more sustainable (and also with some handy healthy benefits for ourselves).

Nicky and Alex consider the whole lifecycle of a housing project and how protecting and enhancing green space in cities is key to nurturing a "sustainable and symbiotic" relationship between the urban and the rural. Density over urban sprawl, building faster and smarter, the green-industrial manufacturing sector, greater quality control, low or zero-carbon materials – this is how we grow cities to be part of a circular, sustainable, and environmental economy. Leaders and innovators must meet the challenges of this century and once again go "beyond" them if we are to succeed.

13

The Habitable City (Part II)

Nicky Gavron and Alex Denvir

How should cities grow? Where should they grow? For whom should they grow? Framed by the climate and ecological crisis, cities must find answers to these challenging questions. By 2050, 2.5 billion more people will live in cities and millions of them will be climate migrants. This urban growth will entail the production of an enormous volume of housing and infrastructure, consuming fast amounts of raw materials and finite resources. The construction sector alone is already responsible for more than 23% of global greenhouse gas emissions. Continuing business-as-usual in this sector threatens to put the world on a fast track towards a global temperature rise of 3°C or more.[1] We will not achieve the scale of production and carbon-free construction without innovation in the way we build our homes.

To address these issues, we need a change of mindset; collaboration between the construction and housing sectors; and bold political leadership. New building technologies and systems open up possibilities for more affordable and sustainable homes, and, indeed, new industrial sectors are already emerging. A net-zero city is a connected city. Homes are part of a system of buildings and neighbourhoods, connected to their surroundings and the global environmental system that we share.

Cities, it is often suggested, face a choice: *up or out* – density or sprawl. Throughout history, cities have grown outwards onto undeveloped land but also faced constraints to that expansion. These are sometimes natural barriers, sometimes policy decisions to prevent sprawl.

Globally, there are many variations of urban growth boundaries, protected open land, and forests on the periphery of cities called the "peri-urban" or "green belt".[2] A green belt does not hinder a city's growth. It channels it into sustainable avenues. In limiting the outward sprawl of a city, these urban growth boundaries further benefit the city's sustainable development by incentivizing urban renewal and the reuse of previously developed land. This in turn spawns innovation and densification that benefits from existing public transport and infrastructure. A denser, compact city that encourages less car

---

1 C40's clean construction declaration, 2020. http://c40.org/clean-construction-declaration.

2 Bishop, P., Martinez Perez, A., Roggema, R., 2020. *Repurposing the Green Belt in the 21st Century*. Chapter 2. UCL Press, London.

The Climate City, First Edition. Edited by Martin Powell.
© 2022 John Wiley & Sons Ltd. Published 2022 by John Wiley & Sons Ltd.

dependency, provides for mixed-income communities living in more walkable neighbour-hoods, and sharing amenities, services, and resources is an inclusive, sustainable city. London – though still a work in progress – is one of the leading examples of this approach.

London's metropolitan green belt remains unparalleled in terms of its variety, scale, and multifunctional and encircling nature. In the UK, nearly half of all citizens live some-where surrounded by a green belt, and as our cities become denser, access needs to be improved and increased, so that more can enjoy the mental and physical benefits such belts offer.

While there are advocates for building on London and the UK's green belts, they are increasingly the envy of planning professionals the world over who recognize not only the resources that have been protected, but also the opportunities offered for a resilient future. Globally, city leaders and planners are increasingly incorporating urban growth bounda-ries into their spatial planning strategies, including in China and Spain, not just to prevent sprawl, but because they are waking up to the climate and ecological emergency and the crucial interrelationship between urban areas and their rural hinterland.

For survival and sustainable urban growth, cities must not only protect their green belts and other peripheral green space but also proactively enhance them. Why? Green belts have a new overarching function in adapting to and mitigating climate change and addressing the ecological crisis. Encircling a city, the woodlands, wetlands, and grass-lands clean and cool the air, sequester carbon, provide continuous wildlife corridors, increase biodiversity, store excess water, and act as a flood defence. They provide new models of horticulture and opportunities for local food production. This is an enormous reservoir of natural capital assets.

The benefits of woodland cover in particular are multiple, and afforestation and refor-estation have become a global priority. Political leaders should seize the opportunity green belts offer for extending tree cover, as well as the economic possibilities from sus-tainable timber production and agroforestry close to towns and cities where the demand is. Timber is a crucial material for the habitable city and, as this chapter will show, we can even "grow" our homes.

Precision Manufactured Housing

Working within the positive constraint of green belts, we can make a city's land a renew-able and recyclable resource, with the future growth of our populations accommodated on urban brownfield land. But how can we make sure that this is affordable, equitable, and sustainable? How can we make better use of the sites available?

A key solution is to harness modern technology to reimagine how homes are designed and rethink the way they are built. Precision manufactured housing (PMH) offers just this.

PMH, modular housing, offsite manufactured housing, and modern methods of con-struction are the various umbrella terms for systems of housing construction in which individual components of a house are made in a factory. There are two typical forms: volumetric (3D units produced and fitted-out in a factory before being transported to site and stacked onto foundations) and panellized (2D flat panels constructed in a factory and transported to site to be fitted together), as well as hybrid systems which combine both.

In traditional or "linear" construction, everything happens on site. With PMH, preparations and foundations are laid on site, while panels or modules are simultaneously being produced in a factory, speeding up the build process dramatically.

The current levels of use of PMH vary greatly around the world. Japan has large industrial companies that have been producing modular homes for decades, with capacity to build 200,000 per year. More than 80% of Swedish homes are factory-built and PMH is popular in Norway and Germany too.

In the UK, PMH has faced the challenge of being associated with the system-build housing of the post-war era, which was often poor quality. The new model of factory-built homes could not be more different: high quality, precision engineered, and eco-efficient with virtually no waste. A cutting-edge industrial sector is emerging. Everything from doors and windows to electrics and plumbing can be made and assembled in a single factory to millimetre tolerances, giving greater control over the quality of the product.

Digitalization of the processes makes it easier for owners and managers to take advantage of advances in building information modelling (BIM), a system that stores details of the building and its different components, allowing for knowledge and tracking of the supply chain and future-proofing the building by enabling more efficient maintenance and refurbishment. This is a huge benefit to those institutional investors and housing providers seeking a long-term income-generating asset, as it reduces lifecycle costs. This could be used in conjunction with a "material passport" that contains data on the characteristics and properties of building materials, to be used when planning reuse and recycling at the end of a building's life (e.g as part of carbon lifecycle assessments, which will be referred to later in the chapter).

PMH works on any development site, of any size. But to achieve dense, sustainable cities, we need to "sweat" all our land assets. The case studies in this chapter demonstrate how PMH is uniquely able to build new homes on constrained sites and towers in tight spaces in London, where high-quality high-density housing would have been unachievable with traditional construction methods.

The speed of delivery means that the business model is well suited to providing the range of rental accommodation that cities need, from private rent to social rent. Schemes can be completed faster; owners can generate income and residents and tenants can move in sooner. This benefits build-to-rent investors who are attracted to the counter cyclical nature of rental homes and want to see returns on investment as early as possible.

Public Housing

Cities across the world already have crises of affordability and significant need for public rented housing to provide affordable homes for families on low and lower–middle incomes. PMH offers the opportunity to build quality new homes, faster and at lower cost (see Cherry Court case study).

Figure 13.1 PLACE/Ladywell, constructed by RSHP for the London borough of Lewisham. (*Source: Morley von Sternberg*)

PMH is also being adopted as an innovative and cost-effective solution for people who are homeless and in acute need of temporary accommodation. PMH homes can be designed to be deconstructed and reassembled on different sites, making them perfect for "meanwhile sites" – sites awaiting longer-term regeneration.

An example of this is PLACE/Ladywell in South East London, designed by Rogers Stirk Harbour & Partners (RSHP), constructed with volumetric timber modules, designed to be reassembled up to five times. The scheme provides 24 sustainable homes for homeless families (Figure 13.1).

This model is being rolled out on a wider basis by the Mayor of London and local authorities. This Pan-London Accommodation Collaborative Enterprise (PLACE) is the first of its kind in the world. Authorities are working collaboratively to pool their "meanwhile sites" and commission hundreds of homes, to address the shortage of temporary accommodation and reduce homelessness.

With growing cities experiencing crises of housing availability, affordability, and homelessness, likely to be exacerbated by climate refugees, PMH allows for efficient use of existing land and, in the recycling of previously developed land, building at greater densities.

Case Study: 101 George Street, Croydon, London

At its completion in 2020, this was the tallest residential modular building in the world, with two terracotta-clad towers of 44 and 38 storeys containing 546 high-quality "build-to-rent" homes (20% of which is for a form of discounted public rent) managed by Greystar, the American real-estate firm.

This is a great demonstration of how PMH can unlock a tight urban location, with modules craned into position on a plot surrounded by roads, a tramway, *and* mainline station. The use of factory-made modules allowed for higher density, built faster, with far less on-site waste and disruption. Crucially, there was also significantly less carbon and air pollution from transport, plant, and construction. As there was no need for cement mixers, waste removal, and follow-on trades, 60% fewer people were required on site. The controlled factory environment vastly reduced the waste produced, of which 97% is recycled, fulfilling circular economy principles. Although each of the 546 volumetric modules arrived by lorry, there were still 80% fewer vehicle movements compared with an equivalent traditional construction.

This is safer, cleaner construction, built at phenomenal speeds; once the initial ground works and core were completed, all the modules (heavy-gauge steel and plasterboard) were in place within eight months.

These towers are an example of what can be achieved with PMH and demonstrate the confidence that large property investment firms have in the sector (Figure 13.2).

Scaling Up Production

PMH offers enormous possibilities for delivering our future homes in quantities that cities need. But there are barriers to this scaling up. Like all manufacturing processes, PMH needs both continuity of demand and volume of supply to justify upfront capital investment. Without a pipeline of projects, factories cannot work at full capacity.

Figure 13.2 101 George Street, Tide Construction and Vision Modular Systems, during and after construction.

Figure 13.2 (Continued)

This is where politicians and leaders must take the bull by the horns to create and incentivize that demand.[3] Decision-makers at national, regional, and local levels must promote PMH using their planning and procurement powers, public land, and funding. PMH can be mandated through public funding; authorities can work in partnership to pool their sites, and rather than one developer procuring a single manufacturer for a large

3 Gavron, N., Devenish, T., Boff, A., Shah, N., Copley, T. 2017. Designed, sealed, delivered. London Assembly, London. https://www.london.gov.uk/sites/default/files/london_assembly_osm_report_0817. pdf.

scheme, many developers or public housing companies can collaborate to procure a number of manufacturers across multiple sites.

Better and faster housing delivery is not the only benefit. PMH factories and their supporting supply chains will create new industrial and manufacturing jobs. While most of the new homes will be built in cities, the manufacturing can be done in their hinterlands – other regions where industrial costs are lower or natural resources are nearby. Significantly, there is also scope for retooling existing factories and retraining their workforce. For instance, in France, in 2008 Ossabois transformed a toaster factory situated in a forest three hours from Paris into a timber module factory, reskilling the 50 staff (70% of whom were women).

As well as rebalancing regional economies, the prize for a politician promoting PMH is a new generation of stable jobs requiring new skills. This has the potential to bring a fresh cohort into the construction industry, including young people and engineering graduates.[4]

Stoking demand, however, without the sector ready to step up with a sufficient volume of homes is a political risk, with the chance of backlash and failure. One route to greater output comes from "vertically integrated" companies such as Vision Modular. The largest globally is Japanese housing manufacturer Sekisui, which produces 9,000 units per year from each of its factories. Unlike the typical developer-led PMH model, a single company is both the developer of the site (overseeing the land, project management, and procurement) and the manufacturer of the homes (owning the factory, creating the modules or panels, and installing them on site). These firms can take a modular-first approach to their sites, securing cost efficiencies through the supply chain and end-to-end control over quality and speed of delivery.

Manufacturers with high-volume output can use a "pattern book" approach whereby architects set the design codes, including building typology and design details. Clients – whether individual owners or housing institutions – can then combine different materials, appearances, and layouts within that code to suit their needs. This is increasingly common now in the UK, Sweden, and Germany. In Japan, with its industrialized approach, design combinations are almost limitless.

Standardization would make PMH an even more attractive proposition to investors and unlock the increased output the sector needs. Currently, every PMH company has its own system, each with its own individual benefits but lacking interoperability. If one manufacturer pulls out, there isn't a compatible alternative. To give certainty to local authorities, developers, and financiers, a degree of cross-compatibility between manufacturers is required to prevent projects stalling or failing.

Politicians and innovators can show the way. The Mayor of London, in partnership with tech-design company Bryden Wood and consultancy firm Cast, has created PRiSM, a web application, which aims to drive interoperability and standardization in PMH systems (Box 13.1).

---

4 Farmer, M., 2016. Modernise or die: The farmer review of the UK construction labour model. CLC, London.

Box 13.1 PRiSM

PRiSM is an app that harnesses the latest digital technology, enabling architects, planners, developers, and landowners to quickly assess the viability of a site for PMH. Users pick a real-world site and specify what kind of development they want. The app will then create an optimized building layout and suggest which manufacturing methods are suitable, which the user can then further tweak (Box Figure 13.1).

Box Figure 13.1 PRiSM. (*Source:* Bryden Wood.)

It offers a rich 3D environment using local geospatial data about roads, site accessibility, transport, and infrastructure – even nearby species of trees. The future plan for the app includes the ability to specify materials and so calculate embodied and sequestered carbon. It is free-to-use, is open source, and can be adapted by any city in the world to conform with its own codes and standards. By creating an open platform, PRiSM will drive manufacturers towards common, interoperable standards that work with the platform. The possibility of access to a larger market incentivizes standardization.

Standardization will not only make it more achievable to build the homes cities need but also make them cheaper. While PMH offers cost savings to landowners and developers in terms of speed and reduced waste, a larger industry would be able to do more for less, providing *more* abundant and affordable housing.

A Materials Revolution

While increasing PMH is important, cities need to think beyond a transformation in *how* they build and change *what they build with*. If we are to meet our climate change targets to be net-zero by 2050, or even 2030 – the ambition of so many cities – there must be a materials revolution reducing reliance on concrete, steel, and iron.

These materials, currently used in the construction of the majority of our homes, contribute 15% of all global greenhouse gas emissions. This is not sustainable in the face of a climate emergency and *must* change. If not, every new home built pumps additional and avoidable emissions into the atmosphere. The new industrial revolution provided by PMH should be matched with a materials revolution building homes with bio-based, renewable materials like wood, hemp, and bamboo – all of which store (sequester) carbon. They are a natural form of "carbon capture and storage", locking CO_2 away from the atmosphere.[5]

In temperate climes, timber abounds. Humans have been building with timber for thousands of years, and most of our cherished heritage buildings have timber as part of their structures. But now, advances in technology, using wood, have created the first significant new construction material since reinforced concrete around a century ago: structural engineered timber.

The most common forms are: cross-laminated timber (CLT) used for slabs and walls; glued-laminated timber (glulam) used for columns and beams; and laminated veneer lumber (LVL) used for particularly demanding structural situations.

This technological advance allows engineered timber to replace the majority of concrete in new multistorey buildings – buildings now largely made from materials that are grown, not mined.[6]

Embodied and Sequestered Carbon

In 2019, the built environment was responsible for 39% of all global carbon emissions, with 11% coming from "embodied carbon" emissions. The replacement of steel and concrete by timber provides the opportunity to substantially reduce the carbon footprint of buildings.[7] Embodied carbon is defined as the carbon emissions released into the atmosphere throughout materials and construction processes associated with different stages of a building's lifecycle. This includes the "product stage" (extraction or supply of raw materials, their transport and processing), "construction process stage" (transport, construction, and installation), "use stage" (materials and processes needed to maintain or refurbish the building, but excluding the "operational carbon" from the energy used by a building), and "end of life stage" (the deconstruction, demolition, recovery, waste processing, or disposal).[8]

The "use stage", from the point of raw material to completion of the product in a factory, is also known as "cradle to gate" carbon; if one includes the additional emissions from transport to site, construction, machinery, and construction waste, this becomes "cradle to site". The true carbon cost of steel and concrete is so high because of the energy required in these stages: from making cement and forging steel, to transporting goods to site and using diesel-powered machinery.

5 Churkina, G., Organschi, A., Reyer, C.P.O., et al., 2020. Buildings as a global carbon sink. *Nature Sustainability, Perspective* 3, 269–276.

6 Dangel, U., 2016. *Turning Point in Timber Construction: A New Economy*, p. 5. https://issuu.com/birkhauser.ch/docs/turning_point_in_timber_constructio.

7 https://www.worldgbc.org/embodied-carbon.

8 RICS, 2017. Whole life carbon assessment for the built environment. RICS, London.

These stages that cover carbon emissions released before the building is in use are known as "upfront carbon". Figure 13.3 illustrates the stages of carbon assessment through the whole lifecycle.

This is where timber comes into its own. It is much less carbon-intensive to produce than other building materials; it can be grown closer to construction sites, reducing transportation emissions; it is lighter and easier to construct, reducing reliance on heavy machinery on-site; and it is much easier to reuse or recycle at the end of a building's life. Research carried out on behalf of the International Energy Agency (IEA) found that using timber instead of concrete, masonry, or steel was one of the most commonly used and most successful carbon reduction strategies. Using timber resulted in upfront carbon reductions of between 27% and 77%.[9]

The case study of Dalston Works demonstrates this well, as it has embodied much less carbon than an equivalent building using concrete. This is also before you subtract the sequestered carbon – that which the trees have absorbed from the atmosphere.

CLT, for example, sequesters on average 831 $kgCO_2e/m^3$ within the product. The "cradle to gate" carbon is just 222 $kgCO_2e/m^3$. So when you subtract the sequestered from the upfront amounts, the sum of carbon is actually saved (rather than released). This therefore makes CLT carbon negative during its use, storing a net amount of 610 $kgCO_2e/m^3$.[10]

During its lifetime a building made with CLT is actually a carbon sink.

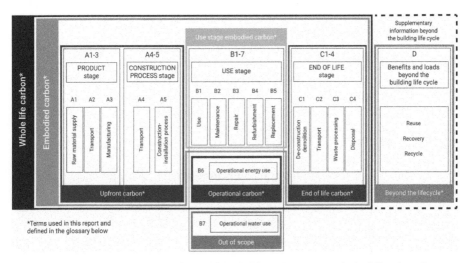

Figure 13.3 World Green Building Council: Whole life carbon stages of a building from European standard EN 15978. (*Source:* World Green Building Council: Whole life carbon stages of a building from European standard EN 15978.)

---

9 Malmqvist, T., Nehasilova, M., Moncaster, A., et al., 2018. Design and construction strategies for reducing embodied impacts from buildings – Case study analysis. *Energy and Building* 166, 35–47.

10 Jones, C., Hammond, G.P., 2019. The Inventory of Carbon & Energy. The ICE Database. Circular Ecology and University of Bath.

Cradle to Grave

To ensure the rapid reduction in the mining and use of polluting materials and finite resources, whole lifecycle assessments (LCAs) including the environmental performance of products (environmental product declarations, or EPDs) must be adopted throughout the industry. LCAs include *all* the whole life carbon stages and are particularly important as they ensure accountability for end of life. When considering timber products and buildings, unless the timber is reused or remanufactured, the carbon stored will be released back into the atmosphere. Regulations, therefore, should follow circular economy principles and mandate reuse and recycling.

A growing number of city leaders are seizing the initiative and committing to the C40 Clean Construction Declaration. This requires LCAs in planning permissions and embeds them into planning policies, processes, and building codes. This declaration commits city leaders to halving embodied carbon in all new buildings, major retrofits, and infrastructure projects by 2030, and striving to achieve 30% by 2025. It also commits them to aiming to procure zero emissions construction machinery by 2025 and for zero emissions construction sites city wide by 2030.[11]

The C40 and Carbon Neutral Cities' Alliance have developed complementary tools to help cities to achieve these targets and incentivize them to go further. And to reach net-zero by 2030 they will *need* to go further. The Carbon Neutral Cities' Alliance has the "City Policy Framework for Dramatically Reducing Embodied Carbon".[12]

Whole LCAs are the realization of the cradle to grave concept; timber can be a perfect example because it is both renewable and recyclable. Figure 13.4 illustrates this.

PMH with Timber

PMH using conventional materials already has lower carbon emissions than traditional building because it reduces waste and lorry journeys. Using timber takes the benefits of building with PMH in cities to another level. As well as the carbon savings, timber is far lighter than steel and concrete, requiring shallower concrete foundations and therefore making it easier to transport and build on constrained urban sites. By combining engineered timber with PMH we will begin to unlock urban development in a way that is even more sustainable.

Timber has other benefits. Studies have shown that exposed wood in buildings slows the heart rate and has a calming effect; it is being used in schools and offices to create happier, more productive work environments for children and adults. It makes for quieter and healthier construction sites too. The construction dust and pollution from traditional construction sites is terrible for local air quality and has harmful health impacts.

---

11 C40's clean construction declaration, 2020. Op. cit.

12 https://carbonneutralcities.org/embodied-carbon-policy-framework and C40's Clean Construction Policy Explorer https://www.c40knowledgehub.org/s/article/Clean-Construction-Policy-Explorer?language=en_US

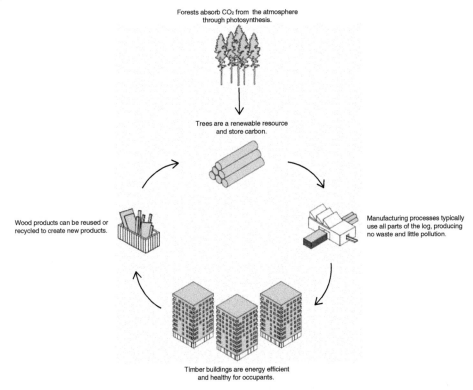

Figure 13.4 The renewable and recyclable capabilities of timber. (*Source:* Dalston Works, Waugh Thistleton Architects.)

Dalston Works and Cherry Court, Bacton Estate

Both these projects in North London were constructed using PMH with CLT and brick cladding. Though different typologies, they demonstrate how CLT can deliver high-density, high-quality housing on very constrained sites.

At the time of completion in 2017, Dalston Works in Hackney, at 10 storeys, was the tallest CLT building in the UK. A mixed-use development, designed by Waugh Thistleton Architects and developed and managed by Regal London, it has 121 residential units and 3,500 m^2 of commercial space (Figure 13.5A).

Bacton Estate Phase 1 was a partnership between the London Borough of Camden and Karakusevic Carson Architects, to build 67 new council homes on the estate, of which 46 are low-cost, low-rise social rented terraced houses (now called Cherry Court).

Railways abut both these projects. In the case of Dalston Works, they also pass underneath, making the use of concrete piles impossible (Figure 13.5B). Instead, it was built on a simple 900-mm-thick raft foundation. Because CLT is 80% lighter by volume than concrete, it made development on this site possible and enabled 25% more homes to be built than a concrete equivalent. Despite the complexity and challenge of the site, the construction time of Dalston Works was still 20% shorter than traditional methods, with 84% fewer vehicle deliveries.

Figure 13.5A Dalston Works. (*Source:* Daniel Shearing.)

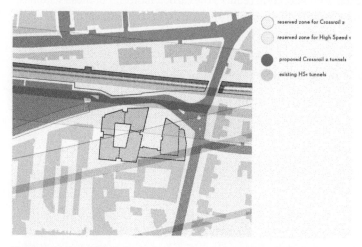

○ reserved zone for Crossrail 2
○ reserved zone for High Speed 1
● proposed Crossrail 2 tunnels
○ existing HS1 tunnels

Figure 13.5B Railway lines and tunnels under Dalston Works. (*Source:* Dalston Works, Waugh Thistleton Architects.)

The environmental statistics of Dalston Works are impressive. Using over 4,650 m³ of timber, the project has absorbed/removed 3,560 tCO$_2$e from the atmosphere. It has been estimated that the building will save 2,400 tCO$_2$e compared to a concrete building of the same size, reducing upfront carbon by 2.5 times (Figure 13.5C).

Cherry Court homes abut a busy railway line. Despite this, the use of CLT means they are utterly soundproofed – there's no noise and no vibrations. Their thermal and energy efficiency cuts utility bills and eliminates fuel poverty. These benefits improve the quality of life for rehoused residents.

Due to the speed of construction, CLT is ideal for regeneration schemes. At Bacton, existing social tenants were central to the development of the project, with workshops

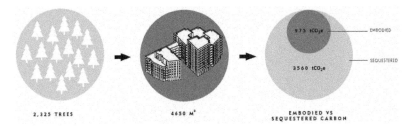

2,325 TREES   4650 M²   EMBODIED VS SEQUESTERED CARBON

Figure 13.5C The timber used in the construction of Dalston Works absorbs more CO_2 than it embodies, making it a carbon negative building. (*Source:* Dalston Works, Waugh Thistleton Architects.)

and events for the community to plan their new homes. To minimize disruption, the 46 social homes were built first, for families to move into, before the rest of the estate regeneration progressed. Construction was completed in 10 weeks and required only 18 lorry loads (Figures 13.5D and E).

While the ease and speed of construction reduces the carbon footprint of these schemes, by far the biggest environmental achievement from using CLT is that these two projects will act as carbon sinks for the duration of their lives.

A Global Transition to Growing Our Homes

Across the world, cities, developers, and architects are recognizing the strength, lightness, and carbon negative properties of engineered timber. There are ground-breaking

Figure 13.5D Cherry Court, Bacton Estate, under construction using CLT panels. (*Source:* Cherry Court, Bacton Estate, Karakusevic Carson Architects.)

Figure 13.5E   Cherry Court, Bacton Estate, completed. (*Source:* ©Tim Crocker, All rights reserved.)

examples of its use in tall buildings in Vancouver and Oslo, both nations with fantastic timber resources. Increasingly, companies are using local timber: Japanese firm Sekisui makes use of local timber in 45% of the homes it builds. Bordeaux, in France, is using local forests to construct timber buildings. In remodelling the south of the city, an engineered timber factory has been set up, catalysing a new regional economy.

As long as there is sustainable forest management (maintaining what is there and replacing what is cut), many countries in the northern hemisphere already have the timber resources they need. Canada has the capacity to harvest 220 million cubic metres of timber per year, enough to house a billion people for a whole lifetime, while sustainable forests in Europe have expanded by an area larger than the size of Portugal over two decades. According to M.H. Ramage, only 25–30% of Europe's forest, managed, utilized, and harvested as it is currently, is needed to house its entire population.[13]

To achieve the mass adoption of engineered timber, it is imperative that this is supported by a strong legal and political commitment to sustainable forest management and robust forest certification schemes.

This transition needs very strong political leadership and development, with changes to building codes, training schemes for construction workers, and expansion of factories.

---

13 Ramage, M.H., Burridge, H., Busse-Wicher, M., et al., 2016. The wood from the trees: The use of timber in construction. *Renewable and Sustainable Energy Reviews* 68(1), 333–359.

It is encouraging that some European and North American countries have adjusted their building codes to allow construction of mid-rise and in some cases high-rise buildings out of wood. France has mandated that by 2022 all new public buildings financed by the French State must contain at least 50% wood or other organic material and that 80% of the 2024 Olympic village should be built in wood. Significantly, France has also introduced a major regulatory change regarding the whole life carbon of a building, giving higher weight to the carbon emitted today than the carbon that will be emitted in the longer term.[14]

Conclusion

Cities will, and must, continue to grow through the twenty-first century. To do so while meeting global climate change targets and providing higher quality places for people to live, there needs to be a technological revolution in the way cities are built and designed and how the relationship of a city with its surroundings and the natural world is conceived. Carrying on building in the same way would lead us down the path to environmental catastrophe.

Density must be preferred to sprawl, in order to maximize benefits from infrastructure and to create sustainable cities. To build faster and smarter, optimizing land, cities must spearhead a new industrial sector in PMH. The opportunity this burgeoning green industrial-manufacturing sector offers is enormous: new construction jobs; significantly reduced build times and costs; greater quality control; cleaner and safer construction processes; better air quality; low or zero-carbon materials; and low construction waste and energy use. Around the world, these benefits are beginning to be realized by leaders and innovators. To meet the challenges of this century and beyond, they must be bolder still.

Protecting and enhancing green space encircling cities is key to nurturing a sustainable and symbiotic relationship between urban and rural areas; green belts have multifunctional uses that improve citizens' lives and address the climate and ecological crisis. These enhancements can create a world where cities can contribute to growing the homes of their future with trees and forests.

Wood, through engineered timber products like CLT, will be the material that unlocks vast gains for the modern city. Its properties are ideally suited to PMH processes: it is light, easier to handle, and abundant. But it is when you look at the sequestered carbon of timber that its use in construction becomes so important. Rather than building homes with steel and concrete, wherever possible we can use timber that has locked CO_2 away. The city of the future will – and the city of today can – be built from homes that are carbon sinks, made from trees. These homes will be in greener, denser cities and be part of a circular, sustainable, and environmental economy.

---

14 https://www.ecologie.gouv.fr/sites/default/files/dp_re2020.pdf.

14

The Resourceful City – Introduction

Nicky and Alex have made the case in Chapter 13 of focusing on density and new construction that considers the whole lifestyle of a housing project. In this chapter, *The Resourceful City*, Conor takes a look at how wasteful our cities have been, the importance of a GHG inventory so we can treat waste as a resource, and what it will take to achieve a circular economy.

In Chapter 13, *The Habitable City*, Nicky and Alex talked about the need for a "materials revolution" in the building sector, and they advocate for the movement away from concrete, steel, and iron towards engineered timber products in order to curb carbon emissions. And as you can see in Figure 14.1, the materials we use to construct the future generations of housing with need to be sourced, manufactured, and installed within an emissions envelope the city can live with.

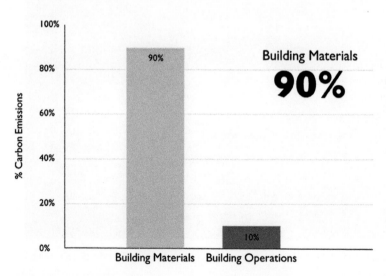

Figure 14.1 Building sector CO$_2$ emissions. New construction, 2015–2050. If things don't change with how we treat embodied carbon, impacts will total 90% of the carbon released from newly constructed buildings between 2015 and 2050. (*Source:* © Architecture 2030, All Rights Reserved.)

The Climate City, First Edition. Edited by Martin Powell.
© 2022 John Wiley & Sons Ltd. Published 2022 by John Wiley & Sons Ltd.

This is a frightening prospect, but along with our "materials revolution" must also come a waste management revolution if we are to have any hope for growing cities to be part of a circular, sustainable, and environmental economy. By looking at developing and developed cities alike, through the lens of the devastating Koshe landfill collapse in Ethiopia, Conor explores what cities are doing and, more importantly, what they *need* to do in order to tackle and transform and diversify waste management and thrive in the future. Figure 14.2 is one of many typical landfills across the world.

Conor looks at the many steps involved in this transformation, beginning with measuring inefficiencies, as "cities can only manage what they can measure". He then goes on to discuss how a city can take a circular approach to resources and the steps it will take to achieve zero waste. He looks at how cities must incentivize material reuse in conjunction with creating recycling systems for the twenty-first century, and, finally, at reducing the costs of waste management in cities.

These models are shifting cities away from a linear consumption model and towards a circular model, keeping resources in use as long as possible, minimizing waste, and protecting natural resources. The World Bank predicts waste volumes will rise 70% by 2050, so the transition will not be an easy one, but it will be a necessary one.[1] In order for us to have success stories in city waste management and to avoid the devastating consequences

Figure 14.2 A typical landfill. Globally we dump nearly 2 billion tonnes of waste. (*Source:* vchal/ Getty Images.)

---

1 World Bank, 2018. Global waste to grow by 70 percent by 2050 unless urgent action is taken.https:// www.worldbank.org/en/news/press-release/2018/09/20/global-waste-to-grow-by-70-percent-by-2050-unless-urgent-action-is-taken-world-bank-report.

of the Koshe landfill collapse, the goal is a completely circular system, or, as Conor describes, "one in which the outputs of certain urban processes are the inputs of other high-value urban processes".

It is imperative that future cities are resourceful. They must use resources efficiently, create a loop system that maximizes resources, and rethink their approach to physical waste through reduction, reuse, and recycling. As Conor reminds us, the Koshe landfill story is not over yet. There is a lot of work still to be done.

14

The Resourceful City

Conor Riffle

Just outside the picturesque capital city of Addis Ababa, Ethiopia, sits a steep mountain of trash. Surrounded on all sides by settlements tucked in among the verdant green of Ethiopia's capital region, the mountain stands out. It towers over the nearby highway and homes. Its odour is overpowering, sometimes causing fainting spells at a nearby school. Parts of the mountain smoke ominously. Birds wheel overhead. This is the Koshe landfill, one of the main storage locations for the trash from Ethiopia's largest city.

In 2017, disaster struck at Koshe. After years of trash piles rising higher and higher on the landfill, one of the towering walls of garbage collapsed. The resulting garbage slide buried a nearby settlement, killing 116 people.

There are thousands of landfills around the world just like Koshe. Some are informal and unmanaged, places where garbage piles up without oversight or safety practices, threatening the lives of those who live nearby or make a living on the landfill. Others are managed and graded, their toxic methane emissions captured, and then eventually closed, covered up, and turned into parks or solar farms. But all of them stand as stark reminders that the primary way that most of our cities deal with waste is the same method pioneered over 2,000 years ago by the ancient Romans – fill a plot of land with garbage until it's full.

It is not just developing cities that struggle to manage their waste. Today, in Rome, the city that invented the modern practices of waste management, the landfill system has reached its breaking point. In 2009, the European Union declared that Rome's main landfill, Malagrotta, could no longer accept waste.[2] This decree ignited nearly a decade of furious efforts to find locations for the 1.7 million tonnes of waste that Rome produces every year. By 2018, the city was so desperate to find enough space to store its waste that the mayor appealed to surrounding cities to open their own landfills to Rome's garbage.[3]

As the Earth's population continues its upward trajectory – the UN projects global population to reach 9 billion by 2050 – the solution to how the Earth's cities manage their waste is becoming even more pressing. The traditional model – the landfill – is

---

2 Di Giorgio, M. 2014. Italy's woeful waste management on trial with Il Supremo trash king. Reuters, May 2014. https://www.reuters.com/article/us-italy-trash-insight/italys-woeful-waste-management-on-trial-with-il-supremo-trash-king-idusbrea4o07k20140525.

3 Zampano, G. 2019. Why is Rome drowning in trash? The Point Guys, February 2019. https://thepointsguy.com/news/why-is-rome-drowning-in-trash.

The Climate City, First Edition. Edited by Martin Powell.
© 2022 John Wiley & Sons Ltd. Published 2022 by John Wiley & Sons Ltd.

environmentally and economically unfeasible in some cities, like Rome, and outright deadly in others, as in Addis Ababa. Future cities, cities that will succeed and thrive throughout the next 100 years or so, are developing new models for dealing with waste. These models – which are detailed throughout this chapter – are shifting cities away from a linear consumption model, in which goods are produced, consumed, and then buried in the ground. Future cities are moving to a circular model, which keeps resources in use as long as possible, minimizing waste and protecting natural resources.

The transition will not be easy. The World Bank predicts waste volumes will rise 70% by 2050, to more than 3.4 billion tonnes per year.[4] But the transition is necessary. Cities around the world are already beginning to reimagine how they think about waste. They are becoming more resourceful – utilizing every precious resource as much as possible and designing systems to minimize waste and inefficiency. The goal is a completely circular system, one in which the outputs of certain urban processes are the inputs of other high-value urban processes – the "Circular Economy". This chapter outlines what resourceful cities are doing and what they will need to do to use the concept of circularity to drive out waste and thrive in the future.

It all begins with measurement.

Measuring Inefficiencies

The first step for cities in managing inefficiencies in the disposal of waste is to identify and measure how much waste cities are actually creating. Cities can only manage what they can measure. The first and most important metric for cities to measure is their greenhouse gas (GHG) emissions. GHG emission inventories tell cities where to focus their efforts to reduce emissions. These inventories provide a map to inefficiencies in the use of energy, the transportation of people and goods, and the management of waste. For example, New York City's early GHG inventories helped the city identify that its buildings were a major source of GHG emissions – primarily old, inefficient buildings that were leaking energy. This information led then-Mayor Michael Bloomberg to launch efforts aimed at improving energy efficiency in these buildings. From 2007 to 2017, these efforts were able to eliminate 7.24 million tCO_2e.[5] As of 2020, more than 800 global cities now disclose emissions and other environmental data using a common reporting framework.[6]

New frameworks for measuring inefficiencies are emerging. For example, some leading cities are taking a consumption-based approach to carbon accounting, which reflects cities' positions as consumers of the world's resources. A consumption-based approach, instead of looking at where carbon emissions are produced, focuses on where the output of those emissions is consumed. When practitioners examine city

---

4 Global waste to grow by 70 percent by 2050 unless urgent action is taken: World Bank report. World Bank, September 2018. https://www.worldbank.org/en/news/press-release/2018/09/20/global-waste-to-grow-by-70-percent-by-2050-unless-urgent-action-is-taken-world-bank-report.

5 Inventory of New York City greenhouse gas emissions. NYC Mayor's Office of Sustainability. City of New York, 2016. https://nyc-ghg-inventory.cusp.nyu.edu.

6 CDP, 2020. Cities. https://www.cdp.net/en/cities.

emissions using a consumption-based approach, new areas of focus come into view. In a consumption-based accounting study of 79 major cities across the globe, a majority of emissions occurred due to goods and services that are imported into cities. That report demonstrated that "over 70% of consumption-based GHG emissions in cities come from utilities and housing, capital, transportation, food supply and government services".[7]

Some cities are explicitly measuring their progress toward creating a circular economy. For example, the EU published the "Indicators for circular economy transition in cities" White Paper in 2019.[8] This report suggests a number of different frameworks and indicators for measuring circularity, including tonnes of waste per household and the ratio of household spending on services versus goods. The report built on pioneering work done by London's Waste and Recycling Board, which outlined in 2015 London's strategy for building a city on circular principles.[9] The city of Amsterdam recently released its own framework for measurement, called the Amsterdam City Doughnut. The Doughnut measures "how the City can be a home to thriving people in a thriving place, while respecting the wellbeing of all people and the health of the planet".[10] This approach informed the City's Circular Strategy.[11]

Each of these measurement approaches can give cities insights into where they can focus to push inefficiencies out of their urban systems. Resourceful cities are transforming each of these areas to make maximum and efficient use of their key resources. The rest of this chapter outlines the strategies that cities are undertaking and must continue to undertake in one of these key sectors: waste management.

Taking a Circular Approach to Resources

In 1979, Dutch politician Ad Lansink introduced into the Dutch parliament a framework for efficiently and productively managing waste. This framework, known as Lansink's Ladder, eventually became the well-known waste hierarchy ("reduce, reuse, recycle"). The waste hierarchy has been adapted for use in various countries, but the principles are broadly similar: when dealing with waste, first attempt to minimize it, then reuse it, then recycle it, then capture its energy, and then, as the last option, put it into a landfill.

---

7 Doust, J., Wang, M. 2018. Consumption-based GHG emissions of C40 cities. C40, March 2018, p. 18. https://www.c40.org/researches/consumption-based-emissions .

8 Indicators for circular economy (CE) transition in cities – Issues and mapping paper (Version 4). Urban Agenda for the EU. European Union, May 2019. https://ec.europa.eu/futurium/en/system/files/ged/urban_agenda_partnership_on_circular_economy_-_indicators_for_ce_transition_-_issupaper_0.pdf.

9 London: The circular economy capital. London Waste and Recycling Board, December 2015. https://www.london.gov.uk/what-we-do/environment/smartlondon-and-innovation/circular-economy.

10 Ibid.

11 Numerous other frameworks exist. See, for example, de Ferreira, F.-N. 2019. A framework for implementing and tracking circular economy in cities: The Case of Porto. MDPI, March 2019, pp. 7–10. https://www.mdpi.com/2071-1050/11/6/1813/pdf.

Circular economy principles supercharge the traditional waste hierarchy. A circular economy: (1) designs out waste and pollution, (2) keeps products and materials in use, and (3) regenerates natural systems.[12] Practitioners in different sectors apply these principles in creative ways. For future cities, adopting circular economy principles means actively structuring municipal operations and economic and social incentives to eliminate the inefficiencies that cause waste.

Future cities are adopting policies that don't just minimize waste – they eliminate it. Cities are promoting initiatives that design reuse into materials from the beginning, allowing them to reuse some materials in a circular loop. Cities are overhauling recycling, turning an expensive and underfunded municipal service into a showcase of efficiency and new technology. These resourceful initiatives together are borne out of necessity – the linear model of waste management pioneered by the Romans is no longer fit for purpose for the twenty-first century. These initiatives will result in future cities that are more sustainable, resilient, and circular.

Moving to Zero Waste

Future cities' first objective is to curb the volume of solid waste that their cities produce. This objective is itself ambitious. The world's largest 27 cities alone account for 13% of the world's total production of waste.[13] Resourceful cities are undertaking a number of strategies to minimize the amount of waste they produce. In 2018, 23 megacities committed to cut the amount of waste generated by each citizen by 15%, reduce the amount of waste sent to landfills and incineration by 50%, and increase the diversion rate to 70%, all by 2030.[14] These cities are undertaking a mix of strategies to reduce their overall volume of waste by 87 million tonnes.

One of the best strategies for cities to reduce waste is to rid their waste streams of the organic waste disposed of by residents. Organic waste comes from discarded food, yard waste, plants, and other organic items. In many cities where this material is disposed of in the regular garbage, organic material makes up 30% or more of the waste stream going to landfill. Yet organic waste is valuable, and landfilling organic waste is counter-productive. Organic waste that decomposes in oxygen-starved environments like landfills creates methane, a potent GHG. And it costs cities money to dispose of organic material in landfill, yet they see no benefit from it as a resource. Leading cities are having success reducing the volume of trash sent to landfill by no longer accepting organic material in their waste streams. For example, in 2009, San Francisco banned residents from disposing of organic material in kerbside bins. Instead, the city required residents to separate their organic waste and place it at the kerbside, where dedicated trucks collect it and take it to

---

12 What is a circular economy? Ellen Macarthur Foundation. https://www.ellenmacarthurfoundation.org/circular-economy/concept.

13 Kennedy, C.A., Stewart, I., Facchini, A., et al. 2015. Energy and material flows of megacities. *Proceedings of the National Academy of Sciences USA* 112(19), April 2015. https://www.pnas.org/content/pnas/112/19/5985.full.pdf.

14 23 global cities and regions advance towards zero waste. C40, August 2018. https://www.c40.org/press_releases/global-cities-and-regions-advance-towards-zero-waste.

a special facility, where it is composted and sold. The city saves money on disposing of waste at its landfill, saves landfill space for the future, and can turn its waste into cash. Stockholm in Sweden was able to turn its organic waste into a productive resource by creating "[t]he world's first large-scale 'biochar' urban carbon sink". With this system, the city collects garden waste and other organics from residents. That waste is turned into biochar, which, once processed, supplies heat to residents.[15]

Organic material, treated correctly, is a valuable resource, one that resourceful cities are eager to use productively. The city of Toronto has developed a truly circular use for organic waste: powering the very trucks that collect it. The city collects organic waste in separate bins from its residents. It then uses this organic waste to produce renewable natural gas (RNG) at its Dufferin Solid Waste Management Facility. The RNG is then used to power waste collection trucks, directly reducing the city's fuel costs and carbon emissions. Over the long term, Toronto seeks to move toward a "closed-loop approach" where all organics collection trucks can be powered by the waste that they collect.[16] This approach towards waste saves money, cuts carbon emissions, and recaptures the potential of once-thought-to-be-wasted goods.

Future cities are employing other strategies to eliminate waste. So-called Pay as You Throw (PAYT) programmes have been successful at directly incentivizing residents to reduce waste volumes. PAYT programmes charge residents or businesses fees for disposing of waste, usually based on the volume of waste: the higher the volume, the more the resident pays. San Francisco has been able to divert 80% of its waste from landfills – the highest rate among major cities in the USA – in part by using PAYT programmes. The recycling bin for San Francisco families is much larger than the bin for trash, creating a direct incentive for families to increase recycling and reduce waste.[17] Other cities like Toronto and many municipalities across Europe utilize PAYT programmes to reduce waste volumes.

Promoting Material Reuse

Another core strategy for reducing solid waste in cities is to promote and incentivize the reuse of materials. Fulfilling the proverb, "one person's trash is another person's treasure", cities are developing resourceful strategies to drive maximum reuse of materials. In Austin, Texas, for example, the city led the creation of the Austin Materials Marketplace, an online marketplace that matches waste producers with those who need raw materials. The Marketplace helped the city's Department of Transportation find a second life for its old parking meters, saving 2,600 pounds of metal and plastic from the

---

15 Bernhardt, H., Zeller, D. 2018. Municipality-led circular economy case studies. C40, December 2018, pp. 126–127. https://www.c40.org/researches/municipality-led-circular-economy. Stockholm Biochar Project, a Mayor's challenge winner, opens its first plant. Bloomberg, May 2017. https://www.bloomberg.org/blog/stockholm-biochar-project-mayors-challenge-winner-opens-first-plant.

16 Turning waste into renewable natural gas. City of Toronto. https://www.toronto.ca/services-payments/recycling-organics-garbage/solid-waste-facilities/renewable-natural-gas.

17 How cities can boost recycling rates. C40, May 2019. https://www.c40knowledgehub.org.

landfill.[18] Vienna, Austria has already diverted over 15,000 tonnes of electronic waste from landfills through its government-funded repair and service centre (RUSZ). The independently run organization allows citizens to bring broken electronics and household appliances for repair, creating jobs, extending the life of common consumer goods, and reducing landfill use.[19]

Cities must become even more resourceful in the future to close the loop in high-impact sectors like construction. Construction and demolition may account for up to twice as much waste volume as that produced by residents in their homes in the USA.[20] Resourceful cities are designing waste out of the construction system entirely. In London, for example, more than half of the capital city's waste comes from construction. Mayor Sadiq Khan recently required all new building projects that meet a certain size to submit a circular economy statement showing how the project will prioritize the reuse of secondary materials and how the project will be designed to encourage reuse of its materials after their full life has been achieved. Amsterdam is working to develop new buildings that have "adaptable functions and systems" that can be repurposed for new uses at the end of their lives. The city is leveraging its procurement processes to lead the way by transforming its own construction projects first. Beginning in 2022, all new urban development in Amsterdam will be based on circular principles.[21]

Houston is also putting these ideas into practice. As a growing American city, it sought to eliminate waste in the construction of new buildings. In 2009 the city opened the City of Houston Building Materials Reuse Warehouse which accepts donations and offers free collection of disused building material. Since its opening, the programme has diverted over 4,000 tonnes of construction waste from landfills and has "given away 90% of diverted construction materials to over 700 non-profit organizations, schools, universities, and government agencies".[22] As more cities incorporate reuse of materials into building codes, the amount of waste that cities produce will fall.

Resourceful cities are working closely with the private sector to encourage new approaches to design and reuse. The city of Brussels has established the Be Circular programme to incentivize circular and sustainable practices among private businesses. The programme provides direct financial support and training for companies in various sectors to adopt recycling and reuse practices. Among its successes, it has helped Brussels divert 91% of its construction and demolition waste to other uses. Additionally, this programme has aided 222 companies in their transition to sustainable practices. These companies include food and retail establishments, construction

18 Bernhardt and Zeller, 2018. Op. cit., p. 110.

19 Ibid., p. 130.

20 The state of the practice of construction and demolition material recovery. EPA, May 2017, p. 4. https://nepis.epa.gov/exe/zypdf.cgi/p100ssjp.pdf?dockey=p100ssjp.pdf National overview: Facts and figures on materials, wastes and recycling. EPA. https://www.epa.gov/facts-and-figures-about-materials-waste-and-recycling/national-overview-facts-and-figures-materials.

21 Policy: Circular economy. City of Amsterdam, p. 67. https://www.amsterdam.nl/en/policy/sustainability/circular-economy.

22 Bernhardt and Zeller, 2018. Op. cit., pp. 66–67.

companies, and companies operating in the waste sector.[23] As cities transition to new systems that maximize reuse of materials, close collaboration with the private sector is essential.

Creating Recycling Systems for the Twenty-First Century

For materials that cannot be kept out of the waste stream, reused, or repaired, cities turn to recycling. Recycling is inherently a circular solution: waste materials are captured, processed, and then sold back to industry as the raw materials for new products. However, recycling programmes are expensive to set up and run. The average cost to run a kerbside recycling programme in the USA can range from $45 to $135 per tonne.[24]

City governments don't control the makeup of waste, but they must dispose of it, whatever it comprises. In most countries, private sector companies produce goods, like bottled soda, for consumption by consumers, but bear no responsibility for ensuring those plastic bottles are recycled properly. For example, there are 1.5 billion plastic bottles sold in the state of Minnesota each year.[25] Minnesota's cities, where more than 70% of the state's population lives, must recycle or landfill most of those bottles.[26] This volume puts enormous strain on municipal waste collection systems – they bear responsibility for capturing, processing, and then reselling these bottles. In the end, less than 25% of these bottles are recycled. Product packaging that is difficult to recycle in traditional recycling facilities, like polystyrene, plastics 3–7, and anything made of two different types of plastic fused together, further hamper municipal recycling efforts. As a result of this system, despite a strong culture of recycling and sophisticated waste infrastructure that includes incineration, approximately 60% of Minnesota's waste ends up in landfill every year.

To help offset the cost and realign incentives to support recycling, resourceful cities, along with regional and national governments, are implementing incentive policies like extended producer responsibility (EPR). EPR assigns a small fee to each producer of recyclable materials. The policy usually then distributes the resulting money to local governments to support recycling and waste collection efforts. Such policies can encourage environmentally friendly design, sustainably sourced materials, and the "use of life-cycle oriented waste and materials management".[27] Such measures increase accountability for producers for the lifetime of their product. Amsterdam, for example, plans to build on the success of EPR programmes for beverages and electronics in the Netherlands by extending similar programmes to important sectors like furniture, retailing, and plastic

23 Ibid., pp. 18–20.

24 The pros and cons of recycling. ThoughtCo, August 2019. https://www.thoughtco.com/benefits-of-recycling-outweigh-the-costs-1204141.

25 Anderson, J.A. 2019. Rethink, retool, then recycle? Next City, January 2019. https://nextcity.org/features/view/rethink-retool-then-recycle.

26 Greater Minnesota: Refined revisited. Minnesota State Demographic Center. January 2017. https://mn.gov/admin/demography/reports-resources/greater-mn-refined-and-revisited.jsp.

27 Environment at a glance: Climate change. OECDOECD, 9 March 2020, p. 2. https://www.oecd.org/environment/environment-at-a-glance/circular-economy-waste-materials-archive-february-2020.pdf.

packaging.[28] These programmes provide much-needed funds for local recycling infrastructure and education, helping to create more effective and efficient recycling programmes.

Future-oriented cities are taking similar action on hard-to-recycle materials like plastics. More than 8.3 billion tonnes of plastic has been produced since the 1950s. Yet only 9% of all plastic waste ever produced has been recycled.[29] Indeed, much of it can be difficult or impossible to recycle with existing municipal recycling infrastructure. In Atlanta, for example, the presence of single-use plastic bags in residential recycling containers can be grounds for dumping the entire container in landfill.[30] Despite the lack of federal policy regulating the use of plastic bags in the USA, at least 150 cities, towns, and counties have adopted policies that charge a fee for or ban single-use plastic bags. That number includes major cities like New York, Los Angeles, Chicago, Boston, San Francisco, Washington DC, Portland, Austin, and Seattle.[31] Cities are taking action to eliminate hard-to-recycle plastic from waste streams to increase the efficiency of recycling programmes, among other reasons.

Well-run recycling programmes can deliver economic benefits for cities. The city of Charlotte, North Carolina, for example, discovered in 2016 that its failure to recycle the maximum amount of material was costing it money. The city identified that 23.6% of materials that end up in landfills could have been recycled with existing programmes. The potential value of the electronics, glass, organics, and other recoverable materials going to its landfills was US$111 million. The city also estimated that if these materials were recycled, 2,000 jobs could be created.[32] With the right programme structures in place, cities of the future can capture more recyclable materials and profit from their reuse.

Reducing the Costs of Waste Management in Cities

One of the largest barriers to reducing waste in cities is the cost of waste management. According to a study by KPMG, the average city worldwide spends US$201 to collect a ton of garbage ($182 a tonne).[33] Considering that large cities can dispose of more than 9 million tonnes of garbage per year, waste management can rank as one of the top costs that

---

28 Policy: Circular economy. Op. cit.

29 Our planet is drowning in plastic pollution. United Nations. https://www.unenvironment.org/interactive/beat-plastic-pollution.

30 Kaufman, L. 2020. The future of recycling is sanitation workers rejecting your bin. Bloomberg, January 2020. https://www.bloomberg.com/news/articles/2020-01-29/the-future-of-recycling-is-sanitation-workers-rejecting-your-bin.

31 National list of local plastic bag ordinances. California Against Waste. https://www.cawrecycles.org/list-of-national-bans.

32 Galadek, E., Kennedy, E., Thorin, T. 2018. Circular Charlotte – Towards a zero waste and inclusive city. City of Charlotte, September 2018, p. 7. https://charlottenc.gov/sws/circularcharlotte/documents/circular%20charlotte_towards%20a%20zero%20waste%20and%20inclusive%20city%20-%20full%20report.pdf.

33 Beatty, M. 2017. Garbage collection. KPMG, September 2017. https://home.kpmg/xx/en/home/insights/2017/09/garbage-collection.html.

cities face. Municipalities in low-income countries spend on average about 20% of their annual budgets on waste management.[34] Garbage trucks are expensive to purchase; landfills are expensive to maintain. But of all of the cost drivers, the cost of the actual collection of waste is often the largest.

As a result, future cities must reduce the cost of managing waste as much as possible. Advances in technology now make possible significant changes in how cities collect waste. For example, fleet efficiency software can now help cities design efficient collection routes. Vehicle tracking can help cities monitor the resources that the trucks collect, ensuring that recyclable materials end up at the recycling station, not at the landfill. In Atlanta, Georgia, for example, the city dramatically lowered its cost of solid waste collection by optimizing its daily residential collection routes. The city saved more than US$780,000 per year in collection costs by partnering with a private technology company to optimize the efficiency of its fleet of garbage trucks. Atlanta adjusted its "solid waste service schedule from four to five days, decreasing the total amount of trash routes per day, and balancing the number of hours driven". These changes not only preserved a significant amount of taxpayer dollars, but also cut GHG emissions, reduced traffic, and increased public health and safety.[35] Cities of the future are incorporating these technologies into their daily operations, ensuring that they can collect municipal solid waste affordably and effectively.

Future cities will continue to drive down the cost of collection with new technologies. A number of forward-thinking cities are beginning to make the shift to natural-gas-powered or even all-electric waste collection fleets. In June 2020, the city of Manchester, UK ordered its first shipment of 27 electric waste collection trucks, which will replace half of its current fleet. This decision was part of the city's larger goal of cutting its carbon emissions in half by 2025 and eliminating all carbon emissions by 2038. These first 27 trucks are expected to eliminate 900 tCO_2e each year. Additionally, the new electric vehicles only cost "marginally" more than it would have to replace outdated trash collection trucks with new diesel trucks. Despite the higher upfront cost of electric trucks, they are expected to result in net savings due to increased efficiency and energy saving.[36]

Artificial intelligence will help reduce the costs of waste collection for cities. Resourceful cities are also finding that they can use their garbage trucks for more than just collecting garbage. These trucks are the only vehicles in a city's fleet that travel up and down every single street at least once per week. With the right equipment and the power of artificial intelligence, these trucks can assess and analyse the streets they drive, sending insights on items like potholes, graffiti, and cleanliness back to the city's

34 Kaza, S., Yao, L.C., Bhada-Tata, P., Van Woerden, F., 2018. What a waste 2.0: A global snapshot of solid waste management to 2050. World Bank, September 2018, p. xii. https://openknowledge.worldbank.org/handle/10986/30317.

35 2019 Environmental, social, and governance report: Toward a future without waste. Rubicon, p. 55p. 55. https://www.rubicon.com/reports/2019/rubicon-esg-report.pdf.

36 Electric dreams: Council makes huge commitment to eco-friendly bin lorries. Manchester City Council, June 2020. https://secure.manchester.gov.uk/news/article/8444/electric_dreams_council_makes_huge_commitment_to_eco-friendly_bin_lorries.

managers in real-time.[37] Furthermore, garbage trucks, like passenger cars and buses, will eventually be autonomous, able to collect trash from homes and businesses without human drivers. The money saved from incorporating advanced technology into garbage collection can be used by the city to invest in other sustainable and cost-saving measures, encouraging further private innovations, boosting the city's economy, and creating jobs.

Cities are exploring ways to relieve themselves of the necessity of waste vehicles altogether. The current waste collection systems in high-income cities involve sending a waste collection vehicle to the door of every household in the city at least once per week. By eliminating the waste collection vehicle, cities can save on vehicle maintenance, fuel, and labour costs, while improving safety. New York City found that at least 26 deaths in the city involved a commercial waste vehicle.[38] In Toronto, the city recently entertained proposals for the redevelopment of its Quayside district. The winning vision proposed the installation of underground pneumatic tubes, designed to carry waste from buildings to transfer stations, limiting or outright eliminating the need for a garbage truck to make multiple stops in a certain area.[39] The proposed underground tube system would be located in the basement of buildings throughout a given neighbourhood and would be able to process just under 1.4 tonnes of waste per day and move waste at speeds of 70 km/h.[40] Resourceful cities are finding ways to reduce the significant costs associated with waste collection.

Case Study

- If the world's cities increased circularity between now and 2050, what would the impact on waste volumes be? How much waste into landfill could we eliminate, and what would be the GHG and monetary savings associated with that be?
- Steps:

  1. Use World Bank statistics (below) for waste disposal and treatment in each region of the world.
  2. Assign each of the 5,600 cities in the world to one of the regions below.
  3. Assume same percentages for cities in each region.
  4. Model 3 scenarios:
     a. Low action: Increase the amount of composting and recycling by 2% every five years, with corresponding drop in landfill and/or open dump. Waste volumes continue to grow at same rate.

---

37 Redling, A. 2019. How Montgomery, Alabama, is using routing software to transform its waste collection. *Waste Today*, April 2019. https://www.wastetodaymagazine.com/article/montgomery-alabama-waste-routing-software.

38 Commercial waste zones. New York City Department of Sanitation, 2018, p. 12. https://dsny.cityofnewyork.us/wp-content/uploads/2018/11/CWZ_Plan-1.pdf.

39 Toronto tomorrow: A new approach for inclusive growth. Sidewalk Labs, 2018, p. 44. https://www.torontopubliclibrary.ca/detail.jsp?Entt=RDM3820301&R=3820301.

40 Ibid., p. 208.

b. Medium action: Increase the amount of composting and recycling by 3% every five years, with corresponding drop in landfill and/or open dump. Waste volumes growth rate falls moderately.

c. High action: Increase the amount of composting and recycling by 5% every five years, with corresponding drop in landfill and/or open dump. Waste volumes growth rate falls significantly (Figure 14.3).

All data are from https://openknowledge.worldbank.org/handle/10986/30317.

East Asia

Landfill: 46%
Recycling: 9%
Incineration: 24%
Composting: 2%
Open dump: 18%
Anaerobic digestion: 0%

Europe and Central Asia

Landfill: 25.9%
Recycling: 20%
Incineration: 17.8%
Composting: 10.7%
Open dump: 25.6%
Anaerobic digestion: 0%

Latin America and the Caribbean

Landfill: 68.5%
Recycling: 4.5%
Incineration: 0%
Composting: 1%
Open dump: 26.8%
Anaerobic digestion: 0%

Middle East and North Africa

Landfill: 34%
Recycling: 9%
Incineration: 0%
Composting: 4%
Open dump: 52.7%
Anaerobic digestion: 1%

Figure 14.3 The significance of composting for the circular economy cannot be underestimated. (*Source:* Marina Lohrbach/Adobe Stock.)

North America

Landfill: 54.3%
Recycling: 33.3%
Incineration: 12%
Composting: 0.4%
Open dump: 0%
Anaerobic digestion: 0%

South Asia

Landfill: 4%
Recycling: 5%
Incineration: 0%
Composting: 16%
Open dump: 75%
Anaerobic digestion: 0%

Sub-Saharan Africa

Landfill: 24%
Recycling: 6.6%
Incineration: 0%

Composting: 1%
Open dump: 69%
Anaerobic digestion: 0%

Conclusion

The story of the Koshe landfill in Ethiopia is not over. After the landfill collapse in 2017, the Ethiopian government and the Japanese government collaborated to bring sustainable landfill practices to Koshe, significantly improving safety. Hundreds of Ethiopians continue to make their living scavenging on its mountainsides, many finding ways to return the waste from the landfill back to productive purposes. But the underlying systems that gave us the Koshe landfill – and the landfills in every country on Earth – are unsustainable for future cities.

Future cities must be resourceful. They must use every resource at their disposal as efficiently as possible. They must move toward a closed loop system that maximizes the life of every resource and material. They must drive inefficiency out of every sector. It begins with measuring waste and identifying inefficiencies. And it continues with cities rethinking their approach to physical waste, moving away from reliance on the one-way system of the landfill, and prioritizing strategies to reduce waste, reuse materials, recycle those materials, and move to a circular economy. As the twenty-first century ages and the Earth's population swells to 9 billion people, this transition is an imperative.

15

The Zero Waste City – Introduction

Conor has given us many examples in Chapter 14 of how cities are moving from a linear model to a circular model made possible by an inventory and some imagination from the key players. His startling statistic that "the world's largest 27 cities account for 13% of the world's total production of waste" gives us the opportunity to focus on a model that works in those cities and will pave the way for smaller ones to follow.

Conor also draws on the role of artificial intelligence to help garbage trucks scan and report potholes and graffiti, and how data and IT can build business cases to start removing unnecessary vehicle trips from the road, as the beginning of "purpose-driven" thinking in our future design. Seth and Eric open this door further in Chapter 18, *The Data City*.

As we now move to Terry and Peter's chapter and what needs to be achieved for a zero waste city, I just want to highlight the projections for waste to 2050. Figure 15.1 makes it clear to me that the anticipated volumes of waste mean we have to achieve a fully circular system, but there will be plenty of volume to make it work financially – something that Colin tackles in Chapter 22, *The Invested City*, and James in Chapter 23, *The Financed City*.

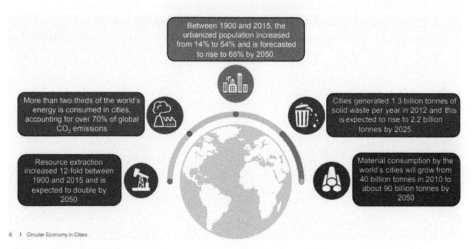

Figure 15.1 A simple illustration of the trend to increased waste through time. (*Source:* Gregory Hodkinson, Hazem Galal, Cheryl Martin, Circular Economy in Cities Evolving the model for sustainable urban future, 2018. Retrieved from: http://www3.weforum.org/docs/White_paper_Circular_Economy_in_Cities_report_2018.pdf.)

The Climate City, First Edition. Edited by Martin Powell.
© 2022 John Wiley & Sons Ltd. Published 2022 by John Wiley & Sons Ltd.

Terry and Peter push Conor's concepts using examples from California. They also focus on the idea that if retailers of electronic goods charge a fee at point of sale for handling the final waste solution, then this fee should guarantee circularity, and we are getting the polluter to pay!

Conor warned us before that there is still a lot of work to be done in regards to waste management in cities. Terry and Peter explore some of the ways we can undertake this work. They begin this chapter talking about the animated children's film WALL-E, which presents to the audience a self-inflicted, apocalypse-style universe in the not-so-distant future, and follows the adventures of one of the robots designed to clean up the mess the human race has brought. It is a beautiful, touching, and often entertaining film, but I cannot help but feel a pang of despair at the not-so-impossible universe it reflects back to us. It reinforces to me that we are destroying our world in order to create food, plastic, fuels, clothing, cell phones, medicines, building materials, packaging, paper, and a lot of other things. At the same time, we discard resources, lack efficiency, fail to reach recycling quotas, rely on waste landfills, and underuse our green technology.

Fortunately, Terry and Peter don't leave us in despair and take on the world of WALL-E as a challenge to find solutions rather than as a destiny. They present "waste" to us as a "resource" that can be embraced and managed rather than dreaded and thrown away. It is a mindset we must adopt, just as Olivia, Nicky, and Alex push for our need to conceptualize and implement a mindset of diverse mobility in Chapters 12 and 13, *The Habitable City*.

This chapter examines the policies, technologies, and finance needed for cities to achieve a zero-waste future and reduce pressures on the natural world to sustain thousands of species, the human race included, and that can also provide jobs for millions. By managing waste as a resource, a city can increase business development opportunities and solutions and, more importantly, save money using resources that are literally under its feet! Terry and Peter also look at a case study in efficiency and climate benefits of zero waste in the Global North and South and the bringing together of developed and developing countries in waste management data sharing and implementation.

It's as simple as *if you've got it – use it*. And use it we must if we are to avoid a WALL-E-style universe in the future. We have everything at hand, and all there is left to do is act.

15

The Zero Waste City

Terry Tamminen and Peter Lobin

In the 2008 animated film "WALL-E" we're introduced to an uninhabitable world where it's clear that 100% of the planet's forests, open spaces, wildlife, and plants have been converted to products used by humans and then discarded into mountainous landfills that resemble the high-rises we once lived and worked in (Figures 15.2 and 15.3).

Figure 15.2 Mountainous landfill from the 2008 film WALL-E. (*Source*: Disney.)

Figure 15.3 Stacked waste bundles today. (*Source*: JenJ_Payless / Shutterstock.)

The Climate City, First Edition. Edited by Martin Powell.
© 2022 John Wiley & Sons Ltd. Published 2022 by John Wiley & Sons Ltd.

The few people who survived this self-inflicted apocalypse have fled to cruise-ship-style space ships awaiting any signs of life on Earth that might indicate the robots, like WALL-E, have succeeded in cleaning up the mess and making the planet habitable once more. I won't spoil the ending of this brilliant film, but it's the beginning that should alarm us, because our world is racing to a future that will make Hollywood look like a prescient oracle.

Already we clear-cut forests, destroy habitats on land and at sea, and pollute 100% of the water we drink and the air we breathe with toxic chemicals and compounds. We do this to create food, plastic, fuels, clothing, cell phones, medicines, building materials, packaging, paper, and a lot of other things that make our lives comfortable.

At the same time, we discard about half of the food produced; are not very efficient in our usage of the energy generated by fuels; and have extremely low recycling rates for the plastics, clothing, electronics, and paper packaging we used ever so briefly. In addition, a majority of construction and demolition (COD) debris waste materials wind up in landfill too.

Cities also tend to be inefficient with water, greenspaces, air quality, intellectual capital, and a wide variety of other valuable resources, and as study after study marks the inexorable loss of forest cover, productive marine habitats, species extinctions, and the billions of humans who go to bed hungry each night, we rarely connect the findings of that research to the growing mountains of waste in landfills, incinerators, and dumpsites, and even the materials collected in recycling centres that, it turns out, will never be put to productive use.

This chapter focuses on how ending the concept of waste, both garbage and energy, saves money, protects ecosystems, and creates new jobs using resources that are literally under our feet. We use examples of energy efficiency solutions that have shown they can be rapidly scaled up to address climate change and used as examples of the "best practices" that can equally be applied to our solid waste challenges.

In places like Denmark, for example, energy is used so efficiently that it produces the highest quality lifestyle on Earth at up to 60% less energy input with little or no human health impacts. Smart cities learning from these models have created thousands of new jobs retrofitting factories, buildings, streetlights, and other municipal infrastructure, and transportation choices. In the vast majority of examples, the energy and maintenance savings that result pay for the improved appliances, lighting, industrial machinery, vehicles, and systems integration/controls. These benefits are derived from enlightened policy, harnessing innovative technology, and bold financing initiatives – the keys to a zero waste city for just about everything that we're not using efficiently today.

So where do we look first for solutions to WALL-E's challenge? We look by seeing "waste" as a "resource" that can be managed. Manufacturers develop global supply chains, at great expense for the raw materials, finished components, and assembly lines to make their products. Cities have already solved that challenge, importing vast quantities of timber, minerals, fossil fuels, energy, and the embedded labour and carbon represented by the harvesting, mining, drilling, generating, processing, and distribution of those materials. We're just not fully utilizing the enormous economic and social value of the "accidental supply chain" that our solid waste and energy sources have created.

Indeed, most cities, consumers, and businesses spend money, utilize space, and provide human capital to dispose of these waste streams in elaborate landfills and highly inefficient incinerators. In developing countries, cities often dump these valuable resources into open pits, roadside piles, and rivers, while their inefficient energy sources contribute to air pollution and human health impacts. All cities contribute significantly to global greenhouse gas emissions with this daily procession of waste – and all of that could end in a matter of a few years, with value-added waste materials paying some of the cost.

The sustainable, climate-resilient, prosperous city of the future will be the one that can connect these dots and harness the opportunity to convert these valuable raw materials into new products, businesses, and jobs, while reducing pollution and destruction of the very natural resources that provide us life in the first place. Though energy production and usage are the largest contributors to climate change, waste and recycling are even more directly under individual control in our day-to-day lives and can therefore have a direct impact on climate change, job creation, equality, quality of life, and resilience to economic, environmental, and even pandemic health disasters.

So with the intent of turning the apocalyptic vision in WALL-E into one of hope, this chapter will consider the policies, technologies, and finance needed to achieve a zero waste future and reduce pressures on the natural world that sustains thousands of species, including our own, and that can provide healthy, dignified, creative, and satisfying jobs for millions. We'll look into where this vision is already a reality and how those examples could be applied in the Global North and South, in communities both formal and informal. Finally, we'll examine the direct link between the zero waste opportunities and rapidly reducing greenhouse gases, while adapting our communities to be more resilient to the climate change-related impacts already being observed.

Policy

Getting to zero waste starts with policy, which has been most effective when following a three-step process:

1) leadership that establishes goals;
2) technically and financially viable standards phased in over time to achieve those goals; and
3) continuous improvement.

Here are two good examples that followed this roadmap, one in waste and one in energy. In 1999, California Governor Gray Davis understood the need to work towards zero waste and approached the problem by setting a goal of reducing the volume of waste being sent to landfills. That same year, the legislature enacted a law mandating state agencies develop and implement an integrated waste management plan outlining the steps to divert 50% of waste entering landfills by 2004, either by reducing waste from various sources or by recycling and productive reuse.[1] The governor and legislature accomplished steps one and two.

Step three took place when I served as Secretary of the California Environmental Protection Agency under Governor Arnold Schwarzenegger in 2004 and was required by

---

1 https://www.calrecycle.ca.gov/stateagency/requirements/lawsregs.

law to certify whether the state had achieved its goal, then to aim higher. Indeed, the data showed we had diverted 55% from landfills on average, with many cities achieving closer to 70%. That gave us the ability to work with agencies, local authorities, and the legislature to understand the composition of the remaining solid waste going to landfills and begin to set goals, policies, and laws to divert those materials. Organics, electronic waste, C&D debris, and other specific wastes were targeted and, as a result, in 2011 Governor Jerry Brown was able to set a new goal of 75% diversion by 2020 (which the legislature enshrined into law). That goal was also achieved.

Another key component of this approach is to make sure goals are both technologically and economically feasible. Electronic waste ("e-waste") provides a good example of how to address a material that is otherwise difficult to recycle. To help achieve its overall waste diversion goals, California banned e-waste from landfills but recognized that there was no effective recycling industry capable of handling the volumes of old phones, computers, TVs, kitchen appliances, and other electronic devices that are constantly facing obsolescence and being replaced.[2]

The state imposed a disposal fee that was charged by retailers when new electronics were sold. The fees were collected by the state and used to pay incentives to e-waste recyclers, giving them the necessary added revenue to handle the most challenging e-waste products and the new increased volumes. This ensured that e-waste would not simply be exported overseas or incinerated, but thousands of new jobs and profitable businesses (and state tax revenues) were created in the collection, disassembly, and recycling of e-waste components.

That example also highlights another aspect of effective policy – the polluter pays. In the e-waste example, we are the polluters when we discard old electronics. Most landfills charge "gate" or "tipping" fees, and of course generators of waste pay to have their output taken away, so, again, the polluter pays, although not always the full cost of dealing with the problem.

The second example of policy leading to zero waste goals addresses energy. Wasting electricity and transportation fuels, by inefficient use of them, contributes to greenhouse gases and other pollutants that threaten human and ecosystem health.

The California Energy Commission (CEC) was established in 1974 to set standards for appliances and buildings to conserve energy.[3] For example, it recognized that some of the highest-consumption, least-efficient use of electricity was by refrigerators. The CEC set standards limiting the sale of refrigerators to only the most efficient ones (and increasing that efficiency over time), saving over half of the energy for the same result.[4] But most people don't buy refrigerators very often, so the CEC provided incentives to consumers to scrap their old models and buy new ones, stimulating the economy, generating tax revenues (which paid for the incentives), and helping consumers save money on energy bills. Most important, the state achieved its energy efficiency goals much faster than if it had waited for the business-as-usual cycle of appliance replacements to take place.

---

2 https://www.greenbiz.com/article/california-becomes-first-us-state-e-waste-law.
3 https://www.energy.ca.gov/about.
4 https://www.eesi.org/papers/view/fact-sheet-energy-efficiency-standards-for-appliances-lighting-and-equipment.

A final example of how policy ignites change is the California Organics Law – California law AB 1826 (2014), which has driven organic waste recycling in the state, and SB 1383 (2016), which supersedes it, has gone even further. AB 1826 set in motion the requirement to begin diverting organics from landfills. SB 1383 accelerates the volume of organic waste being recycled by mandating a reduction from the 2014 level of organics going to landfills in California, 18 million tonnes, or 50% by 2020 and 75% by 2025. Achieving the 50 and 75% mandates will require local governments to speed up the start of food waste recycling programmes. In addition, SB 1383 sets regulations for the procurement of recycled organic products by local governments. The two organic waste products most commonly produced are compost and renewable natural gas (RNG) transportation fuel. It's estimated that over 200 anaerobic digesters will be needed to meet the volume of organic waste diverted from landfills from around the state, creating jobs and state revenues in the process.

Finally, national governments play a role in policy drivers too. The Clinton administration in the USA signed an executive order requiring federal agencies to buy paper stock made of at least 20% recycled content in 1994, as long as its cost was equal to or less than virgin paper, kick-starting the paper recycling industry. Today the US Environmental Protection Agency (EPA)'s Comprehensive Procurement Guideline (CPG) programme is the legacy of this effort and provides cities with a model for procurement of goods with recycled content, helping to develop markets for recycled materials.

Donald Trump recognized the value of reducing waste at one point in his career. Just before he was inaugurated as US President, Leonardo DiCaprio and I went to meet with him to advocate for environmental policy that would also stimulate the economy, especially with programmes like energy efficiency standards and retrofits of buildings, lighting, and appliances. He pointed to the LED lights in his Trump Tower office and the programmable thermostat on the wall and said he appreciated programmes that paid for themselves. History shows he did nothing to advance environmental goals, including those that would have been very good for consumers and the economy, but the encounter proved that reducing waste can appeal to politicians of all ideologies and backgrounds.

One final point about setting policies: Cities learn from each other but also have a tendency to move no faster than the herd. For example, the non-profit C40 works with cities on climate policies and has recruited its members to adopt zero waste goals and share policy options.[5] These collaboratives can establish healthy "competition" among cities, to see which ones can accomplish sustainability and climate goals faster and with more innovation or investment. It also highlights laggards and allows advocates to press their local leaders to follow the best examples. Some comparisons of C40 members show the emphasis is still on recycling, which remains important but has proven to be insufficient as a "zero" waste approach.

The key to achieving zero waste is for state and local governments to change their current line-item-expense approach and begin to view waste as a resource. Reimagining waste and recycling in a way that has long-term benefits for society opens the door to building a zero waste city. The value proposition is simple: maximize the value of local

---

5 https://c40-production-images.s3.amazonaws.com/other_uploads/images/2114_zero_waste_technical_note_final_%28feb_2019_update%29_-_zachary_tofias.original.pdf?1551804802.

resources through coordination among stakeholders, gain economies of scale, and improve operating efficiencies, resulting in greater economic activity, job creation, and a reduced carbon footprint.

Technology

Efficient appliances, lighting, building materials, controls, and other technologies to prevent waste of energy are now ubiquitous and even mandated by regulation in many places. This section will therefore focus solely on municipal solid waste conversion technologies, which are still in their relative infancy in terms of constant improvement, innovation, and deployment.

In few cities today, solid waste is processed efficiently and profitably into raw materials and finished components to be made into new products, including reuse of wood/paper/pulp, glass, metals, plastics, organics (including sewage sludge), tyres, medical/toxic waste, C&D materials, fabrics, and other feedstocks.

By managing the waste as a resource, the opportunity exists to create new jobs to process these materials and still more jobs that can be created by local businesses, farmers, and consumers who reuse them in their supply chains instead of importing raw materials from far away.

In essence, these technologies use relatively cheap feedstocks that have already been purchased and delivered to the city by residents, businesses, and governments, who all willingly pay even more to have these valuable materials taken from their doorsteps, businesses, factories, and civic infrastructure to be disposed of, which has perpetuated a culture of "out of sight out of mind".

There are three technology keys to optimizing the full value of "waste":

1) Mechanical sorting: using the wide variety of options to mechanically sort and decontaminate mixed solid waste, which is up to three times more efficient than separating by homes and businesses into recycling bins versus waste bins.
2) Conversion technologies: the technologies for each type of sorted material to convert it from raw waste into commercial-grade feedstocks for new products, replacing virgin supplies of wood/pulp, oil/plastic, glass, metals, etc.
3) Production technologies: the technologies to utilize those feedstocks for making new finished products.

Mechanical Sorting

After decades of trying, humans are not very efficient at sorting household or industrial wastes into recyclable materials. The USA and EU recover on average just under 40% of what might actually be harvested from wastes (although some individual countries like Germany and Denmark approach 70%). Much of the developing world collects negligible amounts thanks to informal pickers, while the majority of organic and non-organic wastes are thrown into landfills or informal dumpsites. Mechanical sorting, however, is approaching 90% efficiency and has the potential to essentially eliminate the concept of waste by effectively sorting everything into usable materials.

Bulk Handling Systems (BHS) is one company that designs and builds material recovery facilities (MRFs) for mixed waste, aka "dirty MRFs", and to further segregate recycled materials into subcategories, aka "single stream MRFs". MRFs can be highly automated or more dependent on manual labour, depending on relevant budgets for capital expenditure and operational expenses, such as labour costs.

Modern MRFs use a wide variety of technologies to achieve their impressive efficiency, including the following:

- Shredders to break apart bagged waste and large objects, optimizing the material to fit on conveyer belts and screens.
- Conveyors for workers to visually identify and remove objects that are inappropriate for further sorting.
- Sorting robots using artificial intelligence (AI) to identify objects on conveyor belts. These AI robots can recognize recyclables such as cartons, plastic bottles, and containers. After the robot identifies the object, it uses a suction cup to pick it up and place it into the correct bin.
- Optical scanners that recognize objects and can direct them to other conveyors for more specific sorting.
- Eddy currents, magnets, vacuum suction, air jets, screens, and other devices to further separate materials.
- Washers, bailers, and other devices to aggregate each type of waste into more homogenous, usable collection formats.

Conversion Technologies

Once the mixed waste is sorted into individual types, technology exists to convert those materials into feedstocks to replace virgin materials in making new products, either entirely or by some percentage, including:

- Organics: about half of municipal solid waste in most cities around the world is organic.
  - Sources: food (both pre and post-consumer), agriculture, green waste maintenance (trees and lawn cuttings), sewage sludge, and other organic content.
  - Conversion technology: composting has been around for centuries but tends to be located far from urban areas. The primary method used today in cities is anaerobic digestion, essentially the same process as in the human gut, where solids are converted to methane and other usable gases, as well as soil amendments and liquids that can also be used as fertilizers.
  - Challenges: biological decomposition of organic material can be interrupted, can reduce yields of desirable outputs, or can be stopped entirely by the introduction of toxins or by mixing wastes of different types.
- Plastics: recycling of plastic varies widely by country and region, but estimates show not more than 20% of all plastics are recovered.
  - Sources: plastic packaging and beverage containers, fabrics, auto parts, ropes and other fibres, electronics casing and components, tools, and other products made from rigid and soft plastics.

- Conversion technology: primary methods include pyrolysis and other gasification technologies to heat plastics, converting them to gases, which can then be used industrially or can be cleaned, separated, and converted to resins to make new plastic feedstocks. A cheaper conversion process involves shredding, washing, and sanitizing plastic waste of a specific kind so it can be melted or pelletized to become feedstock for making the same product again (example: PET plastic beverage bottles can be processed into flakes or pellets that can make new PET beverage bottles).
- Challenges: each of the dozens of types of plastics require very specific performance characteristics when used for specific products, so mixed plastics can't be used to make clear plastic beverage bottles, for example, but need to be fully sorted to be efficiently utilized in commerce. Moreover, recycled plastics of a given type may lose some of those performance characteristics (clarity, rigidity, tensile strength, etc.) from the original use or the conversion process, so a limited percentage of recycled content can be used in making a new product. "Downcycling" allows for low-value or mixed plastics to be used to make things like floor mats or toys, but those materials become even harder to recycle later.

- Paper/pulp: this is the "good news" category of waste management, because up to 70% is recycled in many countries.
  - Sources: in addition to post-consumer paper and cardboard waste, recycled material comes from paper mill trimmings and other scrap from the manufacture of packaging and paper.
  - Conversion technology: the process generally involves shredding and mixing recycled paper with water and chemicals to break down the lignin, remove inks, and otherwise turn the paper and cardboard into a pulp, which is then heated and further degraded into a slurry. The slurry is screened to separate plastic (especially from plastic-coated papers used for magazines and packaging) and bleached to remove colours/inks, after which it can be processed into new paper products.
  - Challenges: conversion technology, especially if content is repeatedly recycled, creates very short fibres which lose the characteristics necessary for making many kinds of packaging and higher quality papers (although even newspaper requires certain fibre length, strength, and flexibility). As such, virgin fibres are added in varying amounts to finished products.

- Tyres: used tyres are one of only two wastes that cannot practically be reduced, as all other wastes might be under "reduce, reuse, recycle" regimes (the other being sewage sludge). As such, conversion becomes extremely important as billions of tyres already reside in dumps and special-purpose landfills all over the world, creating sources of toxic fires and other hazards.
  - Sources: there are approximately 1.5 billion motor vehicles in the world today and automakers produce another 60 million cars each year, generating roughly 1.5 billion waste tyres annually.
  - Conversion technology: waste tyres are converted with pyrolysis or gasification into oils, gases, and metals, which can then be processed into refined fuels, carbon black (for making new tyres or printer ink, among other things), industrial gases, and any product derived from the recovered steel.
  - Challenges: tyres are notoriously difficult to handle and process prior to feeding the material into conversion technology systems. They come in widely varying sizes and

shapes (after piling waste tyres, many lose their typical rounded shape with even, parallel sides). One solution is to shred them first, but that adds considerable expense and energy inputs. Another is to load them into a pressure cooker, but that is inefficient because a pile of tyres includes up to half of the space being empty.

- Glass: glass is a good alternative to plastic, because containers can more easily be reused, are more sanitary, and don't leach petroleum toxins, and few bottles or other glass products will end up in the environment accidentally, among other benefits.
  - Sources: beverage bottles, food and other storage containers, cooking/glassware, windows, computer/TV screens, consumer goods (e.g. thermometers), and various industrial products.
  - Conversion technology: typically glass is ground into granules or powders, cleaned, and then melted to be processed into new glass products. Another method uses ultrasonic waves to pulverize the glass into more homogenous grains, which are valued as industrial abrasives.
  - Challenges: colour can't be removed from glass, so clear glass is the only recycled material that can be made into new clear glass products. Bottle manufacturers can only use up to 50% of recycled glass due to size to guarantee product quality. Coloured glass is also difficult to use as a major percentage of inputs to new products, because the colour of the waste feedstock may not easily match the desired colour of the new products.

- C&D debris: as more waste is recovered and productively used, C&D is becoming one of the largest remaining components of typical municipal solid wastes.
  - Sources: new and remodelling construction projects; demolition of old buildings; sheetrock; wood; concrete, bricks, and other stone-like materials; replacement of building components, such as carpets, tiles, kitchen cabinets, bathroom fixtures, etc.; paint and coatings; glass; wiring and electrical components; scraps and rejects from building supply manufacturing plants.
  - Conversion: MRFs can be specifically designed to disaggregate mixed C&D waste and separate it into wood, metal, gypsum, glass, and other components. C&D separation also lends itself to manual labour at jobsites to harvest any valuable or reusable objects and feedstocks.
  - Challenges: C&D debris is typically hard to separate. For example, tearing down walls results in wood or metal framing connected to wires, particle board or sheetrock, and coated in paint, wall covering, or other dissimilar materials, including toxic substances such as asbestos and lead. Manual separation is time-consuming and dangerous.

- E-waste: perhaps no other products become obsolete and are replaced so quickly as the countless devices we carry in our pockets and that populate our desktops. Electronic waste has therefore become one of the fastest growing sources of discarded materials globally but includes some of the most valuable recoverable resources.
  - Sources: any electronic appliance, including mobile phones, computers and monitors, printers and other peripherals, televisions, radios, music players and speakers, household appliances, and essentially anything with a battery or a cord.
  - Conversion technology: unlike other forms of sorted waste, e-waste needs a significant amount of manual labour to deconstruct appliances and sort the resulting materials into plastics, glass, metals, and reusable parts (e.g. computer hard drives). Once

separated, the components are recycled as described previously. Electronics contain small quantities of precious metals, but when aggregated from thousands of discarded devices they can add up to a valuable asset for recyclers. In many cases, devices can be refurbished and resold, especially in developing countries where the latest models are too expensive for many consumers.

– Challenges: the countless types and versions of electronic devices present unique challenges to disassemble and harvest valuable components, generally defying mechanical separation. As a result, tonnes of e-waste are sent each year to developing countries where cheap labour and lax safety standards lead to disassembly and recycling by workers, including children, and often leave toxic dumpsites of chemicals used in the process alongside components that have little value outside of a more formal recycling system.

• Metals: metals enjoy among the highest rates of recycling, because they are typically easy to separate from other wastes and a commodity market exists in many cities where even informal workers can sell the feedstock. Sources are ubiquitous and obvious; conversion technology is generally the same as that used with virgin metals (grinders, smelters, and other processors) to make new metals for a wide variety of applications. For example, "according to the International Aluminium Institute's analysis, approximately 75 percent of all primary aluminum ever produced is still in productive use due to its strength, product life and recyclability. Recycling aluminum uses about 8 percent of the energy required to make new aluminum ingot and emits 92 percent fewer greenhouse gases." Figure 15.4 illustrates waste types and the processes for organic waste, construction and demolition waste, electronic waste, and PET plastic waste.

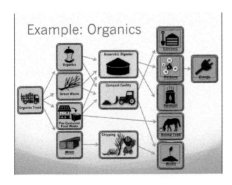

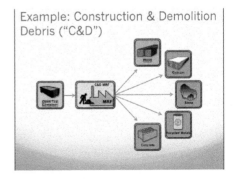

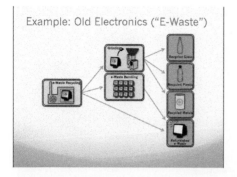
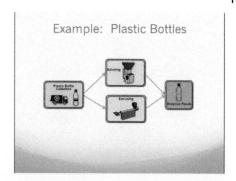

Figure 15.4 Waste optimization by category. (*Source*: Courtesy of author Peter Lobin, author created infographics.)

Production Technologies

As noted previously, conversion technologies convert recovered wastes into feedstocks for making new finished products. In some cases, the recycled material performs the same as virgin feedstocks (e.g. metals), but most recycled content requires specialized production technologies to manufacture new products, generally from a mix of the recycled material with virgin feedstocks (e.g. products made from paper, plastic, and e-waste).

In other cases, the recycled content is not converted to similar products but becomes feedstock for different industrial processes (e.g. pre-consumer food waste can be made into animal feed, which in turn becomes human food, but post-consumer food waste is generally converted to gases and fertilizers, which provide inputs to agriculture, although not human food itself).

One example of conversion technology that generates functional products but which is not always used in that manner is recycled paper used in toilet tissue. Toilet tissue for home use is typically softer than tissues used in public restrooms or office buildings and has other properties that require using virgin wood pulp. As a result, vast swaths of Canadian boreal forest are clear-cut every year and literally flushed down the toilet.[6] Given that these products can be made practical and economically viable (if not always as soft or fragrant) from recycled paper, it is an unconscionable waste of our natural heritage to destroy habitats that protect wildlife and humans in this manner and is therefore another sector of a zero waste future that will require policy solutions before technological ones can be fully realized.

Finance

Without investment for cities and technology providers to build collection, sorting, and conversion technology projects, the goal of a zero waste future would be difficult at best. The challenge to financing this transition is that some recycling and use of waste feedstocks is highly profitable (e.g. metals and certain kinds of plastic), while others are barely

---

6 https://www.theguardian.com/world/2019/mar/01/canada-boreal-forest-toilet-paper-us-climate-change-impact-report.

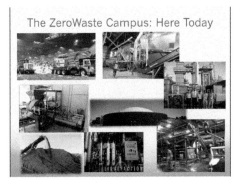

Figure 15.5 The zero waste campus. (*Source*: Courtesy of author Peter Lobin.)

profitable (glass), and still others are profitable in some locations but not financially viable in other places (e.g. waste tyre or sewage sludge conversion).

There are many reasons for this variability, including whether cities or waste generators pay waste collectors a tipping or gate fee for transportation and disposal; whether industries that can use harvested feedstock are nearby and can affordably purchase the recycled materials; and whether volumes of particular wastes are found in sufficient quantity at a given location to generate commercially viable and reliable outputs.

One solution to this challenge that can work in many cities is to vertically integrate – co-locate the sorting, conversion, and value-added manufacturing in the same location, a "zero waste campus" (Figure 15.5). The campus approach can further achieve economies of scale and operating efficiencies by combining the waste and recyclable materials from multiple communities; lowering the cost of transportation; reducing processing costs; increasing market opportunities; and increasing the value of the materials. Larger volumes of materials, if local governments work in concert, will increase market opportunities as well as the options of adding value to the materials – from simply cleaning and selling them to a processor, to processing materials themselves, to manufacturing (converting) products and closing the loop.

Combining the value chain for waste in one place has multiple financial benefits, making otherwise risky or unprofitable projects both reliable and rewarding:

- Transportation costs are reduced or eliminated if sorted waste is taken from an MRF across campus to a conversion facility for each type of material, and from that facility to the end-user.
- Commodity price volatility is reduced, because the MRF is a constant source of raw materials to the conversion facility, so fixed price contracts can be negotiated.
- Higher margins from processing one type of waste versus another can generate predictable, acceptable, risk-adjusted returns for investors when aggregated across all of the different waste streams, feedstocks, and outputs.

A good example of this approach was a project designed by the non-profit R20 Regions of Climate Action for the Kingdom of Bahrain. Bahrain had planned to buy an incinerator from a French company for over US$400 million that would have reduced the volume of waste by 60% and generated electricity. However, to make the economics viable, investors required a tipping fee paid by the municipality of over US$90/tonne; 40% of the volume would remain as ash, still needing landfill disposal; and the resulting cost of electricity would exceed US$16/watt (compared to solar at under $3/watt installed). Instead, a zero waste campus was designed and, based on analysis of the daily waste load, the campus was forecast to be profitable at US$13.5/tonne tipping fee (one-third of the cost that Bahrain's cities are paying today for landfill disposal). New jobs and businesses would be created in the process, and the old landfill could even become an "urban mine" to collect old waste and process it through the new facility.

Zero Waste in the Global North and South: A Case Study in Efficiency and Climate Benefits

Many of the examples described so far are based on projects in the Global "North" (developed countries) but can equally apply to the Global "South" (developing countries).

For example, Brazil's booming economy needed more electricity, but power generators and transmission lines were struggling to keep up. Eletrobras, the country's largest utility, sought projects and technologies to use energy more efficiently, thereby reducing the need to build so many new power plants.

The R20 Regions of Climate Action worked with the city of Rio de Janeiro to replace more than 400,000 inefficient old streetlights with modern LEDs that would save up to 67% of the energy and reduce air and greenhouse gas pollution from that amount of electricity generation, while repaying the cost of retrofits from energy and maintenance savings (Figure 15.6).[7] This payback financing model worked in the USA and Europe, but Brazil faced unique challenges.

Figure 15.6 New LED streetlights in Novo Friburgo, Rio de Janeiro, Brazil. (*Source*: Courtesy of author Peter Lobin.)

---

7 The next generation of street lighting: A feasibility study and action plan for Rio de Janeiro. https://regions20.org/wp-content/uploads/2016/08/Rio-Streetlighting-Action-Plan-ENGLISH.pdf.

First, all LED streetlights were imported and subject to a hefty tariff, making them unaffordable. R20 experts worked with manufacturers to coordinate in-country manufacturing supported by the introduction to dozens of cities with over 1.5 million streetlights, making new facilities in Brazil a good investment and new streetlights very affordable. Second, not all LED streetlights performed well, so the team worked with Eletrobras to develop design and testing standards to ensure high performance in Brazil's unique coastal and interior municipal conditions. Finally, a US$500 million Special Purpose Finance Vehicle was set up by Pegasus Capital Advisors, a US sustainability-focused investment firm, so cities could borrow money to replace all streetlights at once and repay the cost from savings.

In other words, transferring technology from north to south requires more than just sharing experience. Dedicated businesses, government authorities, and NGOs can accelerate the transition to a zero waste, low-carbon economy by rolling up their sleeves and working out the barriers, so that markets can function in one place as they do in others.

The climate benefits of this example are obvious in some ways but subtle in others. Reducing power consumption by nearly 70% for any product, especially one as ubiquitous and beneficial as municipal lighting, will reduce greenhouse gases from energy production dramatically. Less obvious is the embedded carbon in the product itself – the LED streetlight weighed about 5.5 kg, compared to 29 kg for the average fixture it replaced, reducing the need for metal, electrical components, glass, and a host of other materials. LEDs are also suitable for addition of other technologies such as traffic and parking sensors, thereby reducing emissions from traffic and cars circulating on streets looking for a place to park.

Finally, the small, more efficient fixtures generate far less waste when the time comes to replace and recycle them, contributing to a zero waste, low-carbon future.

Conclusion

This chapter was able to highlight the opportunity and challenges of a zero waste city but was only the most basic overview of each topic. By managing waste as a resource, cities can unlock business development opportunities and solutions. A structure that combines collection, recovery, sorting, conversion, and financing is complex, requiring leadership from government officials and business leaders. A public–private partnership is needed to gain the experience, resources, know-how, and capital of the private sector. Local circumstances, geography, population, weather, etc. will vary widely, but the general approach and concepts presented here should provide hope for a future without "waste", and our ability to address climate change before it's too late is as close as our waste and recycling bins.

16

The Resilient City – Introduction

Terry and Peter in Chapter 15 offered us the "zero waste campus" where we can co-locate sorting, conversion, and value-added manufacturing, and we have seen the projected waste growth to 2050 which would make this concept highly financeable as the volumes increase.

How we translate these concepts from the Global North to the Global South relies on the local context being understood, which I advocate again that a city can provide by ensuring it is managed in the right context of its governance structure. The financing opportunity in the circular economy is covered by James in Chapter 23, *The Financed City*.

Let's take a moment to summarize. We have taken some lessons from cities through time, and we have understood some of the "North Star" principles set out in the UN SDGs and trends in electrification, densification, and efficiency of resources and services within our cities that are driving cities to be better. We now need to prepare our cities to handle some of the more alarming trends we are witnessing – increased extreme events, pressures from urbanization, the changing shape of demographics, and the demands from society for higher performing infrastructure.

Richard and Sarah take this emerging but mature concept of resilient cities and use direct interview results from key Chief Resilience Officers (CROs) to guide us through this topic.

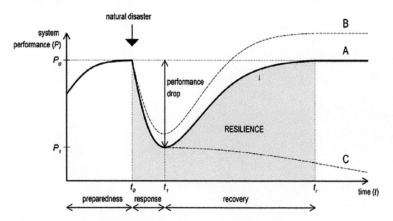

Figure 16.1 The phases of handling extreme events. (*Source*: Koren, D., Kilar, V., & Rus, K. (2017). Proposal for Holistic Assessment of Urban System Resilience to Natural Disasters. IOP Conference Series: Materials Science and Engineering, 245, 062011. Licensed under CC BY 3.0.)

The Climate City, First Edition. Edited by Martin Powell.
© 2022 John Wiley & Sons Ltd. Published 2022 by John Wiley & Sons Ltd.

As you read their chapter, you will get some first-hand accounts from people dealing with some hard situations. I wanted to highlight, in Figure 16.1, the generic concept of handling extreme events – the preparation, the event itself, and the recovery. These phases require different subsets of city actors to collaborate and contemplate how to deal with each phase in real-time. The measure of resilience is how successfully they can operate through these phases.

Throughout the COVID-19 pandemic we have seen turmoil and progress in equal measure, including social unrest; intense climate events with heatwaves, hurricanes, and wildfires; and economic shock. Our cities have been challenged, too, as we all have. This past year has taught us that our cities not only need help to recover but also need the long-term ability to support themselves, to be prepared, and the ability to respond effectively – resilience. In Figure 16.2, you can see some of the factors associated with city resilience.

With these factors in mind, and building upon the work of the 100 Resilient Cities initiative, which aimed to not only assist cities in building resilience to the impact of climate change but also help counter the fallout from economic and social shocks, we now see cities like London and Tel Aviv, at the height of a global pandemic, setting up resilience funds and recruiting CROs.

The pandemic is giving us new insights and providing us with an opportunity to focus harder on resilience more than ever before, proving that resilience planning should be at the heart of any city – pandemic or no pandemic.

Just as Nicky and Alex did in Chapter 13, *The Habitable City*, Richard and Sarah emphasize the importance not only of *why* we need resilient cities but also of *how* we can implement resilient strategies in cities. Using Rotterdam as a prime example of a working resilient city, they examine the various steps the city has taken: through its creation of the CRO role in 2014, its climate adaptation programme, the release of the

Resilience Factors

Effective Policing and Judicial Mechanisms

Provision of Basic Services

Social Cohesion

Strong Community-Government and Inter-Governmental Cooperation

Microeconomic Security & Social Protection

Greater Income & Social Equality

Social Networks Social Support

Figure 16.2 Factors that contribute to city resilience. (*Source*: Conceptualizing City Fragility and ResilienceUnited Nations University Centre for Policy Research, Working Paper 5, October 2016. United Nations University. Retrieved from: https://collections.unu.edu/eserv/UNU:5852/ConceptualizingCity FragilityandResilience.pdf.)

first documented resilience strategy, and its formerly assigning resilience as a governmental responsibility, among other things. Most significantly, Richard and Sarah look at the post-pandemic steps the city has taken, the updating of resilience strategy, and the Seven City projects (the Big 7), which aim to boost the economy and liveability, with a focus on green infrastructure and climate adaptation as well as creating jobs and attracting businesses.

Richard and Sarah identify that there is an important link between infrastructure and resilience. They examine the transformation of the New Orleans data collection networks following Hurricane Katrina. They also present a rare positive outcome of the pandemic through the acceleration of digitalization in cities to meet the demand for accurate data, resulting in quick and decisive action. Just as we moved from the *why* to the *how*, we have shifted away from *theory* and into *action*.

16

The Resilient City

Sarah Wray and Richard Forster

The COVID-19 crisis has focused the minds of local governments in the USA and Europe on the need to build resilience into their recovery strategies.

When the 100 Resilient Cities (100RC) initiative was launched in 2013 by the Rockefeller Foundation, its executive team were keen to emphasize that it was not only to assist cities in building resilience to the impact of climate change but also to help communities counter the fallout from economic and social shocks. While other city networks such as C40 and ICLEI had been leading the way as advocates for action on climate change mitigation and adaptation, it was the 100RC network that reinforced the need for a holistic strategy on resilience.

What this year has shown is that the mission of 100RC has never been more necessary. While The COVID-19 pandemic (the pandemic) has knocked the world sideways, 2020 has also witnessed parallel and compounding crises, including social unrest through the Black Lives Matter protests; intense climate events with heatwaves, hurricanes, and wildfires; and the unfurling economic shocks following the pandemic.

If one positive is to come from the coronavirus outbreak, it is that this perfect storm should be a pivotal moment in changing attitudes to resilience. The focus must be on learning lessons to strengthen communities against future shocks and stresses – whether economic, social, or environmental.

"Having this constant, high level of shocks and stresses together has really brought home to people that you've got to bring that multiple benefit approach into the present," says Lauren Sorkin, Executive Director at the Resilient Cities Network (RCN), which is the organization that evolved out of the 100RC network.

Refocus on Resilience

One lesson that cities have already learned in terms of building resilience has been the need to appoint leadership that can break down silos within local government. Chief Resilience Officers (CROs) are at the forefront of this. Launched and funded as part of the 100RC programme, one of the key roles for city CROs has been to communicate and collaborate across different departments. This is essential to build a coherent strategy that can help cities face up to the challenges posed by climate change, health crises, or an economic downturn.

The Climate City, First Edition. Edited by Martin Powell.
© 2022 John Wiley & Sons Ltd. Published 2022 by John Wiley & Sons Ltd.

The ability to bring multiple strands together is where CROs excel because they "understand people in a different way and they understand the integration between the challenges that the city might face," Sorkin says.

The advent of the pandemic has seen CROs receive greater support in terms of funding, staff resources, and proximity to the mayor. A survey by the RCN, organized with Dalberg Advisors, found that 87% of CROs[1] are involved in their city's pandemic response or recovery effort.

"When city leaders need to make decisions, they want to make sure that they are purposefully building for a resilient future and so they're calling on their CROs for their advice and analysis," Sorkin says.

Despite severe budget challenges globally, cities such as London[2] and Tel Aviv[3] have set up dedicated resilience funds. And several cities have now pledged to recruit a CRO, including Austin, Texas, and a number of states in the USA and regions elsewhere are developing legislation to create resilience officer roles or take on resilience responsibilities. The issue is going up the agenda in fast-growing cities in emerging markets too, as well as at the national policy level. India's Ministry for Housing and Urban Affairs, for example, recently launched the Climate Smart Cities Assessment Framework,[4] which aims to provide a clear roadmap for cities towards combating climate change.

From Why to How

This elevation of resilience is not only at the policy level, and there has been a marked shift towards a more practical approach to deploying resilience measures.

As Arnoud Molenaar, CRO, city of Rotterdam, explains: "We are entering a new phase. So far, we have been busy explaining what resilience is about and why it is important. Now, we are entering a phase that's much more about the how."

Rotterdam is one of the more mature cities when it comes to resilience. It created the CRO role in 2014, broadening the climate adaptation programme which had been in place since 2008. In 2016, Rotterdam launched its first documented resilience strategy.[5] It also has, for the first time, assigned resilience formally as a responsibility of one of its vice-mayors and has appointed a cyber-resilience officer with a dedicated budget. Rotterdam is doubling down on its approach following the pandemic and embedding resilience into all aspects of city planning. "COVID is giving us new insights but also a new trigger to focus even harder than before on resilience," says Molenaar (Figure 16.3).

---

1 https://collections.unu.edu/eserv/UNU:5852/ConceptualizingCityFragilityandResilience.pdf.

2 https://www.london.gov.uk/coronavirus/mayors-resilience-fund-supporting-innovation-resilience.

3 https://telavivfoundation.org.

4 https://www.thehindubusinessline.com/economy/housing-ministry-launches-climate-smart-cities-assessment-framework-20/article32581590.ece.

5 https://www.thehindubusinessline.com/economy/housing-ministry-launches-climate-smart-cities-assessment-framework-20/article32581590.ece.

Figure 16.3 Pond in a Park, Rotterdam. (*Source*: Leonid Andronov/Adobe Stock.)

His team recently began the process of updating the resilience strategy in light of the pandemic and other learnings since 2016, and this work will also feed into the city's mid- and long-term COVID recovery planning.

The 2016 strategy outlines seven qualities of resilience: reflectiveness, resourcefulness, robustness, redundancy, flexibility, inclusiveness, and integrated. These are central to Rotterdam's recovery plan, and the first one in particular – being reflective – is now coming into play.

"If we really want to learn from the situation, then we have to organise on a structural basis, harvest the lessons learned, and harvest the vulnerabilities that have become clear because of COVID," says Molenaar. "When we have a good mapping of this, then we can also focus on these new vulnerabilities. We have to learn from this crisis and make ourselves stronger to be better prepared for the next one."

This is echoed by cities around the world which have pledged to address systemic inequity and prioritize sustainability in their recovery planning.[6] Of 53 respondents from 47 different cities within the RCN that responded to the Dalberg survey, 79% said that enhancing social equity is a top priority for recovery.

Molenaar added: "The challenge is raising awareness about not only coming out stronger from the COVID crisis but also that we are already in a crisis – we can apply the lessons from the pandemic to the economic crisis but also to the bigger wave, which is the climate crisis. It isn't only coming; it's already here."

---

6 https://cities-today.com/green-jobs-public-transport-and-15-minute-cities-top-mayors-agenda-for-covid-19-recovery/.

As part of its COVID economic recovery plan, in June Rotterdam announced seven city projects[7] (the "Big 7") that aim to boost the economy and liveability, with a focus on green infrastructure and climate adaptation as well as creating jobs and attracting businesses. Other key initiatives that are now being accelerated in Rotterdam include the use of a "filter" tool, which is a set of principles to help the city make decisions, with citizens, about which ideas or actions to prioritize.

The leading principle is that initiatives have to contribute to a more resilient society and economy. The city is working with Erasmus University to develop criteria that can be used to judge whether proposals contribute to the resilience goal.

Further efforts to embed resilience thinking throughout all levels of the municipality, rather than it being perceived as the role of the resilience team, include providing training on how to apply the city's "Resilience Scan" and apply it to specific projects.[8] The scan provides a checklist to help people ensure they have taken into account possible shocks and stresses and considered resilience qualities within their plans. So far, this training has been mainly delivered to spatial planners but will be rolled out more widely.

Molenaar comments: "If you think about flexibility in spatial planning related to climate change, you have to take into account a lot of uncertainty and it may be that within 30 or 40 years you have to adjust your plan – for higher-than-projected sea-level rise, for instance. It's important not to make future adjustments impossible and that's a kind of flexibility that's not really common among spatial planners."

Sorkin sees, too, that there is a growing push from cities for practical tools they can use in their recovery planning. In July, the RCN released its Toolkit for Resilient Recovery,[9] developed collaboratively with city members.

"We've seen huge resilience and recovery funds earmarked. The toolkit aims to help cities to prioritise according to multiple challenges," Sorkin says.

Within the network, cities are also collaborating to develop tools and frameworks in key areas including the circular economy, climate resilience, waste management, and prevention of plastic pollution in the ocean.

Delivering the Infrastructure to Support Resilience Planning

If resilience planning has risen to the top of the local government agenda because of COVID – which may assist cities in getting the funding and political support for climate change action at a federal or central government level – a second positive outcome of the crisis has been the acceleration of digitalization in cities.

The pandemic has seen a move away from the vendor-shaped promise of "smart cities" to a need to find out how city governments can quickly get hold of accurate data and make decisions. As with urban resilience, we are seeing a sharp shift from the theory and the benefits to the practical and urgent deployment of measures by government.

---

7 https://www.rotterdam.nl/bestuur-organisatie/stadsprojecten/.

8 https://www.resilientrotterdam.nl/rotterdam-resilience-scan/.

9 https://resilientcitiesnetwork.org.

It was 2005 when Hurricane Katrina devastated New Orleans, killing more than 1,800 people and causing US$125 billion in damage. More recently, pre-COVID, the city has faced everything from a large-scale cyber-attack to the partial collapse of a large hotel under construction, an exploding turbine at a water plant, and countless severe storms.

Katrina had already shown New Orleans that there was an urgent need for the better management of data to build resilience. "We had to get really sophisticated very quickly in how we used information, and that data became a weapon for us and a great resource," says Kimberly W. LaGrue, Chief Information Officer, city of New Orleans.[10] "One of the most important things for us to address at the time was blight. The number of homes that had not been repaired or even started repair work was concerning to the mayor and his administration."

Government officials realized they needed geographic information system (GIS) data, information about properties, and a way to contact owners and to engage with them to start renovating properties. Property databases began to be updated and maintained, which enabled the administration to employ data more into decision-making processes.

"The culture was formed around our need to use information to assist in the city's recovery and to make it a viable city once again," she says. "That then spread throughout city government."

What helped further to contribute to the city's progress was the community's involvement in the use of data.

"They [the community] were not just giving data, they were taking data and creating data for us," she explains. "That accelerated and drove the adoption of other datasets and using data in other areas of the city, including public safety and crime."

LaGrue adds that the technological adoption was built out of need as the community placed demands on the city to assist them in their recovery.

"We weren't just saying as a city, 'Hey here is this cool tool, why don't you use it?'" she says. "It was innovation at its best because they [citizens] were vested."

The experience from Katrina and the culture it engendered within the administration has helped the city to use data to cope with the impact of the pandemic. New Orleans' Mayor, LaToya Cantrell, has consistently said the city is focused "on the data, not the date" when it comes to decision-making around when the city should impose or relax measures. LaGrue says the Mayor's Office has set up a data "war room" to understand exactly what data were required by key players in the pandemic response effort, such as the health department and public safety staff.

"Most important for us was the veracity of the data," LaGrue says, adding that the team would quickly go back to sources where information was incomplete or inadequate.

Using existing tools such as Power BI, New Orleans set up dashboards which track trends and criteria for easing restrictions. One dashboard includes both city and state data on daily and cumulative death rates, test numbers, and positive diagnosis figures. This is overlaid with geographical data to highlight hotspots where further efforts need to be focused. A second dashboard maps the data against baseline milestones set by the city, including the five-day rolling average of new COVID-19 cases and testing, as well as the availability of ventilators, hospital beds, and intensive care spaces.

---

10 https://www.nola.gov/.

LaGrue reveals that the amount of data stored by the city is growing at 15% a year, with the overall budget specifically for this increasing from between 5 and 8% each year.

"Right now we are at a place to catch up, and make a pretty sizeable investment in our infrastructure and growth that we have realised over the years," adds LaGrue.

Investment in hard and soft infrastructure is, according to LaGrue's colleague Ramsey Green, CRO in New Orleans, a key part of building resilience. Since New Orleans saw its first COVID diagnosis in mid-March 2020, Green says the city has executed over US$300 million in infrastructure contracts alone. The city's bid process has gone completely online and 95% of the professional non-field workforce shifted to working at home.

"I believe there is a direct relationship between infrastructure and resilience," says Green. "When the resilience function of our government was started, it was very much a policy ideas shop. Vision is important in city government, especially in a place like New Orleans, but delivery is more important."

Ramsey was appointed as Deputy Chief Administrative Officer for Infrastructure in 2018, following the election of Mayor LaToya Cantrell, and subsequently assumed the position of CRO.

"A big reason I was hired is because the city was so behind in the delivery of infrastructure projects," Green says.

He explained that when he began the role, almost US$1.8 billion in federal infrastructure funding which was provided for the city of New Orleans and the New Orleans Sewerage and Water Board following Hurricane Katrina was largely unspent, despite being approved in July 2016 and earmarked specifically for rebuilding streets, water lines, and drainage.

Another US$300 million in Federal Emergency Management Agency (FEMA) hazard mitigation funds and US Department of Housing and Urban Development (HUD) funds for green infrastructure projects also "sat languishing". Moreover, coordination was an issue, with the same roads often being dug up repeatedly for separate infrastructure projects due to a lack of holistic planning.

"We have aggressively focused our work on building up our in-house ability to honestly and effectively manage the deployment of these funds and the construction of these projects, vastly improving the infrastructure within our city while making a substantial impact on our local economy," Green explains.

Projects underway and set to break ground include the Gentilly area Blue & Green Corridors[11] and Mirabeau Water Garden,[12] which use natural systems to manage water and reduce flooding (Figure 16.4); St Anthony Green Streets,[13] where stormwater solutions are being retrofitted; and the St Bernard Neighborhood Campus,[14] which integrates

---

11 https://nola.gov/resilience-sustainability/areas-of-focus/green-infrastructure/national-disaster-resilience-competition/gentilly-resilience-district/urban-water-projects/blue-green-corridors/.
12 https://www.nola.gov/resilience-sustainability/resources/fact-sheets/mirabeau-factsheet/.
13 https://www.nola.gov/resilience-sustainability/gentilly-resilience-district/st-anthony-green-streets/.
14 https://www.nola.gov/resilience-sustainability/gentilly-resilience-district/st-bernard-neighborhood-campus/.

Figure 16.4 New Orleans City Park – The city is learning how to design in natural drainage into its existing parks and new spaces. (*Source*: SaceyK Photography/Shutterstock.)

green infrastructure and recreational improvements at a school and playground. The Oak Park and St Roch FEMA hazard mitigation projects,[15] which will store up to 30.7 million gallons of stormwater, have also been approved for construction funding.

Green says this delivery culture has also helped New Orleans respond to the pandemic and keep going amid the crisis. Both the state of Louisiana and the city of New Orleans categorized construction as an essential service since the start of the pandemic.

"Construction is a big [way] that we're mitigating ourselves against the impact of future significant, and even minor, storms – putting better drainage lines and green infrastructure in – so we've really accelerated those projects," Green adds.

To date, US$2 million in bond funding has been allocated to the French Quarter pedestrianization concept for pavement repairs, for example, and the construction of bicycle and pedestrian-focused infrastructure is also underway through the Moving New Orleans Bikes (MNOB) programme.[16] Currently, the city is constructing 17.7 km of cycling routes on the West Bank of the Mississippi River in the Algiers neighbourhood.

"We never stopped or slowed down," says Green. "Despite the pandemic, it doesn't mean that the city won't flood; it doesn't mean that the effects that our city faced prior to the pandemic go away so we have to continue to get the work done that we were doing before we got here."

---

15 https://www.nola.gov/resilience-sustainability/areas-of-focus/green-infrastructure/hazard-mitigation-stormwater-projects/oak-park/.
16 https://nola.gov/transportation/moving-new-orleans-bikes/.

Another instance of this has been the launch of the New Orleans' COVID-19 meal assistance programme,[17] which was 75% funded by FEMA, with the rest covered by the city through Coronavirus Aid, Relief, and Economic Security (CARES) Act money. The initiative engaged local restaurants to prepare and deliver food to residents in need. Over 12,000 residents now receive up to two meals per day, and the programme is set to continue for the foreseeable future. Sorkin observes that food resilience programmes have been a key initiative during the pandemic.

"Whereas before [CROs] were looking at strengthening rooftop gardening to provide healthy food and cooling, now they are also thinking about food supply in terms of shoring up their city's nutritional content, and in terms of people not having to travel very far [for food]," she explains.

The meal assistance programme is also an example of the wide-ranging role of the CRO, with Green noting that the pandemic has seen him considering everything from large construction projects to children's playgrounds and the impact of the pandemic on mental health and food insecurity.

"Every day I learn something that amends the way we do this work," he says.

Making the Case for Resilience Funding

Cities are now seeing their digitalization planning as a way of supporting and driving a resilience strategy. Like CROs, the Chief Information/Digital/Innovation Officer is a relatively new post in many cities, and importantly it is also cross-departmental because breaking down data silos is the key to cities being able to improve decision-making.

The convergence of resilience and digital strategies is crucial to cities not only in combatting climate change but also in tackling social and economic shocks. Edinburgh in Scotland unveiled its digital strategy in October 2020[18] with an emphasis on how it can help meet the target of eradicating poverty in the city by 2030. It is the first UK city to set such a target which mirrors that of the city's aim to be carbon neutral by the same year.

But if COVID has shown that cities must adopt a holistic approach to digitalization and resilience planning, it is important that cities have the means to implement this. Smaller municipalities in the USA face challenges with accessing the city-specific data they need from county and state governments, which sometimes lack the technology, resources, and systems to break the information down to this level. This leaves cities trying to work with the data manually or unable to get the granular insights they and their citizens require.

And building the physical infrastructure alongside the data infrastructure which cities need to support resilience planning is ever more challenging in the economic chaos unleashed by the pandemic.

---

17 https://nola.gov/mayor/news/june-2020/city-of-new-orleans-launches-unprecedented-covid-19-meal-assistance-program-in-partnership-with-fema/#:~:text=a%20cost%2dsharing%20collaboration%20with,for%20at%20least%2030%20days.
18 https://edinburgh.gov.uk/news/article/12985/edinburgh-gets-smart-with-new-digital-strategy.

The argument to central government is that resilience not only protects cities from future shocks but also actually provides an opportunity for them to attract wider investment. While mayors have long argued that developing a green economy is an important element of planning for climate change, the real-estate industry has reinforced the need for cities to understand that resilience can also assure a better economic future.

According to the Royal Institution of Chartered Surveyors (RICS), over the next few decades commercial real-estate developers are just as likely to base investment decisions on rising sea levels as on rising interest rates.

"We are being asked by clients more and more about the dynamics of risk and resilience in cities, rather than national economies," says Jeremy Kelly, Director of Global Research at JLL, a commercial real-estate services company.

As factors such as climate change and social unrest exert increasing pressures on cities, traditional metrics for measuring investment risk may shift to incorporating resilience into the decision-making around capital allocation.

While the risk factors are yet to be determined or standardized, and the lack of data on cities is still prevalent, cities and their national governments need to be aware that resilience planning not only is essential to maintain liveable communities but also offers a means to attract investment to guard against shocks.

"[Previously], resilience had been perceived as a risk dealt with through a city's civil contingencies department, but now it is being perceived as an opportunity," says Chris Brooke, Managing Director at real-estate consultant Brooke Husband and Senior Vice-President, RICS. "Because if you can get it right, you could be on the forefront of attracting investment."

Conclusion

The biggest mistake post-COVID would be for national and local governments to let budget shortfalls and short-term economic recovery mask the need to plan for long-standing crises such as climate change and equity. CROs and Digital Officers must work together to change the paradigm and address multiple threats in a way that national governments and the multilateral system have failed to do.

"Everyone's going to be borrowing for stimulus or investment, and that is money that cities would have borrowed for climate action or [to address] the future of work," says Lauren Sorkin of the RCN. "We have one shot to get at those multiple issues, and to recover and achieve climate resilience."

17

The Fragile City – Introduction

In Chapter 16 Sarah and Richard covered a rich tapestry of ideas, showing how embedding resilience into the mindset of everyone will ensure effective resilience planning. Mayor LaToya Cantrell made it clear that "The City is focussed on the data not the date," and perhaps that is decisive action made correctly rather than quickly. This chapter shows us the power of data, and Seth and Eric will take us through this in more detail in Chapter 18, *The Data City*.

Sarah and Richard also showed us the pathway to reconsidering traditional metrics for measuring investment risk. Patricia will expand on this in Chapter 19, *The Measured City*, and Colin and James in Chapter 22, *The Invested City*, and Chapter 23, *The Financed City*.

John will now take us to fragile cities. The nexus between resilience and fragility is an important one; and, as Austin introduced in Chapter 3, *The Emerging City*, when cities emerge and focus on economic success, they can fix environmental blunders as they grow. Resilience cannot be an after-thought in today's emerging cities, as we can see from the worrying global landscape reality in Figure 17.1.

It must be a priority if we are to thrive in the twenty-first century.

In the early stages of the pandemic, COVID-19 was hailed as a "great societal leveller" that could introduce changes to address social, economic, and political inequalities that affect our world today. This has proved not to be true.

As much as the pandemic has highlighted the necessity of resilient cities, it has also, undeniably, exposed their fragility when it comes to income, gender, race, and opportunity, as well as structural factors linked to exposure to violence, poverty, extreme pollution, and natural disasters. With resilience comes the other, less-desirable, side of the coin to be overcome. Half the world's population lives in cities, so the stakes are high, and it is clear that cities registering more equal distribution of income and social services have responded better against the impacts of the pandemic than those with extreme inequality and deprivation. We can see some of the factors associated with fragility in Figure 17.2.

As Richard and Sarah told us in Chapter 16, we must put theory into practice and undertake what John refers to in this chapter as an "urban metamorphosis" in order to overcome this urban fragility. Just as London's devastating cholera pandemic (1846–1860) led to the development of the modern sewage system, we have already seen how the

The Climate City, First Edition. Edited by Martin Powell.
© 2022 John Wiley & Sons Ltd. Published 2022 by John Wiley & Sons Ltd.

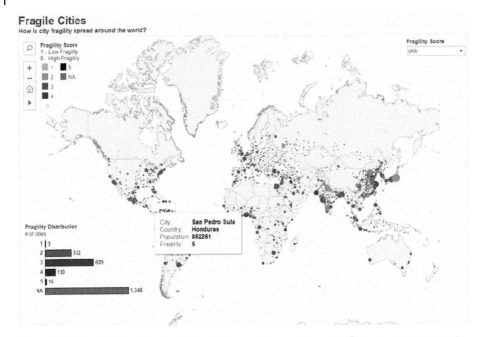

Figure 17.1 Fragile cities: how is city fragility spread around the world? (*Source:* Robert Muggah, 2016. How fragile are our cities?, World Economic Forum. Retrieved from: https://www.weforum.org/agenda/2016/02/how-fragile-are-our-cities.)

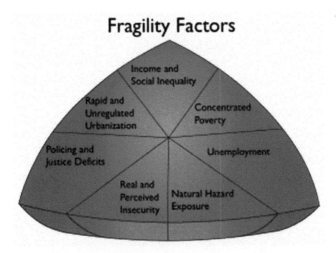

Figure 17.2 Factors that contribute to city fragility. (*Source:* Conceptualizing City Fragility and ResilienceUnited Nations University Centre for Policy Research, Working Paper 5, October 2016. United Nations University. Retrieved from: https://collections.unu.edu/eserv/UNU:5852/ConceptualizingCityFragilityand.)

specific demands of the pandemic have altered the fabric of our cities, through digital tracing, retrofitting city assets for social distancing, responding to healthcare pressures, the shift to e-learning, and providing basic income, among other things.

There are multiple dimensions to urban fragility, so a systems-based approach that identifies factors that interact to shape patterns of risk and resilience is essential to effectively tackle future problems, including addressing deficits, infrastructure, and social and economic inequalities, and enhancing political inclusion, social protections, and access to essential services.

In this chapter John will also touch upon the data divide that Seth and Eric explore in more detail in Chapter 18.

Just as Nick and Leah said in Chapter 7, *The Decoupled City*, it is not good enough just to build back, we have to "build back better". The pandemic has shown us that through effective leadership and governance and a systems-based approach, the virus has been more rapidly contained, as we have seen in cities such as Copenhagen, Seoul, and Taipei. COVID-19 has not been the "great societal leveller", but it has perhaps given us a roadmap to make our cities less vulnerable and to transform fragility into resilience.

17

The Fragile City

John de Boer

There is growing recognition that the cumulative impact of converging environmental, social, political, and economic risks is straining the ability of many cities to deliver essential services to residents in times of shocks and stresses. The COVID-19 pandemic (the pandemic) brought cities around the world to a halt, causing massive disruption and suffering for hundreds of millions of people. It exposed the fault lines in our cities that make them fragile. This includes growing inequalities in income, gender, race, and opportunity, as well as structural factors linked to exposure to violence, poverty, extreme pollution, and natural disasters. This chapter assesses the sources of fragility rooted in our cities and explores approaches that could help cities develop more resilient urban systems, enabling them to function, and even thrive, in times of crisis.

As the pandemic spread across the globe leaving no city unscathed, prominent voices began to portray the pandemic as a potential great societal leveller that could introduce radical changes to address massive social, economic, and political inequalities that characterize our world today.[1] It is still too early to tell whether this prediction will be realized. What is clear, however, is that the pandemic and the lockdowns that have accompanied it have exposed major socioeconomic fault lines in our cities.

Virtually every city, from New York to Paris, Madrid, New Delhi, Baghdad, and Lima, has been affected by the virus. Yet the impact of the pandemic within cities has been highly asymmetrical.[2] Populations living in neighbourhoods characterized by concentrated poverty, a lack of access to public services, low levels of education, high

---

1 Hartog, K., 2020. Black Death historian: 'A coronavirus depression could be the great leveller.' *Guardian*. https://www.theguardian.com/world/commentisfree/2020/apr/30/walter-scheidel-a-shock-to-the-established-order-can-deliver-change. Also see Scheidel who authored the famous "The Great Leveler" which argued how warfare, state collapse, and catastrophic plagues destroyed the fortunes of the rich and introduced radical changes that reduced inequality. Scheidel, W., 2018. *The Great Leveler: Violence and the History of Inequality from the Stone Age to the Twenty-First Century.* Princeton University Press. https://press.princeton.edu/books/paperback/9780691183251/the-great-leveler.

2 OECD, 2020. OECD policy responses to coronavirus (COVID-19). http://www.oecd.org/coronavirus/policy-responses/cities-policy-responses-fd1053ff.

The Climate City, First Edition. Edited by Martin Powell.
© 2022 John Wiley & Sons Ltd. Published 2022 by John Wiley & Sons Ltd.

unemployment, poor infrastructure, and physical insecurity have borne the brunt of this pandemic. The reason links back to the notion that vulnerability and exposure to shocks, pandemics, and disasters in cities is rarely the result of a single risk factor. Rather, as de Boer, Muggah, and Patel described in their paper conceptualizing city fragility and resilience, it is the cumulative effects of overlapping political, social, economic, and environmental risks that drive vulnerability in cities.[3]

Mounting evidence of the risks that populations in urban centres face indicates that the levels and types of vulnerabilities within cities are highly heterogeneous.[4] Policy actors trying to respond and prevent pandemics and other shocks need to take into consideration the hypercontextual factors that shape exposure to risk within cities.[5] To be effective, responses need to be informed by research that disaggregates impact by ethnicity and race, income status, gender, and age, as well as geography, to better understand the distribution of risk and the differential impact that shocks and disasters have within cities and population groups.

Lessons from the Global COVID-19 Pandemic

If there was one lesson to take away from the pandemic for city planners it is that economic geography is a better determinant of contagion risk and mortality than geography. Initial studies that pointed to population density as the main culprit for large urban centres being the epicentres of the outbreak (e.g. New York) have proven to be overly simplistic.[6] As a recent study by the World Bank underscored, "density and infections don't go hand in hand".[7] Low-income neighbourhoods, where people live in more cramped quarters and have jobs that require face-to-face interaction, are at far higher risk of contagion than populations in upper-income neighbourhoods, such as Manhattan, who have more space and can work from home (Figure 17.3). What we have learned is that the virus has had a profoundly differential impact on different groups of people, even within the same city.[8]

3 de Boer, J., Muggah, R., Patel, R., 2016. Conceptualizing city fragility and resilience. United Nations University. https://i.unu.edu/media/cpr.unu.edu/attachment/2227/wp05.02_conceptualizing_city_fragility_and_resilience.pdf.

4 Patel, R., Sanderson, D., Sitko, P., de Boer, J., 2020. Investigating urban vulnerability and resilience: A call for applied integrated research to reshape the political economy of decision-making. *Environment and Urbanization* 32(2). doi: 10.1177/0956247820909275.

5 Ibid., p. 3.

6 Rosenthal, B.M., 2020. Density is New York City's big 'enemy' in the coronavirus fight. *The New York Times*. https://www.nytimes.com/2020/03/23/nyregion/coronavirus-nyc-crowds-density.html. Bozikovic, A., 2020. Cities and the coronavirus: Density isn't the problem. *The Globe and Mail*. https://www.theglobeandmail.com/canada/article-cities-and-the-coronavirus-density-isnt-the-problem.

7 Lall, S., Wahba, S., 2020. No urban myth: Building inclusive and sustainable cities in the pandemic recovery. World Bank. https://www.worldbank.org/en/news/immersive-story/2020/06/18/no-urban-myth-building-inclusive-and-sustainable-cities-in-the-pandemic-recovery.

8 Goldin, I., Muggah, R., 2020. The COVID City. Project Syndicate. https://www.project-syndicate.org/commentary/what-covid19-means-for-cities-by-ian-goldin-and-robert-muggah-2020-08.

Figure 17.3 Manhattan during the COVID-19 pandemic. (*Source:* nycshooter/Getty Images.)

The fact that racial minorities in urban centres across North America have been dispro-portionately affected by the virus is further evidence of this notion. In Canadian cities, Black Canadians have experienced far worse health outcomes related to the pandemic compared to the average Canadian. A report sponsored by the African-Canadian Civic Engagement Council documented that Black Canadians are more likely to report symp-toms and nearly three times as likely to report knowing someone who has died of the virus.[9] This is because they experience multiple risk factors that place them at greater risk of exposure to COVID-19. Black Canadians are much more likely to have jobs that require them to work with people face-to-face (25 times the national average). Black Canadians are twice as likely to take public transportation and twice as likely to report their com-mute as unsafe. Black Canadians are also more likely to experience layoffs or reduced working hours due to the pandemic and to have been negatively impacted financially from the virus. Other studies have documented similar patterns in cities across the USA.[10] The US-based Centers for Disease Control and Prevention has also indicated that many racial and ethnic minority groups are at increased risk of getting sick and dying from COVID-19 due to social determinants of health that have historically prevented them from having fair opportunities for economic, physical, and emotional health.[11]

Low-income neighbourhoods in the cities of developing countries have been particularly hard hit. Many of these places are characterized as slums and informal settlements where

---

9 Innovative Research Group, 2020. Impact of COVID-19: Black Canadian perspectives. African-Canadian Civic Engagement Council and the Innovative Research Group. https://innovativeresearch.ca/wp-content/uploads/2020/09/ACCEC01-Release-Deck.pdf

10 Chen, J.T., Krieger, N., 2020. Revealing the unequal burden of COVID-19 by income, race/ethnicity, and household crowding: US county vs. ZIP code analysis. Harvard Center for Population and Development Studies. HCPDS Working Paper Volume 19, Number 1. HCPDS, Cambridge, MA. Eligon, J., Burch, A., Searcey, D., Oppel Jr. R., 2020. Black Americans face alarming rates of coronavirus infection in some States. *The New York Times.* https://www.nytimes.com/2020/04/07/us/coronavirus-race.html. Bassett, M.T., 2020. Just because you can't afford to leave the city doesn't mean you should. *The New York Times.* https://www.nytimes.com/2020/05/15/opinion/sunday/coronavirus-cities-density.html.

11 Centers for Disease Control and Prevention, 2020. Health equity considerations and racial and ethnic minority groups. https://www.cdc.gov/coronavirus/2019-ncov/community/health-equity/race-ethnicity.html.

people are packed tightly together and suffer from inadequate infrastructure and lack access to essential services. This outcome was not hard to foresee. In April 2020, the World Bank published a study that sought to predict contagion risk hotspots. Its conclusion was that populations living in informal settlements were most at risk. Their limited access to water prevented them from practising proper hygiene, exposure to high levels of air pollution placed these populations at higher risk of developing complications from COVID-19, and the difficulty of maintaining social distance due to overcrowding increased the chances of contagion.[12]

The damage inflicted by the pandemic in cities across Africa and Asia has been stark. It is estimated that more than half of Mumbai's six million slum-dwellers have had the coronavirus.[13] Estimates by the United Nations indicate that the pandemic will push nearly 100 million people back into extreme poverty by 2021, defined as people living on less than US$1.9 a day.[14] Informal workers have been hard hit. According to the UN's International Labour Organization, about 70% of domestic workers (the majority of whom are women) had lost their jobs as a result of the pandemic by June 2020. In addition to the economic turmoil created by this upheaval, the pandemic has provoked a massive exodus of poor urban labourers to go back to their originating villages in developing countries. Tens of millions of the 130 million migrant labourers living in the cities of India have returned to their villages. Major urban-to-rural exoduses have also been recorded in Peru, Mexico, Brazil, Uganda, South Africa, Kenya, Morocco, the Ivory Coast, Gabon, and Zimbabwe.[15]

The cascading consequences created by the pandemic have pushed populations already vulnerable to the brink of survival. It is in zones where political, social, and economic risks converge where suffering has been concentrated. The COVID crisis has clearly exposed how unequal cities can be, with migrants, the poor, women, racial minorities, and isolated older people being hardest hit. Structural inequalities compounded by acute shocks have clearly driven suffering.

From Fragility to Resilience: A Call for Urban Metamorphosis

More than half of the global population lived in cities until the pandemic hit. There has been speculation in some quarters that the pandemic could spell the end of big cities.[16] The pandemic saw hundreds of thousands of people flee urban coronavirus hotspots for

12 Bhardwaj, G., Esch, T., Lall, S.V., et al., 2020. Cities, crowding, and the coronavirus: Predicting contagion risk hotspots. World Bank Group. http://documents1.worldbank.org/curated/en/206541587590439082/pdf/cities-crowding-and-the-coronavirus-predicting-contagion-risk-hotspots.pdf.

13 Biswas, S., 2020. India coronavirus: 'More than half of Mumbai slum-dwellers had Covid-19.' BBC News. https://www.bbc.com/news/world-asia-india-53576653.

14 Canadian Broadcasting Corporation, 2020. Coronavirus: United Nations warns of widening gender poverty gap due to pandemic. https://www.cbc.ca/news/world/coronavirus-covid19-world-sept2-1.5708944.

15 Saunders, D., 2020. The pandemic's unnoticed toll: Hundreds of millions of missing city dwellers. *TheGlobeandMail*.https://www.theglobeandmail.com/opinion/article-the-pandemics-unnoticed-toll-hundreds-of-millions-of-missing-city.

16 Florida, R., 2020. Does COVID-19 spell the end of big cities? Munk Debates. New Geography. https://www.newgeography.com/content/006724-does-covid-19-spell-end-big-cities-munk-debates-with-guests-joel-kotkin-and-richard-florida. Kleinman, M., 2020. Will Covid-19 spell the end of our cities? *The Telegraph*. https://www.telegraph.co.uk/news/2020/05/19/will-covid-19-spell-end-cities.

rural areas. While such a prognosis is premature, it will certainly change the design of many cities and how people live in them. The virus is forcing many mayors and urban planners to rethink certain aspects of urban life, especially density and mobility.[17]

Pandemics have dual-edged effects on some cities. The cholera outbreak, which devastated London in 1850, led to the development of the modern sewage system that separated waste and drinking water.[18] What will be the legacy of the pandemic on our increasingly urbanized world?

With cities focused on short-term health and economic priorities, most urban leaders are only at the earliest stages of redesigning them for the long-term. Some analysts predict that we may be entering a post-urban economy where the digital economy will deconcentrate city centres where goods and services are typically concentrated. Other researchers insist that urban density will continue being the driver of innovation and economic growth but that our cities must be reconfigured to be more sustainable, socially cohesive, and better prepared to tackle future pandemics.[19]

Cities preparing for a future with COVID-19 need to be mindful not just of the immediate challenges but deep trends as well. This pandemic has thrown up urgent new demands related to ensuring mass testing and digital tracing, retrofitting city assets for social distancing, responding to healthcare pressures, shifting to e-learning, and providing basic income. It has also accelerated deeper trends such as the digitalization of retail, remote working arrangements, food production localization, the pedestrianization of cities, and climate-resilient and people-centred city design.

The push for change is not just a social-justice imperative but also an economic one. Cities are in significant financial strain. Early estimates suggested that megacities such as Mumbai have sustained economic losses of between US$1 and 2 billion a week in revenue since March 2020.[20] By May 2020, US cities had collectively lost more than US$130 billion in revenue since lockdowns went into effect mid-March.[21] In Canada, cities called on the federal government for tens of billions of dollars in financial aid as they grappled with the virus's cascading impact.[22] A full recovery, assuming this is even possible, could take

17 Shoichet, C., Jones, A., 2020. Coronavirus is making some people rethink where they want to live. CNN. https://www.cnn.com/2020/05/02/us/cities-population-coronavirus/index.html.

18 Shenker, J., 2020. Cities after coronavirus: How Covid-19 could radically alter urban life. *The Guardian*. https://www.theguardian.com/world/2020/mar/26/life-after-coronavirus-pandemic-change-world.

19 Loh, T., Love, H., Vey, J., 2020. The qualities that imperil urban places during COVID-19 are also the keys to recovery. The Brookings Institution. https://www.brookings.edu/blog/the-avenue/2020/03/25/the-qualities-that-imperil-urban-places-during-covid-19-are-also-the-keys-to-recovery. Muggah, R., Katz, R., 2020. How cities around the world are handling COVID-19 – And why we need to measure their preparedness. World Economic Forum: Agenda. https://www.weforum.org/agenda/2020/03/how-should-cities-prepare-for-coronavirus-pandemics.

20 Nahata, P., 2020. Covid-19: 'Maximum city' Mumbai braces for maximum economic hit. Bloomberg. https://www.bloombergquint.com/business/covid-19-maximum-city-mumbai-braces-for-maximum-economic-hit.

21 Navratil, L., Van Berkel, J., 2020. Lobbyists for Minnesota cities, counties converge on Washington for help. *StarTribune*. https://www.startribune.com/lobbyists-for-minnesota-cities-counties-converge-on-washington-for-help/570724662.

22 Patel, R., Blouin, L., 2020. Cities grappling with COVID-19 costs call for federal, provincial co-operation for emergency aid. Canadian Broadcasting Corporation. https://www.startribune.com/lobbyists-for-minnesota-cities-counties-converge-on-washington-for-help/570724662.

years, and the long-term repercussions of the virus on city life are difficult to predict.[23] While wounded by deficits and loss in revenue, cities will survive the epidemic, and the smartest among them will take this opportunity to build back better.

If this pandemic is to live up to expectations as a great leveller, public authorities and municipal leaders will have to make decisions that proactively shape a more inclusive urban future. How cities manage risks that shape vulnerability to shocks and crises in cities and invest in factors that reduce these vulnerabilities will determine the extent and pace of recovery as well as exposure to future risk.

Cities that proved more fragile to the pandemic had coping capacity deficits, often with interlocking systemic challenges where the state of basic service provision and access, including public health, security, and social support, were challenged. These pockets of fragility have long been areas of extreme inequality and deprivation. Evidence clearly points to the fact that cities characterized by higher levels of material and social deprivation in income were most exposed to the effects of COVID-19. This finding is consistent with studies on the impact of infectious diseases and mortality rates dating back to 2013.[24]

In contrast, cities registering more equal distribution of income and social services (including in terms of access to public health) fared better. Just as inequality can lead to fragility, greater equality can protect against the risks that cities face. Greater resource equity (or less income inequality) has been proven to associate with higher levels of resilience to numerous types of shocks.[25]

As de Boer, Muggah, and Patel have outlined, "resilient cities are those that have been able to activate protective qualities and processes at the individual, community, institutional and systems level to engage with hazards and stresses to maintain or recover functionality in times of crises".[26] This includes investing in characteristics and actions that reduce exposure and limit vulnerability by minimizing pre-existing risks and enhancing adaptive capabilities in times of crisis.

Required are public policies and investments that address the full spectrum of risks that contribute to various negative outcomes to the health and wellbeing of urban populations. These approaches need to take into consideration urban complexity that emerges from diversity and constant change in cities. As Patel et al. have argued, a systems-based approach that identifies factors that interact to shape patterns of risk and resilience is essential to effectively tackle the multiple dimensions of urban fragility. This would include addressing deficits in the urban economy, infrastructure, and services, social and

23 *The Economist*, 2020. London may have gone into a Covid-accelerated decline. https://www. economist.com/britain/2020/05/23/london-may-have-gone-into-a-covid-accelerated-decline?cid1=cust/ ednew/n/bl/n/2020/05/21n/owned/n/n/nwl/n/n/ME/478899/n. Roy, Y., 2020. Consultant: Full recovery from virus' economic impact will take NY 3 years. Newsday. https://www.newsday.com/business/ coronavirus/cuomo-state-budget-boston-group-1.44114268.

24 Thurston, G.D., Ahn, J., Cromer, K.R., et al., 2016. Ambient particulate matter air pollution exposure and mortality in the NIH-AARP Diet and Health Cohort. http://ehp.niehs.nih.gov/1509676. Cesaroni, G., Badaloni, C., Gariazzo, C., et al., 2013. Long-term exposure to urban air pollution and mortality in a cohort of more than a million adults in Rome. http://ehp.niehs.nih.gov/1205862.

25 Sherrieb, K., Galea, S., 2010. Measuring capacities for community resilience. *Social Indicators Research* 99(2), 227–247. doi: 10.1007/s11205-010-9576-9. https://www.researchgate.net/ publication/226732284_measuring_capacities_for_community_resilience.

26 de Boer et al., 2016. Op. cit., p. 3.

economic inequalities, and governance by enhancing political inclusion, social protection, and access to essential services. Public authorities also need to be held accountable for ensuring that their public policies actually help neighbourhoods grapple with the compounding risks that they face.[27]

In the digital age, addressing the digital divide in cities will also be critical. For the majority, digital technologies have not helped them cope during this pandemic. Sixty-one percent of the world's employed population (2 billion workers) are in the informal economy and were more likely to be exposed to health and safety risks associated with the pandemic.[28] In the USA, due to the nature of their jobs, less than 30% of workers could work from home. The digital divide is one of the the many inequalities exposed by the pandemic.[29] Less-digitally connected countries and cities are falling behind. Focused investments that help address the digital divide will also help address social inequalities, particularly in a world where schools, businesses, and government services have gone digital.

Conclusion

Ultimately, addressing urban fragility to pandemics and other shocks will require effective governance and leadership. A clear lesson emerging from the COVID-19 pandemic has been that where there is leadership and public–private coordination, the virus has been more rapidly contained. This is evident in the cases of cities such as Copenhagen, Seoul, and Taipei.[30] If cities are to deliver on the idea of building back better, they need to move to a logic that prioritizes accessibility to services and social, economic, and political inclusion.

The pandemic has clearly illustrated that the devastating impact of converging and compounding risks is most profoundly felt among populations in marginalized neighbourhoods with limited access to services and resources. COVID-19 has not proven to be the great leveller. Instead, it has exacerbated socioeconomic and health-based inequalities. It is imperative that we use this opportunity to tackle the socioeconomic determinants of risk in our cities.

---

27 Patel et al., 2020. Op. cit., p. 6. doi: 10.1177/0956247820909275.

28 OECD, 2020. Tackling coronavirus (COVID-19): Cities policy responses. https://read.oecd-ilibrary.org/view/?ref=126_126769-yen45847kf&title=coronavirus-covid-19-cities-policy-responses.

29 Ibid.

30 Muggah, R., Ermacora, T., 2020. Redesigning the COVID-19 city. National Public Radio. https://www.npr.org/2020/04/20/839418905/opinion-redesigning-the-covid-19-city.

18

The Data City – Introduction

In Chapter 17 John showed us the inequality the COVID-19 pandemic has exacerbated, as well as the 100 million people pushed back into extreme poverty. But he also laid out important lessons that can be learnt from disasters in improving the resilience of a city.

John referenced London's cholera pandemic (1846–1860) and how this led to the city's investment and innovation of a brand-new sewage system. In 1854, at the height of a new cholera epidemic in the Soho district around Broad Street, John Snow, a physician, discovered the connection between contaminated water and cholera by plotting the course of the cholera outbreak on a map he drew himself, as you can see in Figure 18.1.

He discovered that all the victims had used the same water pump, so he removed the handle from the pump and the Broad Street epidemic ended. Overall, it took 20 years of data collection to discover, analyse, and prove this finding – *cholera was a water-borne disease.*

At this time the Thames River was little more than an open sewer system, with much of the city's waste running freely through the streets and thoroughfares and into the river. In short it was a total health hazard. London's new sewage system was built by Sir Joseph Bazalgette and consisted of a network of enclosed underground brick main sewers to intercept sewage outflows and street sewers to intercept the raw sewage. At a time when London's population totalled 1 million people, Bazalgette made the ingenious decision to build it for 4 million people, "ensuring it will never need to be expanded". Nowadays, London's sewers manage approximately 1.25 billion kilogrammes of excrement each year; London's population is now 8 million people; and we have eroded green spaces and put in more hard surfaces, so on a rainy day the overflow goes into the river. Consequently, there is now a new super sewer being bored under the Thames.

The point I would like to outline, however, is that Snow took 20 years to collect the data he needed, whilst nowadays this can be done in 20 minutes if we understand the problem that we are trying to solve, hastening the progress of innovators like Bazalgette.

The Climate City, First Edition. Edited by Martin Powell.
© 2022 John Wiley & Sons Ltd. Published 2022 by John Wiley & Sons Ltd.

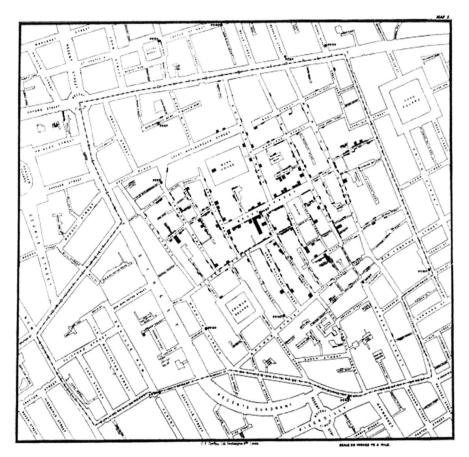

Figure 18.1 A map of the Soho district around Broad Street where John Snow recorded data to prove the link between contaminated water and cholera. (*Source:* John Snow / Wikipedia Commons / Public Domain.)

As we build back better after the pandemic, we must address "inclusion" by reducing the digital divide. Data are about tenacity, determination, freedom, and "imagination", and cities need to collect the right data, analyse it, and apply it to the context of the problem they are trying to solve.

In this chapter Seth and Eric now "warn" us about the abundance of data we can now mine, collect, and analyse, emphasizing the importance of collecting the right data.

Ensuring a just, equitable, and safe future for humanity in the fight against climate change is embedded in data and centralized in the city. In a world that is seemingly awash with the stuff, we somehow don't have enough to tackle the climate demands we need to meet. There are major gaps and failures that need to be overcome if we are to hold the planet to 1.5°C of warming and achieve the UN-appointed SDGs.

Cities are central to this endeavour, with city government's being the most present and accountable form of government in the world. Seth and Eric see cities as "convenors" due to their ability to enforce standards and incentivize. They will examine how cities can place themselves in a more constructive role within the data ecosystem by avoiding the mistakes of the past and looking at city climate action and refocusing public–private partnerships. They will also look at the tenants of a revitalized procurement and platform pivot for the twenty-first century and the required design parameters to ensure a successful pivot.

Seth and Eric tell us there is good reason to be optimistic about the long-term ascent of data-centric techniques and the potential for collaboration between public and private sectors. However, due to the increasing threat the climate crisis poses, it is also clear that immediate action is, once again, preferable to theory and conjecture.

18

The Data City

Seth Schultz and Eric Ast

In a book about cities and climate change, this may be the single most important chapter. The road that we must travel in the remainder of the twenty-first century to ensure a just, equitable, and safe future for humanity will be paved with data. This view is based on three key understandings – the climate crisis requires transformative action within cities by 2030; effective global action against climate change requires quality local data; and that quality local data require city government leadership and partnership.

At a time when we are awash in more data than ever in the history of our planet, the ability for the right people to access the right data at the right time, with a clearly framed problem in a site-specific way, still eludes us at the speed and scale required to combat climate change. Therefore, in order for humanity to effectively combat the climate crisis, we must reimagine the role that cities play in the collection, dissemination, and use of data.

We say reimagine because this is not a new problem. Cities, companies, and governments have been trying to harness the power of data for well over a century. While massive strides have been taken in terms of our ability to collect, analyse, and maintain data, there remain major gaps and failures that need to be overcome in order to effectively meet, manage, and maintain the transformation that cities around the world must undertake if we are to hold the planet to 1.5°C of warming while also achieving the United Nations' Sustainable Development Goals (SDGs).

This re-imagining is to recognize cities around the world for what they are – convenors. This may sound simple, but it is essential. At a time where we need more investment, innovation, implementation, and accountability than ever before, cities are not only the best convening mechanism but also where the "rubber meets the data road" to implement this change. If we think about this in terms of data, they are platforms. Effective platforms have two key characteristics: the ability to enforce standards and a mix of carrots and sticks to incentivize participation. For cities, procurement is the mechanism for both. Thinking about cities in this way is what we call the "procurement and platform pivot". This is born of an honest assessment of what has and has not changed throughout the

The Climate City, First Edition. Edited by Martin Powell.
© 2022 John Wiley & Sons Ltd. Published 2022 by John Wiley & Sons Ltd.

Box 18.1 Tenets of local data supporting collective global action

- Access to quality, granular data quickly when lives and livelihoods are at stake.
- Identification of specific policy impacts when many overlapping initiatives are taking place.
- Comparison of data between geographies to determine effectiveness of policies.
- Aggregation of data across geographies to understand collective conditions and impact.

---

century-old battle to bring empiricism to the halls of government, acknowledgement of the strengths and weaknesses of city government, recognition of the critical nature of evidenced-based and data-driven policymaking, and the speed and scale of action that is required. Box 18.1 identifies four key tenets of local data.

City governments are the most present, tangible, and accountable form of government throughout the world. Their mayors do not spend vast amounts of their time in remote places. They are responsible, day in and day out, for answering clear and evident questions that directly and immediately affect residents. Are the streets safe? Is the snow cleared? Has the garbage been collected? City governments own, regulate, or otherwise influence all major sources of global greenhouse gases (GHG) – the electricity that flows underneath city streets into buildings and vehicles, the buildings whose heating and cooling keep residents comfortable at nights and workers productive during the day, the infrastructure that moves public transit and private vehicles, the food that flows through distribution centres to grocery stores and restaurants, and the goods entering ports and making their way to stores.

By saying that local data require city government, we are not suggesting that cities alone shoulder the burden for all data collection and management. A city can never do everything on its own, but it needs to be part of everything, following the adage that you don't have to be great to start, but you have to start to be great. In order to instigate the data revolution required at the local level, cities need to position themselves in their most constructive role within the data ecosystem.

In determining that role, and ultimately navigating the role data can play within the larger context of governance and policymaking, it's important to consider a number of factors. First, acceptance of the limitations of empiricism in everyday social and commercial activity is important. When a private company buys an advertising spot during the Super Bowl, can they pinpoint the return on investment (ROI)? Do we crunch the numbers when choosing a friend or a partner, or heavily consider proximity to family in choosing where to live? We live in a world of limited and imperfect evidence, and where evidence is operating in concert with more complex moral and ethical value frameworks. We do our best to apply the most fitting evidence where we can, *and acknowledge when we cannot, or when other values are paramount*. We should expect no more, and no less, from our cities. In the words of sociologist William Bruce Cameron, "Not everything that counts can be counted, and not everything that can be counted counts."

Second, an understanding that transformative data work is failing everywhere is necessary. Over two decades into the Internet of things (IoT)-fuelled "smart cities" era, with tens of billions of dollars spent and hundreds of billions more anticipated,[1] the movement's vision is now characterized by leading firms, less as futuristic transformation and more as modesty and incrementalism.[2] For those looking for greener pastures, over two-thirds of leading private corporations say that they are failing in efforts to create data-driven organizations as well.[3] Building and maintaining the technology, talent, and culture necessary to successfully integrate data into the DNA of a business is brutal work. It takes a village, from the top to the bottom of organizations. There are many success stories with firms dominating their industries with data-led transformations, but they are the exception, not the rule. The grass may in fact be greener in the private sector, but it is not an Eden. These across-the-board struggles tell us that even in an optimistic scenario where the public sector can achieve data parity with the private sector, it would not be sufficient in the time required.

Third, some problems are very well suited for empirical approaches, and systemic city problems, like combating climate change, are not generally among them. City problems are most often complicated social problems. They rarely feature the benefit of an unambiguous objective function and must instead balance jobs, security, public health, equity, and environmental concerns, among other factors. More discrete, more technically focused problems – how to optimize a supply chain, how to serve more relevant advertisements, how to maximize runs scored per nine innings or goals scored per 90 minutes – are considerably better suited for data-driven techniques than the messy problems that cities tackle every day. With city problems it's better to view data as a spectrum and ask "How data-driven can this be?" rather than "Is it data driven or not?".

Fourth, day-to-day life and the accompanying activities relevant to climate change do not sit atop a clean, contained, machine-data environment. The data that cities seek in informing strategy and action are fragmented, owned by others, and survey-reliant or simply may not exist. Some sectors are better equipped than others to overcome these challenges. For example, electricity consumption data tend to be good, as such data are essential to the business of utility companies, but may be hard to access by cities. Cities have overcome this issue by purchasing their local electric utilities or forming larger-scale aggregation schemes, as was the case with Melbourne (Melbourne Renewable Energy Project) and San Francisco (CleanPowerSF; Figure 18.2). Others like Tokyo (Emissions Trading System) have enacted legislation that necessitates annual disclosure of data, but these solutions suffer from scalability issues and leave many cost and operations burdens with the city itself. Data on consumption-based emissions (what citizens purchase, their diets) are considerably more fragmented, are less consistently tracked, and have a murkier line of sight to accurate assessment.

These factors help us to contextualize the role of data and choose battles wisely. There exists a considerable amount of variety in the nature of problems that could be solved with data-driven techniques, and our goal should be to make sure we don't hold our hammer and eye everything simply as a nail.

---

1 https://www.idc.com.
2 https://www.govtech.com/data/a-new-smart-city-model-is-emerging.html.
3 https://hbr.org/2019/02/companies-are-failing-in-their-efforts-to-become-data-driven.

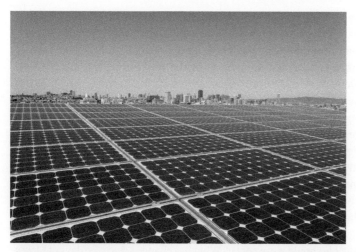

Figure 18.2   CleanPowerSF, San Francisco's renewable energy programme, is an example of city activities that can claim greater data autonomy. (*Source:* Tom Penpark/Shutterstock.)

Avoiding Mistakes of the Past: Government and Data – A Century-Long Journey

"It is essential ... in order to avoid personal and political pressure ... [performance] could be reduced to mathematical grades ... [assessment] determined by a science precise enough to give everyone ... the exact rewards he deserved."

This description of a performance programme would look perfectly at home in an email inbox full of newsletters discussing the latest data-driven techniques applied in a city or the newly formed team that will bring better information to bear across an administration. However, these words aren't in an email. They are in a book written in 1974 about events taking place in 1914.[4] They are about a young Robert Moses, the man who would eventually become the controversial and towering "master builder" whose four decades in New York politics reshaped a state and defined an era of destructive, auto-centric American urbanism. In 1914 Moses was an idealistic young reformer, and he spent the first few years of his career attempting to implement a more empirical system for assessing the performance of city employees. It was a system that would be familiar to anybody today if one were to substitute his data-collecting index cards for a web-based form.

Moses was left in tatters by this effort. He brought an empirical slingshot to a political gunfight. He was up against a corrupt political machine and bureaucracy that easily sidestepped, obfuscated, and frustrated him and his system into oblivion. He was fired, and his career in public service appeared to be over. Moses turned this episode into a historically tragic lesson of political Machiavellianism. We must turn this episode into a long-game view of the relationship between information and politics.

---

4 https://en.wikipedia.org/wiki/The_Power_Broker.

In 1960, almost 50 years after Robert Moses's crusade inside the halls of city government, Robert McNamara became President John F. Kennedy's Secretary of Defense. McNamara was a Harvard Business School graduate and a Second World War veteran of the Army Air Force Office of Statistical Control. Over the 15 years prior to his appointment as Secretary of Defense, McNamara had risen through the management ranks to become president of the Ford Motor Company, the first person outside of the Ford family to claim that post. He took an aggressive data-centric approach to managing the Vietnam war. This was an approach that served him incredibly well at Ford, the pinnacle of American industrial enterprise at the time. Every deployment, casualty, and bomb dropped would be neatly quantified, reported, and analysed. The USA would manage the war with a degree of operational and empirical sophistication that had never before been seen in the history of the world. McNamara's data often told him things were going well (Figure 18.3). History has told us a different story.

So what is the same today? What is different? Why should we expect greater success this time around, and how quickly should we expect it? What has changed since the days of Moses and McNamara are certainly the technologies applied, both in computing, the IoT, and the overall transparency and accountability of city government. In 1914 New York City government did not even have a budget, and pre-progressive era systems of patronage and vote-buying were rampant. While this is no longer the case in most developed countries, this is still the case in countless city governments around the world, particularly in the developing world, where data are more critical than ever. As such, what has not changed is the underlying nature of how the proverbial political sausage gets made and the suitability, or lack thereof, of empirical methods towards complicated, multifaceted political problems.

Figure 18.3 Computers at the Combat-Operations Center at the headquarters of the North American Air-Defense Command. (*Source:* Bettmann/Getty Images.)

The lessons we glean from these episodes and the constraints of data-driven work are not that the introduction of data into the public sphere is hopeless, but rather it is nuanced, and we should set our expectations for achievement and approach accordingly. One cannot take the square peg of Ford Motors, or Facebook, or Google and find a perfectly fitting home in government. It is a different animal. It is understandable that our view of what is possible is painted heavily by the private sector, or academia, and by what is considered to be on the cutting edge, but we should take a good look at how sharp that edge truly is and how far the problems we seek to solve are from it.

City Climate Action: The Urban Context

Cities around the world are investing considerable resources in gathering climate-relevant data and answering critical questions, as shown in Box 18.2. However, progress can be expensive and slow. Cities are able to compile GHG inventories, the backbone of climate strategies, no more than once per year, and often only every two or three, limiting constructive cycles of iteration and improvement. Risk assessments are typically completed even less frequently, and while the broad set of hazards a city faces might not change at a high speed, the nature of hazards is highly location specific, and who lives and works in vulnerable areas, especially in developing cities, changes rapidly. In a world now measured in nanoseconds of transaction times in financial markets, this is glacial.

These difficulties are driven by a number of factors. City governments are balkanized. The nature of city government work demands this. If a private conglomerate owned a

---

Box 18.2   Urban climate data and questions of focus

Key data

GHG inventory:
- Electricity generation and consumption
- Building energy utilization
- Public and private transportation
- Waste disposal
- Food consumption

Risk assessment:
- Risks present in a city (e.g. flooding, extreme heat)
- Specific geographies associated with risks
- Estimated costs associated with risks

Primary questions

What are our climate change challenges?
- Where are our emissions coming from?
- What hazards do we face?
- What should we prioritize?

How can we take action at speed and scale?
- How can we effectively evaluate policies in the near and long terms?
- How should we operate and implement policies?
- How can we best leverage open data and local stakeholders?
- How can we lean into uncertainty, risking failure and using data as our guardrail?

How can we communicate most effectively?
- How can we keep citizens engaged, involved, and supportive?
- What's the best approach to providing policymakers with data, tools, and processes?

healthcare company, a security firm, and a ridesharing app, we wouldn't expect their data to seamlessly integrate, and we wouldn't expect efforts to combine data in a joint effort to go smoothly. Cities often have dozens of departments and agencies. They operate semi-autonomously for valid reasons, but these legacy arrangements are poorly suited for tackling systemic issues like climate change. City governments are also bureaucratic, as all complex organizations are. Cities employ thousands and tens of thousands of people. In the private sector, we call it "governance". The systems of keeping thousands of people aligned, productive, and accountable are imperfect and ever-evolving, and cities have an additional layer of broadly positive anti-waste, anti-corruption, and transparency considerations to meet, further complicating these objectives. Additionally, large quantities of critical city data are locked away in private companies. Cities directly oversee a number of aspects of city life, but they also rely on a vast ecosystem of private firms. This includes direct support and outsourcing of work, like the contracting of waste services, as well as indirect private activity that exists within the boundaries of the market, like the operation of private residential housing. The default for private ownership over an activity is private ownership over the data that comes with it. Finally, cities are not operationally tuned for data. Cities face all of the same challenges that have the majority of private companies failing in the journey to become more data-driven, in addition to others. Cities cannot pay as much for in-demand talent as private companies, they cannot dangle large paydays in the form of stock options, and they do not have the same cultural cache. This work is already an uphill battle, and cities have a slightly higher angle. Things are improving in city halls around the world as there is more investment in data and analytics and a growing number of Chief Data Officers, but we can reasonably set expectations for the scale and trajectory of this activity based on business as usual.

It's only with a deep understanding of the specific data relevant to the problems being solved and the system in which that data are scoped, collected, and used that a path forward can be cut. There has been major progress and accomplishments that data-driven governments have achieved over the last few decades. There are more people in government plying data craft, greater commitment to transparency, and greater understanding of the biases and weaknesses present in data-driven work. The ultimate question isn't whether or not there is progress, but if it is of sufficient scale and speed to confront the climate crisis.

Refocusing Public–Private Partnerships to Power a Data Revolution

Similar to the challenges laid out above with regard to data, cities have been grappling with surging populations, budget gaps, infrastructure needs, and the pressures of climate change and the fourth industrial revolution. One of the modern mechanisms to address this is the public–private partnership (PPP). First emerging in the UK and becoming widely popular by the 1990s, PPPs are now worth hundreds of billions of dollars globally.

PPPs involve a long-term contract between a public party, like a city – the procuring authority – and a private party for construction or rehabilitation, finance, and operation of an infrastructure asset in which the private party bears significant risk throughout the life of the contract. This approach has been used for over 40 years in a diverse range of markets across Europe, the Americas, Asia, Oceania, and Africa. A PPP contract typically lasts for more than 20 years after procurement, through construction and operations.

There are multiple reasons that PPPs have become popular. They help a city fund projects that they don't have the resources to do on their own. They bring in capacity and expertise that often local governments do not have. And they allow cities to move debt off their balance sheet, thus allowing them to borrow more money. Cities from Hong Kong (Land Value Capture for Mass Transit Investment) to Manila (Water Infrastructure) and London (Transit Modernization) have all employed the PPP model, to varying degrees of success.

Given the need to also align national governments and cities through more ambitious nationally determined contributions (NDCs) to GHG reduction, it is more important than ever to consider national budget strategies, pipelines, and priorities. In times of financial stress which the world is now facing amid the COVID-19 pandemic, governments focus on developing budget strategies that identify economically feasible and affordable PPPs to match national development goals and economic stimulus. This is leading to requirements that line ministries and subnational agencies coordinate project pipelines with national pipelines (often controlled by national treasuries and ministries of finance) in order to prioritize projects meaningfully.

The need to create public and private sector collaboration mechanisms with blended funding is more important than ever due to the investment needed in cities around the world over the coming decade. Estimated to be roughly US$3.5 trillion per year, the amount of planning, design, construction, and operations and maintenance in cities is staggering. Not to mention the fact that early-generation PPPs that were started during the early 1990s are now maturing, with few governments focused on the date when the project contracts would expire, and the responsibilities for the operations and maintenance would be handed back to them. And this was the case before the pandemic struck the world, further exacerbating government budgetary constraints.

However, PPPs are controversial, to say the least. There is a wide range of opinions, research, and case studies making the cases ranging from highly successful to catastrophic. After more than 30 years of innovation, testing, and evaluation, two things are clear. Traditional PPPs still have a long way to go in terms of delivering the expectations and returns expected, and cities need a mechanism for leveraging private expertise and investment now more than ever. So what happens next?

A Path Forward: Political Power, Procurement, and Platforms

What happens next is a new PPP for the twenty-first century, the procurement and platform pivot. There are three key tenets of this shift. The first is for cities to leverage their power of procurement towards data ownership and standardization. This requires that cities take ownership of establishing what data should be collected. Inventor Charles Kettering famously remarked, "A problem well stated is a problem half-solved." Cities are in the best position to state problems well and determine what data are necessary to answer them. This model requires a meaningful shift in the economics of deals, with cities no longer able to use data as a form of de-facto payment. This approach, which keeps up-front costs down, shares much in common with consumers receiving low- or no-cost technology for the hidden price of their private data. It's an approach that is penny-wise and pound foolish.

It is also an approach that creates a damaging cycle of dependence and data poverty, particularly in the Global South where starting data quality can be poor. As we have seen occur in Lagos, Nigeria, poor data quality leads to tenders that require data ownership as payment to offset the risk caused by the data limitations themselves. As the contract is implemented, the private company uses the data, which they now solely own, to identify what value exists within the system. When the contract is up and the lion's share of value has been extracted, the city is left in a weaker position than where it started. They have little data, have no improved capacity, and now face the "damned if you do, damned if you don't" choice between being in the dark with their existing provider or going back to tender to firms who now recognize the low-hanging fruit has already been picked.

The second tenet is to use open data platforms to empower the local civic, research, and entrepreneurial ecosystem and vertically integrate data. Every city has access to a robust web of NGOs, universities, civic organizations, and philanthropies. The totality of these organizations is in a stronger position to aggregate and utilize climate data than any specific city can be. By focusing on relationships with these entities, scoping work appropriately, and ensuring the right questions are being asked, cities can dramatically increase analytic productivity. Cities have already set a strong precedent for platformization. There currently exist no fewer than 10 private and open-source technologies specifically designed for the collection and dissemination of data. These platforms are in use by hundreds of governments around the world. This technology is ready, and cities have proven an ability to leverage it. These platforms can also serve to immediately satisfy the data needs of national and international aggregation and of crisis scenarios.

The third tenet is for cities to legislate and regulate where possible, *but use data to empower allies in the absence of authority*. Sometimes cities have the authority to collect desired data. In 2006, New York City mandated that its famous yellow cabs install GPS, which dramatically increased the city's understanding of for-hire vehicle activity. Often, cities do not have the direct capability to capture the specific data they seek, or they need additional public support to do so. In these cases they can use what data they have in order to create political pressure in favour of disclosure. Less than 10 years after requiring yellow cabs to install GPS, Uber and Lyft transformed the New York City's taxi market, and the data that came with it, leaving the city in the dark. Use of the data the city did have, in particular relating to street safety, played an important role in New York City's ability to apply political pressure to Uber and Lyft and ultimately to acquire the new data.

We have also seen the utilization of city power play out in the more recent world of micro-mobility. Bird, which famously burst onto the scene in 2018 with an "ask forgiveness not permission" approach to scooter placement in cities, has evolved to share data in the face of increased political and regulatory scrutiny. The company has gone as far as offering a platform, GovTech, specifically designed for city governments to access and analyse data and exercise authority over where scooters can operate.[5] As the fastest start-up in the gig-economy to reach a billion dollar evaluation, it is notable to observe their pivot to a platform provider. For those sceptical of industry-led data-sharing efforts, there is also the Mobility

---

5 https://datasmart.ash.harvard.edu/news/article/lessons-from-leading-cdos-966.

Data Specification (MDS). MDS is an open-source, non-profit-led standard which originated in Los Angeles and is now used by over 90 cities worldwide[6] to ensure consistency and interoperability across platforms and service providers. Standards like these could be mandated during procurement processes, helping to eliminate the issue of vendor lock-in.

With 400,000 pedestrians and cyclists killed every year due to road traffic,[7] these data blind spots go beyond climate consequence. There are thousands of people alive today in Bogotá, Colombia, a user of MDS, because the city reduced road traffic deaths by over 60% from levels in the 1990s through a series of reforms including Bus Rapid Transit and improved pedestrian and cycling infrastructure. In a world where there is a 5× difference in the rate of road traffic fatalities between the best and worst performers, greater access to data can help avoid the unnecessary deaths of hundreds of thousands of people simply walking or riding their bikes in the cities they call home.

Required Design Parameters in Order to Ensure a Successful Pivot

The context of any city action is a political one. The axiom "culture eats strategy for breakfast" is complemented by the idea that "politics eats data for lunch". It is therefore necessary not just for individual practitioners to be politically attuned, though minimally politicized, in how they apply their craft, but for the entire set of stakeholders within the city data ecosystem to recognize this dynamic. In order to ensure the critical elements of the political power structure as it pertains to the use of data in cities is understood and incorporated into a procurement and platform pivot, it is essential to understand the following characteristics that drive the manner in which cities can best play to their strengths.

First, city legislative and regulatory power is varied and limited. There is no global continuity in how city governments function. We know that cities tend to exercise more power over transportation infrastructure and building regulations than they do over energy and waste treatment.[8] Given the need for comparable, consistent data across cities, this variability means we cannot depend on direct control alone to meet global data needs.

Second, cities exercise significant power through the procurement process. Cities spend. The demands on that spending are vast, but it is estimated that government procurement accounts for 10–15% of the GDP of economies worldwide.[9] Leveraging this system to determine the conditions of how data are treated is a key strength of cities.

Third, cities operate within a rich technological and research ecosystem. Be it the private sector, NGOs and the third sector, or academia, cities have the opportunity to work with a litany of local and global organizations that can offer constructive services and partnerships towards their goals. Recognition of these characteristics, and working within them, is essential to successful design and implementation of future strategies.

---

6 https://www.wto.org/english/tratop_e/gproc_e/gproc_e.htm.

7 https://www.bird.co/blog/bird-announces-new-govtech-products-and-team-cities-primary-customer-for-new-offerings.

8 https://www.openmobilityfoundation.org/faq.

9 https://www.who.int/violence_injury_prevention/road_safety_status/2018/en.

Conclusion

The climate crisis presents a unique challenge to humanity and a unique challenge for the intersection of data, cities, and worldwide political action. There's good reason to be optimistic about the long-term ascent of data-centric techniques within the political sphere and the potential for collaboration between public and private sectors around the globe. However, due to the immediate nature of the climate crisis, a clear-eyed view of our current trajectory dictates that bold action towards accelerating action is necessary.

If one has spent a long time on a river with relatively stable current, it is more reasonable to conclude that they are on a very long river with a way to go, and not that there is a waterfall soon ahead. We should not expect this watershed moment in the near-term if we follow the current playbook. We need a pivot. A procurement and platform pivot.

19

The Measured City – Introduction

In Chapter 18 Seth and Eric opened us to the importance of using the right data to solve the right problem. The fact we can extract data from disparate systems and disparate infrastructure types into a common analytical space allows us to prove our business cases with far more certainty. The success rates of projects and programmes can be dramatically improved.

In this chapter Patricia shows us the importance, as our cities grow, of measuring with more like-for-like comparisons. Patricia's team developed the world's first ISO standard for cities. This standard means that cities can really monitor against their UN SDGs and ensure they are moving in the direction of their North Star.

We live in a highly connected world with measurement of GDP, GNP, and other national income, and with investment and monetary measures globally standardized and comparable in order to ensure well-informed data. Within our national systems, cities have become critical sites with unrivalled population growth, investment, invention, prosperity, and climate mitigations, to name a few examples. Yet globally standardized and comparable measurement has lagged at city level.

Patricia examines the reasons behind why we should focus on globally standardized measurement in cities. And how it can improve overall performance, drive economic development, and create a "peer-to-peer" learning network across cities, which can in turn foster friendly comparison and a culture of innovation in cities. Currently, cities are measuring the same things but just in very different ways, resulting in uneven data. Cities across the world are facing the same problems, whether it be climate change, poverty alleviation, health pandemics, or cultural tolerance, among other things, and as a consequence even and reliable data are vital.

Patricia details the three standards for sustainable, smart, and resilient cities and the accompanying performance indicators in ISO 37120 *Sustainable Development of Communities – Indicators for City Services and Quality of Life*. The ISO 37120 series is the world's first international standard for cities and city data. She also discusses the evolving world of international standards and the drive to enable and implement standardized city data and how cities are using these data to target six areas: economic development planning; monitoring the progress of SDGs; results of infrastructure investment; tracking climate change; smart city agendas; and shock recovery and resilience.

The Climate City, First Edition. Edited by Martin Powell.
© 2022 John Wiley & Sons Ltd. Published 2022 by John Wiley & Sons Ltd.

Patricia's chapter, *The Measured City*, fosters more informed decision-making and strategic action that propels success for the city and its residents. Just as Sarah and Richard interrogated what it will take to make a city truly resilient in Chapter 16, *The Resilient City*, it is clear that the standardized data gained are essential for a more resilient and prosperous future in cities.

19

The Measured City
Patricia McCarney

We increasingly live in a highly connected world. In the recent past, such a statement would be attributed to national governments connecting through trade, security, and global economic monetary policy. Today we have measurement at national level with Gross Domestic Product (GDP), Gross National Product (GNP), and other national income, and investment and monetary measures have been globally standardized and are comparable in order to ensure these global, state-to-state relations are well informed and data driven. However, also in the recent past, cities have been rising in stature and performance as critical sites in this highly connected world. Cities are sites not only for dominant population growth and global migration, but also critical sites where investment, invention, prosperity, climate mitigation, security, and social wellbeing can either succeed or fail. Globally standardized and comparable measurement, so valued at national level to drive data-informed global relations, has, however, lagged at city level, despite this emergence of cities as crucial players on the global stage and as they enter into their own global networks and relations with other cities: hence the title here – *The Measured City*.

Why should we focus on globally standardized measurement in cities? What global standards exist for city data that drive and enable "the measured city"? Why is "the measured city" so important for cities today, and how are cities embracing global standards to propel their success?

Advancing the Measured City: Why Should We Focus on Globally Standardized Measurement in Cities?

Cities are a defining phenomenon of the twenty-first century, both in demographic and economic terms. For the first time in history, the majority of the global population lives in cities. With cities now responsible for greater than 70% of global GDP,[1] cities are quickly becoming economic powerhouses, taking centre stage in the development and prosperity of nations. Alongside these global demographic and economic trends comes a new set of challenges for city leaders on the ground. City leaders are being tasked with a wider and

---

1 The top 780 global cities already produce almost 60% of all world economic activity. Oxford Economics, 2018. *Global Cities: The Future of the World's Leading Urban Economies to 2035*. Oxford Economics Group, Oxford.

The Climate City, First Edition. Edited by Martin Powell.
© 2022 John Wiley & Sons Ltd. Published 2022 by John Wiley & Sons Ltd.

deeper set of challenges – from crime prevention, to more efficient mobility, to creating healthier environments, to security and emergency preparedness. In order to attract investment and drive economic development, cities need indicators to measure their performance in delivering services and improving quality of life. In addressing global challenges and opportunities for sustainability and prosperity, the need for globally comparable city data has never been greater. The ability to compare data across cities globally, using a globally standardized set of indicators, is essential for comparative learning and progress in city development. City metrics equip city leadership for intelligent decision-making, strategic planning, and policy change, and, overall, guide more effective city governance.

Now more than ever, a stable and sustained trajectory for economic development in cities is dependent upon effective management and evidence-based policymaking. Cities need data – to drive economic development, inform investment decisions, benchmark progress, and, moreover, drive a culture of innovation in their cities – but not just any data. City leaders worldwide want to know how their cities are doing relative to their peers. City leaders need to talk to each other for peer-to-peer learning – hence the need for globally comparable and standardized city data has emerged as a pressing priority for success. If cities are measuring in the same way, in simple terms with "apples to apples" data, then city leaders have the ability to compare data across cities within the same region, across the same country, and with other cities across the globe. Being equipped with a globally standardized set of indicators is essential for comparative learning and progress, regardless of size or level of economic development.

There is no question that most cities have extended datasets inside their city halls. However, these data, until recently, have never been standardized, enabling cross-city comparisons. Even cities within the same state or country are measuring city services differently, often with different definitions of what is being measured, according to different methodologies on how to measure and also according to different urban boundaries. Most cities produce metrics and measure their core service delivery and their quality of life across almost all of the same themes, but the problem has been that there has been no standardized way that these data are being collected and reported. This holds true for all sorts of data points – across transit, recreation, education, housing, safety, air quality, and other city services, and including quality of life data that affect a city's planning and economic policy. Definitions on what is being measured and methodologies on that measurement are uneven and differ despite very similar objectives. To put a point on this – cities are measuring the same things but in very different ways.

This unevenness in the data creates an inability to compare and an inability to improve based on peer cities' experiences. Without standardization, it is difficult for mayors to know how they are doing, for example on the overall safety of their citizens, when some cities define violent crime rates according to different definitions, or they measure emergency response time according to different methodologies such as "from the time of the emergency call" versus "from the time the emergency vehicle is dispatched". Hence a basic, fundamental question posed by city officials or citizens such as "How safe is my city?" is not easily answered. It is not a question of insufficient data, just a lack of data that are standardized and verified for a sound context and comparability.

Similarly, without comparable data, economic development officers lack information on how to accurately and effectively promote their city. Economic development in cities is by nature propelled by comparative data. Moreover, senior levels of government lack credible comparative city data to make informed decisions on investments locally, this being particularly acute in large infrastructure spending by national and regional (provincial/state) governments. Accurate and trusted city-level data that are comparable in cities across a country or region support both local leadership in making informed requests for infrastructure investments and upper-level government leadership in making informed decisions on where infrastructure investments are most necessary. Once allocated and according to accurate and comparable baselines, tracking year-over-year progress of the results and outcomes of these investments also builds efficiencies in municipal management and transparency for citizens.

Standardized indicators leading to a common base of city-level data enables city leaders to measure their performance and compare and monitor progress relative to other cities. In addition, comparable city-level data can help build collaboration and understanding by fostering information exchange and sharing of best practices across cities. Comparative analysis and knowledge sharing are vital in the face of rapid urbanization and the associated demand for larger scales of infrastructure investment and the evolving and more complex task environment in delivery of city services. In addition, the existence of standardized indicators and this common base of city-level data are pivotal if cities are to better confront and respond to the emergent global challenges – including climate change and the associated demand for sustainability planning, as well as city resilience and emergency preparedness. Now, in 2020 and 2021, confronted by the COVID-19 pandemic, the need for data and globally comparative analytics has never been more critical. Standardized measurement in cities across the globe is at the core of ensuring economic recovery, health recovery, and the development of a more resilient business continuity strategy in cities into the future, as well as a more resilient healthcare strategy in cities going forward. In this current crisis we are witnessing challenges ranging from rising inequality in cities to global supply chains. Globally standardized indicators designed in the past for national-level exchange are now more than ever required specifically at city level. City-level data that are standardized and comparable are essential for comparative learning and fundamental to the strategic planning discussions being spearheaded by city leaders whether inside city halls, across states or provinces, within national territories, or with global peers.

As global challenges increasingly find expression in the world's cities – whether climate change, poverty alleviation, cultural tolerance, health pandemics, global financial crises, local business continuity, or global risk and conflict – cities are sites where these global challenges are most symptomatic, where the greatest concentrations of individual citizens and communities are affected, and where informed responses can be most strategic. In addressing these challenges and the global opportunities for economic growth and prosperity, the need for globally comparable data, strategic analytics, and comprehensive knowledge on cities has never been greater.

Cities are expanding and changing, and not just in terms of their population and spatial form but also in terms of their social, economic, and political spheres of influence. Cities, regardless of their size, have evolving labour markets, real-estate markets, financial and business markets, and service markets. Cities are centres of innovation where information and communications technology converge to inform alternative futures, highly altered urban form, and varied economic trajectories. Cities equipped with internationally standardized and comparative data are able to drill down into comparative case studies to understand and, most importantly, learn from other cities. City-to-city learning, propelled by data-informed conversations, supports innovation.

High-quality, standardized city-level data, generated by cities, are at the core of efforts to plan and build a smart and sustainable city system worldwide. Information and communications technology (ICT) and high-quality data allow city and systems managers to gain clear insights on how to optimize performance and create efficiencies of rapidly evolving and complex systems. These increasingly complex functions demand more comprehensive and data-driven planning and a more robust data-informed governance framework to ensure integrated service delivery for prosperous, sustainable, and inclusive futures for their citizens. When indicators are well developed and soundly articulated, they can drive learning and ultimately policy change for smarter urban development. This learning is dependent on the presence of appropriate information with the capacity to change society's behaviour.[2]

The dynamic economic transition of cities in both a national and global context together with the emergence of a whole new paradigm of connected cities driven by ICT and city data movements frame the sustainable and smart city agendas as countries and cities seek new paths to meet massive scales of new infrastructure investment, the circular economy, and climate-sensitive investment in assets and design, while building transformative innovation platforms for better performance and improved quality of life in cities for the post-2020 era.

What Global Standards Exist for City Data that Drive and Enable "the Measured City"?

There are three international standards now published and being implemented in cities across the globe that drive and enable "the measured city". The evolving world of international standards has only very recently begun to address the need for standardization in cities.[3] International standards bodies, such as the International Organization for Standardization (ISO), the International Electrotechnical Commission (IEC), and the International Telecommunication Union (ITU), have only since around 2013 started to address the pressing cities agenda, with new work ranging from smart grid, to smart city infrastructure, to international telecommunications and management systems.

As part of a new series of international standards being developed for a holistic and integrated approach to sustainable development in communities under ISO TC268 Sustainable Development of Communities, the first international standard was published on 15 May 2014 by ISO in Geneva. ISO 37120 *Sustainable Development of Communities –*

---

2 Hezri, A.A., Dovers, S.R., 2006. Sustainability indicators, policy and governance: Issues for ecological economics. *Ecological Economics* 60(1), 86–99.

3 McCarney, P., 2015. The evolution of global city indicators and ISO37120: The first international standard on city indicators. *Statistical Journal of the International Association for Official Statistics* 31(1), 103–110.

Indicators for City Services and Quality of Life[4] became the first international standard for cities and, more specifically, the first international standard for city data.

ISO 37120, compared to other international standards, had its own unique development path within ISO. Most international standards are developed within ISO and then tested and marketed for public consumption. The development path of ISO 37120 was the opposite of this. At least 75% of indicators within ISO 37120 were tested, prioritized, and reported on by some 200 cities globally,[5] led by a group of Canadian scholars and data scientists in Toronto. This catalogue of key performance indicators (KPIs), with definitions and methodologies developed by city leaders in Canada and across the globe, was brought forward by this group of Canadians to ISO in 2011. In 2012, a new ISO Working Group was established and led by Canada within ISO[6] that developed this Draft International Standard (DIS), and ISO 37120 was published two years later, in 2014. The involvement by cities in developing ISO 37120 is extremely important as they are the ultimate adopters of this new ISO standard. It was cities who directed the prioritization of these KPIs and advised on the drafting of standardized definitions and methodologies that best fit their city measurement frameworks.

Throughout the ISO development process the indicators, definitions, and methodologies in ISO 37120 were further fine-tuned by incorporating over 300 comments from participating experts. New indicators were added to the various drafts for consideration by member country experts. Following a series of international ISO ballots and commentary on multiple drafts of this standard, ISO 37120 was successfully balloted and was published by ISO in 2014.

ISO recognizes that, given the strong correlation between economic growth and urbanization, ISO standards are becoming increasingly important. ISO is continuing to lead on the development of many new standards aimed at helping cities develop more intelligent and sustainable infrastructure, supporting better city management and planning, and making cities better places to live. This work within ISO on cities and communities is now moving forward to address the key challenges arising on climate, sustainable development, and in particular health and resilience in cities. On the latter, in the case of healthcare, McCarney and McGahan argue that standardized city indicators developed in partnership with city leaders to establish a common, accepted methodology for measuring health and other urban conditions can unlock an understanding of each city's unique and shared health challenges and thus enable cross-city learning.[7]

The evolution of the work in ISO in building KPIs for cities has continued since 2012, and following the 2014 publication of the first ISO standard for city data – ISO 37120 – this

---

4 ISO, 2014. *ISO 37120: Sustainable Development of Communities – Indicators for City Services and Quality of Life*. ISO, Geneva.

5 This network of cities that developed and tested the initial set of indicators that eventually led to the development and publication of ISO 37120 was the Global City Indicators Facility (GCIF), a network of cities headquartered in Canada. The GCIF was dissolved in 2014 when ISO 37120 was first published and replaced by the World Council on City Data (WCCD).

6 McCarney, 2015. Op cit.

7 McCarney, P., McGahan, A.M., 2015. The case for comprehensive, integrated and standardized measures of health in cities. In: R. Ahn, T. Burke, and A. McGahan (eds), *Innovations, Urbanization and Global Health*. Springer, New York.

work has propelled the creation of a series of standards on city data, now globally referred to as the *ISO 37120 Series*. This evolution was, once again, propelled by cities. City leaders already adopting ISO 37120 were expressing their need for more indicators, particularly in building datasets to support their work on the rising agendas around smart cities and resilient cities.

About the ISO 37120 Series

The ISO 37120 Series of international standards contains the first set of global standards published by ISO that are devoted to building city-level data that are globally comparable. The ISO 37120 Series includes three international standards: ISO 37120 – *Indicators for Sustainable Cities*,[8] ISO 37122 – *Indicators for Smart Cities*,[9] and ISO 37123 – *Indicators for Resilient Cities*.[10] The series includes KPIs for cities distributed across 19 themes, all prioritized by cities to create a globally standardized baseline of information and ensure accurate year-over-year performance. These indicators in the ISO 37120 Series are distributed and clustered across 19 themes, as shown in Figure 19.1.

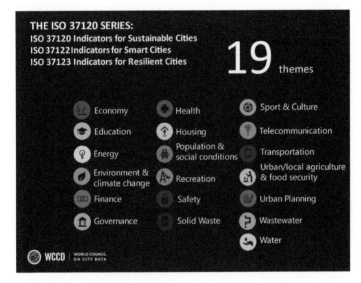

Figure 19.1 Thematic grouping of indicators. (*Source:* WCCD ISO 37120 SERIES ON CITY DATA, World Council on City Data (WCCD).)

---

8 ISO, 2014. *ISO 37120: Sustainable Development of Communities – Indicators for City Services and Quality of Life*. ISO, Geneva.

9 ISO, 2019a. *ISO 37122: Sustainable Development of Communities – Indicators for Smart Cities*. ISO, Geneva.

10 ISO, 2019b. *ISO 37123: Sustainable Development of Communities – Indicators for Resilient Cities*. ISO, Geneva.

Across the Three Standards in the ISO 37120 Series

Each of the three ISO standards is being implemented by the World Council on City Data (WCCD)[11] to support cities in reporting data in conformity with the definitions and methodologies contained in this series of ISO standards. The ISO 37120 Series contains 276 KPIs in total, all developed within ISO since 2012, voted on through global ballot by some 30 member countries between 2012 and 2019, and published between 2014 and 2019.

Cities across the world are stepping up to become ISO certified by the WCCD. The WCCD is working with over 100 cities in 35 countries, building a platform for high-calibre city-level data that are globally comparable for the first time. Cities are being ISO certified by the WCCD under ISO 37120 – Indicators for Sustainable Cities (128 indicators), ISO 37122 – Indicators for Smart Cities (80 indicators), and ISO 37123 – Indicators for Resilient Cities (68 indicators). Each of the three standards shares this common set of 19 themes that inform the strength and interlinkage of these KPIs across the ISO 37120 Series. This thematic mapping of the series' KPIs across each of the themes creates a data framework designed to help position cities to lead in sustainable, smart, and resilient city development globally with ISO standardized, comparative, and independently verified city data.

In response to the successful passage and publication of ISO 37120, the WCCD was launched, with its headquarters in Toronto, Canada. The WCCD was created to facilitate the adoption and implementation of ISO 37120 for cities worldwide as of 2014 and, working with national standards bodies in several countries, developed the certification and audit protocol for city certification under this new international standard. The web-based reporting mechanism with a certification protocol was built by the WCCD and includes an ontology to support the international standard. An ontology includes machine-interpretable definitions of basic concepts in the domain and relations among them. In her research on developing ontologies for sustainability indicators, Lida Ghahremanloo[12] outlines the steps in designing an ontology to systematically represent knowledge about a set of indicators. The first step is to understand the set of indicators and their main features. The second step consists of designing a systematic approach to represent the key information of the indicators.[13]

This WCCD ISO city indicator framework and data-reporting platform serves to support cities of all sizes globally that are adopting the ISO 37120 Series, as they seek to build a more data-driven sustainable, smart, resilient, inclusive, and prosperous future. The series provides cities with quantitative, globally comparable, and independently verified local-level data, enabling any city, of any size, to measure and compare its social, economic, and environmental progress internally year over year, and also in relation to other

---

11 The WCCD was incorporated with the mandate to establish the ISO certification protocol for cities globally, support cities in adopting ISO 37120, assist cities in reporting data in conformity with the definitions and methodologies of ISO 37120, create an audit protocol for this reporting, set up a third-party verification scheme, and create the ISO 37120 certification tiered system.

12 Ghahremanloo, L., 2012. *An Integrated Knowledge Base for Sustainability Indicators. Australasian Computing Doctoral Consortium.* RMIT Melbourne, Melbourne.

13 A similar approach was used in the development of the ISO 37120 Series ontology.

peer cities locally and globally. Cities like Boston, Los Angeles (Figure 19.2), Dubai, and Melbourne that were first certified by the WCCD in 2014 and 2015 created a solid baseline of data and now have five years of data to measure year-over-year progress and to develop predictive analytics.

About ISO 37120: Indicators for Sustainable Cities

ISO 37120 includes 104 KPIs – all prioritized by cities to measure performance on city services and quality of life. These KPIs are distributed across the themes set out in Figure 19.1. The WCCD developed the certification and audit protocol for cities to adopt ISO 37120 and report data in conformity with this standard.

These metrics are modest and "do-able" and, for the most part, are already being collected by cities across the world, just not in a standard or "apples to apples" way. Designed as core performance measurements for *city* leaders, these 104 KPIs serve as a base framework for "the measured city" upon which the other two ISO standards for city data are built.

About ISO 37122: Indicators for Smart Cities

Since around 2010, the concept of a smart city has been increasingly embraced by cities globally. Beyond the city level, many national governments and international organizations have also adopted smart city development as a key policy priority. However, until now, there has been a lack of global coherence around this concept, in particular in terms

Figure 19.2   The new vision for Los Angeles is built on a measurable baseline. (*Source:* Ron and Patty Thomas/Getty Images.)

of two fundamentals: a general lack of clarity on the definition of what a "smart city" truly is; and a lack of internationally standardized indicators to measure progress, drive smart city investment, drive city-to-city learning, and create tools for year-over-year bench-marking. The recent standard published by ISO in 2019, ISO 37122, is designed to help cities better adopt quantitative, citizen-focused metrics for the rising smart city agenda and support smart city planning with year-over-year benchmarking and impact reporting. The standardized data generated by cities reporting data in conformity with ISO 37122 and being certified facilitate city-to-city learning, encourage city solutions to travel globally, and foster smart city innovation.

ISO 37122 contains 80 KPIs with standardized definitions and methodologies. These KPIs are distributed across the themes set out in Figure 19.1. The WCCD developed the certification and audit protocol for cities to adopt ISO 37123 and report data in conformity with this standard.

As defined in ISO 37122, a smart city is one that increases the pace at which it provides social, economic, and environmental sustainability.[14] Smart cities respond to challenges such as climate change, rapid population growth, and political and economic instability by fundamentally improving how they engage society, apply collaborative leadership methods, work across disciplines and city systems, and use data information and modern technologies to deliver better services and quality of life to those in the city (residents, businesses, visitors), now and for the foreseeable future, without unfair disadvantage of others or degradation of the natural environment.

About ISO 37123: Indicators for Resilient Cities

Cities are increasingly confronted by shocks which include extreme natural or human-made events which result in loss of life and injury, and material, economic, and/or environmental losses and impacts. These shocks can include floods, earthquakes, hurricanes, wildfires, pandemics, chemical spills and explosions, terrorism, power outages, financial crises, cyber-attacks, and conflicts. A resilient city is able to manage and mitigate ongoing human and natural stresses in a city relating to environmental degradation (e.g. poor air and water quality and wastewater management), social inequality (e.g. chronic poverty, exclusion, and housing shortages), and economic instability (e.g. rapid inflation and persistent unemployment) that cause persistent negative impacts in a city.

The concept of a resilient city is being embraced by cities globally. Beyond the city level, many national governments, private institutions, and international organizations have also adopted resilient city development as a key policy priority. With cities growing in population and density, the level of risk associated with environmental shocks has risen. However, until now, there has been a lack of global coherence around this concept, in particular in terms of two fundamentals: a general lack of clarity on the definition of what a "resilient city" means and how to measure it; and a lack of internationally standardized indicators to quantify and strategically measure progress, drive resilient city investment, drive city-to-city learning, and create tools for year-over-year benchmarking. ISO 37123 defines a resilient city as a city able to prepare for, recover from, and adapt to shocks and stresses.

---

14 ISO, 2019a. Op. cit., p. 2.

ISO 37123 contains 68 KPIs with standardized definitions and methodologies. These KPIs are distributed across the themes set out in Figure 19.1. The WCCD developed the certification and audit protocol for cities to adopt ISO 37123 and report data in conformity with this standard. The WCCD, in partnership with the United Nations Disaster Risk Reduction (UNDRR), developed the core set of KPIs in ISO between 2016 and 2019. ISO 37123 was published in 2019, and the two organizations are working together now to operationalize and coordinate the global effort to build high-calibre, independently verified, and globally comparable city data for stronger and more resilient cities.

The availability of ISO 37123 standardized metrics for resilience will assist cities to prepare for, recover from, and adapt to shocks and stresses. When equipped with high-calibre and standardized data, cities can efficiently plan for, and recover from, shocks such as floods, ice storms, fires, earthquakes, pandemics, and many other stresses city leaders are increasingly being tasked with.

Why Is "the Measured City" so Important for Cities Today, and How Are Cities Embracing Global Standards to Propel Their Success?

"The measured city" fosters more informed decision-making and strategic action that propels success for the city and its residents. Cities equipped with globally standardized data are able to quantitatively assess their performance and measure progress over time, which is vital for continuous programme and service improvement and cost-effectiveness. Data are moving fast; in a global context of big data and the information explosion, the ISO 37120 Series is being embraced by cities as a way to build a reliable foundation of core knowledge for city decision-making, while enabling comparative insight and global benchmarking. Cities across the world are stepping up to become ISO certified by the WCCD. The WCCD is working with over 100 cities in 35 countries, building high-calibre city-level data that are globally comparable for the first time.

The particular value of municipal data as generated by the WCCD and the ISO 37120 Series includes a growing list of insights. In particular, these are data that are: globally standardized in conformity with the ISO 37120 Series of international standards for city data; considered trusted in that the data are outside of government (ISO certified and hosted on the WCCD global platform) and also independently and third-party verified; and regularly reported (ISO indicators are annually reported).

These data are creating data-driven municipalities and incentivizing performance across departments and divisions inside city halls, while facilitating peer city exchange locally and across the globe with "apples to apples" data to ensure city-to-city lessons and insights.

As cities embrace these three global standards and develop these high-calibre data, they are increasingly using these data to propel their success. Specific uses of the data are varied across cities in the growing global network. Naturally, and at the broadest level, cities are using these data to build more effective and transparent governance, to inform decision-making in policy and city management, and in particular to improve performance in service delivery. Cities are able to use these data to help inform budget and better target their spending, as well as improve their city's credit and bond ratings. City leaders are using their ISO certified data for international benchmarking and are, in the process, establishing comparative targets. Similarly, they are using the data for local benchmarking with neighbouring municipalities for comparative learning and sharing of informed practice across cities. Many cities are currently working with the WCCD in mapping KPIs across the ISO 37120 Series to support their strategic plans. Finally, at this broad level, cities are using this high-calibre and trusted data to leverage funding and recognition in international entities and with senior levels of government.

More specifically, cities are putting these data to use in six targeted areas.

First, cities are using this trusted, high-calibre, and comparative city-level data to inform and propel economic development planning. Previously, economic growth and development have been closely associated with the performance of nations. But the acceleration of urbanization has strengthened the weight of urban areas, making cities much more important to national and global economies. The prosperity of nations and regions is, more than ever, dependent on the economic performance of cities. The data informed by the KPIs in the ISO 37120 Series are helping cities to demonstrate investment attractiveness. For example, a core asset for cities in building prosperity is the skilled and educated workforce that the city produces within its region and also that it attracts in from outside. For instance, a city could look at its education indicators in an effort to understand how the education system and skills base contributes to the ability to propel business and attract investment in the city, compare it to that of peer cities, turn to peer groups or target cities in the network with strong performance in the relevant indicators to learn alternative models and practices, and pose key policy questions such as, how is our high school completion rate relative to our peers? How does our class size (student:teacher ratio) affect success? How many university degrees do we have per 100,000 population in our city and region? How do we look in terms of this comparative advantage? Cities of all sizes that are part of this growing global network of data-driven cities are using globally standardized data to attract investment, drive job creation, and direct economic development opportunities.

Second, cities are using their WCCD ISO certified data to monitor their progress on the Sustainable Development Goals (SDGs). Once equipped with globally standardized data, cities are able to track their progress in a quantitative sense on each of the 17 SDGs. Such a quantitative tracking has been challenging for cities and nations alike, and more qualitative measurement has been the norm. The WCCD, in 2016, undertook a detailed quantitative mapping exercise of ISO KPIs for cities with regard to all of the 17 global goals. This was based on an early recognition that to be effective and propel success on this global agenda, the SDGs, all 17 themes, needed to be localized for city action. Measurement is at the core of this success, and measurement on each and every one of the SDGs is critical since cities are pivotal leaders for sustained action. As a result of this early recognition and mapping, the WCCD network of cities and the data platform of reported, ISO certified data offer member cities the opportunity to effectively localize the SDGs and be equipped with globally comparable data to calculate progress towards the 2030 Urban Agenda of the United Nations.

Third, cities are using these standardized municipal-level data to make intelligent choices in infrastructure investment with measurable results. City leaders need good data to plan modern public infrastructure. Mayors are increasingly asking, how are we doing relative to our peers? Equipped with comparative data, they are positioned to set specific priorities and to formulate data-driven requests for infrastructure investments from federal and state governments. For example, a city is using these data to underscore a weakness in its wastewater treatment facility and is formulating an infrastructure funding application based on these ISO certified data. Others are citing data-informed deficiencies in their transit systems or their waste treatment plants, and still others are using the indicators to assess energy consumption in their public buildings to inform retrofit and efficiency investments. Cities across the Global South are using these data in a similar way to formulate "asks" and drive discussions with international funding agencies, bringing these certified data to bear. Likewise, senior levels of government are recognizing the importance of these certified and trusted municipal-level data to both inform their infrastructure spending in and across cities, and then to showcase the impact of federal and state infrastructure spending to citizens and communities across the country.

Fourth, cities are using their WCCD ISO certified data to track climate change performance. Cities are estimated to be responsible for 75% of global CO_2 emissions,[15] with transport and buildings being among the largest contributors, and city-level data are essential for a more calculated action. The ISO 37120 Series includes a suite of KPIs on emissions (e.g. GHGs, NO_2, SO_2, O_3), on energy (e.g. end-use energy consumption per capita in cities and percentage derived from renewables, energy consumption in streetlighting), and on mobility services (e.g. kilometres of public transit, commute times and ridership, modal split, automobiles per capita, kilometres of bike paths). The full remit of infrastructural and service decisions that city managers make today will be critical in setting the global emission trajectories for the future. The data being generated under the ISO 37120 Series are helping to improve cities' ability to act and move the world towards better climate performance management.[16] These data are helping cities to track progress on their local and the global climate agendas and to develop a data-inspired framework for sustainability planning.

Moreover, poorer urban households are usually more vulnerable due to weaker structures, less-protected city locations and building sites, and lack of resilient infrastructure to withstand climate damages. Similarly, the relation between urban health and climate change risks is particularly heightened under conditions of urban poverty in cities. When basic infrastructure is inadequate, existing conditions of poor sanitation and drainage and impure drinking water are further stressed under conditions of extreme weather events and flooding, leading to the transmission of infectious diseases, which puts poor urban households at high risk. This situation is worsened under circumstances of higher

---

15 UNEP, 2020. https://www.unenvironment.org/explore-topics/resource-efficiency/what-we-do/cities/cities-and-climate-change#:~:text=at%20the%20same%20time%2c%20cities,being%20among%20the%20largest%20contributors.

16 Hurth, V., McCarney, P., 2015. International standards for climate- friendly cities. *Nature Journal* 5, December 2015.

densities in urban areas.[17] The ISO 37120 Series provides cities with the necessary data to address these core vulnerabilities to climate risk in their cities.

Fifth, cities are embracing more data-driven planning and data-informed smart city agendas through the data generated by the ISO 37120 Series. With large infrastructure deficits and climate-related challenges, sustainable urban growth is dependent upon effective data-driven planning and management for smart city investments and evidence-based policy-making. Cities need data to measure their performance and timeliness in delivering services and improving quality of life. The ability to compare data across neighbouring cities locally and other peer cities globally, using a globally standardized set of indicators, is essential for comparative learning and progress in city planning and smart city development.

Urban space can be a strategic entry point for cities in driving the smart cities agenda. Physical planning and urban design professionals are pivotal actors in overcoming the path-dependent historical trajectory of urbanization, a trajectory that has to date been marked by sprawl and unplanned expansion. Some of the most innovative global cities are incorporating these data to drive more informed and smarter investment that contributes to an increased quality of life, and moreover creates an urban space that is healthier, cleaner, and, almost inevitably, more attractive and therefore more prosperous.

The first cities now certified under the standards in the ISO 37120 Series are using these data to drive a more informed planning process. Increasingly, a key role for cities and their planning functions lies in offering efficient and cost-effective city services and time-effective mobility and transport. For example, comparative transportation data in the series indicate what proportion of a cities' residents commute to work outside of a personal vehicle. Comparative data help decision-makers plan for and target potential improvements by learning from other cities' strategies. Citizens and city leaders alike are questioning how their city compares to other peer cities on, for example, the kilometres of bike paths and lanes per 100,000 population or on the kilometres of light passenger public transport system per 100,000 population – all core indicators in the ISO 37120 Series. City leaders, citizens, and planners alike are increasingly asking, how safe is our city? How clean is my air? How does our level of safety compare to our peers? How green is my city? The WCCD ISO data help answer these questions and inform land-use planning and strategic planning for smarter and more investable cities.

Sixth, city leaders are using the ISO 37120 Series to ensure their cities are less vulnerable to external shocks and chronic stresses and to direct recovery. The risks that cities are now facing as a result of the COVID-19 pandemic, climate change, and natural disasters, the pressing shortfalls in urban water and sanitation services, and the deteriorating quality of air and water in city environments are being experienced in a context of intense urban growth of cities that increasingly manifests deepening income inequities and socioeconomic exclusion. A growing international focus on resilient cities and how to ensure cities are less vulnerable to external shocks and chronic stresses and how they can recover quickly from such shocks is a core agenda item for the decades leading up to 2050.

---

17 McCarney, P., Blanco, H., Carmin, J., Colley, M., 2011. Cities and climate change: The challenges for governance. In: C. Rosenzwieg, W.D. Solecki, S.A. Hammer, and S. Mehrotra (eds), *Climate Change and Cities: First Assessment Report of the Urban Climate Change Research Network* (pp. 249–269). Cambridge University Press, Cambridge.

ISO 37123 – Indicators for Resilient Cities, the newest ISO standard published as part of the ISO 37120 Series, defines the resilience of cities as a city's ability to prepare for, recover from, and adapt to shocks and stresses.[18] The core capacity in any city's resilience to external shock includes the presence of effective institutions, governance, and infrastructure that are resilient in cities, especially in those cities that are increasingly facing extreme weather and other natural disasters. Cities are increasingly vulnerable to heavy winds, hurricanes and tornados, wildfires, heatwaves of higher intensity and longer duration that augment already existing urban heat island effects,[19] rainfalls of higher intensity and longer duration, flooding of coastal and riverside city settlements, seismic events and earthquakes coupled with tsunamis, and, most currently, the rapid spread of viruses triggering health pandemics worldwide. While all cities and their inhabitants are at risk, the poorest cities and the most vulnerable populations are most likely to bear the greatest burden of the storms, flooding, heatwaves, and other impacts anticipated to emerge from global climate change.[20] A resilient city is also able to manage and mitigate ongoing human and natural stresses in a city relating to environmental degradation (e.g. poor air and water quality), social inequality (e.g. chronic poverty and housing shortages), and economic instability (e.g. rapid inflation and persistent unemployment) that cause persistent negative impacts in a city.[21] A city's resilience will be enhanced by reducing social and structural inequalities,[22] since vulnerability is exacerbated by rapid urbanization, expanding slums, ineffective land-use planning, and poor enforcement of building codes. Resilience strategy and planning in cities requires addressing pre-existing social, political, and economic factors that make poor and marginalized people more vulnerable in the first place.

Resilience planning and preparedness is often lacking quantitative data. Too often, city resilience plans are based on qualitative assessments, narrow asset management values, and anecdotal evidence. ISO 37123, coupled with ISO 37120, includes KPIs that help cities undertake a quantitative assessment of the extent of a city's vulnerability, according to a set of social, geophysical, and economic indicators. This assessment includes such measures as per capita income; the city Gini coefficient to measure inequality; the percentage of the city's population living in poverty and slums; a demographic profile of vulnerable populations including older people and children; population size and density; physical profiles and mapping of the urban territory; a city's location in high-risk zones (seismic, flooding, hurricane, etc.); and physical form, climate type, and availability of resilient infrastructure and services including, for example, stormwater systems, construction and building code enforcement, hospital beds and emergency response. These data are supporting city leaders in resilient city development and also strengthening knowledge for a

---

18 ISO, 2019b. Op. cit., p. xi.

19 Urban heat island effect is a condition whereby cities tend to be hotter due to the absorption of heat by concrete and other building materials and due to the loss of greenspace and removal of natural vegetation that support cooling.

20 Ibid., p. 250.

21 ISO, 2019b. Op. cit., p. xi.

22 Jabareen, J. 2013. Planning the resilient city: Concepts and strategies for coping with climate change and environmental risk. *Cities* 31, 220–229.

more informed and strategic engagement with the insurance industry and ratings agencies, which has fundamental implications for city budgets and household and commercial risk assessments.

Conclusion

Data being generated across all ISO standards but particularly the ISO 37123 Series are supporting cities and helping communities to recover and strengthen their resilience to pandemics, natural hazards, and the shocks and stresses related to global climate change. These data are the essential starting point for a more resilient and prosperous future.

20

The Smart City – Introduction

In Chapter 19 Patricia showed us that cities are measuring the same thing but in different ways. This is probably the most dangerous scenario for comparison, as it will lead cities to divert their resources to places that may not need improvement and, conversely, away from places that need additional attention.

Importantly, we can focus on datasets for sustainable cities, resilient cities, and smart cities which, usefully, also fall into the domain of someone's responsibility.

The measurement protocol will allow cities to drive economic development planning further, monitor SDG progress, make informed infrastructure choices, measure their progress toward climate actions, start using IoT and data analytics to drive incremental value from infrastructures across the city, and improve their resilience awareness and planning.

In this chapter Noorie takes the concept and applies a layer of technology choices to demonstrate the outcomes that can be achieved. We will deal with the scaling up for communities across the world in Chapters 22, *The Invested City*, and Chapter 23, *The Financed City*, but with standard measurement protocols across cities it means we can apply the same technology scaling principles across cities and demonstrate technology pathways that optimize local jobs, local environmental outcomes, and local social outcomes at the individual city level.

Currently, 50% of the world's population lives in urban areas, and in less than 10 years' time it will be 60%, creating unprecedented challenges for cities and local government, with current urban infrastructure not equipped to handle the forecasted growth. As a consequence, technologies will have to take the lead in solving these new challenges and in driving climate action in cities. This idea of a "smart city" not only is of a digital and connected city but also will use technologies to modernize infrastructures, enhancing city operations and improving the lives of its citizens.

Noorie examines how technologies enable smarter, greener, and more prosperous cities by looking at the main indicators for improved city life. She looks at how the environment and health can be optimized through the reduction of carbon emissions and improved air quality, technologies that can help drive down congestion figures, and safety in regards to both crime rates and the reduction of vehicular and pedestrian accidents. She also looks at how the adoption of predictive analytics can increase resilience, how we can increase the quantity of local jobs, and how to increase our operational efficiency.

The Climate City, First Edition. Edited by Martin Powell.
© 2022 John Wiley & Sons Ltd. Published 2022 by John Wiley & Sons Ltd.

Every city is unique, so whilst it is impossible to generalize the climate impact of this "smart" technology, Noorie recognizes the significance of the differing factors that can affect a city's emissions. She provides a detailed case study of electric vehicles for carbon emissions reduction, a comparison of five different typologies that show how different city characteristics affect the city emissions after replacing 20% of all cars on the road by electric vehicles.

Finally, she looks at how the solutions for urban sustainability are profoundly interconnected and, more importantly, outlines how we need them all in order to solve the climate crisis. As Noorie writes, "sustainability and resilience are complementary". Once again, we must go "beyond" sustainability and implement technologies that build resilience, just as we have heard from Sarah and Richard in Chapter 16, *The Resilient City*, and Patricia in Chapter 19, *The Measured City*. The goal is not simply to sustain but to sustain the advancement of economic prosperity, business success, environmental integrity, and human wellbeing, despite external threats. That is a "smart city".

20

The Smart City

Noorie Rajvanshi

Before 2030, 60% or more of the world's population will be living in urban areas.[1] The growth of urbanization is creating unprecedented challenges for cities and local governments that will result in profound effects on the quality of life for city residents. Current urban infrastructure and accompanying services that cities provide are not equipped to handle the forecasted urban growth. In 2012, urban areas of the world (cities with population greater than 0.5 million) were responsible for around 47% of global carbon emissions. Thinking very carefully how cities should be built and managed, as well as which technologies should take the forefront in solving the new challenge is key to driving climate action in cities.

The concept of a "smart city" extends beyond a digital and connected city. A future city, irrespective of its geography and economic class, will use technologies to improve the lives of its citizens both by modernizing infrastructures (e.g. for generation and transmission of energy, creating more efficient and liveable buildings or optimizing transportation) and by using information technology to enhance city operations and services.

Over seven years of work at Siemens has shown[2] that no matter where the city is located, its climate or socioeconomic standing, when it comes to climate action cities share their priorities for long-term sustainability, and it is believed that deep carbon reductions will be achieved by:

- decarbonizing the electricity grid, by adopting 100% renewable electricity;
- reducing energy use through increased efficiency, for example in buildings and transportation; and
- electrifying everything, from heating to transportation.

---

1 United Nations, Department of Economic and Social Affairs, Population Division, 2019. World urbanization. https://population.un.org/wup/publications/files/wup2018-report.pdf. Prospects: The 2018 Revision (ST/ESA/SER.A/420). United Nations, New York.

2 Technology pathways for creating smarter, more prosperous and greener cities – a blueprint for creating jobs, reducing carbon emissions and improving air quality in cities across North America. 2018. https://assets.new.siemens.com/siemens/assets/public.1551137070.16acf802-36d2-4699-8d25-32d09779c3c9.cypt-wp-f2.pdf.

The Climate City, First Edition. Edited by Martin Powell.
© 2022 John Wiley & Sons Ltd. Published 2022 by John Wiley & Sons Ltd.

All three of these actions need to be backed by hundreds of technologies. For example, reduction in energy usage in transportation means more investment in public transit, which in turn could mean more buses and trains that run on cleaner fuel and with higher efficiency but also with increased capacity and at an optimized schedule to serve the maximum number of passengers. Or grid decarbonization means not only replacing existing fossil fuel powered plants with ones running on renewable fuel but also implementing software and hardware technologies to help understand when, where, and why electricity is being used; to provide transparency to consumers about pricing; and to do predictive analytics for grid maintenance. Transparency in usage and production patterns would also help identify prosumers (producers and consumers) of electricity and enable them to sell and buy electricity.

Although advances in technologies are taking place every day, almost all of these actions needed for future zero-carbon cities are powered by existing technologies that have been proven to work on a large scale. In an analysis done by Project Drawdown,[3] one of the major insights has been that we can reach drawdown by mid-century if we scale the climate solutions already in hand. In the original version of this analysis published in 2017, approximately 50 of the proposed 80 solutions are technology based and if implemented at full possible scale would reduce over 15 Gt CO_2 annually worldwide.

Technologies Enable Smarter, Greener, and More Prosperous Cities

In a smart city, technology is an enabler and not the end game that will improve the lives of people living in cities. The primary indicators of improved city life – environment and health, reduced congestion, improved safety, improved resilience, more local jobs, and improved operational efficiency – are all linked to technologies. Most of these technologies also positively impact more than one indicator, as shown in Figure 20.1.

Environment and Health

Improving air quality and reducing GHG emissions in cities has been directly linked to an improvement in health resulting in lower mortality rates and hospitalizations. Studies have shown direct linkages between street-level traffic-related air pollution and risk of cardiovascular diseases in older people.[4] Technologies across transportation, buildings, and energy sectors can drive down emissions through (1) a shift in the energy generation mix from non-renewable to renewable energies (e.g. photovoltaics, or PV) or improving

---

3 Hawken, P., 2017. *Drawdown: The Most Comprehensive Plan Ever Proposed to Reverse Global Warming.* Penguin Books, New York.

4 Alexeeff, S.E., Roy, A., Shan, J., et al., 2018. High-resolution mapping of traffic related air pollution with Google street view cars and incidence of cardiovascular events within neighborhoods in Oakland, CA. *Environmental Health* 17, 38. doi: 10.1186/s12940-018-0382-1.

Figure 20.1 Technology driving positive impact on the indicators of improved city living. GHG, Greenhouse gas; AQ, air quality; FTEs, full-time equivalents.

efficiency of existing fossil fuel powered plants (e.g. through combined heat and power); (2) improvements in energy efficiency in buildings and transportation by replacing existing technologies such as inefficient motors; and (3) a modal shift in transportation (e.g. by adding a metro line a city could move passengers away from single-occupancy vehicles).

Congestion

Technologies can help drive congestion down by increasing the usage of public transit modes through increased frequency and capacity. Technologies such as signalling systems that use communications-based train control can increase the flexibility of scheduling and in turn reduce the headway, ensuring more frequent trains, which would make them more attractive for passengers as their wait time would be significantly cut down. Also, implementing low emission zones or congestion charging zones or enforcing managed lanes on highways that would optimize traffic by charging an occupancy-based toll are some of the technologies that can push travellers to use public transit. For example, in Singapore's congestion zone, in which every vehicle containing less than three people was charged US$2 per day on weekdays, car trips decreased by 70% and bus ridership increased by about 20%.[5] Similar trends have been observed in Milan, Stockholm, and London (Figure 20.2) with varying degrees. New York City will be the first US city to implement a congestion charging fee starting 2022, and this could have significant impact on the average speed of traffic, as well as a reduction in car travel.[6]

5 Rahman, S.M.R., Kabir, E., Mannan, M. R., 2015. A scoping review of congestion pricing: Past work and its implementation benefits. *Port City International University Journal* 1851120791(01773225500), 74.

6 Glynn, A.C., 2007. Report to the New York City Traffic Congestion Mitigation Commission. New York State Department of Transportation, New York. Goodwin, P., Dargay, J., Hanly, M., 2004. Elasticities of road traffic and fuel consumption with respect to price and income: A review. *Transport Reviews* 24(3), 275–292.

Figure 20.2 Ultra low emission zone (ULEZ) in London, introduced in 2019 to improve air quality in the city. (*Source:* Alena/Adobe Stock.)

Safety

Both aspects of safety – reduction in crime rates and reduction in vehicle and pedestrian accidents – are made significantly more achievable through technologies. A study in New York City neighbourhoods has shown that lighting reduces outdoor night-time index crimes by approximately 36% and reduces overall index crimes by approximately 4% in affected communities.[7]

Intelligent street-lights not only are more energy efficient but also can collect data on which intersections are most dangerous and need to be redesigned for pedestrian safety.[8] The sensors installed in these smart street-lights could in future also detect certain sounds to automatically alert police to dangerous situations, by recognizing the sound of broken glass or a car crash, for instance.[9] These features might benefit cities beyond safety measures and can help optimize traffic flows by communicating with local transportation departments; for example, the ability to monitor intersections and note when traffic backs up could be useful information that might one day be used to adjust traffic signals.

Adoption and Resilience

Technologies alone cannot create resilient communities but nevertheless play a very important role. Modernization of city infrastructure will leverage better data, more transparency, and better predictability through connectivity of energy and transportation infrastructure. Modernizing the grid has emerged as a high priority for ensuring a resilient electric future. The implementation of innovative software capable of understanding

---

7 Chalfin, A., Hansen, B., Lerner, J., Parker, L., 2019. Reducing crime through environmental design: Evidence from a randomized experiment of street lighting in New York City. No. 25798, NBER Working Papers, National Bureau of Economic Research, https://econpapers.repec.org/repec:nbr:nberwo:25798.

8 https://spectrum.ieee.org/computing/it/san-diego-installs-smart-streetlights-to-monitor-the-metropolis.

9 https://statetechmagazine.com/article/2020/01/power-smart-street-lighting-smart-cities-perfcon.

when, where, and why electricity is being used helps to cut down on waste. Energy stakeholders are also capable of performing predictive analytics for more efficient grid maintenance.

Simulations using digital twin can be used for resilience planning. These models using data and simulation tools can project potential outcomes in the wake of disasters and can help city officials identify gaps as well as generate scenarios for investments.

Strengthening urban energy systems through hybrid grids that combine mini/micro-grids and energy storage and can operate in "island" mode (generating power independently in the event of a grid power outage) could restore power to critical infrastructure in as short a time as possible. Utilities can better manage peak demands with technologies such as distribution management technologies, distributed power generation, virtual power plants, and microgrids. Energy storage equipment (e.g. batteries and electric vehicles) can provide additional power at times of abnormal peak demand or shortages in supply, to help in maintaining energy supply to consumers. Combined with a microgrid, battery storage creates a hybrid power grid that can stand-alone in the absence of a central grid.

Analysis of the New York City region after Superstorm Sandy[10] showed that a "full investment" scenario that includes technologies for flood protection and undergrounding plus smart grid investment would require the city to spend approximately US$3 billion over a 12-year period and would produce value through reduced outages and transmission and distribution losses, reduced disruptions to priority consumers, and improved system energy efficiency. The financial value of these improvements might be as high as US$4 billion.

Jobs

Behind successful implementation of all of these technologies are people. Installing and maintaining technologies like PV on rooftops and electric chargers on streets, or retrofitting buildings with more efficient controls will create local jobs. These direct jobs will be available for varied skill levels and accompanied by more indirect or induced full-time equivalents (FTEs), which could range from additional retail and restaurants to parts manufacturing or fabrication facilities. In an analysis done by Siemens for the city of Los Angeles[11] it was found that implementing 19 infrastructure technologies across the energy, building, and transportation sector could not only help LA get to its climate goals but would create approximately 1.8 million local jobs over by 2050. Roughly 35% of the projected jobs will require special skills and training, and LA must continue to invest in higher education to graduate more engineers than any other metro area in the region, to have the solar power technicians, transportation engineers, and building mechanics needed to install and maintain LA's infrastructure.

---

10 Toolkit for resilient cities: Infrastructure, technology and urban planning. June 2013. https://rpa.org/work/reports/toolkit-for-resilient-cities.

11 Climate LA: Technology pathways for LA to achieve 80 × 50 in buildings and transportation. https://assets.new.siemens.com/siemens/assets/api/uuid:1e3d9de1-b4a1-4612-a795-5b7f16c43a05/version:1573494431/tech-pathways-resilient-la-online-version-lores-110219.pdf.

Operational Efficiency

Connected technologies have an added advantage of a data layer. These data can help everyday applications and technologies become smarter and learn from past usage patterns and trends. This information is also useful in predicting future events, especially failures. Key to predicting events is data and connectivity. Without information you can't predict failures or optimize performance. Traditional predictive maintenance can reduce downtime on equipment by up to 45% while reducing maintenance costs by up to 30%.[12]

Predictive analytics can help cities not only with traditional tasks such as remote building monitoring to reduce energy consumption[13] or transportation network planning and optimization[14] but also with unexpected problems such as pest control, specifically rodents in Chicago[15] and Washington, DC,[16] or predict disease outbreak areas by forecasting areas at risk for mosquito-borne disease, rodent infestation, and poor food quality.

Diagnosis of existing and past failures is key to predicting future cases, which in turn is essential for prescriptive analytics. Prescriptive analytics uses machine-learning algorithms to enhance the advantages provided by predictive analysis by giving applications the ability to process large amounts of data at greater speed and giving them the ability to self-learn.

Climate Impact of Technologies

Quite a lot of research has been done to evaluate the maximum potential of creating zero-carbon cities using both technologies and policies.[17, 18] All these studies agree that different types of cities require different mitigation strategies, but technologies are almost as important as policies for effective climate action. Different city typologies based on income, density, and travel preferences influence how effective a certain technology or policy will be in a city. As an example, a study of 274 cities around the world[19] has shown

---

12 EERE, Operations & maintenance best practices – A guide to achieving operational efficiency. August 2010. https://www.energy.gov/sites/prod/files/2013/10/f3/omguide_complete.pdf.

13 Smart buildings: Using smart technology to save energy in existing buildings. https://www.aceee.org/sites/default/files/publications/researchreports/a1701.pdf.

14 Zhu, L., Yu, F.R., Wang, Y., Ning, B., Tang, T., 2019. Big data analytics in intelligent transportation systems: A survey. *IEEE Transactions on Intelligent Transportation Systems* 20(1), 383–398. doi: 10.1109/TITS.2018.2815678.

15 Using predictive analytics to combat rodents in Chicago. https://datasmart.ash.harvard.edu/news/article/using-predictive-analytics-to-combat-rodents-in-chicago-271.

16 Mayor Bowser highlights citywide efforts to reduce rodents. https://mayor.dc.gov/release/mayor-bowser-highlights-citywide-efforts-reduce-rodents.

17 Global typology of urban energy use and potentials for an urbanization mitigation wedge. https://www.pnas.org/content/pnas/112/20/6283.full.pdf.

18 Estimating the national carbon abatement potential of city policies: A datadriven approach. https://www.nrel.gov/docs/fy17osti/67101.pdf.

19 Global typology of urban energy use and potentials for an urbanization mitigation wedge. https://www.pnas.org/content/pnas/112/20/6283.full.pdf.

that in high-income economy, mature cities, higher gasoline prices combined with compact urban form can result in savings in both residential and transport energy use, whereas in low-income cities, compact urban form (resulting in higher population density) accompanied by transport planning can encourage low-carbon emission patterns for travel. Findings from a report from the Coalition of Urban Transition (CUT[20]) echo these analyses and quantify the impacts of technically feasible, widely available mitigation measures to reduce GHG emissions from cities by almost 90% by 2050. CUT's work shows that 42% of the abatement potential lies in cities in non-OECD regions with emerging economies and populations lower than 750,000.

Every city is unique – located in different climate regions, with its own compact form or density, its people moving around the city in a unique way, building homes and using energy that is different than even a neighbouring city. And we cannot generalize impacts of the same technology on different cities. This means that an effective carbon mitigation strategy in one city might not be as impactful in a different city. The infographics in this section demonstrate this point through the impact of the adoption of electric cars on CO_2 emission reduction.

Case Study: Electric Vehicles for Carbon Emission Reduction

A comparison of five different typologies shows how different city characteristics affect city emissions after replacing 20% of all cars on the road by electric vehicles. These five typologies:

- high-income economy, low density, car-centric;
- middle-income economy, high density, transit-centric;
- high-income economy, high density, transit-centric;
- middle-income economy, low density, transit-centric; and
- high-income economy, high density, car-centric

represent the majority of the urban areas of the world. The biggest missing typology is cities from emerging economies, not included due to lack of data when this analysis was completed. The data for this analysis come from a tool called the City Performance Tool (CyPT[21]) developed by Siemens. CyPT was developed to help cities make informed infrastructure investment decisions, identifying which technologies from the transport, building, and energy sectors best fit a city's baseline in order to mitigate CO_2e emissions, improve air quality, and add new jobs in the local economy. CyPT has been used in over 40 cities worldwide.

A *high-income economy, low density, car-centric (HILD-car)* city (Figure 20.3) is one that can be best characterized as urban sprawl. With just over 2,300 people in a square

---

20 Climate emergency, urban opportunity. https://urbantransitions.global/en/publication/climate-emergency-urban-opportunity/.

21 City performance tool by Siemens. https://new.siemens.com/global/en/products/services/iot-siemens/public-sector/city-performance-tool.html.

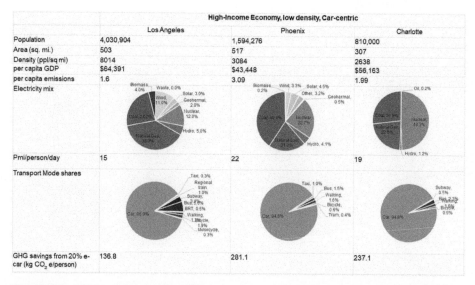

Figure 20.3 HILD-car. Ppl, People; Pmi, person miles; e-car, electric vehicle; BRT, bus rapid transit.

kilometre in this city there is plenty of space. The primary mode of transportation is personal cars. Most of these cities have an average of one to two vehicles per household. Within this typology the characteristics that most impact the GHG savings from electric vehicles are the percentage of renewable energy in the electricity generation mix and percentage of travel by cars.

A *middle-income economy, high density, transit-centric (MIHD-transit)* city (Figure 20.4) is very compact with over 7,700 residents in just 1 km². The average resident travels over 27 km per day, and with only 36% of the mode share from private cars, public transit and non-motorized travel are prevalent. It is no surprise that GHG saving potential per capita of electric cars is less than half compared with HILD-car cities, even with high penetration of renewables in the electricity grid. These compact cities also have 20% lower per capita emission rates compared to HILD-car cities, and this aligns with an analysis of 120 cities which found that a 10% increase in density correlates with a 2% decrease in per capita carbon emissions.[22]

The two examples of *high-income economy, high density, transit-centric (HIHD-transit)* cities (Figure 20.5) used in this analysis are very different. These two cities sit in two opposite geographical regions and have a very big difference in population. What puts them both in the same category for this analysis is the percentage of residents using a car as their primary mode of travel – average 32%. Surprisingly, the more compact city does not win the GHG savings war here, mostly because 10% fewer people use cars in Seoul compared to Copenhagen to begin with.

22 Annex 5 – Relationship between urban density and urban greenhouse gas emissions. https://urbantransitions.global/wp-content/uploads/2020/04/climate-emergency-urban-opportunity-methodological-annexes.pdf.

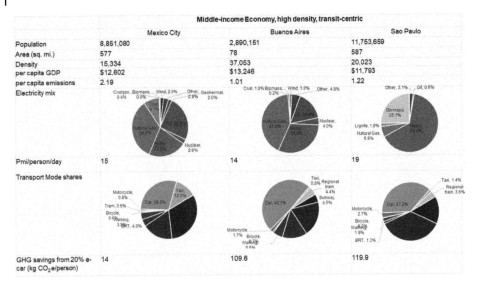

Figure 20.4 MIHD-transit.

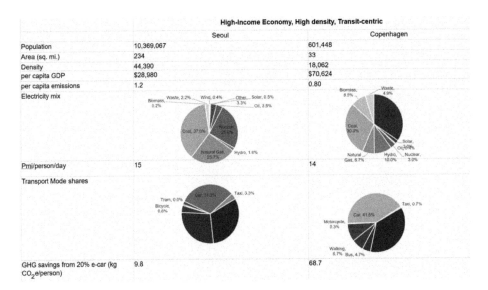

Figure 20.5 HIHD-transit.

The typology of *middle-income economy, low density, transit centric (MILD-transit)* cities (Figure 20.6) is a rare one. Cities with fewer than 1,150 people per square kilometre, such as Ningbo in China, are usually designed for personal car travel, so any city with less than 50% mode share by car which is not compact would have its own set of challenges to make public transit effective.

Finally, the *high-income economy, high density, car-centric (HIHD-car)* cities (Figure 20.7) are most common in OECD countries. Examples such as Washington, DC or the one

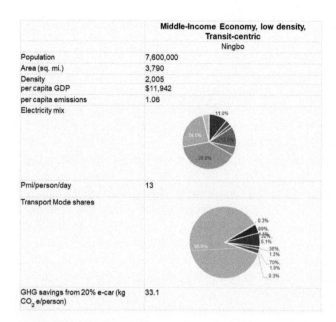

	Middle-Income Economy, low density, Transit-centric Ningbo
Population	7,600,000
Area (sq. mi.)	3,790
Density	2,005
per capita GDP	$11,942
per capita emissions	1.06
Electricity mix	
Pmi/person/day	13
Transport Mode shares	
GHG savings from 20% e-car (kg CO₂ e/person)	33.1

Figure 20.6 MILD-transit.

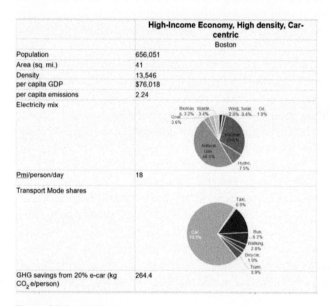

	High-Income Economy, High density, Car-centric Boston
Population	656,051
Area (sq. mi.)	41
Density	13,546
per capita GDP	$76,018
per capita emissions	2.24
Electricity mix	
Pmi/person/day	18
Transport Mode shares	
GHG savings from 20% e-car (kg CO₂ e/person)	264.4

Figure 20.7 HIHD-car.

used in this analysis of Boston are densely populated but still have between one and two cars per household. These cities will greatly benefit from electrification of the private fleet as that is the primary mode of transportation here.

If we can take away a single aspect from the multitude of infographics in this section, that would be that the impact of a technology on carbon emission reduction in a city is dependent on many variables and it is important to consider as many as possible before making implementation decisions.

Interconnected Technologies

Solutions for urban sustainability are interconnected, and we need all of them to solve the climate crisis. Although we would like to cling to the idea of a "silver bullet" or that one big thing that will solve our climate issues, this problem is too complex for that to be possible. A whole ecosystem of solutions is needed to make our modern cities sustainable, prosperous, and liveable.

Technologies that can be combined and that complement will be the ones that will have the highest impacts and are capable of solving the most complex problems. For example, in order for electric vehicles to have the highest benefits in improving air quality and reducing GHG emissions, we need a grid that is powered entirely by renewables. In turn, a 100% renewable grid needs technologies that can modernize the grid by implementing software to incorporate alternative sources of energy (like rooftop PV panels or microgrids, which use the grid as back-up).

The CUT report[23] evaluated the contribution of various mitigation scenarios to creating zero-carbon cities by 2050. Of the total 15.5 billion metric tons of CO_2 reduction achievable in cities by 2050, 58% would come from adoption of more efficient homes and commercial buildings and 18% from cleaner transportation, but both of these strategies are only half as effective without cleaner electricity, making a clear case for the interdependencies of technologies.

An example from a CyPT analysis[24] for the city of Los Angeles showed that five technologies that make up an eMobility strategy for the city if implemented at scale in 2050 would consume over 2 GWh (Figure 20.8). This is 1,663% more electricity used for transportation in LA than today! This number makes an excellent case for upgrading the existing grid and making sure our electricity generation and distribution system is able to handle the stress electrified transport would bring by 2050.

Some additional technologies like distribution management technologies would also enable smoother integration of new loads into the grid. These tools would ensure availability of a minimally sufficient amount of electricity where and when it is needed. It can include smart metering, which is the digital monitoring of consumption data shared bidirectionally between consumer and energy provider. When combined with active load tools, the energy provider has some control over consumption and can use this to mitigate peak demand. For example, such systems can delay or reduce electricity consumption of non-essential functions during peak times. This means that electric vehicles could charge

---

23 Ibid.

24 CyPT analysis for Los Angeles: Climate LA, technology pathways for LA to achieve 80 × 50 in buildings and transportation. https://assets.new.siemens.com/siemens/assets/api/uuid:1e3d9de1-b4a1-4612-a795-5b7f16c43a05/version:1573494431/tech-pathways-resilient-la-online-version-lores-110219.pdf.

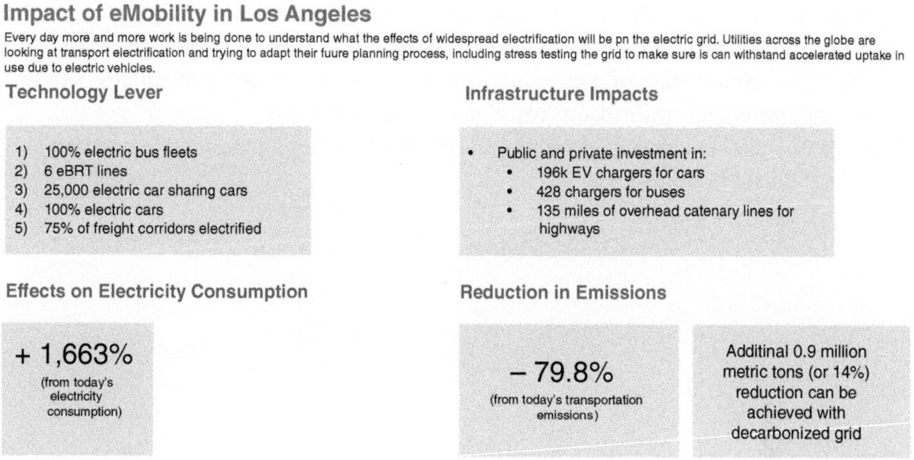

Impact of eMobility in Los Angeles

Every day more and more work is being done to understand what the effects of widespread electrification will be pn the electric grid. Utilities across the globe are looking at transport electrification and trying to adapt their fuure planning process, including stress testing the grid to make sure is can withstand accelerated uptake in use due to electric vehicles.

Technology Lever

1) 100% electric bus fleets
2) 6 eBRT lines
3) 25,000 electric car sharing cars
4) 100% electric cars
5) 75% of freight corridors electrified

Infrastructure Impacts

- Public and private investment in:
  - 196k EV chargers for cars
  - 428 chargers for buses
  - 135 miles of overhead catenary lines for highways

Effects on Electricity Consumption

+ 1,663%

(from today's electricity consumption)

Reduction in Emissions

− 79.8%

(from today's transportation emissions)

Additinal 0.9 million metric tons (or 14%) reduction can be achieved with decarbonized grid

Figure 20.8 The impact of eMobility in Los Angeles.

gradually throughout the night, rather than at maximum speed immediately after being plugged in during the evening after work.

Distributed energy resources (DERs) are smaller-scale generation, mostly renewables that could be generated close to consumer locations, such as PV on residential rooftops or on car parks of commercial or retail space. DERs can be set up as small microgrids that can be operated independently of the central grid if needed. Microgrids reduce transmission losses compared to the centralized systems, offer resilience against grid failure, and can enable new types of users, aka prosumers. Prosumers are consumers who are also producers of (renewable) energy and who use energy in a smarter and more efficient manner. Prosumers require microgrids to be optimized for bidirectional energy flow or to be able to handle variable amounts of electricity, such as from batteries in electric vehicles.

Conclusion

It is becoming apparent through daily events that the time to invest in resilience is now. Resilience is the ability of cities and its private citizens, organizations, or systems to prepare for, respond, recover from, and thrive in the face of hazards. The goal is to ensure the continuity and advancement of economic prosperity, business success, environmental quality, and human wellbeing, despite external threats. In the face of the current climate crisis, becoming resilient is the only way for cities to remain economically competitive and attractive for business growth in this globalized world.

Sustainability and resilience are complementary and overlapping concepts. Resilience planning helps broaden the goals set by a city's long-term climate action plan and embraces the turbulence of daily life. Resilience is about learning to live with the spectrum of risks that exist at the interface between people, the economy, and the environment, and maintaining an acceptable stability or equilibrium in spite of continuously changing circumstances. Resilience also addresses the interdependencies between systems and minimizes unforeseen "gaps" in risk management.

The rise of digital technology is helping systems everywhere to anticipate, predict, and respond to failures in the system.

21

The Just City – Introduction

In Chapter 20 Noorie showed us pathways with existing technology that can transition us to better health and air quality outcomes, reduced congestion, and safer communities, all while building resilience across critical city systems. By being able to quantify the job creation, we can begin to plan education programmes and job brokerage services that can match the local city employment profile to the future needs generated by this transition to "the climate city".

As a vivid example of how far we have gone so far in the book, Noorie's case study has mapped the idea of "Electric vehicles for CO_2 reduction" to cities, accounting for their economic status, their density, and the modal-share of the city – meaning we can describe a pathway to deliver the Paris Agreement commitments for every unique scenario. "The dangerous thing about knowing the truth is you may have to act on it!"

Noorie also points out the necessity to move towards more distributed energy generation: off-grid and islanded energy centres that can give communities energy independence, resilience, and affordability. This will also provide a valuable twin-tracked strategy to put more renewable energy, faster, into our energy mix.

We have a three-part section on *The Just City*. Hayley leads off with an all-encompassing view of the world's greatest environmental killer – air pollution. In Part II Jane and Matt augment Hayley's section with a case study on what a low-cost monitoring network for air quality could mean for cities everywhere. And in Part III Jenny tells the very sad story of Ella. The focus that Ella's story brings, not just to London but to cities everywhere, in maintaining healthy cities gives us cause for optimism.

Part I

It is estimated that 80% of city dwellers breathe air the World Health Organization (WHO) has deemed unsafe.[1] The greatest risk a city dweller faces is not becoming a victim of a crime but of air pollution. Hayley examines the extent to which this sometimes invisible

---

[1] https://www.who.int/airpollution/data/cities-2016/en/#:~:text=more%20than%2080%25%20of%20people,cities%20are%20the%20most%20impacted.

The Climate City, First Edition. Edited by Martin Powell.
© 2022 John Wiley & Sons Ltd. Published 2022 by John Wiley & Sons Ltd.

Death rates from air pollution, 2019

Death rates are measured as the number of deaths per 100,000 population from both outdoor and indoor air pollution. Rates are age-standardized, meaning they assume a constant age structure of the population to allow for comparisons between countries and over time.

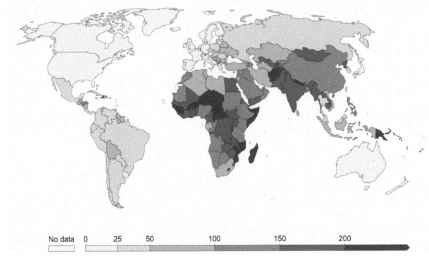

| No data | 0 | 25 | 50 | 100 | 150 | 200 |

Figure 21.1 Death rates from air pollution, 2019. (*Source:* IHME, Global Burden of Disease.)

enemy affects our cities as she explores the human cost of air pollution, the staggering figures associated with premature deaths totalling 7 million per year, and the disproportionate effect on minority groups.[2] Some of this we can see reflected in Figure 21.1.

Hayley offers us some inspiration, however, by looking at the benefits of air quality, the results of which we have started to see as a result of the lockdown measures the pandemic have precipitated, as well as the importance of measuring and monitoring air pollution levels. She then goes on to talk in more detail about how we can tackle vehicle emissions, looking specifically at the Cheonggyecheon Expressway in Seoul.

Many of the solutions Haley outlines were successful in part due to the efforts to mobilize support and to overcome this urban mindset that the climate is a faraway intractable challenge for future generations. It is not. It is now. Luckily there is nothing more motivating than one's health.

Part II

While Hayley outlines the problem, Jane and Matt go into the specifics, which means ... more data! Looking at London specifically, they introduce us to Gerald Nabarro, a champion of the Clean Air Act of 1956 and the monitoring system that was put in place following the tragic consequences of the Great Smog. Chapter 18, *The Data City*, Chapter 19, *The Measured City*, and Chapter 20, *The Smart City*, have all argued that data are crucial if we are to effectively combat air pollution.

---

2 https://www.who.int/news-room/fact-sheets/detail/ambient-(outdoor)-air-quality-and-health.

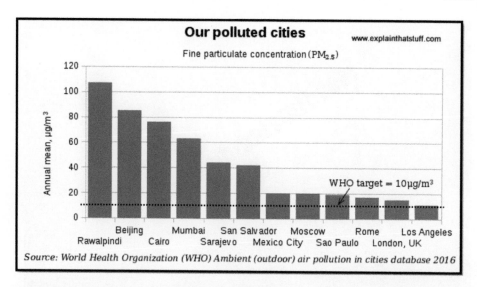

Figure 21.2 The annual mean PM2.5 for a range of cities. (*Source:* https://www.greenerjobsalliance. co.uk/courses/air-pollution-a-public-health-emergency/module-1-causes-health-impacts-of-air-pollution/.)

Jane and Matt look at how air pollution is measured and how data have enabled actions that have improved air quality in London over recent years despite the city's growing population. They also show us that London's network is, unfortunately, an anomaly, with many cities lacking the required time, money, and expertise to implement robust networks. As we can see in Figure 21.2, the toxic air pollution in London is modestly good in relation to other cities. Monitoring allows directed action to hotspots that can have tremendous improvements across the city.

They introduce lower-cost attempts to monitor air quality as well as the attempt to overcome these challenges made by the Environmental Defense Fund that launched the Breathe London pilot project in 2018 and has since seen impressive results.[3]

Once again, data lie at the crux of the future of cities.

Part III

Finally, Jenny relays the human cost of air pollution (Figure 21.3) and the moral imperative we have to do better through Ella's story. Ella died in 2013 at the age of nine after developing asthma at the age of seven due to exposure to excessive air pollution. Her death certificate lists her cause of death as air pollution, the first time air pollution has ever been listed as a cause of death, although it has attributed to the tens of thousands of premature deaths that occur in the UK every year. Jenny looks at the problems air

---

3 https://www.breathelondon.org.

Figure 21.3 Heavy traffic on a London Street, coupled with emissions from buildings, creates toxic hotspots. (*Source:* magicbones/Adobe Stock.)

pollution pose, where we currently are in dealing with that problem, the failure of local leadership to enact affirmative action, and, most importantly, what we need to do in order that we don't add to the problem but make steps to change it.

Ella's life ended far too early; her death identified a problem we all knew existed but did not do enough about, inspiring positive action to further improve our air quality. How unfortunate it is that it takes a child's death certificate to propel this action. But as Jenny points out, it perhaps will be Ella's death that sets the new air pollution agenda. The listing of air pollution on Ella's death certificate could lead to cleaner air and better health for many others.

The Just City is one of collective responsibility and appropriate and responsive action.

21

The Just City (Part I)

Hayley Moller

There is a prevailing perception that living in cities means greater risk of being the victim of a crime. But depending on where you live, the greatest risk of death when living in a city isn't crime.

It's air pollution.

Around the world, 80% of city dwellers breathe air the WHO has deemed unsafe.[1] This is true in the developing world – the ochre skies above New Delhi and the infamous Beijing smog come to mind – and in the developed world: from London to Sydney, particulate matter from the wear of brakes, tyres, and roads poses an invisible threat. Though the pollutants – and their relative visibility when gazing out over a skyline – differ, the threat remains high in cities around the world.

Nowhere are the interlocking challenges of public health, climate change, and economic inequality more obvious than in the air quality of our cities.

Human Costs of Air Pollution

The global health effects are staggering. Air pollution kills 7 million people every year,[2] accounting for 12% of all deaths annually,[3] and making it the top environmental risk factor for premature death worldwide.[4]

Low- and middle-income countries suffer the greatest toll,[5] but even in Europe, estimates suggest that particle pollution from human sources takes 9 months off the average life expectancy.[6] In the USA, more people die each year from car exhaust than car

---

1 https://www.who.int/airpollution/data/cities-2016/en/#:~:text=more%20than%2080%25%20of%20
people,cities%20are%20the%20most%20impacted.

2 https://www.who.int/news-room/fact-sheets/detail/ambient-(outdoor)-air-quality-and-health.

3 https://www.who.int/news-room/fact-sheets/detail/the-top-10-causes-of-death.

4 Fuller, G., 2018. *The Invisible Killer: The Rising Global Threat of Air Pollution – and How We Can Fight Back.* Melville House, London.

5 https://www.who.int/news-room/fact-sheets/detail/ambient-(outdoor)-air-quality-and-health.

6 Ibid.

The Climate City, First Edition. Edited by Martin Powell.
© 2022 John Wiley & Sons Ltd. Published 2022 by John Wiley & Sons Ltd.

accidents.[7, 8] In New York City alone, particle pollution causes more than 3,000 deaths and some 6,000 emergency room visits each year.[9]

The good news is that aggressive strategies to clean up air pollution can have significant and measurable health benefits. For instance, China's ambitious air quality efforts increased average life expectancy by 2.5 years in just 5 years.[10]

The primary cause of air pollution mortality is long-term exposure to PM2.5 – particulate matter of 2.5 microns or less in diameter, or approximately 1/40th the width of a human hair. This fine particle pollution causes both chronic and acute respiratory diseases, including asthma, and can penetrate deep into human lungs and enter the bloodstream, causing strokes, heart disease, and lung cancer.[11]

In addition to particle pollution, ozone (O_3), nitrogen dioxide (NO_2), and sulphur dioxide (SO_2) pose health threats.[12] These pollutants can stunt lung growth in children, reduce fertility, increase dementia,[13] and increase the likelihood of premature, underweight, or stillborn babies.[14]

And while WHO offers "safe" thresholds for these pollutants, they can cause adverse health effects at any concentration – which is why it's so important to address them.[15]

The chronic health impacts associated with air pollution carry a heavy economic cost – from missed days of work, to healthcare costs, to lost productivity. In 2018, particle pollution from fossil fuels alone was responsible for 1.8 billion days of work absence, 4 million new cases of asthma in children, and 2 million premature births.[16] The economic costs are devastating, estimated at US$5.7 trillion, or 4.4% of global GDP, in 2016.

Poor air quality causes further economic trouble by discouraging tourism and making it harder to recruit and retain top talent for the local job market. Each day spent in Delhi, India, for instance, translates to 2 hours less off your life each day you spend there – an inconvenient fact for job recruiters and tourism bureaus.[17]

When we talk about air pollution, the scale of the discussion matters significantly. Air quality is determined by local, regional, and even global factors, since recent studies have shown that particle pollution can travel vast distances – a dust cloud from the Taklamakan Desert in China, for example, made a complete circuit of the globe in 13 days.[18]

But air quality is also hyperlocalized. Even in a city with high pollution controls that is situated within a similarly well-regulated region, specific neighbourhoods can have vastly different air quality.

---

7 https://www.sciencedirect.com/science/article/abs/pii/S135223101400822x.

8 https://www-fars.nhtsa.dot.gov/main/index.aspx.

9 https://www1.nyc.gov/assets/doh/downloads/pdf/eode/eode-air-quality-impact.pdf.

10 https://www.ted.com/talks/angel_hsu_how_china_is_and_isn_t_fighting_pollution_and_climate_change#t-437694. (Original source is Air Quality Life Index, 2018).

11 https://www.who.int/news-room/fact-sheets/detail/ambient-(outdoor)-air-quality-and-health.

12 Ibid.

13 Smedley, T., 2019. *Clearing the Air: The Beginning and the End of Air Pollution.* Bloomsbury, London, p. 13.

14 https://www.nytimes.com/2020/06/18/climate/climate-change-pregnancy-study.html.

15 https://www.who.int/news-room/fact-sheets/detail/ambient-(outdoor)-air-quality-and-health.55

16 https://energyandcleanair.org/wp/wp-content/uploads/2020/02/Cost-of-fossil-fuels-briefing.pdf.

17 https://www.bloomberg.com/news/articles/2015-01-26/mr-president-world-s-worst-air-is-taking-6-hours-off-your-life.

18 Fuller, 2018. Op cit., p. 104.

As a result, air pollution does not affect everyone equally. All too often, poor communities and communities of colour bear a disproportionate burden of both air pollution and climate change.[19] Lack of political power to oppose nearby chemical facilities and power plants, limited housing and employment options, minimal access to healthcare, and systemic discrimination all play a role in perpetuating this dangerous dynamic.

In my home country of the USA, studies consistently show that the risk of death from air pollution for Black Americans is significantly higher than any other population.[20, 21] Black children in the USA are twice as likely as white children to have asthma, and they are 10 times more likely to die from it.[22]

While the COVID-19 pandemic (the pandemic) gave residents of some cities clean air and blue skies for the first time in recent memory, the pandemic has also made systemic societal inequities highly visible. A respiratory disease, COVID-19 visits much more serious illness and higher death rates among those with already burdened respiratory systems. A much-discussed study showed that COVID-19 patients living in areas with higher particle pollution before the pandemic were much more likely to die than those who breathed cleaner air.[23]

It is perhaps unsurprising, then, that since the early weeks of the COVID-19 pandemic, majority-Black counties had three times the rate of COVID-19 infections and nearly six times the death rates of majority-white counties in the USA.[24] At the same time, Black people were four times more likely to die from COVID-19 than white people in the UK.[25]

Co-benefits of Good Air Quality

The good news is that tackling air pollution, especially in cities, not only is a fast-track to positive and measurable public health, and economic and equity outcomes, but also provides major benefits from an emissions standpoint, since many of the worst air pollution sources are also major greenhouse gas emitters. The same emissions from power plants, cars, and factories that produce immediate health impacts locally also cause long-term warming globally.

The pandemic illustrates this overlap. In the first months of July 2020, lockdowns, travel restrictions, and shuttered industry not only caused global emissions to fall by nearly 5%, but also decreased fine particle pollution by nearly 4%.[26]

---

19 https://www.unenvironment.org/news-and-stories/story/air-pollution-hurts-poorest-most.

20 https://www.pnas.org/content/116/13/6001.

21 Ibid.

22 https://www.cdc.gov/asthma/asthma_stats/asthma_underlying_death.html.

23 https://www.nytimes.com/2020/04/07/climate/air-pollution-coronavirus-covid.html.

24 https://www.washingtonpost.com/nation/2020/05/06/study-finds-that-disproportionately-black-counties-account-more-than-half-covid-19-cases-us-nearly-60-percent-deaths.

25 https://www.ons.gov.uk/peoplepopulationandcommunity/birthsdeathsandmarriages/deaths/articles/coronavirusrelateddeathsbyethnicgroupenglandandwales/2march2020to10april2020.

26 https://www.eurekalert.org/pub_releases/2020-07/uos-sei070820.php.

Furthermore, given the inequities inherent in the reality of who breathes air pollution, addressing air pollution can also help alleviate some of the systemic injustices that limit opportunities for poor communities, those living in informal settlements, and – especially in the USA – communities of colour. Thoughtful, systemic, and science-based solutions to air quality issues can lock in long-term health and economic benefits for urban residents while contributing to decarbonization and a cleaner future for all.

The history of air pollution measures is littered with examples of well-meaning decision-makers inadvertently worsening health outcomes. In the early 2000s, for instance, much of Europe put greater restrictions on nitrogen oxides. Unfortunately, this approach actually increased levels of nitrogen dioxide, the single worst pollutant in the group, thereby exacerbating air quality overall.[27]

This offers a cautionary lesson for cities: a solution is not tenable if it solves one problem while aggravating another. A piecemeal approach to tackling the climate crisis in cities may reduce emissions, but at what cost? As we implement climate and sustainability solutions in our cities, we must evaluate their impact on real people – especially those most vulnerable.

The United Nations' Sustainable Development Goals recognize that we cannot make real progress on any one while ignoring the others. That is exactly the type of comprehensive approach we must apply in our efforts to address urban air pollution. And to do so, we must transform how we think about our cities.

Addressing an Invisible Adversary

Approaching the central question of what all (or most) cities can do to tackle air pollution is challenging, since cities face very different circumstances globally. Not only do different cities face variations in sources, pollutants, and meteorological conditions, but they also face different realities of the air pollution controls by the regional and national governments under whose jurisdictions they fall.

Broadly, cities in the Global North have long histories of monitoring, researching, and addressing air pollution. London, for instance, began seriously working to improve air quality after the London Smog of 1952, which killed 12,000 people.[28] Los Angeles, once panned for its iconic smog, has been able to greatly improve its air quality over several decades in collaboration with the highly effective California Air Resources Board.[29]

In many cases, these cities were able to identify pollutants from car exhaust as causing some of the worst health effects. In a classic case of treating the symptom and ignoring the root cause, however, cities like London and Los Angeles narrowly focused on cleaning up tailpipe emissions. Had they recognized and addressed the greater systemic challenge, we may not have so greatly prolonged our reliance on gasoline- and diesel-powered vehicles,

---

27 Fuller, 2018. Op cit., pp. 146–147.

28 Fuller, 2018. Op cit.

29 Smedley, 2019. Op. cit., pp. 189–190.

a major source of greenhouse gas emissions.[30] Meanwhile, cars remain a top pollution source globally.

With respect to another major air pollution source – fossil fuel combustion for energy – the good news is that a clean energy transition is underway globally, and cities are working to accelerate this trend. In the USA alone, more than 200 cities have established 100% clean electricity commitments.[31] As more cities adopt these mandates, the reliance on fossil fuels – and the associated pollution – will continue to wane.

There are, of course, other sources of air pollution to consider in the Global North, including agricultural ammonia (especially in North America), construction zones, manufacturing facilities, aircraft and airports, and volatile organic compounds (VOCs) from chemical products like paints, pesticides, and perfumes.[32]

In the Global South, the picture is much different, with a completely discrete set of air quality challenges.

Some 3 billion people in low- and middle-income countries rely on wood, straw, dung, and other biomass for cooking. At-home wood burning – classified as indoor air pollution – caused 2.85 million premature deaths in 2015, with the highest impact among women, children, and older people.[33] These traditional cooking practices are also responsible for somewhere between 2 and 5% of annual greenhouse gas emissions globally, making them a prime target for action.[34]

As we expand access to electricity and fuel – in line with the UN Sustainable Development Goal of universal access to clean energy – decisions about fuel sources will be a matter of life and death for many.

The mayor-friendly adage "if you can't measure it, you can't manage it" is highly applicable in the case of air pollution. In the Global South, inadequate measurement is a major impediment to improving air quality. The WHO ambient air pollution database, which catalogues air quality data from more than 4,000 settlements worldwide (Figure 21.4), includes 2,626 European settlements, but just 41 African settlements. Of the 930 settlements included from the Americas, just 170 are located in low- and middle-income countries.[35]

A first step for any city hoping to improve its air quality is to install monitoring equipment to determine the contours of air quality in that city. Are there high levels of particle pollution from cooking fuel? Spikes of ozone from cars or agricultural activities? The signature of regional fossil fuel combustion? The fingerprints of different sources will appear, giving scientists and policymakers critical information to inform effective next steps.

International frameworks like the BreatheLife Cities 2030 campaign or the C40 Cities Climate Leadership Group's air quality network can help foster access to technical knowledge and provide tools and resources particularly for countries in the Global South.

---

30 Interview with Iyad Kheirbek, Program Director, Air Quality at C40 Cities Climate Leadership Group. 16 July 2020.

31 https://innovation.luskin.ucla.edu/wp-content/uploads/2019/11/100-Clean-Energy-Progress-Report-UCLA-2.pdf.

32 https://www.scienceintheclassroom.org/research-papers/whats-major-source-urban-air-pollution.

33 Fuller, 2018. Op. cit., p. 181.

34 https://drawdown.org/solutions/improved-clean-cookstoves.

35 https://www.who.int/airpollution/data/aap_database_summary_results_2018_final2.pdf?ua=1.

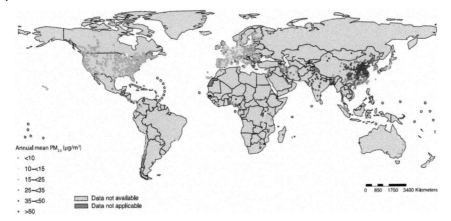

Annual mean PM₂.₅ (µg/m³)
<10
10–<15
15–<25
25–<35
35–<50
>50
Data not available
Data not applicable

0 850 1700 3400 Kilometers

Figure 21.4 Location of the monitoring stations and PM2.5 concentration in more than 4,000 human settlements, 2010–2016. (*Source:* https://www.who.int/airpollution/data/AAP_database_summary_results_2018_final2.pdf.)

Low-Hanging Fruit: Tackling Vehicle Emissions

As noted throughout this book, it's not enough to address a symptom of a larger systemic problem. When it comes to air pollution, the most effective solutions are the ones that transform city spaces and people's daily lives. Because air pollution touches so many interconnected issues, from health to racial justice, we must think about it not as a siloed problem to be solved, but as an opportunity to build cleaner, healthier, more inclusive cities.

Good air quality solutions should be designed to:

- reduce key air pollutants without exacerbating others;
- reduce greenhouse gas emissions that cause climate change;
- improve public health and decrease associated economic costs;
- decrease inequality and be applied equitably; and
- increase economic opportunities, particularly for the most vulnerable populations.

In considering solutions that meet the above criteria and that are also applicable to most cities around the world, one blight rises above the rest as being a constant and significant – but relatively easy to tackle – source of air pollution: vehicles.

While some might argue that cars have been the focus of most air pollution controls throughout history, and we may have even already reached "peak car", I think it's exactly those factors that make vehicles such a prime target. We should capitalize on the existing trend to capture the greatest gains by designing cars out of our lives.

One of the earliest ways that cities – especially in Latin America – sought to curb vehicle emissions was through vehicle restrictions, which typically ban cars on certain days according to their licence plates.[36]

Popular evolutions of vehicle restrictions include congestion pricing and low emission zones. In these schemes, older vehicles must pay a fee to enter or are restricted altogether

---

36 Fuller, 2018. Op. cit., p. 194.

from accessing a particular region of the city – usually the city centre. By the start of 2015 there were more than 200 low emission zones in 12 European Union countries.[37]

These zones can be highly effective in reducing local air pollution and inducing scrappage of older vehicles, accelerating trends already underway. Concerns that highly polluting vehicles will simply skirt the zone have not proven true.[38] In order to ensure that these zones are most effective, however, they must be designed around the realities of the city's traffic, include vehicles like large freight trucks and buses, and be enforced around the clock. They must also cover a large enough area to have a measurable impact.[39] Finally, these zones should be deliberately sited to reduce harmful air pollution for a city's poorest and most vulnerable residents.

In Europe diesel vehicles remain dominant, with three-quarters of all the diesel vehicles in the world driven on European (including UK) roads.[40] To address diesel emissions, the UK and some European cities have simply banned diesel vehicles. However, there is scepticism about the effectiveness of this action, since some fail to factor in buses, trucks, and heavy vehicles.[41] Even Paris's ban on both diesel and gasoline vehicles by 2030 won't eliminate vehicle-related emissions: electric vehicles aren't zero emissions if the electric grid remains powered by fossil fuels, and even electric vehicles create unregulated particle pollution from tyre-, brake-, and road-wear.[42] To mitigate these issues, cities pursuing vehicle bans should also consider committing to 100% clean electricity, while lowering speed limits and opting for pavements with the least potential for wear.

One perennial challenge with all of these efforts to restrict vehicle pollution is freight. Since deliveries are typically handled by private companies and originate outside city limits, city planners often neglect to factor them into their policies. This is an oversight: in New York City, freight vehicles (mostly heavy diesel vehicles) account for just 5% of vehicle miles travelled but are responsible for half the health effects of air pollution.[43] This is also a massive equity issue, since freight vehicles often pass through low-income neighbourhoods.[44]

The number of delivery vans in cities has skyrocketed in recent decades as citizens and retailers alike become increasingly reliant on online shopping. These on-demand deliveries result in incredible transport inefficiency: in 2014, an analysis showed that 39% of vans driving around London were less than one-quarter full.[45] There are few examples of cities tackling this issue, but there is massive potential to increase efficiency by deduplicating trips and deploying fully packed, electric delivery vehicles through a centralized distribution system.[46]

---

37 Ibid., p. 190.

38 Ibid., pp. 192–193.

39 Ibid., p. 190.

40 Ibid., p. 156.

41 Ibid., p. 100.

42 Ibid., p. 203.

43 Interview with Iyad Kheirbek. Op. cit.

44 Ibid.

45 Fuller, 2018. Op. cit., p. 201.

46 Ibid., p. 201.

Case Study: Seoul – The Cheonggyecheon Expressway

One city in particular devised an incredibly elegant solution to choking congestion, which may serve as inspiration to cities around the world.

In 2001, the Cheonggyecheon Expressway, which carried an average of 160,000 cars[47] into the heart of the city centre each day, was a nightmare. The city was under pressure to deal with the increasing traffic congestion and concerns about the structural integrity of the expressway, which had been built 30 years prior by paving over an intermittent creek.[48]

In urban planning, the Braess paradox describes an interesting phenomenon: expanding or building new roads does not have the expected effect of alleviating traffic – it actually makes traffic worse. Drivers flock to the new roads, increasing numbers of trips and no longer avoiding peak congestion times. As it turns out, however, the Braess paradox also works in the reverse: reducing road capacity alleviates traffic – often permanently reducing the number of cars in a particular location.[49]

So instead of adding lanes or establishing a congestion fee, the city of Seoul came up with a simple solution: to remove the offending expressway entirely. In its place, the city built a 9-km linear park and revitalized stream,[50] fed by treated water from the Hangang River.[51] The city established terraces around the water passageway, complete with a pedestrian park full of indigenous plants, ramps for the disabled, fountains, lights, and street furniture (Figure 21.5).[52]

Meanwhile, the city established 98 km[53] of designated bus-ways to support commuters, implemented a car restriction policy,[54] and increased parking fees.[55]

The results were stunning. Critics warned the project would exacerbate the congestion issues,[56] but the number of vehicles entering and exiting the area fell by nearly half.[57]

In the four years following the conversion, levels of coarse particle pollution fell 21% near the project. Nitrogen dioxide density and VOC concentration fell 17% and as much as 65%, respectively.[58]

The ancillary benefits were numerous: public transit accessibility increased by 13.4%, and local property values increased by more than twice the city's average rate. The

---

47 https://www.theguardian.com/environment/2006/nov/01/society.travelsenvironmentalimpact.

48 https://issuu.com/rujak/docs/lifeanddeathofurbanhighways_031312 (pp. 27–31).

49 Fuller, 2018. Op. cit., pp. 197–199.

50 https://www.theguardian.com/world/2014/mar/13/seoul-south-korea-expressway-demolished.

51 Information provided by Director Kwon and Hyejung Yeo, Environmental Policy Division Seoul Metropolitan Government. 18 August 2020.

52 https://issuu.com/rujak/docs/lifeanddeathofurbanhighways_031312 (pp. 27–31).

53 Information provided by Director Kwon and Hyejung Yeo. Op. cit.

54 https://issuu.com/rujak/docs/lifeanddeathofurbanhighways_031312 (pp. 27–31).

55 https://www.nytimes.com/2009/07/17/world/asia/17daylight.html?_r=2&pagewanted=all.

56 https://www.kdevelopedia.org/themesub.do?thememainid=13.

57 https://issuu.com/rujak/docs/lifeanddeathofurbanhighways_031312 (pp. 27–31).

58 Ibid.

Figure 21.5 Left: The elevated freeway prior to the 2003 start of construction on the stream restoration project. Right: The restored Cheonggyecheon Stream after construction was complete in 2005. (*Source:* Left: Seoul Metropolitan Government. Right: Binh/Adobe Stock.)

conversion brought down average summer temperatures by 3.6°C[59] and reduced the urban heat island effect by as much as 8°C near the project.[60]

The project reduced odour and noise pollution and created a new natural habitat that fostered enormous increases in the numbers of fish, bird, plant, and insect species,[61] while providing better natural stormwater management than city sewers.[62] The new park has become a popular destination for locals to engage in recreation and participate in cultural events like the lantern festival. And it has become a must-visit for tourists, with many tens of thousands of visitors per day.[63]

One dimension setting this project apart from other urban revitalization projects was the massive public buy-in necessary to execute the project. The mayor who executed the project ran with the demolition of the expressway as a key part of his candidacy, and he sought representation from diverse voices throughout the planning process, convening a Cheonggyecheon Citizens Committee[64] that met hundreds of times over the course of two years.[65] The success of the project sparked a spate of urban revival projects that have transformed the city of Seoul over recent years, including the removal of at least 15 additional expressways in Seoul.[66, 67, 68]

---

59 https://www.theguardian.com/world/2014/mar/13/seoul-south-korea-expressway-demolished.

60 https://issuu.com/rujak/docs/lifeanddeathofurbanhighways_031312 (pp. 27–31).

61 Ibid.

62 https://www.cnu.org/what-we-do/build-great-places/cheonggye-freeway.

63 https://issuu.com/rujak/docs/lifeanddeathofurbanhighways_031312 (pp. 27–31).

64 Ibid.

65 https://www.nytimes.com/2009/07/17/world/asia/17daylight.html?_r=2&pagewanted=all.

66 https://www.cnu.org/what-we-do/build-great-places/cheonggye-freeway.

67 https://www.landscapeperformance.org/case-study-briefs/cheonggyecheon-stream-restoration.

68 Ibid.

The above benefits aren't exclusive to Seoul. Permanently reallocating major vehicle arteries in cities to pedestrians and cyclists can have massive health, economic, and climate benefits in cities around the world, while providing vulnerable populations with much-needed greenspace.

In fact, more walking and cycling opportunities not only boost public health – one study put the value of health benefits from cycling 3 miles (4.8 km) instead of driving at €1,300 per year[69] – but also create more opportunities for small local businesses to thrive.[70, 71]

Of course, it's important to be clear-eyed about the challenges and limitations of such projects. The Cheonggyecheon project was criticized by some as promoting gentrification,[72] and while the city took steps to help relocate displaced street vendors, more could have been done for the project to proactively alleviate equity and poverty issues. Furthermore, the project ran over budget, raising concerns about how cities in the Global South – or those whose coffers have been badly depleted by the COVID-19 pandemic – can afford similar plans.

Cities can build on the lessons from Seoul for even more comprehensive and systemic benefits by further addressing equity; promoting the uptake of electric vehicles and pursuing ways to minimize their air pollution impact; and tackling emissions from freight vehicles within city limits.

Specifically, cities should select areas to convert road capacity to pedestrian space in parts of the city where air pollution is highest, so as to maximize health and economic benefits for poor and marginalized groups. Cities may also consider incorporating additional services into these projects, like cooling or heating facilities for populations experiencing homelessness. This is a particularly bright opportunity to improve access to services in cities with large informal settlements, like Karachi, Cape Town, Nairobi, and Rio de Janeiro.[73]

Conclusion

Many of the solutions outlined in this chapter were successful at least in part because of the efforts to mobilize public support.

While the climate crisis can feel to urban dwellers like a faraway (and potentially intractable) challenge for future generations, air pollution offers an immediate – and often visible – threat that is also personally motivating: one's health.

As indoor air pollution sources play a larger role in overall air quality, particularly in the Global South, public education about the health dangers of such activities and clear, comprehensive education about alternative options will be absolutely critical to addressing the problem.

---

69 Fuller, 2018. Op cit., p. 208.

70 Ibid., p. 207.

71 Smedley, 2019. Op. cit., p. 167.

72 https://www.tandfonline.com/doi/full/10.1080/10630732.2013.855511?scroll=top&needaccess=true.

73 Fuller, 2018. Op cit., p. 213.

And in countries like the USA and Australia, where the climate crisis is highly politicized, talking about solving air pollution issues is a way to address climate change without having to talk about climate at all.

Furthermore, cities may be able to capitalize on positive public sentiment around the benefits of highly visible, large-scale projects to enact further air pollution and climate policies.

With travel patterns disrupted given the ongoing COVID-19 pandemic and public transport in crisis, cities have more leeway to make significant infrastructure changes to the built environment.

Indeed, several cities, including Paris and New York, have already closed commercial streets to traffic in order to provide space for recreation or for outdoor restaurant dining.[74] Bogotá established some 80 km of "emergency" bike lanes at the start of the pandemic, while lowering speed limits across the city.[75] These measures are a step in the right direction, and public support for such measures now may be more easily converted into support for making these solutions permanent features of our cities.

One thing is certain – we have no time to waste. Now is the time for city leaders to take bold action on air pollution. The future of our health, climate, and equality of our cities depends on it.

---

74 https://www.bloomberg.com/features/2020-city-in-recovery.
75 Ibid.

21

The Just City (Part II)

Jane Burston and Matt Whitney

From Air Pollution to Data Revolution: How London's Fight for Clean Air Is Based on Data

Born in London in 1913, Gerald Nabarro was elected to the British parliament in 1950. Motivated by the Great Smog of London in 1952, when the smoke from coal burning combined with cold, windless weather to envelope the city in a thick smog that killed thousands of people in a matter of days, he quickly became a leading advocate for clean air. He made impassioned speeches on the topic in parliament and introduced the Bill that led to the Clean Air Act of 1956, legislation that is widely seen as the first example of a national government taking meaningful action on air pollution.

Nabarro had had enough of the lack of action taken by successive British governments on air pollution. In one rousing speech he described how, on a visit to the UK city of Nottingham in 1257, Queen Eleanor of Aquitaine had needed to leave due to the "obnoxious atmosphere", and joked that the timing until the introduction of his Clean Air Bill – almost 700 years later – was "normal progress for parliamentary democracy".[1] Determined not to see events such as the Great Smog unfold again, Nabarro told parliament that "we mean business today, after a lapse of several hundred years".

In championing the proposed legislation, Nabarro drew upon important data to demonstrate the urgency of the matter. "An experiment was carried out a few years ago with the growing of a lettuce in the open air," Nabarro said. On the outskirts of a city, where soot had fallen at the rate of 15 tonnes per km^2 per annum, the lettuce grew to 140 g. But in the middle of that industrial area, which measured a soot fall of 188 tonnes per km^2 per annum, "the lettuce grew to a size of only 44 grams". It may not be the most scientific of experiments, but it shows how important observation and evidence are a precursor for ambitious action.

The Clean Air Act passed in 1956 restricted use of the most polluting coals and provided generous funds to help people upgrade their home heating. Whilst making substantial progress, it didn't solve the problem; we may not suffer the same "pea-souper" smogs that once plagued London, but pollution isn't only harmful when it's visible. The gases and tiny particles that are produced from our modern engines, construction, heating systems,

---

1 https://api.parliament.uk/historic-hansard/commons/1955/feb/04/clean-air-bill.

The Climate City, First Edition. Edited by Martin Powell.
© 2022 John Wiley & Sons Ltd. Published 2022 by John Wiley & Sons Ltd.

restaurants, and agriculture are an invisible killer, causing thousands of deaths a year in London even today[2] and many millions around the world.

Solving our air pollution problem will not be easy, but it will be worth it. Clean air means healthy citizens, a more productive workforce, and lower healthcare costs. Clean air also means reducing the risk of climate change. The causes of climate change are often the same as the causes of air pollution, so making the changes needed to reduce air pollution often reduces greenhouse gases too.

Critical to achieving this future are data. Data are a common theme in this book: data have the power to make visible the invisible, demonstrate that decisions have consequences, and show the path from problem to solution. In short, data are a fundamental requirement for progress.

But collecting data on air pollution is a challenge. Air pollution, despite the huge damage it does to our health, makes up tiny fractions of our atmosphere. Microscopic particles of dust and soot called particulate matter are measured in micrograms, or millionths of a gram. Toxic gases like nitrogen dioxide are measured in "parts per billion", a measure that tells you how many molecules of nitrogen dioxide there are for every billion molecules in the atmosphere. Air pollution is also highly localized, varying significantly even between individual streets and blocks. Forming a thorough understanding of pollution, such as where it is coming from and where concentrations are highest, therefore requires a network of monitors across a city that are able to operate for long periods in the relentless conditions of the great outdoors.

Thanks to the groundwork of the Clean Air Act, London is a leading city in its approach to monitoring air pollution. Decades of investment have resulted in a monitoring system that provides information about historic, current, and predicted future levels of multiple pollutants. The monitoring system is made up of a network of high-quality "reference-grade" monitoring stations that continually measure a range of pollutants at over 120 different locations.[3] The data are fed into models to help epidemiologists understand the health impact of pollution and provide policymakers with an understanding of the impact that proposed policies will have on cleaning up the air.

These data have enabled actions that have gradually improved air quality in the city over recent years (Figure 21.6), despite London's growing population. However, whilst there has been progress, there is still a long way to go: 99% of London still has pollution of particulate matter above levels recommended by the WHO.

London's dense network of monitoring is, unfortunately, the exception and not the rule (Figure 21.7). Some cities have set up similarly robust networks to London's, but many have not, largely because of the significant time, money, and expertise that it takes to do it well. Even national governments can struggle to find the means to measure air quality. Astoundingly, half of the world's national governments, representing 1.4 billion people, produce no air quality data at all. Of those governments that do measure air quality, it is not always the case that they publish the data, for all to see and use.[4]

---

2 https://www.standard.co.uk/futurelondon/theairwebreathe/3800-deaths-in-london-caused-by-air-pollutant-study-finds-a4345831.html.

3 https://www.london.gov.uk/sites/default/files/air_quality_in_london_2016-2020_october2020final.pdf.

4 https://openaq.org/assets/files/2020_opendata_stateofplay.pdf.

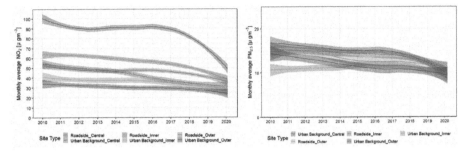

Figure 21.6 Trends in NO$_2$ (left) and PM2.5 (right) in London between 2010 and 2020 (the data series ends just before the start of COVID-19-related impacts). (*Source:* AIR QUALITY IN LONDON 2016-2020, London Environment Strategy: Air Quality Impact Evaluation, October 2020. Retrived from: https://www.london.gov.uk/sites/default/files/air_quality_in_london_2016-2020_october2020final.pdf.)

Figure 21.7 The London skyline. Cities often conceal their pollution to the naked eye, building the case for more widespread monitoring. (*Source:* Cavan Images/Adobe Stock.)

This pervasive lack of data is a barrier to action. Despite decades of research, satellite data, and scientific campaigns providing overwhelming evidence of the causes, impacts, and necessary actions needed to address air pollution almost everywhere on Earth, decision-makers will oftentimes want locally generated data on which to make decisions.

To solve the air pollution problem, we need to urgently reduce the cost and effort involved in providing compelling information to decision-makers. Not doing so will result in the issue continuing to be ignored, millions dying prematurely, and many millions more living in the misery of highly polluted cities.

Fortunately, the last few years have seen the introduction of new, smaller, and lower-cost sensors that aim to reduce the barriers to monitoring air quality. Anybody can now go online and buy a personal air quality monitor for a few hundred dollars. Those wanting something a little more sophisticated, such as researchers, NGOs, and governments, can turn to low-cost monitors that cost anything from hundreds to a few thousand dollars, but still significantly less than the tens of thousands invested by governments in the high-quality reference-grade monitors (not to mention the many thousands more invested in the labs and technicians needed to maintain them).

The potential of this revolution in low-cost monitoring is undeniable. Billions of people live in places where there simply isn't the capacity or funding to build a monitoring network anything like that of London's. Cheaper monitoring – if it works – would mean more data, opening the potential for near-ubiquitous, hyperlocal, and real-time information that can help our future cities to live up to the title of being "smart".

But we're not there yet. As sensors have reduced in size and cost, compromises have been made in their accuracy and reliability. Those challenges in measuring tiny concentrations of pollution in harsh outdoor environments haven't gone away, and as a result often even the most sophisticated low-cost monitors can be notoriously unreliable, making measurements that are significantly divergent from the reference monitors that governments have traditionally relied upon.

To test out how to overcome these challenges, an international consortium led by the Environmental Defense Fund launched the Breathe London pilot project in 2018.[5] The project uses three innovative monitoring approaches – personal exposure monitors, mobile monitors, and low-cost monitors – against London's established monitoring system to test and improve the technology and help its rollout to other cities globally.

In just two years the Breathe London pilot revealed an incredible level of detail about London's air pollution problem, despite the city already having had one of the most sophisticated air quality monitoring systems in the world. It demonstrated that all three monitoring approaches will have an important role to play in helping cities around the world to understand and act on air quality.

The data collected by the project have been used to analyse the impact of ambitious policies including the ultra-low emission zone (ULEZ), under which a levy was introduced to deter the most-polluting vehicles from entering Central London. Levels of harmful nitrogen oxides from diesel cars – which the policy aims to reduce – were found to be on average 23% lower within the zone, showing its effectiveness but also demonstrating the need for the policy to be expanded alongside financial support to those that need to replace their polluting vehicles.[6]

Combining the data with models has provided a breakdown of pollution sources across London. On Shaftesbury Avenue in Westminster, for example, on average three-quarters of pollution comes from vehicles, which can be further broken down to analyse the individual contribution of cars, taxis, vans, and buses (Figure 21.8).

5 https://www.breathelondon.org.

6 https://www.edfeurope.org/news/2020/19/08/diesel-car-pollution-higher-neighbourhoods-outside-city-centre.

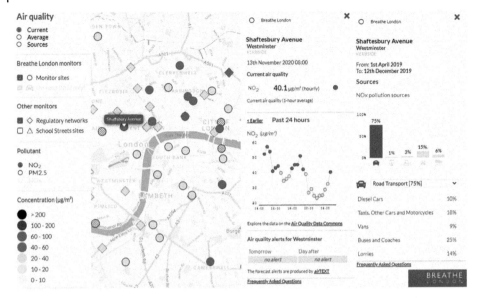

Figure 21.8 The interactive Breathe London data portal from the pilot phase, showing current air quality and average sources of pollution, in this case for a monitor in Shaftesbury Avenue, Westminster. (*Source:* https://www.breathelondon.org.)

What about those streets that don't have a monitor? Even with cheaper sensors it will always be impractical to install equipment on every street across a city the size of London. Instead, the Breathe London team equipped Google StreetView vehicles with specially adapted reference-grade monitors to analyse air quality as the cars move about the city. After a year of driving, covering approximately 40,000 km and taking measurements every 1 second, the team created an incredibly rich understanding of hyperlocal pollution across the city's roads. It revealed stark differences in pollution within only a few tens of metres, with major through-roads found to be on average 50% more polluted than neighbouring backstreet roads (Figure 21.9).[7] This demonstrates the potential of combining air quality data with existing mapping to help individuals avoid pollution, much like we use live maps now to avoid traffic jams.

But perhaps the most surprising outcome of Breathe London has been the breadth of people the project has engaged along the way. Homeowners and concerned parents frequently approached the project asking for a monitor to be placed around their location, and councils have used the data as justification for taking new measures to reduce pollution in their boroughs.[8]

---

7 Google Earth Story.

8 https://www.ealingtimes.co.uk/news/18674235.brent-council-responds-edf-europe-wembley-pollution-data.

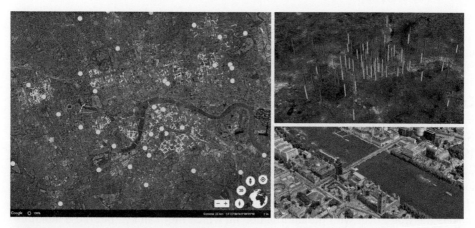

Figure 21.9 Mobile monitoring data of air pollution in London, showing concentrations in Central London (left), trends in average concentrations across the day (pictured is 11 am) (top right), and a zoomed-in view of pollution in the streets around Parliament (bottom right). (*Source:* Google LLC.)

Special focus has been put on providing schools with monitors, and an additional 250 schoolchildren became air quality scientists themselves by carrying personal monitors that measure their exposure across the school day, informing students, parents, and teachers about the steps they can take to reduce exposure and avoid contributing to the problem.[9] Schools across the capital have regularly approached the team wanting to take part in the project, and the Mayor of London announced a new scheme to support all London schools to drive further reductions in pollution based on Breathe London data.[10] It shows the power of data to engage and motivate people to improve their communities and provide policymakers with the evidence they need to act.

The static monitors deployed in the project were operational through the first of London's coronavirus-related lockdown, providing an unexpected opportunity to understand the impact of a full city lockdown on air pollution levels. Monitors in Central London showed an average reduction during restrictions of 20–24% for nitrogen dioxide and up to a 37% drop during commuting hours. Across Greater London, the monitors registered up to a 17% drop. While we never want to see those measures put in place again, it demonstrates that immediate action on air pollution can bring immediate results.

The project was not without its challenges. The expected issues of reliability and inconsistent data quality emerged and were mitigated by developing innovative machine learning and statistical tools to detect failing monitors and aid their calibration.

There were also many unexpected issues. High buildings around London meant that many monitors could not be solar powered as was initially intended. London's dense traffic meant that plans for taking mobile measurements from the road had to be reduced.

---

9 http://blogs.edf.org.

10 https://www.london.gov.uk/press-releases/mayoral/mayor-unveils-plans-to-reduce-toxic-air-at-schools.

And installation of the monitors themselves – especially where power had to be drawn from lampposts – involved negotiating permissions separately with London's 32 individual boroughs, making an already ambitious project timeline even more challenging.

Conclusion

Nabarro and his fellow Members of Parliament who voted in the Clean Air Act in the mid-1950s could only have dreamt of seeing the level of detail provided by Breathe London data. Yet although we have moved on a long way from weighing lettuces, we still have a long way to go. The Breathe London pilot project demonstrated the clear potential that cheaper and more accessible monitoring has to democratize data and provide decision-makers the world over with the information on which they can act. But these monitors are not the panacea, and a lot still needs to be learned, and shared, about how to use them to make the most of their potential.

What is in no doubt is that more and better data will be an essential part of the future city and an enabler for a world where everyone can breathe clean air. Let's just hope it happens, as Nabarro put it back in 1955, without a "lapse of several hundred years".

21

The Just City (Part III)

Jenny Bates

Ella

16 December 2020 changed the way air pollution is seen. A London inquest stated that Ella Roberta Adoo Kissi-Debrah had died, aged nine, "of asthma contributed to by exposure to excessive air pollution".[1] Ella lived close to one of London's busiest roads, and this was the first time that air pollution had ever been stated to be a cause of death on a death certificate. It was the culmination of years of truth-seeking by her mother Rosamund.

For decades we have known that modern air pollution, even though largely invisible, is bad for health. But the more it is researched, the more medical evidence there is of how dangerous it is. We know how short-term exposure to air pollution triggers heart attacks and strokes as well as asthma attacks; and how long-term exposure can cause respiratory and cardiovascular diseases. Outdoor air pollution has been put in the same category as smoking as a cause of lung cancer by the WHO.[2]

We know that it impacts the most vulnerable people hardest, including the young whose lungs can develop smaller,[3] older people, and the most disadvantaged who tend to live near main roads.[4] It should be a wake-up call for drivers that people in vehicles can be exposed to worse pollution than those walking or cycling the same road.[5]

Ella's inquest was important because finally a face and a name has been attached to one of the thousands of early deaths that are attributed to outdoor air pollution in London each year[6] and tens of thousands in the wider UK.[7] It was a first – the family's solicitor at

---

1 https://www.innersouthlondoncoroner.org.uk/news/2020/nov/inquest-touching-the-death-of-ella-roberta-adoo-kissi-debrah.

2 https://www.iarc.who.int/wp-content/uploads/2018/07/pr221_e.pdf.

3 https://www.kcl.ac.uk/news/air-pollution-restricting-childrens-lung-development.

4 https://www.healthyair.org.uk/am-i-at-risk.

5 https://www.healthyair.org.uk/healthiest-transport-option-video.

6 https://www.london.gov.uk/what-we-do/environment/pollution-and-air-quality/health-and-exposure-pollution.

7 https://www.gov.uk/government/publications/nitrogen-dioxide-effects-on-mortality/associations-of-long-term-average-concentrations-of-nitrogen-dioxide-with-mortality-2018-comeap-summary.

The Climate City, First Edition. Edited by Martin Powell.
© 2022 John Wiley & Sons Ltd. Published 2022 by John Wiley & Sons Ltd.

the inquest, Jocelyn Cockburn, said: "Air pollution has never been cited as a potential cause of death and therefore never investigated in relation to the death of anyone in this country or, so far as I am aware, worldwide."[8]

The Problem

The inquest was quite clear that air pollution had been a "significant contributory factor to both the induction and exacerbations of her asthma". And the coroner linked it to the fact that during the course of her illness between 2010 and 2013 Ella was exposed, largely from traffic emissions, to nitrogen dioxide (NO_2) and particulate matter (PM in excess of WHO guidelines.

Shockingly, levels of the toxic gas NO_2 are still over the national and EU legal standards (which are the same as the WHO's), when they should have been met originally from 2010. London in 2020 was still failing these limits, along with a further 32 of the 43 Air Quality Zones into which the UK is divided for air quality purposes.

The most health damaging, fine PM2.5 pollution, which can get deep into the lungs and into the bloodstream, is also a huge issue in London. While the UK is meeting current national and EU legal standards, the WHO guidelines, which are more than twice as stringent for this pollutant, are not being met in London and many other places across the UK.[9] Virtually all of London suffers with pollution above the WHO PM2.5 guideline,[10] and this needs to be addressed.

But these limits don't even represent "safe" levels of pollution. The WHO has found health effects below its guideline level and the UK and EU's current legal levels for NO_2, and below even its guideline levels for PM2.5s. The WHO says for PM2.5s that "no threshold has been identified below which no damage to health is observed". And there are other pollutants of concern with varied sources.

Where Are We?

We have reached the point where air pollution is still a terrible problem in London but is now firmly up the agenda for Londoners, the media, and politicians, as well as campaigners. Air pollution hasn't been such a high-profile issue since London's "pea-souper" smogs in the 1950s and "acid rain" in the 1990s. It has taken a long while to get to this level of awareness on the modern air pollution problem, and it begs the question why more action was not taken before, and what action is needed now.

Recently, air pollution has been gaining momentum as an issue along with climate change, which has many of the same sources and solutions, particularly in relation to transport. The co-benefits of action on one for the other are enormous, and solutions need

---

8 https://www.hja.net/news-and-insights/hja-in-the-news/civil-liberties-human-rights/landmark-ruling-today-excessive-levels-of-air-pollution-contributes-to-death-of-ella-adoo-kissi-debrah.

9 https://uk-air.defra.gov.uk/library/annualreport.

10 https://www.london.gov.uk/press-releases/mayoral/new-imperial-study-on-mayors-aq-policies.

to be win-win. Climate change may ultimately be a far bigger threat to the planet, with its own health implications, but air pollution is often of more immediate relevance to people's health and lives.

Resurgence

The London Air Quality Network's monitoring work since the 1990s has given us good information on the UK capital's air pollution problem, and there have been academics working on the issue, along with professionals and others, for years – but there hasn't been enough public awareness or action.

Campaigning has been needed to get media attention and help focus politicians' minds – and this has ramped up again it seems from the 2000s, and more so in the 2010s. Also, hard targets and legislation have been key to holding politicians to account and are always needed – as is adequate funding to put in place measures needed.

I woke up to the problem after I'd moved to Greenwich in South East London in the early 1990s and got involved in proposed new road river crossings in the area as a volunteer activist with Friends of the Earth, and later as a staff member covering London campaigning. In the early 2000s it became clear that Mayor Ken Livingstone's proposed new Thames Gateway road bridge would have resulted in more traffic pouring onto local roads and creating a new breach of UK and EU legal limits for NO_2.

There was a public inquiry at which the inspector agreed that this was unacceptable, and the scheme was eventually scrapped by Mayor Boris Johnson in 2008.[11] Unfortunately, he then pursued other East London road river crossing ideas. One of these, the four-lane Silvertown road tunnel, is still being taken forward by Mayor Sadiq Khan, and would still worsen air pollution already over legal limits for some.

Around that time in the 2000s, Clean Air London was started by an ex-banker who was looking to clean up the air ahead of the 2012 London Olympics, and campaigning lawyers Client Earth were seeking to use the law. The fact that London and the UK had failed to meet the 2010 deadline to comply with national and EU NO_2 legal limits focused attention, and a "Healthy Air Campaign" coalition was set up in 2011. This brought together both environmental and transport organizations, as well as big health charities.[12]

A series of events helped build momentum. In 2014, media attention was captured when a serious air pollution episode carrying red dust from the Sahara Desert blew into the UK (although the main source turned out to be agricultural emissions), and in 2015 "Dieselgate" exposed how car makers were fiddling air pollution data. Friends of the Earth and others were supporting "citizen science" monitoring of NO_2 pollution with tubes people put up in their own neighbourhoods.[13] Client Earth took the government to court for failing to meet the NO_2 targets, winning a landmark case at the Supreme Court in 2015 and ultimately and finally forcing the government to produce a much-improved NO_2 plan in 2017. This required many cities to consider clean air zones (CAZs) to restrict

---

11 https://en.wikipedia.org/wiki/Thames_Gateway_Bridge.

12 https://www.healthyair.org.uk/partners.

13 https://friendsoftheearth.uk/who-we-are/our-history.

the dirtiest vehicles from the most-polluted areas, though the first wasn't in place until 2021. Meanwhile, London Mayor Sadiq Khan was planning for the introduction in 2019 of an ultra-low emission zone (ULEZ[14]) for Central London, which is to be enlarged in 2021, though many campaigners argue it needs to be bigger still and cover all London.

Momentum to improve air pollution has gone hand in hand with that to address climate change. London Mayor Ken Livingstone built on his ground-breaking Congestion Charge of 2003 with a science-based target for cutting London's climate emissions. Nationally, the Climate Change Act of 2008 added legal force to climate initiatives, and following the 2015 Paris Agreement on climate change, the UK put a "Net-Zero by 2050" target into law in 2019. Since then, David Attenborough, Extinction Rebellion, and Greta Thunberg's school strikes have further raised awareness. The UN "COP26" climate talks in Glasgow in November 2021 are expected to take this further. All this helps deliver cleaner air.

Cleaner and Fewer

The ULEZ in London and CAZs, as well as measures to restrict traffic at drop-off and pick-up times at schools, are important steps forward. But people also need incentives, like a scrappage scheme to enable them to switch from "dirty" to "clean" vehicles, or preferably away from private vehicles altogether. This can be helped with better incentives to join a car club or get a rail season ticket or e-bike loan. Ending the sale of diesel and petrol cars and vans is being brought forward to 2030 in the UK, and electric vehicles (EVs) need to be supported with charging infrastructure. The switch from fossil-fuelled vehicles will result in lost government revenue from road tax, but road user charging is seen as a solution.

While dirty vehicles are being progressively restricted and phased out, this is not enough – EVs and other "clean" vehicles are only part of the solution. We need fewer vehicles on the roads, or fewer miles driven – at least a 20% cut in car miles by 2030 for climate reasons,[15] and also for air pollution reasons because while "clean" vehicles may have no exhaust emissions, they produce PM air pollution from brake and tyre wear.

And to support less vehicle use, alternatives need to be properly supported. It must be made easier and safer to walk and cycle, that is to use "active travel", and there must be more affordable (or even fare-free) public transport.

Don't Add to the Problem

All these measures are important, but planning is also needed to reduce the need for people to have to travel unnecessarily, and to avoid traffic-generating developments like car-dependent housing, new road building, or airport expansion. London has benefitted from

---

14 https://www.london.gov.uk/press-releases/mayoral/dramatic-improvement-in-londons-air-quality.

15 https://policy.friendsoftheearth.uk/insight/radical-transport-response-climate-emergency.

much good transport investment but has also suffered from bad proposals. As well as the various road river crossings proposals, there have been plans to expand London City Airport in the east and Heathrow in the west. All would generate new extra traffic and have been fought on air pollution and climate grounds.

A third runway at Heathrow, meaning more road traffic as well as flights, would inevitably make air pollution worse for Londoners. Ultra-fine PM air pollution from Heathrow has been found even in Central London.[16] A third runway was opposed by the Conservative party leadership in 2009, but now the government has backed such plans which would allow 50% more flights. Legal challenges have been made on air pollution grounds, but a case brought by Friends of the Earth on how climate issues, including the Paris Agreement, have not been properly taken into account won at the Court of Appeal. It was later overturned at the Supreme Court, but it was established that a third runway would have to be assessed against the more stringent climate targets in place at the time planning permission was applied for, which means that significant obstacles for Heathrow expansion remain.[17] Meanwhile, the Committee on Climate Change which advises the government on getting to Net-Zero emissions has said that there should be no net airport expansion – that is, if a Heathrow third runway went ahead, capacity should have to be reduced elsewhere.

That such huge, pollution-generating schemes keep coming forward is also a failure of planning assessment and guidance. Alternatives are often inadequately looked at, and the current National Policy Statement for National Networks, covering large road schemes, even allows air pollution already over legal limits to be further worsened and new breaches. This is madness.

Conclusion

The COVID pandemic and lockdown has given people a glimpse of how cleaner air feels. It has resulted in councils unexpectedly giving more road space to active travel. This is how cities should be developed. And COVID must itself be a driver for action. Living with bad air may worsen COVID-related illness and outcomes,[18] and there are now many more people living with breathing difficulties, such as those recovering from Long COVID.[19] Symptoms of "shortness of breath and cough" associated with NO_2[20] sound uncannily like some of those of COVID. This should be a wakeup call to everyone.

NO_2 air pollution has been a prime driver of action on air pollution since the early 2010s, and there is much more that has to happen to bring down NO_2 levels. However, it may well be that PM pollution drives action more in the future. Whereas legal action had forced a separate NO_2 plan, PM is one of the pollutants included in a wider Clean Air

---

16 https://www.kcl.ac.uk/news/pollution-from-heathrow-detected-in-central-london.

17 https://friendsoftheearth.uk/climate/heathrow-expansion-remains-very-far-certain-says-friends-earth.

18 https://www.theguardian.com/world/2020/aug/13/study-of-covid-deaths-in-england-is-latest-to-find-air-pollution-link.

19 https://www.nhs.uk/conditions/coronavirus-covid-19/long-term-effects-of-coronavirus-long-covid.

20 http://www.londonair.org.uk/londonair/guide/whatisno2.aspx.

Strategy of 2019. While transport sources are important for PM2.5, domestic wood and coal burning is the biggest primary source, with secondary PMs from agriculture also an issue.

With the UK having left the EU (aka Brexit) at the end of 2020, the UK won't benefit from ongoing strengthening of EU air pollution standards, but new UK legislation, starting with the Environment Bill, gives an opportunity to redress the UK's current inadequate PM2.5 standards. Legally binding requirements to meet WHO guideline levels for PM2.5, by 2030 at the latest, must be enshrined in the Environment Bill (and for levels to be updated as the WHO revises its guidelines).

City mayors from across the country are calling on government to adopt those WHO guidelines in the Environment Bill.[21] London has been showing the way and (following Manchester) has itself signed up to the WHO's Breathe Life initiative[22] committing to just those requirements – showing that cities can be, and often are, ahead of national government. There is much to do to address PM air pollution, but achieving WHO guidelines is within reach, even in London (with support from national government), according to a Greater London Authority report.[23]

Politicians should look not only to the cost to human health of air pollution but also to the economy. Air pollution is estimated to cost the UK economy £20 billion a year,[24] but the Confederation of British Industry has shown that meeting the WHO PM2.5 guideline could result in a £1.6 billion a year benefit to the UK.[25]

But perhaps it will be the tragic death of little Ella that sets the new air pollution agenda. Her mother, now a WHO Advocate on Health and Air Quality, insists that the UK must commit to the WHO PM2.5 standards in the Environment Bill. The listing of air pollution on Ella's death certificate could lead to cleaner air and better health for many others. It could lead to a better London.

---

21 https://www.theguardian.com/environment/2021/jan/27/uk-mayors-boris-johnson-tougher-air-pollution-targets.

22 https://www.london.gov.uk/press-releases/mayoral/every-londoner-is-exposed-to-dangerous-toxic-air.

23 https://www.london.gov.uk/sites/default/files/pm2.5_in_london_october19.pdf

24 https://www.rcplondon.ac.uk/projects/outputs/every-breath-we-take-lifelong-impact-air-pollution.

25 https://www.cbi.org.uk/articles/what-is-the-economic-potential-released-by-achieving-clean-air-in-the-uk-1.

22

The Invested City – Introduction

In Chapter 21 Hayley showed us how a bold vision can be delivered to overcome the air pollution challenge. Jane and Matt showed how a monitoring network can provide the vital evidence to effect change. This will allow hotspots to be targeted, and in time whole cities will live within an envelope of emissions, both air pollutants and greenhouse gases that no longer impact our wellbeing. Jenny described the story of Ella and the listing of toxic air pollution on her death certificate. Ella died in 2013.

I lived in New York with my family when it was the epicentre of the COVID-19 pandemic. A study from Harvard[1] showed that if the amount of PM2.5 (the fine particulate matter that gets absorbed into our lungs) were reduced in New York City from 12 μg to 11 μg for a period of the last 20 years or so (since around 2000) there would have been significantly fewer COVID deaths. The paper provides a precise number because we can apply the same methodology that WHO uses. At the height of lockdown, the level in New York City was 4 μg. To sustain that going forward all we have to do is run our vehicles with zero tail-pipe emission – electric vehicles, electric buses – and to continue the modal shift away from cars. There will still be particulate matter from tyre and brake wear and this needs to be tackled, but the most progressive cities have accepted this challenge too.

We have covered a lot of ground and seen examples of what is possible to achieve a sustainable future in our cities. The next two chapters look at what it's going to take to accelerate sustainability trends to redirect, scale up, and mainstream that investment. In Chapter 7, *The Decoupled City*, Nick and Leah identified the areas where low carbon measures could cut emissions in urban areas by 90% by 2050. In Figure 22.1 the investment and the returns will offer a significant job creation opportunity for every city.

In this chapter Colin gives us insight into some of those areas where investment needs to be redirected and the importance of environmental, social, and governance (ESG) criteria in making our organizations more transparent and accountable in what they do and as a lever to drive more responsible behaviour and thus more responsible investing.

The Invested City starts by examining sustainable capitalism and accelerating sustainability trends. Colin examines this "Sustainability Revolution" driven by global population growth, increased pollution, resource constraints, the climate crisis, rising inequality, and

---

1 Wu, X., Nethery, R.C., Sabath, M.B., Braun, D., Dominici, F., 2020. Air pollution and COVID-19 mortality in the United States: Strengths and limitations of an ecological regression analysis. *Science Advances* 6(45), eabd4049.

The Climate City, First Edition. Edited by Martin Powell.
© 2022 John Wiley & Sons Ltd. Published 2022 by John Wiley & Sons Ltd.

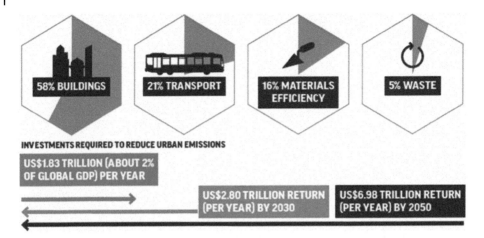

Figure 22.1 Technically feasible low-carbon measures could cut emissions from urban areas by almost 90% by 2050. (*Source:* Climate Emergency, Urban Opportunity, 2019. Coalition for Urban Transitions.)

poverty. It is a global revolution with the "magnitude of the Industrial Revolution and the speed of the Digital Revolution", and cities lie at the heart of it.

Colin looks at cities as financial centres and as the mainstream of sustainable investing. Despite the logic of sustainability, most financial assets are not managed with sustainability in mind. Although there has been great progress in mainstreaming ESG criteria across financial markets, the level of understanding of the risks and opportunities associated with the Sustainability Revolution as well as issues of climate and inequality remains inadequate across our financial system. Colin looks at the initiatives currently underway to ensure that ESG integration and sustainable capitalism are practised in financial markets.

As we head towards our 2030 net-zero goals, sustainable innovation has remained at the forefront of this movement. Colin looks at how cities are not only centres for innovation but also excellent places for investors to focus on. He analyses examples of sustainable investments, such as Neom, a projected smart city in Saudi Arabia, as well as the specific infrastructure required for sustainable success: net-zero buildings and natural infrastructure; personal mobility and public transit; clean energy, storage, and grids; and the benefits of locally sourced, healthy, and accessible food.

Cities are central to the financial system itself but also for new technologies and innovations that are tested and invested in for future sustainability. They have the highest need and consequently hold the key to the solutions. The responsibility is immense, but it once again proves that cities are well equipped to become global leaders in the fight against climate change.

22

The Invested City

Colin le Duc

Sustainable Capitalism and Accelerating Sustainability Trends

We appear to be in the early stages of a systemic, secular, multidecade transition to a sustainable economy. This transition is being driven by a combination of factors. Global population growth, increased pollution, resource constraints, the climate crisis, rising inequality in the developed world, and poverty are driving the need for change. Technological innovation and consumer demand are accelerating this transition. Many refer to this transition as the "Sustainability Revolution", which has the magnitude of the Industrial Revolution and the speed of the Digital Revolution. Companies and cities who lead this transition by offering sustainable products and services are well positioned for the long term. Sustainable solutions are manifesting in material ways across significant industries all around the world. Every industry is ripe for disruption. Cities are at the epicentre of many of these trends.

The present economic system has delivered global and economic growth benefits for a significant portion of society since the mid-nineteenth century. However, evidence is mounting of the system-wide unintended negative consequences of this historical economic growth. Consequences such as the climate crisis, extreme inequality in access to opportunity, education, healthcare, and wealth, mass biodiversity extinction, unparalleled levels of plastics in our oceans, local air pollution, widespread compromised emotional and mental wellbeing, the obesity epidemic, and deforestation are all signs of the fundamentally unsustainable trajectory we are on.

The very foundations of our economy depend on social cohesion, environmental stability, and sustainable access to resources, all of which are increasingly at risk due to the unintended consequences of our significant population growth and resource-intensive economy. Since 1970, the global population has more than doubled. and global real GDP has grown more than five-fold, driving increased demand on natural resources. By 2050 it is anticipated that 2.5 billion more people will become urban residents, which will further increase the strain on natural and social capital. The need for change is presenting unparalleled investment opportunities across all sectors; high quality, economic growth across

The Climate City, First Edition. Edited by Martin Powell.
© 2022 John Wiley & Sons Ltd. Published 2022 by John Wiley & Sons Ltd.

income categories and over varying time horizons and geographies can only be achieved if natural resources remain readily available; affordable energy is abundant; infrastructure is reliable; and citizens have access to healthy food, clean air and water, affordable medical care, quality education, and fair wages. These factors, each a hallmark of sustainable capitalism, enable – albeit do not guarantee – vibrant societies and stable operating environments. Investors actively seek out these conditions, as risk is better understood and relatively more predictable.

Moreover, it may not be possible to optimize land, labour, and capital in the absence of these conditions. The investment community increasingly understands the need to confront these realities as the stakes are so high across the whole financial system. After all, the majority of the world's 100 largest economic entities are no longer nations or cities but companies themselves. And the 10 biggest corporations now manage more wealth than most nations in the world combined. The best investors understand the enlightened self-interest inherent in sustainable investing and sustainable capitalism.

Pursuing a more inclusive form of growth ensures the benefits of capitalism are shared more widely. If we do not do this, we run the risk of still more popular rejection of capitalism and its associated liberties – which could lead to even worse outcomes in the future. This need not be a defensive exercise. Indeed, if it is pursued thoughtfully, it can instead prove to be inspirational. One of capitalism's real strengths is its flexibility to respond to a society's expressed needs. At different times and in different places, capitalism has evolved and changed, as have our attitudes toward it. Consequently, there is no reason we should consider the particular form of capitalism in place today in the USA and Western Europe to be the best possible form of capitalism for human development. To the contrary, better forms of capitalism may lie ahead, and many investors are now embracing their responsibilities in manifesting these advances by allocating capital more sustainably and advocating for a more sustainable financial system.

Recently, the COVID-19 pandemic has disrupted all aspects of normal life, in particular in cities. Enormous health, economic, and social costs have ensued. It has also brought home the urgency of a decisive shift to sustainability. Many long-term sustainability trends are accelerating and some new ones are emerging. Cities are at the forefront of these trends. The implications of these new dynamics for action on climate and inequality are profound and investors are taking note. It is also increasingly clear how vital the role of partnerships is in raising ambition on sustainability, building on the experience of the on-going collective responses to the pandemic.

Companies have a critical role to play in building back better. Cities are critical enablers of this and arguably at the forefront of how to do this. How the investment community is contributing to the recovery and building an inclusive, sustainable economy will determine which companies and investments are successful long term. Moreover, the extent to which investors embrace sustainability will determine which geographies thrive long term and which cities emerge as leaders in the post-pandemic world. It is clear that a change in investor ethos is necessary to find a sustainable pathway out of this immediate crisis and out of the longer-term secular crises of climate and inequity.

The pandemic is revealing new insights into our shared sustainability challenges. The interlocking nature of our climate and ecological and social crises has become obvious.

Additionally, the vulnerability and lack of resilience of our societies and cities to "grey swan" shocks are quite clear. The depth and nature of economic and social inequalities has been dramatically exposed. Our dependence on institutions and cooperation for prosperity and stability, at a time when mutual trust has collapsed, has been brutally revealed. 2020 has proved to be a profoundly difficult time for many people and has added to the riskiness of investing, which often leads to less capital availability and a negative spiral of lack of investment.

As investors focused on long-term value creation and sustainability, we collectively need to lift our eyes to the horizon. The effects of the pandemic will be with us for some time, but severe health impacts might be stemmed sooner with an effective vaccine or treatment. There is, however, no vaccine for the climate crisis. Addressing racial, gender, LGBTQ, and economic inequality likewise requires both urgent and sustained investments across multiple fronts.

The acceleration of sustainability trends provides important glimpses into the future of cities: how we live, how we work, where we want to live, and what we want, and need, to consume. Beyond the short-term impacts of the on-going pandemic, there is a sense of new political and social realities forming that provide a generational opportunity. The action and momentum triggered by events in 2020 will likely prove a powerful catalyst for sustainability and sustainable investing.

Could some companies and governments waver in their commitments to sustainability, blown off course by the short-term impact of the pandemic and the economic crisis? This is certainly a risk. Yet for many investors and other organizations, the lesson from recent events is that sustainability and resilience are two sides of the same coin.

The pandemic has revealed investors' and companies' commitment to sustainability in a way that is difficult to see in normal times. Many businesses have stepped up, working with others to find solutions with unprecedented speed and taking a broad view of stakeholder value when it mattered most. This redefining of what is possible must now be taken into other shared challenges for our societies, not least for climate.

Our quality of life and civility of society now depend on our collective efforts to ensure a safe, resilient, and sustainable recovery. Strikingly, to limit global temperature rise to 1.5°C, greenhouse gas emissions need to fall by 7.6% per year by 2030 – more than the drop expected in 2020 due to the pandemic. Clearly, this means pulling all available levers. Yet there is also a new sense of perspective. The net investment required for climate action is much smaller than the economic cost of the pandemic, and, unlike the pandemic, it represents a major economic and social opportunity. The costs of inaction on climate change meanwhile are so large they are difficult to comprehend.

Governments must chart the course for a green recovery through their industrial policies and green deals. The EU, for instance, sees its Green Deal as the "motor of the recovery". There is too little of the global cooperation that was so crucial in pulling the world out of the depths of the global recession in the late 2000s.

Tremendous strides have been made in recent decades in tackling extreme poverty, improving human health, and developing a raft of clean energy technologies and affordable digital solutions. The task now is to ensure these gains are protected, accelerated, and felt by all. Cities are at the forefront of the Sustainability Revolution, and investors are increasingly embracing sustainability as best practice.

Cities as Financial Centres (Mainstreaming Sustainable Investing)

Decisions about how the world's financial capital is allocated are mostly made in cities. Financial centres such as London, Frankfurt, Singapore, New York, Shanghai, Nairobi, and Brasilia rely on the unique features of cities in bringing together talent that is attracted to the challenges of managing our financial capital and market systems. Central banks, insurance companies, pension funds, investment banks, stock exchanges, family offices, rating agencies, investment advisors, asset managers, high net worth individuals, foundations, and corporations all have roles to play in determining what future we will inherit by virtue of their financial and capital allocation decisions. The questions are whether this capital is being allocated sustainably and how to evolve our capitalist system to being more sustainable.

Currently, despite the overwhelming logic of sustainability, most financial assets are still not managed with sustainability in mind. There has been enormous progress in mainstreaming ESG criteria across financial markets in recent years. All asset classes of investments are embracing sustainability – including public equity, private equity, fixed income, commodities, real-estate, infrastructure, and government bonds. However, the level of understanding of the risks and opportunities associated with the Sustainability Revolution and fundamental issues of climate and inequality remains remarkably inadequate across our financial system.

Encouragingly, a huge variety of initiatives are underway to ensure that ESG integration and sustainable capitalism become accepted best practice in financial markets. The most powerful way to influence investors is for them to experience financial losses or gains due to sustainability. Plenty of examples exist as to stranded assets (e.g. US shale oil companies) and huge positive returns (e.g. electric vehicle companies) from companies on the right side of the sustainability equation, and it is now widely accepted that ESG data help investors optimize risk and return.

Beyond this somewhat Darwinian way of demonstrating the financial case for sustainable investing, many other initiatives exist to bring forth sustainable capitalism: changes to listing rules of public companies, enhancements to corporate reporting guidelines, new types of shareholder agreements, an evolved understanding of fiduciary duty, impact measurement and tracking, formalized ESG training for financial market professionals, central bank mandates to include the management of systemic climate risk, the establishment of carbon pricing and trading, the explosive growth of green bonds, and the increased prominence of green banks. These are just a subset of examples of the breadth of initiatives that many people are working on to ensure the increased sustainability of our financial system.

In particular, the financial sector is starting to take climate-related risks and alignment with the Paris Agreement seriously. Although there is a long way to go, this is a critical ingredient for tackling the climate crisis. The pandemic shock will only strengthen the growing emphasis on managing systemic climate risk. Climate-focused investor groups are developing a coherent plan from the finance and investment sector in time for climate negotiations in November 2021. It is unlikely that the growing momentum will be derailed by the pandemic and the economic crisis.

Investors also need to demonstrate their commitment through their investment practice. This could include: (1) assessing the implied temperature rise of investment portfolios and the implications for Paris-Agreement-aligned investing; (2) asking companies invested to set science-based targets and to double down on climate action in light of the pandemic; and (3) encouraging companies to focus on their long-term strength and resilience.

Getting sustainable investment and ESG right requires a joined-up approach with a wide set of stakeholders involved. The sector must be accountable for delivering on climate and other crucial social and environmental commitments. As part of this, ESG data, a key ingredient for how the sector is developed, need to be more forward-looking and more closely connected to companies' real-world impact.

The ripple effect of ESG integration will be felt across the financial markets as more sustainable investments get funded and capital is allocated away from sunset industries, countries, cities, and outdated business models that no longer serve society's real needs. It is increasingly accepted that sustainable investment helps optimize risk and return. The question now is whether this shift can be applied and adapted to all capital in the economy. We believe the EU taxonomy will help with eradicating "greenwashing", as will the new focus from the US Securities and Exchange Commission (SEC) on this topic.

Furthermore, the recent government interventions to recover from the enormous impacts of the pandemic have revealed the importance of public intervention in economies. This will serve as a dress rehearsal for dealing with the climate crisis. It is obvious that free markets alone can't solve massive, collective challenges like poverty, climate, or pandemics. Governments are needed. This is exemplified by the €750 billion EU Recovery Plan,[2] as well as the multi-trillion dollar recovery stimulus plans established in the USA. Many of these plans have a "green" recovery dimension at their heart. City, state, and country budgets haven't been sufficiently robust to handle the costs of these crises, and free market business hasn't solved things by simple "market forces". Huge investments from the public sector, in addition to strong policy guidelines, are required for us to fulfil the promise of a fair and sustainable economy for all.

Cities as Investments (Sustainable Innovations)

Cities are typically excellent places for investors to focus on, especially as test beds for new innovations. Investors like risks that are understandable and relatively predictable. Cities often offer a stability that many more regional areas may not. Further, cities are dense centres of demand, so it's easier to test and distribute new products and services. Investing is about taking calculated risk, and predictability is highly sought after. Investing to transition cities to becoming more sustainable presents a significant investment opportunity driven by innovative technology, social policy, and consumer preferences.

Recently, innovation has contributed to significant cost down-curves and performance improvements in core sustainable innovations such as solar photovoltaics (PV), LED lighting, genomic sequencing, supercomputing processing power, onshore

---

2 https://ec.europa.eu/info/live-work-travel-eu/health/coronavirus-response/recovery-plan-europe_en.

and offshore wind, electric vehicle batteries, robotics, artificial intelligence, and data storage. These unprecedented breakthroughs are enabling new business models to propagate sustainable products and services and disrupt very large traditional markets.

In addition, consumers are demanding better, more environmentally and socially considerate products. Companies with truly sustainable solutions are benefitting from a shift in demand by consumers away from incumbent, unsustainable products and services. Cities are often early adopters of many of these new technologies and business model innovations. Sustainable innovation, products, and governments can often offer a framework for investors to get behind in a way that countries cannot easily do. This could be in relation to regulations enabling the testing out of new products and services, as was the case with ridesharing, for example. Online grocery pioneer Ocado initially only served densely populated areas in the UK, to begin with in London, in order to be able to do sufficient home delivery drops per van per day to be economic. This early market allowed the company to refine its systems and reduce costs, which subsequently enabled it to profitably expand its geographic footprint across more rural parts of the UK.

Entirely new cities are being built from scratch with sustainability firmly in mind as a founding design principle. The UAE and Saudi Arabia have, for example, very actively courted international investors to help realize these visions. The visions of these new sustainable cities demonstrate the types of technologies and innovations that investors are focused on.

As per Masdar City's own statements:

> Masdar City is one of the world's most sustainable urban communities, a low-carbon development made up of a rapidly growing clean-tech cluster, business free zone and residential neighbourhood with restaurants, shops and public green spaces.
>
> Masdar's philosophy of urban development is based on the three pillars of economic, social and environmental sustainability. Masdar City is a "greenprint" for the sustainable development of cities through the application of real-world solutions in energy and water efficiency, mobility and the reduction of waste.
>
> Masdar City is open to the public and welcomes tourists, residents, students, academics, entrepreneurs, business leaders and investors to its collaborative environment, which encourages people to live, work and play sustainably.

A newer city ambition is that of NEOM in Saudi Arabia, a key part of the 2030 Vision for the country to accelerate its transition beyond economic reliance on fossil fuels. The vision statement of NEOM is a further example of the sustainability aspirations that can be manifested through a new city:

> NEOM is an accelerator of human progress that will showcase the future of innovation in business, livability and environmental sustainability.

NEOM is a bold and audacious dream. It is a vision of what a New Future might look like (in fact, NEOM means, "new future"). It's an attempt to do something that's never been done before and it comes at a time when the world needs fresh thinking and new solutions. NEOM is being built on the Red Sea in northwest Saudi Arabia as a living laboratory – a place where entrepreneurship and innovation will chart the course for this New Future. NEOM will be a destination, a home for people who dream big and want to be part of building a new model for sustainable living, working and prospering.

NEOM will include towns and cities, ports and enterprise zones, research centers, sports and entertainment venues, and tourist destinations. It will be the home and workplace for more than a million citizens from around the world.

An exciting variety of innovative sustainable solutions are being invested in and across existing and new city and urban landscapes all around the world. These touch on essential areas such as energy, mobility, buildings, healthcare, food, and wellbeing.

Net-Zero Buildings and Natural Infrastructure

The city's built environment has an enormous carbon footprint today and is ripe for improvement. Commitments to net-zero buildings are becoming widely regulated, thus providing investors with a framework within which to evaluate all types of low carbon technologies and solutions to invest in. The wellbeing of people within buildings is a further focus area, with solutions for air quality, noise control, and mental health offering investment opportunity.

Within the residential sector, smart thermostats were deployed in nearly 20% of US homes by the end of 2019, and prices for LED lightbulbs dropped about 95% from 2008 to 2019. Many companies are vying to provide the leading smart operating system of the home. The construction industry has, for a variety of reasons, been one of the last to digitize. There is significant investment opportunity for efficiency improvements through better software and planning, potentially leading to lower inventory, lower waste, lower downtime, and more stable jobs throughout the sector. The construction industry is responsible for a significant amount of waste that results in landfill – for example, in the UK, one-third of all waste to landfill is from the construction sector. New construction materials such as timber to replace concrete and steel are also an exciting area for investors to focus on. Retrofitting buildings with insulation and net-zero solutions are also recipients of much investment.

Greening city infrastructure has become widely accepted as socially beneficial and is integrated into city planning. Investors are focused on green real-estate, tree planting, parkland development, integrated green communities, and various adaptations to climate impacts via the build-out of various forms of natural infrastructure.

Our collective experience indoors during lockdown is a reminder of how this push for net-zero buildings is part of a wider conversation about quality of life and wellness at home, in the workplace, and across our lived experience in cities.

Personal Mobility and Public Transport

Cities are dealing with the unintended consequences of their success, notably the challenges of pollution and congestion and how to move people and goods effectively around the highly complex urban environment. Investors have become increasingly interested in technologies such as electric vehicles, autonomous vehicles (Figure 22.2), ridesharing, swappable battery systems for scooters, bike sharing, emerging ideas like vertical take-off and landing (VTOL) vehicles, delivery drones (Figure 22.3), electric buses, bus rapid transit systems, and emerging notions such as solar roads that recharge cars while they drive over them (Figure 22.4). All these types of innovations are enabled by digital connectivity and the maturation of cyber security. Investors are able to express their theses on sustainable innovation for mobility in cities by investing in both the enabling infrastructure and the very specific sustainable mobility solutions.

Technological and business innovations are vastly improving the efficiency of the transport systems and supply chains that move people and products globally. Within transport, the convergence of autonomous, connected, and electric transport has the capability to transform the efficiency of and improve flow in next-generation cities. Autonomous vehicles may be deployed as part of large fleets in some cities within the 2020s. The adoption of electric vehicles is already growing exponentially, with the

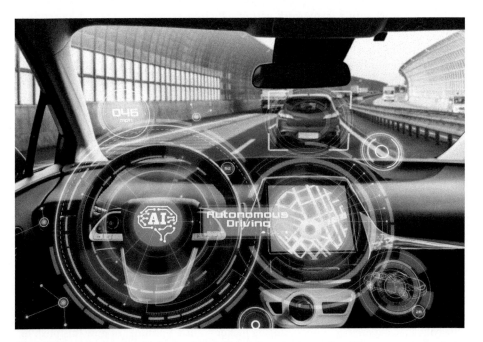

Figure 22.2 Autonomous driving in cities will be a long orchestration. (*Source:* metamorworks/ Adobe Stock.)

Figure 22.3 Delivery drones could offer street level relief. (*Source:* NASA.)

global stock of electric cars surpassing 5 million vehicles in 2019, up from fewer than 100,000 in 2011. Electric cars, scooters, and buses are increasingly attractive to consumers as the price continues to decrease: average electric-vehicle battery costs have dropped from around US$1,000/kWh in 2010 to around US$200/kWh in 2019, with further declines expected to follow. Moving goods more efficiently through global supply chains can bring enormous resource productivity improvements. Shipping and logistics providers are beginning to leverage advanced technology to enable more efficient services and provide visibility. About a quarter of global third-party logistics companies are now using advanced analytics and data-mining tools in their supply chain to better track and reduce downtime in the flow of goods. How these innovations are uniquely implemented within a city environment is a major focus area for investors.

Additionally, existing and new public transit systems are of interest for investors. Investing is typically done via public–private partnerships and is thus quite complex and usually the domain of highly specialized infrastructure investors. One benefit of the revamping of existing public transit systems and also the building of new public transit is that it offers investors an opportunity to invest significant amounts of capital in a relatively low-risk way. The impact of the pandemic on how the public engages with public transit systems is a major consideration as to where demand will manifest in mobility systems and consequently how investors will allocate capital.

Figure 22.4 Solar road in Dubai. (*Source:* World Traveler / Getty Images.)

Clean Energy, Storage, and Grids

BP's statistical data indicate that renewable energy generation (excluding hydro) accounted for 10% of global power generation in 2019, up from approximately 1% 10 years before. Renewables are the fastest-growing source of electricity, contributing almost half of the growth in global power generation in 2019. We have witnessed the challenging cycles since around 2005 in clean energy, and investors often seek to avoid capital-intensive businesses with regulatory exposure, such as project-development businesses. Many investors prefer to focus on more capital-light market opportunities such as off-grid energy, consumer-energy engagement, grid integration, virtual power plants, energy efficiency, and energy storage.

A US Department of Energy survey of clean energy prices found that from 2008 to 2019 the total cost of land-based wind energy in the USA fell by 60%. The survey shows that the cost of distributed solar PV energy fell by 54% and utility-scale solar by 64% over the same period. Once the combined cost of solar plus storage declines below the cost of gas, many commentators anticipate a powerful inflection point in the adoption of solar as the principal electricity generation form.

Cities are obviously huge consumers of electricity and as such are major sources of emissions. The opportunity to integrate distributed energy generation across cities varies according to the city design. However, investors are allocating significant capital to clean, pollution-free electricity as well as ways to reduce energy demand. Many billions of dollars are being invested in solar, wind, battery storage, and grid enhancements that enable higher integration of renewables.

Local, Healthy, and Accessible Food

City food systems depend on entire global food supply chains but also highly localized sources of nutrition. Innovations in enabling technologies such as LED lighting, sensors, protein extraction, and new food delivery models have given rise to investment opportunity in solutions such as vertical, urban farming, fresh food delivery models, and new consumer-facing brands propagating non-animal-based proteins.

Moving our agriculture and food systems in a sustainable direction offers plenty of investment opportunity. This includes healthier, local food and better agricultural production systems which drive higher yield. An important trend that has attracted investor interest relates to improved environmental intelligence. Data-enabled decision-making in agriculture is becoming the norm. A majority of the agricultural market in North America has been estimated to benefit from some form of precision agricultural technique (such as aerial vehicles, GPS, satellite imagery, or soil sampling). Advances in biology that are displacing the use of chemistry in agriculture and food production are another important development and driver towards more sustainable operations. According to the US Environmental Protection Agency, approximately 10% of US greenhouse gas emissions were derived from the agriculture sector in 2019.

The largest source of emissions was agricultural soil management through activities such as fertilizer application and other agricultural practices that increased nitrogen availability in the soil. Another important source was enteric fermentation (livestock such as cows producing methane as part of their digestive processes). Individual consumption habits are also changing and may contribute to the transition to a more sustainable economy. Alternative proteins (such as those produced from nuts and legumes) are growing and diversifying rapidly. Plant-based meat alternatives are disrupting traditional business models and have been recipients of hundreds of millions of dollars of invested capital.

Conclusion

This chapter has touched on how cities are at the forefront of the transition to a more sustainable form of capitalism. Cities are hubs for both the financial system itself and how new technologies and innovations are tested and implemented.

Cities play a crucial role in mainstreaming sustainable investing and enabling ESG factors to be fully integrated into capital allocation decisions. Additionally, in areas of critical societal needs such as building, transport, food, and energy, cities are incubators of new, innovative, sustainable models that can be tested and perfected to become mainstream solutions.

23

The Financed City – Introduction

In Chapter 22 Colin was clear on the role cities have to play to enable ESG factors to be fully integrated into capital allocation.

In *The Financed City*, James answers the question of why cities matter so much, by looking at them as complex interrelated systems, central to a population's wellbeing, and examining their pre-existing and ongoing commitment to a sustainable future and net-zero action. Most significantly, James looks at ways of mobilizing finance for low carbon circular cities by redirecting and scaling up investment to make cities fit for the future in the context of climate change.

He examines why the current way of doing things needs to change so that finance is "not only focused on financial returns but is also purpose-led", aware of the upcoming tragedy of climate change rather than discounting it. He looks at how public, private, and city finance is also changing, with 130 banks, accounting for one-third of the global banking sector, signing up to align their businesses with the Paris Agreement goals alongside the work of the Cities Climate Finance Leadership Alliance.

In Figure 23.1 we can see a move to the municipal green bond market – those cities that have embraced measurement and greenhouse gas (GHG) inventories and produce clear pathways to reducing their risk are benefitting.

James also explores how cities can create a new paradigm for financing low carbon circular transition through policy, the market, citizen intervention, creditworthiness, and combining funding and financing measures.

The goal is to create a more viable and equitable system to help enable a net-zero goal. James details three of the key enablers that will support this system to enable finance and funding to flow towards low carbon resilient cities of the future:

1) devolved decision-making (city-level creditworthiness, citizen involvement);
2) addressing consumption and waste (circular economy);
3) a vision-led approach (ambition alongside clear measurable goals).

James ends his chapter with a rallying call to action. Policymakers at national, city, and local level must play their part, as must citizens as voters, consumers, and investors. Businesses, whether they be established or start-ups, and investors, whether they are banks or asset owners, are also responsible. The climate emergency requires urgent action

The Climate City, First Edition. Edited by Martin Powell.
© 2022 John Wiley & Sons Ltd. Published 2022 by John Wiley & Sons Ltd.

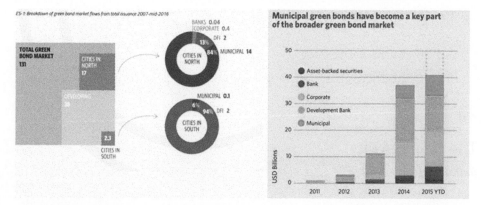

Figure 23.1 Green bond market flows and the role of municipal green bonds. DFI, Development Finance Institution. (*Source*: HOW TO ISSUE A GREEN CITY BOND, THE GREEN CITY BONDS OVERVIEW. 2015, Climate Bonds.)

and cities have a significant role to play. Just as Colin wrote in Chapter 22, *The Invested City*, cities are at the forefront of innovation and investment that means they are best placed to confront the problem at hand. With a new financial paradigm, investment must combine social purpose with risk-adjusted returns.

23

The Financed City
James Close

Why Cities Matter

Seventy percent of greenhouse gas emissions come from cities, and over 50% of the world's population live in cities.[1] They are the foundations of our modern society and economy. As a result, cities are central to managing the transition to a low carbon, resilient future. The focus must be on redirecting and scaling up investment to make cities fit for the future in terms of addressing climate change and also mobilizing that finance to create low carbon cities using circular economy principles.

Cities will need to transition from their historic trajectory of high carbon development to address climate change. Cities are well equipped to make this transition because they are dense, homogenous, and concentrated in terms of both population and infrastructure. Their long-term plans need to be informed by a compelling vision of the future.

Net-zero carbon cities are an important aspiration. However, the net-zero carbon aspiration will not be sufficient by itself. Cities also need to reduce consumption-based emissions by adopting circular economy principles so they can eliminate their contribution to climate change. Figure 23.2 illustrates the increasingly important role of cities over time to reduce consumption-based emissions.[2]

Cities Will Remain Central to Our Collective Wellbeing

Until the COVID-19 pandemic (the pandemic), cities were assumed to be the engine of the economy and the driver of future economic growth. Their concentration of people and emissions suggested that cities are a key building block of a low carbon transition whilst building resilience and creating a pathway to sustainable growth. The pandemic has shifted the trajectory, accelerating trends toward homeworking, internet shopping, and shifts in commuting. How these trends play out is uncertain. Whatever role cities play after the pandemic, they will remain central to our collective wellbeing.

---

1 https://www.c40.org.

2 http://www.lse.ac.uk/granthaminstitute/wp-content/uploads/2019/09/financing-inclusive-climate-action-in-the-uk_an-investor-roadmap-for-the-just-transition_policy-report_56pp.pdf.

The Climate City, First Edition. Edited by Martin Powell.
© 2022 John Wiley & Sons Ltd. Published 2022 by John Wiley & Sons Ltd.

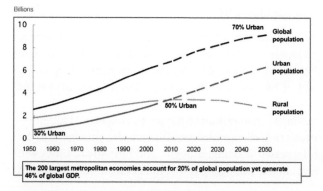

Billions

70% Urban — Global population

Urban population

50% Urban

Rural population

30% Urban

1950 1960 1970 1980 1990 2000 2010 2020 2030 2040 2050

The 200 largest metropolitan economies account for 20% of global population yet generate 46% of global GDP.

Figure 23.2 Role of cities over time to reduce consumption-based emissions. (*Source*: Modified from LSE cities based on United Nations World Urbanizations prospects, 2007 Revisions.)

The most successful cities will find ways to be great places to work, live, and engage in recreation. Those that are not low carbon and resilient will lose comparative advantage to those that are.

Cities Are Complex, Interrelated Systems

Vision-led development in cities started at the earliest times of civilization in Ancient Greece with the city states of Athens, Sparta, Corinth, and Troy and the founding of the Roman Empire in Rome. Cities as engines of development continued through the Industrial Revolution in Manchester in the 1700s to the leading global cities of the last century – New York, London, Tokyo, and Paris. These clusters of enterprise, culture, and ideas are joined in this century by cities in emerging markets including Shanghai, Beijing, Mumbai, and Sao Paulo.

At the heart of the success of – and challenges for – these cities is the concept that the concentration of economic activity promotes development and growth. These economies of agglomeration create clusters which offer economies of scale and network effects. There is a gain in efficiency which comes from having many firms and workers in close proximity to one another. The benefits of the exchange of ideas and knowledge with the attraction of cultural and social engagement create dynamic living and working conditions.

There are disadvantages too. Spatially concentrated growth may create problems of crowding and traffic congestion. It is the tension between economies and diseconomies that allows cities to grow but keeps them from becoming too large. These tensions manifest themselves in the interplay between policy, the market, and the behaviour (and choices) of citizens. Finance responds and shapes these forces and enables visions and aspirations to come to life.

Cities' complexity also arises from the relationship between people, space, and infrastructure. The word citizen relates to cities and citizenship – around participation in public life – and is central to the organization of cities. Identity and culture are parts of the system that make successful cities dynamic and ever-changing.

Cities Are Already Committing to a Sustainable Future

Many cities across the globe have made commitments to achieve net-zero emissions. Along with 77 countries and 10 regions, over 100 cities have committed to net-zero carbon emissions by 2050.[3] These ambitions are supported by organizations representing cities, including C40, the EU Covenant of Mayors, and the ICLEI – Local Governments for Sustainability global network.

The level of action by cities on climate change is underpinned by enlightened self-interest. Whilst contributing to the problem of climate change, cities are also vulnerable to the impact of climate risk through increased flooding, drought, and heat stress. Cities will also need to adapt to the changing climate through investment in climate-smart, resilient infrastructure.

There Are Significant Benefits from these Actions

Climate commitments and actions can also drive innovation and increase efficiency. There are many co-benefits coming from reduced noise, congestion, and pollution, with the potential for better health outcomes coming from improvements in physical and mental health. Furthermore, in the competition for capital and talent, a firm and consistent commitment to creating low carbon, liveable cities can create an attractive environment for skilled labour and entrepreneurial activity.

Net-Zero Is Not Enough

C40 Cities, in its recent report "The Future of Urban Consumption in a 1.5°C World",[4] has set out a pathway for cities to contribute to achieving the more ambitious goal of the Paris Agreement. This pathway identifies the importance of accounting for consumption-based emissions.

Figure 23.3 shows that total estimated consumption-based emissions for the 94 member C40 Cities is 4.5 $GtCO_2e$, whilst those associated with traditional territorial accounting (i.e. those that occur within the boundary of the city, region, or country) are 2.9 $GtCO_2e$.

According to the report, the average per capita impact of urban consumption must decrease by 50% by 2030 and 80% by 2050. For example, in London, Londoners are currently responsible for an estimated 12 tonnes of consumption-based CO_2e emissions per person per year, and so by 2030 this must fall to around 6 tonnes per person and by 2050 to around 2.5 tonnes to hit the targets of the Paris Agreement. At the same time, production of energy needs to be decarbonized through the rapid roll-out of renewables and electric mobility. Energy efficiency makes the challenge easier by reducing the demand for energy.

The Ellen MacArthur Foundation has published a report by Material Economics[5] showing that 45% of global CO_2e emissions arise from the management of land and the production of goods (mainly the production of cement, steel, plastic, and aluminium)

---

3 https://sdg.iisd.org/news/77-countries-100-cities-commit-to-net-zero-carbon-emissions-by-2050-at-climate-summit/.

4 https://www.c40.org/researches/consumption-based-emissions.

5 https://www.ellenmacarthurfoundation.org/publications/completing-the-picture-climate-change.

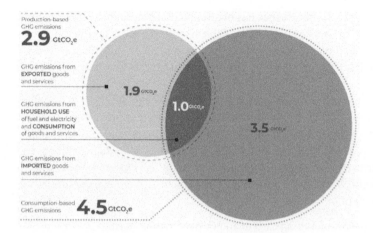

Figure 23.3 Consumption-based emissions versus territorial emissions for C40 Cities.

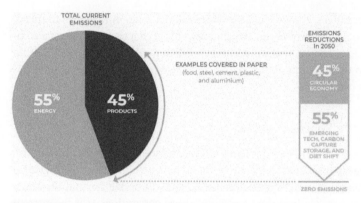

Figure 23.4 Forty-five percent of CO_2e emissions are from products and agriculture.

(Figure 23.4). These emissions are harder to abate and will require a combination of applying circular economy principles to reduce production of new materials and increase recycling and emerging technology to rapidly decarbonize production.

Taken together, these two reports set the aspiration for the cities of the future. Similar to the co-benefits of reducing the production of emissions, addressing consumption-based emissions provides additional benefits from the creation of local jobs, the strengthening of communities, and reduction of waste.

Financing the Transition

The Current Way of Doing Things Needs to Change

Financing the transition to low carbon, resilient, and circular cities exposes the arguments around the way in which long-term economic and environmental development can be combined. The existing model that has evolved is based on a form of capitalism that may have reached the limits of its success. The current regulatory and institutional systems are inadequate and open to abuse. This form of capitalism may have served us adequately in the past, but it needs to change significantly to address the challenges of the future.

The transition from the exiting model presents three broadly defined choices. The first choice is to muddle through with the current inadequacies of the capitalist system. The second choice is to adopt the "degrowth" model where there is a predisposition to a lack of investment which could degrade the quality of infrastructure and reduce potential for economic and jobs growth. The preferred choice is to adapt the current system with a clearer link to the benefits for all citizens and the environment. This choice requires more effective regulation including:

- the pricing of externalities including carbon and environmental harm;
- less concentrated and more equitable returns; and
- greater community involvement and engagement.

Under all these alternatives, finance will flow in a rational way in response to incentives, with the perceived balance between risk and reward. Like it or not, there is a competition for capital. Cities are competing with each other for investment, and, within cities, projects have to demonstrate their relative attractiveness against social, environmental, and economic criteria.

The Need to Address the Tragedy of the Horizon

Mark Carney, the former Governor of the Bank of England and current United Nations special envoy for climate action and finance, referred to climate change as the tragedy of the horizon.[6] The catastrophic impacts of climate change will be felt beyond the traditional horizons of most actors – imposing a cost on future generations that the current generation has no direct incentive to fix.

The business cycle, the political cycle, and the horizon of technocratic authorities, like central banks, who are bound by their mandates are too short. Once climate change becomes a defining issue for financial stability, it may already be too late. Earlier action will also mean less costly adjustment.

The historic way of deploying finance has to change, so that finance is not only focused on financial returns but also purpose-led. Finance needs to be deployed to create a better, more sustainable, and more equitable world. This imperative has always been relevant to public funding, and a transition is underway for private finance.

Public, Private, and City Finance Is Changing

An international group of 27 institutional investors has signed up to delivering on a bold commitment to transition our investment portfolios to net-zero GHG emissions by 2050.[7]

---

6 https://www.bis.org/review/r151009a.pdf.

7 https://www.unepfi.org/net-zero-alliance/.

Representing nearly US$5.0 trillion assets under management, the United Nations-convened Net-Zero Asset Owner Alliance shows united investor action to align portfolios with a 1.5°C scenario, addressing Article 2.1c of the Paris Agreement. In addition, 130 banks, accounting for one-third of the global banking sector, have signed up to align their businesses with the Paris Agreement goals.

In April 2019, finance ministers from more than 20 countries launched a new coalition aimed at driving stronger collective action on climate change and its impacts.[8] The newly formed Coalition of Finance Ministers for Climate Action endorsed a set of six common principles, known as the "Helsinki Principles", that promote national climate action, especially through fiscal policy and the use of public finance.

The Cities Climate Finance Leadership Alliance,[9] a multilevel and multistakeholder coalition, is aimed at closing the investment gap for urban subnational climate projects and infrastructure. The alliance provides a platform to convene and exchange knowledge among all relevant actors dedicated to urban development, climate action, and/or financing. Alliance members include public and private finance institutions, governments, international organizations, NGOs, research groups, and networks that represent most of the world's largest cities. Its members also represent the main market players in city-level climate finance.

Cities Can Create a New Paradigm for Financing the Low Carbon Circular Transition

Successful cities will need to win the competition for capital by making their city attractive for investment. The nature of the complex interrelated system cannot be described simplistically, but it can be characterized by the convergence of and interrelationship between policy, the market, and citizens:

- *Policy* is a deliberate system of principles to guide decisions and achieve rational outcomes. It signals future direction and enables implementation. In cities, it generally comes from a political mandate and is influenced by the expectations and aspirations of citizens and the real or implied preferences of the market.
- *The market* in its broadest sense comprises the relationship between buyers (which include public organizations and citizens) and sellers (which include businesses and citizens as suppliers and employees). These relationships are fixed in transactions and price signals.
- *Citizens* are at the heart of this dynamic through their preferences and voting intentions, and they send signals in all the choices they make directly and indirectly.

A more explicit relationship between these forces and a purpose-led approach to the deployment of finance will create this new paradigm. Cities have the potential for large-scale experimentation where successful implementation can be rapidly scaled up. They can also be building blocks for an economic, political and social system that is better suited for the challenges of the twenty-first century.

---

8 http://pubdocs.worldbank.org/en/646831555088732759/fm-coalition-brochure-final-v3.pdf.

9 https://www.citiesclimatefinance.org/.

Creditworthiness Is at the Heart of this New Paradigm

There is a direct relationship between resilience, investment, and risk. Strong and effective city institutions enhance the reputation of cities and increase their relative attractiveness for investment as they compete for capital.

A city's creditworthiness is dependent on the sovereign context. It is often complicated by the funding agreement with the national government, the volatility of its revenues, and restrictions on its ability to borrow. Sometimes these constraints are appropriate and rational, but often they are a function of history and political. There are often significant ranges between cities in the same country. For example, as rated by Moody's sub-sovereign rating:[10]

- Canada's cities range between Aaa and Aa2.
- The Czech Republic's cities range between Aa3 and A3.
- Mexico's cities range between Baa1 and B3.
- Switzerland's cities range between Aa1 and Aa3.

Those cities on a relatively low rating compared to peers in their countries can materially reduce their borrowing cost by improving their rating to the level of the highest rated city. Cities can improve their creditworthiness by:

- *strengthening financial performance* through effective revenue collection, efficient running of operations, and sound financial controls;
- developing an *enabling legal and regulatory, institutional, and policy framework* for responsible subnational borrowing through reforms at the national level;
- *improving the "demand" side of financing* by developing sound, climate-smart projects that foster green growth;
- *improving the "supply" side of financing* by engaging with private sector investors.

Box 23.1 shows the range of city creditworthiness for selected cities and their countries.

Combining Funding and Financing

The role of funding and financing is central to this transition. In funding, the public sector provides a specific amount of money for a specific purpose (e.g. to a project, usually free of charge or interest-free) with no expectation of repayment. In financing, someone (usually one or more financial institutions) provides an amount of capital (debt or equity) to a project with the expectation that it will be repaid with interest.

Cities that optimize the alignment between public funding and private financing, which comprises of wholesale, national, and international financing with retail community investment, will be more effective in attracting investment.

These tasks are complicated by the need to address a series of complex trade-offs, which include:

- *national versus local priorities* as defined by policy, tax-raising powers, and the distribution of benefits and investment in cities;
- *risk allocation and pricing of risk* between the public sector and private actors;
- *degree of concessionality* of public funding and the role that it can play in financing higher-risk activities such as risks relating to regulation and development;

---

10 https://www.moodys.com/research/sub-sovereign-ratings-list--pbc_124094.

Box 23.1 Range of city creditworthiness for selected cities and their corresponding country

City	City rating	Sovereign rating
Buenos Aires	Caa3	Caa2
Ottawa	Aaa	Aaa
Toronto	Aa1	Aaa
Vienna	Aa1	Aaa
Berlin	Aa1	Aaa
Budapest	Baa3	Baa3
Rome	Ba1	Baa3
Milan	Baa3	Baa3
Merida	Ba1	A3
Mexico	Baa1	A3
Barcelona	Baa1	Baa1
Madrid	Baa1	Baa1
Sao Paolo	Ba2	Ba2
Rio de Janeiro	Ba3	Ba2

(*Source*: Data from Moody's Sub-Sovereign Ratings, 17 July, 2020.)

- *short-term versus long-term horizons* for investment and planning and how these horizons create uncertainty and optionality.

Furthermore, broadening the source of public funding and finance can come from national and local tax revenue-raising powers. Innovative and hypothecated instruments such as green bonds can be new sources of capital. Project finance, corporate finance, and social investment all have key roles to play in combining public funds and private finance effectively.

As cities look to upgrade their infrastructure and deploy smart, low carbon technologies, paying for those projects presents challenges relating to budgetary constraints. Cities can identify business models that can help to attract private financing in order to make the introduction viable and financeable.

Matching projects to the most appropriate financing requires a deep understanding of the project, its potential cash flows, the range of financing options available (locally and internationally), and available procurement methods to government in order to deliver.

Harnessing the benefits that come from effectively combining public funding and purpose-led private finance needs to be at the heart of the redesign of the current system so that it becomes more viable and equitable.

Creating a More Viable and Equitable System

Opportunities for Investment

Cities already deploy significant amounts of public and private capital for investment. Work by the Global Commission on the Economy and Climate[11] suggests that, by 2035,

---

11 https://newclimateeconomy.net/.

roughly US$93 trillion of infrastructure designed to be low-emission and climate-resilient will need to be built globally. According to a report by the Brookings Institute[12] based on work by the Commission's analysis, more than 70% of this infrastructure will be built in urban areas, at a cost of between US$4.5 and $5.4 trillion per year. The value of infrastructure required in urban areas by 2035 could be greater than the US$50 trillion value of all the infrastructure in the world today.

Given that financing and funding cities will be the majority of money flow to 2035, like the rest of the economy cities will need to shift the flow of money to finance future sustainable infrastructure. This presents a major opportunity for investment for all investors irrespective of whether they are asset owners, asset managers, or individuals.

Three Key Enablers

Three of the key enablers that will support the creation of a more viable and equitable system to enable finance and funding to flow towards low carbon resilient cities of the future include:

- increasing the effectiveness of *devolved decision-making*;
- shifting the focus towards *addressing consumption and waste*; and
- taking a *vision-led approach* which focuses on the long-term development of the city with a view to creating great places to live, work, and play in a way that supports cities' comparative advantage.

Increasing the Effectiveness of Devolved Decision-Making

Devolving decision-making and, with it, creditworthiness to a city level and, where appropriate, to a more local level can improve prioritization, increase citizen involvement, and identify and manage risk more effectively. Many decisions are best made at a city or local level and will be most effective when working symbiotically with national and international policy frameworks.

The role citizens play can also contribute, despite the fact that local democracy often has inadequacies. For example, municipal governance can create policy gridlock. Legislators who differ on how best to spend city money will compromise, which can result in suboptimal decisions and higher perceived and actual risk.

Shifting the Focus to Consumption and Waste

Even if cities hit their "net-zero" targets, it will not be sufficient if the goods and materials consumed continue to come from high carbon imports and if there is unnecessary waste. Most cities have significant scope 3 emissions – emissions that are imported from other sources through the consumption of food and materials. Cities need to promote circular

---

12 https://www.brookings.edu/research/driving-sustainable-development-through-better-infrastructure-key-elements-of-a-transformation-program/.

economy business models which help to address consumption-based emissions and have the additional benefit of reducing waste that goes into the system.

Whilst much of the low carbon investment will need to be directed towards infrastructure, these flows of investment need to be combined with the promotion and adoption of business models that help reduce consumption-based emissions:

- *products as services* by selling access to products while retaining ownership of assets;
- *renewable inputs* by shifting to using secondary materials as the inputs for products;
- *recovery of value at the end of life* through effective recycling, including composting and anaerobic digestion;
- *product life extension* through maintenance, designing for durability, reuse, and remanufacture of products and components;
- *sharing economy* by sharing assets (e.g. cars, rooms, appliances) via sharing platforms.

Box 23.2 explains the circular economy and its role.

Transitioning to a circular economy[13] requires the scaling up and rolling out of the business models described above. Large corporations can make this transition, and, in the new paradigm, there are opportunities for purpose-led corporate finance to be oriented towards start-ups and small- and medium-sized enterprises. The acceleration of the development, deployment, and scaling up of circular economy businesses will increase sharing, repairing, reusing, and recycling through the adoption of their innovative business models and principles.

Box 23.2   Role of the circular economy

Many investments in low carbon energy production are now competitive with other forms of higher carbon infrastructure even before externalities and social benefits are included. This creates a clear pathway for financing an energy transition in cities.

The current economy is linear, which means that things are made with virgin raw materials, used, and then thrown away. The circular economy, in contrast, keeps products and materials circulating within the economy at their highest value for as long as possible through reuse, recycling, remanufacturing, delivering products as services, and sharing.

A circular economy approach not only is more resource efficient but also protects businesses from fluctuating commodity prices. It provides an opportunity to develop a more stable operating environment for manufacturers, retailers, and consumers and can respond to the increasing interest from business leaders, employees, and other stakeholders in a new, positive vision for the economy, delivering tangible benefits for society and the environment. The circular economy provides a new, more equitable and sustainable direction for the economy while delivering tangible positive outcomes. Profitable circular companies of all sizes can create jobs with opportunities for upskilling, and they can also contribute to resilient, biodiverse ecosystems.

---

13 https://www.c40.org.

A Vision-Led Approach

Given the complex systems underpinning cities and their development, finding the right balance between a clear vision with specific, measurable, and time-bound aspirations of a low carbon future with the organic development of the city in a national and an international context is important.

According to C40 Cities, a city's vision should outline the main features and benefits of becoming an emissions neutral and climate-resilient city by 2050 and should include a commitment to take transformational and inclusive action in key sectors (e.g. energy, buildings, transport, and waste). The commitment should specifically endorse the Paris Agreement.

Essential components include a written commitment from the mayor or city leader to begin implementing transformational and inclusive action to deliver an emission-neutral and climate-resilient city by 2050, consistent with the objectives of the Paris Agreement. Cities can go further with a signed legislative commitment with cross-political and/or multisector support for delivering the plan.

The declaration of a climate emergency is shown in Box 23.3.

Box 23.3   Declaring a climate emergency

A number of UK cities have now passed motions declaring a "climate emergency", including Edinburgh, London, Leeds, Oxford, and Sheffield.

In some places, for example Oxford, this declaration has been accompanied by an agreement to create a citizens' assembly to consult on the council's response. Some cities are planning their own pathways to net-zero. In March 2019, for example, Leeds City Council adopted a science-based carbon roadmap which laid out targets for the city to reach net-zero emissions by 2050.

London is also acting on climate change;[14] 28 boroughs and the Mayor of London have now passed climate emergency declarations. In December 2019, London Councils' Transport and Environment Committee and the London Environment Directors' Network (LEDNet) published a Joint Statement on Climate Change which includes a series of seven joint ambitions.

Conclusion

The climate emergency requires urgent and thoughtful collective action. Cities have a significant role to play. Their contribution comes from rethinking the vision for the cities of the future. Mobilizing funding and financing with the objective of creating low carbon, resilient, liveable, productive cities is central to enabling the vision. With a new paradigm, investment needs to combine a social purpose with adequate risk-adjusted returns.

---

14 https://www.london.gov.uk/what-we-do/better-infrastructure/london-infrastructure-plan-2050#acc-i-43211.

The transition will require the global financial system to prioritize managing climate risks and channel capital towards cities. As Mark Carney indicates, the transition will require "a massive reallocation of capital".

Cities are conducive to system-wide innovation that can drive transformation through new ways of engagement, revised regulation, and new business models. These actions can create new financial markets and instruments that support the transition more effectively. All actors need to play their part.

Policy makers must set the long-term trajectory to facilitate optimal trade-offs:

- At a *national level*, policy must take account of the disproportionate role for cities in addressing climate change.
- At a *city level*, policy must focus on the comparative advantage of the city in a low carbon future.
- At a *local level*, policy can support grassroots change and the promotion of new ways of operating that support community action, local jobs, and circular economy business models.

Citizens must play their part so that they can realize the benefits:

- In their role as *voters*, citizens can engage in politics at all levels by influencing politicians, making clear their priorities, and casting their votes accordingly.
- In their role as *consumers*, citizens can ensure that they make choices that signal their commitment to reducing consumption, supporting local activity, and promoting new business models.
- In their role as *investors*, individuals, as the ultimate savers and beneficiaries, can make sure that their decisions actively support a low carbon transition, including through physical as well as social resilience, and help to reconnect the financial system to the real economy, particularly in terms of local development.

Businesses, collectively as the market, need to prepare for and take advantage of these changes:

- *Established businesses* can transition to low carbon, circular economy business models to ensure they create the market and avoid falling behind.
- *Start-ups* can respond to market opportunities and, if they have access to capital, rapidly scale up their business models, creating high-quality new jobs.

Investors can protect themselves from downside risks and benefit from new opportunities:

- *Asset owners*, such as pension funds, insurance firms, and foundations, can apply a city lens to setting investment beliefs, signalling expectations throughout the financial system, and driving accountability for results.
- *Asset managers* can allocate capital across different asset classes (including public and private equities, fixed income, and private debt, as well as real assets such as infrastructure and real-estate) to support city action.

- *Banks* can set the trajectory through committing to being purpose-led and facilitating the transition at all levels.

A concerted and collective effort can accelerate the transition to low carbon circular cities and provide opportunities for capital and investors. Optimizing the funding and financing of the cities of the future is urgent and important.

24

The Adapted City – Introduction

The following chapter will now connect this city story to the "risks" that both Colin and James referred to in Chapters 22 and 23 respectively. The inevitable impacts of climate change have already altered the planet. We have to adapt our cities to this new normal.

Adam takes us through the changes that we will have to make. Cities are already on the front lines of the climate crisis, dealing with the catastrophic impacts of climate change-driven heatwaves, coastal storms, droughts, and wildfires. But it's not just climate change that is putting people at risk; it is also decades of poor urban planning and design. The shape and form of our cities will help decide how well we can withstand today's extreme weather and what lies ahead. "*The Adapted City*" looks at the practical actions cities around the globe are taking to adapt to rising temperatures, too much water, and too little water – and the changes needed to scale up these actions to address the urgent reality of the risks we face. This chapter outlines six key principles for mayors and city leaders to embrace to protect their residents from climate change and highlights several case studies that provide a roadmap for urban leaders on how to accelerate the breadth and scale of their work.

The World Economic Forum's 2020 Global Risk Report listed extreme weather and climate action failure as the biggest risks facing the world, with all five of the top risks likely to occur on the survey related specifically to climate change.[1] Adam rightly points out that climate change is "changing the rules". Environments we thought we had conquered are becoming increasingly erratic and dangerous. We have lost control and we need to adapt cities quickly if we are to overcome these challenges.

Adam looks at how cities can adapt to heat not only by reducing the number of automobiles on the streets but also by planting trees and increasing green spaces, much like the revitalization of the Cheonggyecheon Expressway in Seoul which Hayley discussed in Chapter 21, *The Just City*. Adam also explores cool strategies such as shading spaces, water features, cooling roofs, and the strategic use of the public right of way to decrease the heat absorption of dark surfaces.

---

1 Franco, E.G., 2020. The global risks report 2020. World Economic Forum, Geneva.

The Climate City, First Edition. Edited by Martin Powell.
© 2022 John Wiley & Sons Ltd. Published 2022 by John Wiley & Sons Ltd.

We all know how flooding can paralyse a city, as we have seen countless times in cities all over the world, with one-third of Bangladesh underwater in 2020 due to excessive flooding. Adam examines how we can adapt to water and decrease not only the human cost but also the economic and financial damage it can bring. The options available to us include increasing permeable surfaces through investment in wide green infrastructure and changing behaviour by financing green stormwater infrastructure and raising the price of stormwater.

Adapting to climate change is no longer a choice. The question is not if cities need to adapt to climate change but how they will adapt to climate change. Just as our human bodies have evolved over millions of years to optimum environmental condition, so must our cities. Fortunately, many of the solutions offered go hand in hand with reducing our carbon emissions, keeping the 2030 net-zero goal close.

24

The Adapted City

Adam Freed

Throughout history, much of mankind's time and energy have been dedicated to trying to conquer nature. So far, this has been a challenge we have largely been able to master – or fool ourselves into thinking we have overcome. But climate change is changing the rules. The environments we thought we bested are becoming increasingly erratic and dangerous.

The World Economic Forum's 2020 Global Risk Report, which surveys more than 1,000 global business, political, and non-profit leaders to identify their biggest concerns, listed extreme weather and climate action failure as the biggest risks facing the world.[1] In fact, the top five risks in terms of their likelihood to occur are all related to climate change, including extreme weather (1st), climate action failure (2nd), natural disasters (3rd), biodiversity loss (4th), and human-made environmental disasters (5th). As our climate risks increase – driven by climate change and a century of poor urban design and planning – cities need to adapt to a shifting environmental baseline and more severe and frequent natural hazards.

The past few years offer a stark example of the challenges we face. In 2018, Cape Town, South Africa, home to 4 million people, came within days of "Day Zero", when the city's reservoirs where expected to run dry during an unprecedented drought. Later that year, two category 5 hurricanes ripped through the Caribbean just weeks apart, decimating the islands in their path. 2019 was the second hottest year on record on the planet (only behind 2016).[2] This trend is set to continue. In October, Phoenix, Arizona set a new record with 143 days of triple-digit temperatures.[3] Over the same period, more than 8,000 wildfires in California have burned over 4 million acres of land (more than 4% of the state's total area), earning 2020 the distinction of the largest wildfire season in the state's modern history.

---

1 Franco, E.G., 2020. The global risks report 2020. World Economic Forum, Geneva.

2 National Oceanic Atmospheric Administration, 2020. 2019 was 2nd-hottest year on record for Earth say NOAA, NASA. NOAA press release, 15 January 2020.

3 Associated Press Wire Service, 2020. Phoenix breaks record with 144th day of triple digit heat. Associated Press, 14 October 2020.

The Climate City, First Edition. Edited by Martin Powell.
© 2022 John Wiley & Sons Ltd. Published 2022 by John Wiley & Sons Ltd.

The CDP estimates that cities experienced US$140 billion in damage annually between 2005 and 2014 as a result of climate change. Absent lower emissions or adaptation actions, by 2050 US$158 trillion in urban assets will be at risk from climate events.[4] The shape and form of our cities will help decide how well we can withstand the extreme weather that lies ahead.

But this is not a book about doom and hopelessness. It is about the practical actions cities are taking to tackle the climate crisis today and how they can be replicated in other cities to achieve transformative change. From New York City to Durban, Copenhagen to Wuhan, and Paris to Medellín, cities are adapting to rising temperatures, more intense rainstorms and coastal storms, sea level rise, and more frequent and longer droughts. This chapter highlights several of these actions, with a focus on heat and intense rainstorms, and shows how cities can move from pilots to systemic change. Cities are scaling up their implementation by taking multipronged strategies which include significant public investments to catalyse action and provide proof of concept, new standards and codes to ensure widespread adoption from public and private actors, and public engagement and education to change individual behaviours.

The Intergovernmental Panel on Climate Change points out that as cities work to combat climate change, "the tools are at hand. We possess the material basis and policy solutions for transformations and systems changes."[5] What is required is the political will and funding to implement them.

The good news is that many of the things we need to do to adapt our cities can also reduce our GHG emissions (and vice versa), if designed well and as part of a strategic plan. And they can help improve our air quality, increase park access, support sustainable modes of transportation, create jobs, and generally make cities more livable and equitable.

So how do we make these changes a reality?

To adapt our cities to climate change – and do so at a scale and pace that can address the urgent reality of the risks we face – cities and their leaders need to embrace six key principles.

First, they need to recognize that adaptation is not an optional course of action or a "Plan B" solution. Cities need to act now. The decade since 2010 has made this abundantly clear. Climate adaptation is not just about protecting us from future risks, it is about responding to and planning for the realties we face today. The World Economic Forum calls for adaptation "to be given urgent priority, not only to prepare for the possibility of very dangerous levels of climate change in the future, but also to eliminate the resilience deficit we face today".[6] In taking action, cities need to act in what Rushad Nanavatty, a Senior Principal at the Rocky Mountain Institute, calls a "crisis relevant timeframe". Investments and transformative change of urban infrastructure need to happen within the next decade, not over the course of the next century.

---

4 CDP, 2020. Cities at risk. https://www.cdp.net/en/research/global-reports/cities-at-risk.

5 Bazaz, A., Bertoldi, P., Buckeridge, M., et al., 2018. Summary for urban policy makers: What the IPCC special report on 1.5°C means for cities. C40 Cities Climate Leadership Group, London. doi: 10.24943/SCPM.2018.

6 Franco, 2020. Op. cit.

Second, urban leaders need to recognize that our increased exposure to climate risks is not just driven by climate change, but also by decades of poor urban planning and design. The choices made have shaped our cities and put our residents at risk. A recent study found that development in eight US coastal states is occurring at a faster pace in flood zones than in other areas of those states.[7] This is directly putting people and property at risk – even if sea levels were not rising and coastal storms becoming more frequent and intense. Similarly, the abundance of dark asphalt and rooftops in cities and lack of trees, which are the result of deliberate design choices, contribute to cities being up to 11°C hotter than their surrounding regions. The World Resources Institute warns that "urban development that is blind to climate risk is increasing exposure to climate hazards in cities".[8]

Third, cities need to use data to understand and quantify the risks they face. This will help build political and public support for climate actions and help target investments to communities most at risk from climate hazards. When discussing the risks of climate change, Sir Nicholas Stern said, "we were gambling the planet".[9] Just like a first-time gambler in Las Vegas, many cities have a fundamental misunderstanding of the odds. This is driven by the inherent challenges of getting people to understand and value avoided loss, as well as a lack of accessible, understandable, local data to drive planning and investment decisions. Many cities lack basic flood maps or are using outdated information to direct development and infrastructure investments. Prior to Super Storm Sandy, New York City was using federally developed flood maps that were based on 30-year-old data.[10] Cities need up-to-date data to draw the proverbial (and in some cases literal) line in the sand about where development and infrastructure investments can occur and to what standards it should be built.

According to the CDP, there is a significant gap in understanding risks by cities. Of the 530 cities that reported climate information to the CDP in 2018, only 46% completed vulnerability assessments; and only 25% identified vulnerable populations within their assessments.[11] Former New York City Mayor Michael Bloomberg repeatedly says, "You can't manage what you don't measure." Not having climate risk data enables cities to ignore the risks they face or take shots in the dark when trying to reduce them.

Fourth, cities need to target their adaptation actions to address systemic inequities – including historically (and sometimes deliberately) uneven infrastructure investments and service delivery. Taking action on real and significant environmental justice issues in cities

7 Flavelle, C., 2019. Homes are being built the fastest in many flood-prone areas, study finds. *New York Times*, 31 July 2019.

8 Chu, E., Brown, A., Michael, K., Du, J., Lwasa, S., Mahendra, A., 2019. Unlocking the potential for transformative climate adaptation in cities. Background Paper prepared for the Global Commission on Adaptation, Washington, DC.

9 Stern, N., 2009. The global climate deal: Climate change and the creation of a new era of progress and prosperity. New York Public Affairs, New York.

10 City of New York, 2007. Panic 2030: A greener, greater New York. City of New York, New York.

11 CDP, 2020. Cities at risk. https://www.cdp.net/en/research/global-reports/cities-at-risk.

is an equity and adaptation action. Ani Dasgupta, Global Director of WRI Ross Center for Sustainable Cities, and now Head of WRI points out that "done carefully, transformative adaptation can put cities on a stronger, safer path that offers opportunity and a higher quality of life for all".[12] A 2020 report by WRI draws the connection between adaptation and equity. "Transformative adaptation reorients urban climate action around addressing entrenched equity and climate justice challenges. It focuses on systemic changes to development processes that improve people's quality of life, enhance the social and economic vibrancy of cities, and ensure sustainable, resilient, and inclusive urban futures."[13] Given the increasing focus on equity by mayors and city residents, explicitly linking these goals and actions is a key way to build critical public support for climate action and maximize the benefits gained through policy changes and infrastructure investments.

To do this, cities need to engage residents – particularly the most vulnerable and exposed communities. The scale of change needed to adapt to climate change requires new land-use zoning, building codes, development patterns, infrastructure design, and funding streams (in additional to individual behavioural change). All of these elements will require sustained public and legislative support across multiple political cycles. The most successful and impactful actions will also need to be rooted in local knowledge and designed to address social, economic, and environmental inequities – not just climate risks. Strong community engagement will help achieve these goals.

Fifth, city leaders need to recognize that the scale of climate risks we face – and the systems that need to be adapted to respond to them – require new governance models to facilitate metropolitan- and regional-scale planning and investments. Cities are key players in the fight against climate change. But they cannot do it alone. As climate risks span beyond city boundaries and the infrastructure systems cities rely on cross political jurisdictions, they need to find ways to enable broader metropolitan-level collaboration between large cities and surrounding regions. The Nature Conservancy found that cities occupy 2% of the planet's land surface but draw water from 20% of the land.[14] To solve urban water problems, cities need to think (and act) beyond their political borders. For mayors, this means stepping up to take climate leadership, but also bringing others along within their metropolitan region and natural resource sheds. Within cities it also means breaking down the traditional bureaucratic silos that prevent transformation of outdated forms of infrastructure.[15]

Finally, cities and urban leaders need to recognize that every action, whether intentional or not, is a climate action and will lower or increase their risk. Creating new

---

12 World Resources Institute, 2020. https://www.wri.org/news/2019/10/release-new-research-finds-climate-adaptation-opportunity-transformative-change-cities.

13 Chu et al., 2019. Op cit.

14 McDonald, R.I., Shemie, D., 2014. Urban Water Blueprint: Mapping conservation solutions to the global water challenge. The Nature Conservancy, Washington, DC.

15 In New York City, for instance, piloting green stormwater infrastructure in the public right of way required forming partnerships between the City's Department of Environmental Protection, which funded stormwater management; the Department of Transportation, which managed the city's roadways; and the Department of Parks and Recreation, which had the knowledge to maintain greenery.

parking areas of dark asphalt, building housing far from transit, paving over wetlands, or allowing development in coastal areas without adequate flood protections will increase a city's climate risk. Helsinki Mayor Jan Vapaavuori says that in Helsinki "every action is a climate action".[16] This is true in every city. Cities can choose to adapt or perpetuate a status quo that puts their residents and economies at risk.

Given space constraints, this chapter does not cover all of the climate risks facing cities or all of the solutions that need to be (and in many cases are being) deployed to adapt to climate change. It only covers extreme heat and more intense rainfall. Climate risks such as sea level rise, coastal storms, and droughts are not discussed in detail. Nor are secondary impacts such as wildfires, shortened lifespans for infrastructure, lower bond ratings municipal debt, food shortages, climate migration, ocean acidification, or changes in disease vectors. These all pose significant risks for cities. Much more space is also needed to adequately discuss the regressive nature of climate risks, which impact poorer communities the hardest. Addressing this issue is paramount to successfully tackling climate change and making more equitable cities (and society).

Adapting to Heat

Extreme heat kills more people a year than all other natural hazards combined. This is particularly troubling for cities, which can be 12°C hotter than surrounding regions due to the urban heat island effect.[17]

The impacts of heatwaves, which the International Panel on Climate Change (IPCC) predicts will continue to become more frequent and intense, can be devastating. A 2003 heatwave in Europe is estimated to have killed close to 70,000 people, many of them vulnerable and senior residents in cities.[18] In 2010, 55,000 people died in Russia during a heatwave, including 700 people in Moscow a day.[19] In 2018, more than 30,000 people were admitted to the hospital for heatstroke over a two-week period in Japan.[20] As temperatures rise, the Natural Resources Defense Council estimates that 150 Americans will die every summer day due to extreme heat by 2040, with almost 30,000 heat-related deaths in the USA annually.[21]

---

16 Remarks at the Bloomberg Green Summit, 16 September 2020.

17 Urban heat islands are urbanized areas that have higher temperatures than surrounding areas, creating figurative islands of heat. Dark surfaces, such as buildings and roads, absorb the sun's heat, increasing surface temperatures, while exhaust from cars and air conditioners can also raise air temperatures. Non-urbanized areas, in contrast, are cooled by natural features such as forests and waterbodies.

18 Robine, J.-M., Cheung, S.L.K., Le Roy, S., et al., 2008. Death toll exceeded 70,000 in Europe during the summer of 2003. *Comptes Rendus Biologies* 331(2), 171–178.

19 Wallace-Wells, D., 2019. *The Uninhabitable Earth: Life after Warming.* Tim Duggan Books, New York, p. 41.

20 Tan, R., 2018. Death toll climbs as Japan wilts under a record-breaking heat wave. *Washington Post*, 24 July 2018.

21 Barkley, T., Constible, J., Knowlton, K., Shafer, W., 2018. Climate change and health: Extreme heat. Natural Resources Defense Council, New York.

Since the early 1980s, there has been a 50-fold increase in the number of "dangerous heatwaves" occurring across the globe. Globally, nine of the ten hottest years on record have happened in since the early 2000s. A recent report estimated that 1% of the planet is a "barely livable hot zone" today. As temperatures continue to rise, by 2070 almost 20% of the planet, home to one-third of the world's population, could be a hot zone.[22]

Climate change is not the only factor making cities hotter. The last century of urban planning and design has turned our cities into ovens. Urban growth driven by the rapid adoption of automobiles has led to cities with increasing dark impermeable surfaces such as parking areas, asphalt streets, and rooftops with far less green space and trees. This, combined with more cars and air conditioners that exhaust hot air, leads to an intensifying urban heat island effect. Even within cities, temperatures can vary greatly. Several recent studies have demonstrated that historically redlined neighbourhoods in the USA have significantly lower tree canopy cover than other neighbourhoods, contributing to higher temperatures.[23]

As design is contributing to rising temperatures in cities, it also provides the means to reverse this trend. Cities can reduce the impact of heat by providing more shade, adding greenery and water, lightening the colour of surfaces, and adapting building design. Cool roofs, light-coloured pavement, and increased tree planting could reduce heat-related deaths citywide by more than 25%.[24] Research from the University of Wisconsin found that covering blocks with 40% tree canopy coverage can lower temperatures by as much as 5.5°C.[25] The key is finding ways to implement these solutions at scale to have a significant impact on air temperatures.

Several cities are showing how this can be done through ambitious programmes that blend significant public investments, new regulations and codes, and public engagement.

Planting Trees

Milan has developed a multipronged strategy to increase tree cover in the city and broader metropolitan region. In 2019, Mayor Giuseppe Sala and partners launched the ForestaMi campaign to plant 3 million trees in the metropolitan region by 2030. A key focus of this work within the municipality's boundaries is planting trees in the hottest parts of the city, particularly in "grey" areas such as pavements and along roadways. This work, which I am supporting through my work at Bloomberg Associates, is guided by heat maps that

22 Xu, C., Kohler, T.A., Lenton, T.M., Svenning, J.C., Scheffer, M., 2020. Future of the human climate niche. *Proceedings of the National Academy of Sciences* 117(21), 11350–11355. doi: 10.1073/pnas.1910114117.

23 Hoffman, J., Shandas, V., Pendleton, N., 2020. The effects of historical housing policies on resident exposure to intra-urban heat: A study of 108 US urban areas. *Climate* 8(1), 12. doi: 10.3390/cli8010012.

24 Walker, A., 2019. Our cities are getting hotter – And it's killing people. *Curbed*, 12 July 2019.

25 Ziter, C.D., Pedersen, E.J., Kucharik, C.J., Turner, M.G., 2019. Scale-dependent interactions between tree canopy cover and impervious surfaces reduce daytime urban heat during summer. *Proceedings of the National Academy of Sciences* 116(15), 7575–7580. doi: 10.1073/pnas.1817561116.

utilize satellite-derived land surface temperatures to identify neighbourhoods that pose the greatest heat risks for residents (Figure 24.1). In addition to directly planting trees, the city incorporated these heat maps into its updated territorial land-use plan (the Piano di Governo del Territorio, or PGT), which requires increased blue and green infrastructure in public infrastructure and private development in areas with higher levels of heat. ForestaMi also includes robust public engagement and awareness campaigns to attract partners, raise money, and build support for the programme.

Medellín has an ambitious plan to create a series of green corridors throughout the city to cool the city and bring in cooler air from the surrounding mountains. These corridors will add more than 4.5 million m² of public space and greenery to the city by 2030. Each intervention that is part of the plan, which includes pedestrianized streets, neighbourhood parks, and revitalized waterways, will include vegetation in at least 40% of its area (Figures 24.2 and 24.3). These corridors will provide shade and draw in cooler air from surrounding mountains to help lower temperatures. Through this programme, the city has found a way to leverage the collective impacts of individual projects to have a citywide impact.

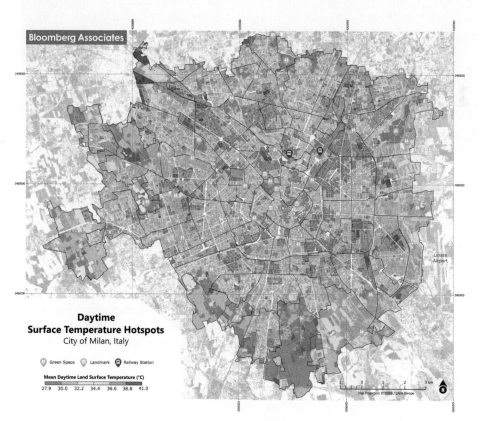

Figure 24.1 Map developed for Milan using satellite data to identify neighbourhoods in the city with the highest surface temperatures. These data are being used to target cooling interventions. (*Source*: Bloomberg Associates.)

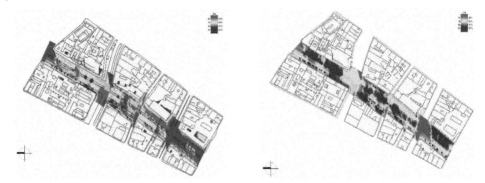

Figure 24.2 Thermal imagery of a street in Medellín before and after construction of a green corridor, showing the cooling benefits of this intervention. (*Source*: City of Medellín.)

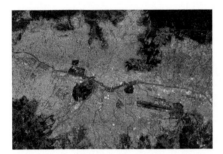
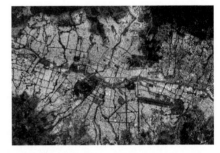

Figure 24.3 Construction of Medellín's green corridors will add 4.5 million m² of public space throughout the city. (*Source*: City of Medellín.)

Creating Shade

In areas where extensive open space or parks do not exist for large-scale tree planting, cities are creatively using their public assets to provide shading. The city of Paris launched an effort to transform some of its 700 school courtyards into "urban oases" to provide shaded spaces for children during school days and to the public during non-school hours. Compared to the current design, the renovated schoolyards are expected to generate a 10% decrease in surface temperatures and a 1–3°C decrease in daytime air temperatures. Students are engaged in helping to redesign their school's schoolyard and take part in monitoring the results of the work. In the first two years of the programme, Paris has transformed 45 schoolyards with an investment of more than US$16 million.[26]

---

26 Sitzoglou, M., 2020. The OASIS Schoolyards project Journal No. 1. The Urban Lab of Europe, Paris.

Adding Water Features

In addition to greenery, water features can be used to help cool urban areas. Water features like stepwells in India and public fountains (both decorative and drinking) in European cities are common urban design features that can help heat-stressed urban residents cool off and rehydrate. Some cities are deploying pop-up strategies to provide cooling opportunities where larger capital budgets are not available. New York City allows residents to install city-approved spray caps on fire hydrants, with city approval, to create temporary "cool streets". This effort complements the 650 spray showers and misting stations that exist in parks and other public spaces across the city.[27]

Improving the Public Right of Way

Cities must also strategically use their public right of way, which can represent up to 30–35% of total land area in cities, to decrease the heat absorption of dark surfaces.[28] In Los Angeles, parking occupies more than 17 million m^2 of land, as much as nearly 1,400 football fields.[29] Los Angeles is piloting the use of reflective coatings to increase the solar reflectance of its streets. This can reduce surface temperatures by up to 8°C.[30] This method can be adapted to other surfaces besides roads, such as surface parking areas. If successful, the city's roadway and parking area standards could be modified to require higher albedo (lighter coloured) surfaces.

Paris has pledged to remove half of the 140,000 parking spaces in the city, which will free up 65 hectares of space for greenery. Accomplishing this ambitious goal requires several interrelated policies to reduce dependence on personal vehicles through expanded bike lanes, investments in transit, new roadway regulations, and changes to parking and land-use regulations. David Belliard, Paris's Deputy Mayor for Transport, says the goal of this work is to "make greenery in the ground and prepare [Paris] for periods of heatwave which will be more and more frequent".[31]

Cooling Roofs

In conjunction with ground-level cooling efforts, New York City's Cool Roofs programme is working to transform the city's 14.8 million m^2 of rooftops with a light-coloured coating. New York City's building code requires all new and substantially reroofed buildings to be

---

27 Kashiwagi, S., 2020. NYC to open cooling centers, spray showers, cool streets, Island included. *Staten Island Advance*, 9 July 2020.

28 Scruggs, G., 2015. How much public space does a city need? *Next City*, 12 January 2015.

29 Peters, A., 2017. See just how much of a city's land is used for parking spaces. *Fast Company*, 20 July 2017.

30 Elder, J., 2019. If you can't take the heat, get into the street. Bloomberg Associates, Medium, 13 August 2019.

31 Henry, C., 2020. 'Nous allons supprimer la moitié des places de stationnement à Paris', annonce David Belliard. *Le Parisian*, 20 October 2020.

"cool roofs". This, combined with a city programme that uses volunteers to coat the roofs of non-profits and public housing for free, is contributing to the creation of more than 90,000 m^2 of cool roofs in the city every year. Coating all eligible rooftops in New York City could mitigate the urban heat island effect by up to 1°C.[32]

Cities must find ways to deploy these strategies quickly. Currently, 354 major cities have an average maximum summertime temperature at or above 35°C. By 2050, scientists predict this could more than double to 970 cities, containing 1.6 billion people.[33] As temperatures rise, we cannot simply air condition our way out of this problem. Today, more than 10% of global electricity use is dedicated to air conditioning and cooling.[34] If we continue on a business-as-usual pathway, with rising temperatures and incomes, this could increase to 30% of global electricity use – causing more excess heat in cities and making it more difficult to reduce GHG emissions. Mayors and city leaders must act now to adapt the urban design and form of their cities to the realities of an age of extreme heat.

Adapting to Too Much Water

Surface flooding caused by intense rainfall is another critical risk for cities that is increasing as a result of climate change and urban design choices. Impermeable areas, such as streets, parking areas, and rooftops, prevent water from being able to be absorbed into the ground. Instead it remains trapped on the surface and can cause severe flooding.

This flooding can cripple a city. In August 2007, a severe thunderstorm in New York City dropped 43 mm of rain in Central Park in just one hour, causing massive flooding and shutting down the city's subway.[35] The frequency and intensity of these types of events are increasing in many regions as a result of climate change. Houston, Texas experienced three 1-in-500-year floods not centuries or even decades apart but in just three years.[36]

Cities in Asia and Southeast Asia will be hit particularly hard by the increase in intense rainfall, putting significant risk on a region where extreme rainfall events are already the norm. 2020 was a sign of what is to come. Earlier that year, one-third of Bangladesh was underwater due to extreme rainfall and rising rivers.[37] In China, 55 million people were

---

32 City of New York, 2010. Mayor Bloomberg coats one millionth square foot of white rooftop as part of NYC Service Cool Roofs Initiative. City of New York, press release 23 October 2010.

33 Wallace-Wells, 2019. Op cit. p. 47.

34 Burabyi, S., 2019. The air conditioning trap: how cold air is heating the world. *The Guardian*, 29 August 2019.

35 Barron, J., 2007. A sudden storm brings New York City to its knees. *New York Times*, 9 August 2007.

36 Ingraham, C., 2017. Houston is experiencing its third 500 year flood in 3 years: How is that possible? *Washington Post*, 29 August 2017.

37 Cappucci, M., 2020. Severe flooding displaces millions in China and Bangladesh as the monsoon wreaks havoc. *Washington Post*, 31 July 2020.

impacted by floods, including 3.8 million people that were evacuated, in just the first half of 2020 as 53 rivers hit or approached historic high water levels.[38]

These changes in extreme short-duration rainfall events will have significant impacts on urban drainage systems and cities. The IPCC warns that given the design of urban infrastructure, these changes may have non-linear impacts on cities. Extreme rainfall changes in the range of 10–60% may lead to changes in flood and combined sewer overflow (CSO) frequencies and volumes between 0 and 400%.[39] McKinsey and Company estimates that a 1-in-100-year flood in 2050 in Ho Chi Minh City "would likely do three times the physical damage as a similar storm today and deliver 20 times the knock-on [economic] effects".[40]

Unlike coastal flood risks, cities do not need to be in coastal areas, or even along rivers, to face surface floods risks. Over 98% of cities in China with more than 1 million people (641 of 654 cities) are already exposed to frequent flooding. In 2018, more than 250 of these cities experienced floods that resulted in more than €46.9 billion of damage.[41]

Cities have several options to reduce surface flood risks. They can build bigger sewer and drainage systems to manage increasing amounts of water over shorter periods of time,[42] construct underground storage tanks, elevate buildings and infrastructure, channel stormwater out of developed areas, and/or enhance the permeability of the built environment to replicate the regulating functions of natural areas. In many cases, given the density of cities, limited space for construction, finite financial resources, and scale of the risk, cities must use an "all of the above" approach.

Increasing Permeable Surfaces

Cities are increasingly looking to distributed green infrastructure to reduce the impacts of intense rainstorms and flooding, given the expense of traditional grey infrastructure, such as underground storage tanks or larger sewer systems. These strategies, including bioswales, green roofs, enhanced parkland, and constructed wetlands, not only reduce flood risks but also help cool cities, improve air quality, support biodiversity, increase property values, and provide recreation opportunities.

To get to a scale to reduce widespread flood risks, cities need to engage the three key levers previously discussed: significant public investments, regulatory changes to standards and codes, and public engagement to achieve individual behavioural change.

---

38 Pike, L., 2020. China's summer of floods is a preview of climate disasters to come. *Inside Climate News*, 17 August 2020.

39 Revi, A., Satterthwaite, D.E., Aragón-Durand, F., et al. (eds), *Climate Change 2014: Impacts, Adaptation, and Vulnerability. Part A: Global and Sectoral Aspects. Contribution of Working Group II to the Fifth Assessment Report of the Intergovernmental Panel on Climate Change* (pp. 535–612). Cambridge University Press, Cambridge.

40 Woetzel, J., Pinner, D., Samandari, H., et al., 2020. Can coastal cities turn the tide on rising flood risk? McKinsey Global Institute, 20 April 2020.

41 Zevenbergen, C., Fu, D., Pathirana, A. 2018. Transitioning to sponge cities: Challenges and opportunities to address urban water problems in China. *Water*, 12 September 2018.

42 A significant challenge in doing this is retrofitting existing sewer systems in cities, where increasing capacity of large networks of underground systems is an expensive and disruptive task.

One of the biggest and most ambitious examples of a system-wide green infrastructure approach to flooding is in China. In 2015, the national government launched the Sponge Cities Initiative with the aim of restoring cities' "capacity to absorb, infiltrate, store, purify, drain, and manage rainwater and [regulate] the water cycle as must as possible to mimic the natural hydrological cycle".[43] The stated goal of the programme is to capture or reuse 70% of the rainwater that falls on 80% of urban land by 2030. The Chinese Ministry of Housing and Urban–Rural Development estimates that this effort will cost approximately €13.4–21.1 million per km^2.[44]

As part of the national programme, initiatives are underway in 30 cities, including Wuhan, Beijing, Tianjin, Shanghai, and Shenzhen. Much of this work aims to replace the water regulating services that had been provided by natural systems prior to urban development. Wuhan had been home to more than 100 lakes in its central core in the 1980s, which helped store and control rainfall. Today, only 30 lakes remain.[45]

Within Wuhan, more than 228 projects have been launched as part of the Sponge City Initiative, spanning actions to retrofit schools, streets, parks, and rooftops to capture water. To date, more than ¥11 billion has been spent to retrofit 38.5 km^2 of the city – an area more than half the size of Washington, DC.[46]

But there are some limitations to this model. While the breadth of the ambition is large, most sponge city projects are only built to manage a 1-in-30-year flood, which leaves a lot of risk still unmet. New approaches are important, but they must be tied to an appropriate design standard to reduce risks.

It is also unclear how this model will be scaled across the 600 cities in China with more than 1 million people.[47] Finding a way to do this is critical to China's urban success – and could offer important lessons for cities around the world.

Financing Green Stormwater Infrastructure

One promising route to scale is demonstrated in Washington, DC, where the city has found a way to leverage private investments in new market-rate developments to build green stormwater infrastructure in flood-prone neighbourhoods.

But cities must put the price on stormwater high enough to force people to change behaviours and make the economics of green stormwater infrastructure attractive to developers and property owners. Several cities in the USA, such as Philadelphia and Detroit, have adopted separate stormwater charges based on impervious areas in the hopes of incentivizing green stormwater infrastructure, only to find that their charges were too low to change behaviours.

---

43 Zevenbergen et al., 2018. Op cit.

44 Zevenbergen et al., 2018. Op cit.

45 *Bloomberg News*, 2020. After 500 years trying to tame fatal floods, China builds sponge cities. *Bloomberg News*, 13 August 2020.

46 Jing, L., 2019. Inside China's leading 'sponge city': Wuhan's war with water. *The Guardian*, 23 January 2019.

47 Zevenbergen et al., 2018. Op cit.

The overall cost facing cities to meet the challenge of increased surface flooding is huge. The Asian Development Bank estimates that US$800 billion is needed between now and 2030 to build grey and green infrastructure in Asia alone. But the cost of doing nothing may be greater. China suffered US$25 billion in flood damages through the first three-quarters of 2020.[48]

Conclusion

Since 2010 we have seen that adapting to climate change is not a choice for cities. It is a necessity. But it doesn't need to be an additive burden. Thoughtful and inclusive adaptation actions can help address systemic issues facing cities – from poor-quality housing, to economic inequalities, to lack of green space – and weather-related hazards. The question is not whether cities need to adapt to climate change or when, but how they choose to do so in a way that delivers clear benefits to residents – particularly their most vulnerable citizens – today.

The solutions and cities highlighted in this chapter can provide concrete examples of the types of action cities can take to protect their residents and economics. They also illustrate the multifaceted approach that is required to get to scale to provide real protection. The critical step is for city leaders to recognize that, like Mayor Vapaavuori in Helsinki, every action they take in their home cities is a climate action, whether intention or beneficial, or not.

---

48 Coca, N., 2020. Flooded Asia: Climate change hits region the hardest. *Financial Times*, 6 October 2020.

25

The Open City – Introduction

Chapter 24, "*The Adapted City*", looked at the practical actions cities around the globe are taking to adapt to rising temperatures, too much water, and too little water – and the changes needed to scale up these actions to address the urgent reality of the risks we face. Adam gave us six key principles for mayors and city leaders to embrace to protect their residents from climate change.

I offer you a startling statistic, which is shown in Figure 25.1 – by 2050, anticipated growth in air conditioning (AC) is set to increase four-fold. This is a consequence of growing populations, increasing middle classes, and a warming planet. The update of renewable energies can support this growth, but the importance of natural canopy to manage the heat island effect will be critical.[1]

The minimum amount of green space per person recommended by the WHO is 9 m². For example, Barcelona falls below this threshold, and according to Figure 25.2, so does Tokyo at just 3 m². Paris exceeds it at 11.5 m². It is no mystery why Berlin, New York City (NYC), London, and Rome regularly appear in the top 20 of liveable cities; regardless of costs of real-estate and transport, the provision of high-quality green spaces contributes to the equation of success.

Peter now moves the debate to an "open city", which is spatially diverse and generous and celebrates its public spaces, its parks, squares, and streets. Open cities are places where citizens meet, exchange goods and ideas, debate, linger, play, and celebrate. This is where the civic life of a democratic society takes place. You can judge the health of a city by its open spaces. Public space is not a commodity, and the market will not provide it (except under very limited conditions). It is public – that is, communally owned and maintained for the use and enjoyment of all. It needs to be protected, managed, and cared for. Where it is lacking it needs to be provided, not as a luxury but as a necessity for urban living.

The importance of public spaces is not to be underestimated, as they face increasing threats. Peter explores the problems posed by private and intermediate space in cities to deliver a truly open city. Most significantly, he looks at the growing risk present in the monestarization of public space by city authorities and the accompanying restrictions

---

1 https://www.buildinggreen.com/newsbrief/study-says-urban-heat-islands-worsen-smog.

The Climate City, First Edition. Edited by Martin Powell.
© 2022 John Wiley & Sons Ltd. Published 2022 by John Wiley & Sons Ltd.

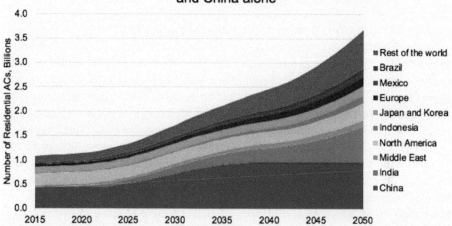

Figure 25.1 The fourfold increase of residential air conditioning by 2050 as a result of growing populations, a rising middle class, and rising temperatures. (*Source:* From "Diesel and electricity consumption for households gaining cooling access in hot countries in the baseline scenario, 2020-2050", IEA. All Rights Reserved.)

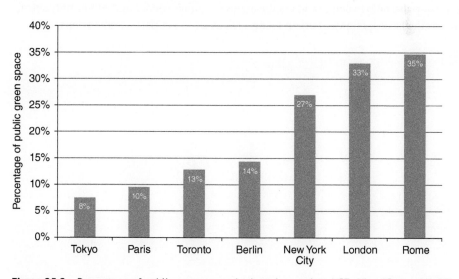

Figure 25.2 Percentage of public green space in densely populated G7 cities. (*Source:* Healthy people, healthy planet, The role of health systems in promoting healthier lifestyles and a greener future. OECD, 2017. Retrieved from: www.oecd.org./health/health-systems/Healthy-people-healthy-planet.pdf.)

posed by security, policing, and by-laws. There is no question public spaces must be regulated, but it is paramount that the citizen must also have licence to enjoy them. Peter interrogates the true function of a public space in conjunction with its architectural design.

Peter details the theory and, more importantly, the practice of public space provision. He examines how mass car ownership has distorted urban strategies and led to the diminishment of public space that has transformed the fabric of cities. He looks at how this has been challenged through the work of people like Jane Jacobs in New York City and her focus on the radical idea that city centres should be designed around the needs of people. He also looks at Jan Gehl's work in Copenhagen which led to a 2–3% parking reduction per year, as well as how mayors across the globe have implemented public space as part of their strategies for urban renewal. He explores ways this interest in public, or "open", cities is continuing to be developed further by a new generation of pragmatic urbanists through "small adjustments rather than grand interventions", in the push towards environmentally sound peopled landscapes.

As we saw in Chapter 24, successful cities are implementing adaptive strategies where agility and resilience are seen as the new hallmarks of stability. Increasingly situated at the overlapping points of urban programmes, Peter outlines strategies for new public spaces in the city and how they have led to innovation and experimentation.

Peter writes, "the contract between the city and nature is there to be redefined". Public spaces can redefine our cities for an environmentally sustainable future in a century where global temperatures are set to rise and extreme climate events become more frequent. They also help to define and inspire a sense of public service and community spirit, as well as the basic qualities of urban living such as clean air, parks, and gardens.

25

The Open City

Peter Bishop

> *There is perhaps no need of the poor of London which more prominently forces itself on the notice of anyone working among them than that of space. ... How can it best be given? And what is it precisely which should be given? I think we want four things. Places to sit in, places to play in, places to stroll in, and places to spend a day in.*
>
> Octavia Hill (1838–1912)[1]

Roberto Rossellini set his 1945 film *Rome Open City* in the context of the withdrawal of occupying forces and the handing of the city back to its inhabitants. At Design for London we adapted the term[2] to reflect our desire to engage on strategies to make London open and inclusive to all. Under our alternative definition, an "open city" was not just a city designed around the needs of all of its citizens but also one that was outward looking, tolerant, and inclusive. An open city allows individuals from different cultures, nationalities, and backgrounds to become citizens, to participate in the affairs of the city, to build their lives, achieve their ambitions, and have a sense of shared belonging. An open city is essentially a democratic city, one that has shared values that apply to all of its citizens. London in the twenty-first century, like New York in the twentieth, displays this condition. As a design team, the term open city provided us with a touchstone by which to judge our work.

An open city is spatially diverse and generous and celebrates its public spaces, its parks, squares, and streets. Public space constitutes those areas of the city that are commonly owned, publicly managed, and accessible to all. They are places where citizens meet, exchange goods and ideas, debate, linger, play, and celebrate. They are places of protest and in a last resort a default place of refuge for those who are homeless. This is where the civic life of a democratic society takes place. Public space also includes streets, which are the arteries of movement in the city, and herein lies a potential conflict of priorities. Public spaces and streets are the complex neural network of the city. Where priorities are tilted

---

1 Octavia Hill, a friend of Ruskin, campaigned for improved housing conditions for the urban poor. She also recognized the importance of open spaces that were easily accessible to the urban population: "the life-enhancing virtues of pure earth, clean air and blue sky".

2 This was in 2006. Design for London was the mayor's architecture studio. We cannot claim that we recoined the term, and others were beginning to use it around the same time: for example, the 2007 Rotterdam Biennale also used the term "open city".

The Climate City, First Edition. Edited by Martin Powell.
© 2022 John Wiley & Sons Ltd. Published 2022 by John Wiley & Sons Ltd.

towards (in Europe a minority) of car users, this network can become damaged and with it the health of the city. In many ways one can judge a city by its streets and public spaces. It is, for example, extraordinary to compare the investment in new highways to the atrocious condition of pavements in many cities, particularly in the developing world. This is a matter of choice, one that favours the most affluent over the lives of the many. Judge a city, and its political values, by its pavements.

Nature and Importance of Public Space

The theories behind public space are well documented, and the importance of good public space is now generally accepted by most urban practitioners. Increasingly, the simple division between public space and private space is becoming blurred by intermediate spaces or "privileged" space,[3] where the public has access under particular terms including payment. Under this definition, cinemas, shopping malls, theme parks, hotel lobbies, bookshops, and cafes have become important intermediate spaces that also nurture a rich civic life. Oldenburg[4] talks about "third places" that can "host the regular, voluntary, informal, and happily anticipated gatherings of individuals beyond the realms of home and work". These places are neutral ground and have no formal membership requirements. They are in essence convivial and comfortable spaces. In contrast, "corporate" space, places where public access is allowed under certain conditions (including dress code and social background), offer another aspect of urban life. These include the plazas and street networks of office and shopping precincts that are becoming a ubiquitous part of city life. While recognizing the important contribution that these spaces make to the variety of city living, they are not truly inclusive and cannot be a substitute for publicly owned and managed public space. Entry into them and their use is not a matter of right and is not open to all. Their growth in the twenty-first century (often encouraged by public authorities that are strapped for cash and do not wish to maintain them) is potentially divisive. The actual or implied restrictions of use posed by gated residential communities and overmanaged corporate plazas inexorably declare parts of the city "out of bounds" for certain sectors of society.

The threat to the free and open use of public space in the city is not restricted to the private sector. A growing risk is present in the monetarization of public space by city authorities and restrictions posed by security, policing, and by-laws. Publicly owned and managed areas of a great city are of course regulated by law, but within these parameters the citizen should have licence to use and enjoy them. Good public spaces are permissive; they enable rather than prescribe behaviour. It is no accident that autocratic and dictatorial regimes design public space for the control and regimentation of their populations often on a scale that crushes the individual. It is not, however, scale that determines the intrinsic value of public spaces, but function. St Peter's Square in Rome might be

---

3 Term coined by John Worthington of DEGW.

4 Oldenburg, R., 1999. *The Great Good Place: Cafes, Coffee Shops, Bookshops, Bars, Hair Salons and Other Hangouts at the Heart of a Community*. Marlowe & Co., New York.

monumental, but Bernini's design has created a contained space that acts as a grand setting for the basilica, a place for the culmination of pilgrimage, and as a place for formal gathering. Many cities have similar formal spaces that fulfil a national or political function. These are the places that we gather in to celebrate national events, sporting successes, and seasonal festivities. Although important, these grand formal spaces are of course not the most important day-to-day spaces in the life of the city. Liveable cities are based on a mosaic of public spaces – from the large municipal park to the recreation ground, play ground, and small sitting areas ("pocket parks") (Figure 25.3).

The spatial variety, the inherent "messiness", of a city like London stems from the incremental way in which it developed. The London "square" was a real-estate construct to increase the value of surrounding housing. The royal parks were planned, but often for private rather than public use. The metropolitan parks and recreation grounds of the late nineteenth and early twentieth century were a response to the social and health needs of a growing urban population, and larger areas such as Epping Forest, Hampstead Heath, and Highbury Fields were safeguarded through the acts of enlightened individuals and authorities. An important part of London's spatial morphology is its green belt. This blurs the edges of the city and the countryside – a Swiss cheese of a city where outward expansion has captured not only outlying settlements but also farmland, river valleys, heaths, marshes, and commons. London is particularly fortunate in the extent of its public spaces

Figure 25.3 Altab Ali Park, Whitechapel. A space to be enjoyed by all. (*Source:* Author Peter Bishop.)

and the diversity of its public spaces. Names tell us a lot about their different functions, as well as their origins and legal status:

- Hyde *Park*
- Highbury *Fields*
- Wimbledon *Common*
- Shepherd's Bush *Green*
- Kensington *Gardens*
- Cavendish *Square*
- Failop *Plain*
- Hackney *Downs*
- Hackney *Marshes*
- Hampstead *Heath*
- Epping *Forest*
- Highgate *Cemetery*
- Paddington *Recreation Ground*
- Tufnel Park *Playing Fields*

The function of public spaces is complex, and so should be their design. If this sounds like an obvious point to make then it is surprising how few planners, architects, and designers understand public space function. A computer-generated image of happy people drinking coffee, wielding shopping bags, and of children trailing balloons really is not good enough. Public space has to be designed around function – and function that is achievable given the immediate context, surrounding uses, footfall, and the demographic and cultural characteristics of its users. Great public spaces are open stages that invite the individual to use them for purposes that suit them rather than the designer.

Theory and Practice of Public Space Provision

In a European context, much older traditions of the square, piazza, the common, and the municipal park have always been staples of urban design before the advent of mass car ownership. Although mass car ownership had only really been a factor in European cities from the middle of the twentieth century, planning policies centred on the car have distorted urban strategies and led to the virtual destruction of large swathes of the fabric of many cities. Arterial roads and one-way streets severed communities, altering urban streets from commercial and social places to hostile traffic corridors.

A new interest in public realm and urban landscape emerged in the 1960s from a number of practitioners in North America and Europe. Jane Jacob's work in New York[5] focused on the then-radical idea that the centres of cities should be designed around the needs of people, including the concepts of mixed uses and streets that were walkable and safe. She had organized local opposition to protect neighbourhoods from comprehensive "slum" clearance programmes, and in particular attempts by Robert Moses to drive major roads through inner city neighbourhoods.

---

5 Jacobs, J., 1961. *The Death and Life of American Cities*. Random House, New York.

Jan Gehl's work in Copenhagen is another early example of a reappraisal of the role of the car. A programme of pedestrianization had already started with the pedestrianization of the main shopping street, Stroget, in 1962. Gehl's work commenced with studies of public space in 1968 that concluded that such spaces were popular and valued. The amount of car-free streets in the city centre increased six-fold from 1962 to 2000. Importantly, the impacts of pedestrianization programmes and public reactions were monitored over the years by the School of Architecture at the Royal Danish Academy of Fine Arts. This provided a research foundation to continue the initiative. Subsequently, parking was reduced by 2–3% per year as city streets and spaces were progressively turned over to cyclists and pedestrians.

Gehl went on to apply his methodology in other cities, including Melbourne. In the middle of the 1980s the city centre was in decline, but under Gehl's proposals, new housing was built and streets were pedestrianized. Between 1994 and 2004 there was, in consequence, a 39% increase in pedestrian footfall in the city centre and the night-time economy doubled.[6]

Barcelona under its Mayor Pasqual Maragall used public space as a central part of his strategy for urban renewal in the runup to the 1992 Olympics, a strategy taken up by a later mayor, Dr Joan Clos. In his foreword for the book *Global Public Space Toolkit* for UN Habitat, Clos writes: "Public spaces contribute to defining the cultural, social, economic and political functions of cities. They continue to be the first element to mark the status of a place from a chaotic and unplanned settlement to a well-established town or city".[7]

In the UK much of this thinking was encapsulated in the work of the Urban Task Force,[8] which fashioned a set of urban planning values that included environmental responsibility and social wellbeing. It identified "the ecological imperative" as one of the key drivers for changes to urban thinking. Public spaces were seen as not only a "good thing" but also an essential ingredient of urban strategy if the city was to be both environmentally responsible and resilient. In a century where global temperatures are set to rise by at least 2°C and where extreme climate events will become more frequent, public space becomes increasingly important in ameliorating high temperatures and flooding. In denser and more compact cities, open spaces become ever more important if the urban population is to have places for relaxation and recreation (Figure 25.4). The contract between the city and nature is there to be redefined. The Final Report of the Urban Task Force defined urban spaces as a key element of community cohesion, describing their importance thus: "Safe, well managed and uncluttered public spaces provide the vital 'glue' between buildings and play a crucial role in strengthening communities."[9] The report recommended that local authorities establish public-realm strategies that would set out a clear

---

6 Gehl, J., Gemzøe, L., 2000. *New City Spaces, Copenhagen*. Danish Architectural Press, Copenhagen.

7 UN-Habitat, 2016. *Global Public Space Toolkit: From Global Principles to Local Policies and Practice*. UN-Habitat, Nairobi. https://www.local2030.org/library/82/Global-Public-Space-Toolkit--From-Global-Principles-to-Local-Policies-and-Practice.pdf (accessed 14 March 2020).

8 The Urban Task Force, chaired by Richard Rogers, was set up in 1997 by the Blair Government with a remit to rethink urban policy. Urban Task Force, 1999. *Towards an Urban Renaissance – Final Report of the Urban Task Force*. HMSO, London.

9 Ibid., pp. 57–59.

Figure 25.4 Water, play, and delight – open space on London's South Bank. (*Source:* Author Peter Bishop.)

hierarchy of open space provision, management, and maintenance. This was turned into practice in many of the programmes promoted by Design for London.[10]

In the 1980s a group of thinkers and activists began to "rediscover" some of the fundamental concepts of the European city – the street, square, and public space. This approach, led by individuals such as Aldo Rossi, saw the emergence of an urbanism based on the citizen and democratic ideals (values largely ignored by the functionalism of twentieth-century modernism). Colin Rowe further developed the idea of cities as collaged and superimposed places[11] where successful places are the result of a ceaseless process of fragmentation and superimposition over successive generations. The city is a place of constant adaptation and never forms a completed project. Post-war functionalism (zoning and pedestrian and vehicle segregation) was rejected in favour of new urban forms

---

10 Design for London 2006–2013 was the mayor's architectural studio led by Peter Bishop.

11 Rowe, C., 1978. *Collage City*. MIT Press, Cambridge, MA.

centred around the characteristics of historic European urban typologies. This approach developed themes of "careful urban renewal" and "critical reconstruction" that incorporated principles of community participation and engagement.

The new interest in open spaces within and around the city was further developed by Denise Scott Brown and a generation of British urbanists and architects including the Smithsons and Colin Ward. This strand of urban thinking recognized and celebrated the "stuff" of cities – elements that were wider than the conventional building blocks that had been the focus of many mainstream European urban theorists. These elements included industrial sheds, motorways, and shops, as well as elements as disparate as the retail shed, the football pitch, and the cemetery. Their interest also recognized the primacy of the street and both formal and informal urban spaces. This was the perspective of the culture of places – "found" elements in the urban fabric that the urban designer could choose to work with rather than obliterate (Figure 25.5). Deeply embedded in this thinking was an antithesis towards urban design as a mechanism for pursuing "neatness" – theirs was an exploration of the gaps in the planning system that might be filled with new design strategies – to create a "culture of the ordinary".[12]

Figure 25.5 Skateboarding on London's South Bank – a "found" space in the city. (*Source:* Author Peter Bishop.)

---

12 White, D. (ed.), Wibert, C. (ed.), Ward, C., 2011. *Autonomy, Solidarity Possibility: The Colin Ward Reader*. AK Press, Edinburgh.

This approach to urbanism is grounded in careful examination of the ways in which the city functions. It stresses the importance of survey, mapping, and documentation, as well as a thorough understanding of urban form and function. This is an urbanism based on small adjustments rather than grand interventions that are framed by the structure of the city, by its motorways, its suburbs, and its infrastructures. It is an urbanism of negotiation – of understanding that process was a key part of design and that the perfect plan was always likely to be side-lined by the reality of the situation. It was not incompatible with European mainstream thinking at the time, but it did add a dose of pragmatism.

Peopled Landscapes

Around the same time as interest was renewing in city spaces, new attention was applied to urban landscapes. Interest in the relationship between the city and its surrounding countryside was not new of course. There had been a rich history of thought going back to the nineteenth century that had manifested itself in the creation of green belts and buffer zones. Sir Patrick Abercrombie's 1944 London Plan sought to address this by containing London's sprawl within a literal "green belt" of undeveloped land. As such it was intended to meet both the agrarian and recreational needs of the London region.[13] The plan envisaged that the green belt would be connected by "green wedges and parkways" to the central areas of London. Within London, the plan proposed that all existing open spaces should be protected from development, that a variety of open spaces be established, and that a series of parkways should be created to allow residents to walk between the major open spaces, unimpeded by traffic.

In 1977 Oswald Mathias Ungers joined Rem Koolhaas at the Berlin Summer Academy for Architecture and their collaboration led to the book *The City in the City. Berlin: A Green Archipelago*,[14] published the following year. The treatise contained proposals to reconfigure Berlin as a set of "islands-in-the-city". This green archipelago sought to exploit urban depopulation as an opportunity to rethink the nature of urban life. Ungers proposed a set of "urban islands" in Berlin – centres where the nature of urbanity might be reinforced while the rest of the city returned to open nature, becoming a "natural lagoon". The urban environment would thus become a series of interconnected fragments surrounded by open and accessible countryside.[15]

Interest in London had been taken up by David Goode at the Greater London Council (GLC), who in the early 1980s mapped the ecological significance of large areas of London. His work was continued by the London Ecology Unit (LEU), which continued its work until it was incorporated into the Greater London Authority (GLA) in 2000.[16] The LEU

---

13 Abercrombie, P., 1944. *Greater London Plan, 1944*. University of London Press, London.

14 Ungers, Os.M., Koolhaas, R., 2013. *The City in the City – Berlin: A Green Archipelago: A Manifesto (1977) by Oswald Mathias Ungers and Rem Koolhaas [with others]*. Edited by F. Hertweck and S. Marot. UAA Ungers Archives for Architectural Research. Lars Müller Publishers, Zurich.

15 Schrijver, L., 2008. *OMA as Tribute to OMU: Exploring Resonances in the Work of Koolhaus and Ungers*. Journal of Architecture, 13 (3).

16 Although disbanded in 2003 its work informed the 2004 and subsequent London Plans.

developed the concept of "ecological deficiency units" – areas of the city that were more than a kilometre from areas of ecological value. Ken Livingstone, London's mayor at the time, was personally interested in ecology, and this became a theme within the 2004 London Plan. Others were interested in the strategic nature of public spaces in London. Peter Beard was part of a group of thinkers at the Architectural Association that included Lisa Fior, Florian Beigel, and Julian Lewis. They were interested in the "grittiness" of urban landscapes – leftover spaces (*terrain vagues*[17]) that hosted ordinary experiences of people's daily lives. The London Green Belt and its associated metropolitan landscapes were "a great conceptual space without density". These were spaces "of slackness, wildness and escape" and a necessary counterpoint to Richard Roger's notion of the compact city. They explored these largely neglected urban spaces, and one of the results of their thinking was Marianne Christiansen's conceptual piece "Picnics in the Green Belt".[18] This saw the green belt as a neglected space, in many ways a void around the city. They postulated that access and use of the green belt offered an opportunity to incorporate it into the spatial and social life of the city.

Two of Design for London's programmes explored the relationship of public spaces within the city and how these might link outwards to the surrounding countryside. The Mayor's 100 Public Spaces (subsequently renamed the Mayor's Great Spaces programme), set out to create new public spaces within the city. These often opportunistic schemes sought to "carve" new spaces out of the existing fabric of the city: for example, by road closures and the amalgamation of otherwise forgotten and underused spaces into new civic parks. The other programme was the East London Green Grid. In this the many pieces of abandoned and leftover spaces in east London were mapped, analysed, and then linked together by small interventions to create a grid of interconnected space that ultimately joined the open countryside beyond London's boundaries. The Green Grid was a walking and cycling network, an ecological and environmental resource, and a place for recreation and leisure. It also mitigated against flood risk and ameliorated the high summer temperatures of the densest and poorest part of London.

Strategies for New Public Spaces in the City

It is considerably easier to recognize a good place than to understand the factors that make it so. Designing good places is more than playing with building blocks or elements out of a catalogue. Elements that contribute to successful places include the interrelationship of different activities and the provision of physical and social infrastructure. Successful cities are responding with adaptive strategies where agility and resilience are seen as the new hallmarks of stability. City government is redefining itself as strategist, enabler, and facilitator rather than sole provider. While bureaucracies may at times be

17 The term *terrain vagues* was first introduced by de Solà-Morales i Rubió, M., 2008. *Las Formas de Crecimiento Urbano*. UPC, Barcelona.

18 Brearley, M., Christiansen, M., Lewis, J., 1996. *Picnics in the green belt*. www.east.com.uk (accessed July 2020).

efficient deliverers of services, they are seldom suited to dealing with the so-called wicked issues that require interdisciplinary working. This weakness became increasingly problematic in the final decades of the twentieth century. The new issues facing cities were less tangible and discrete and far more interconnected. Economic restructuring, crime and social marginalization, disparities in health, and demographic change began to dominate the agenda. Ageing and ethnically diverse populations pose far more complex challenges around social cohesion. While the agendas might be changing, the components of the city stay more or less the same. It is the function and use of these building blocks that is coming under scrutiny. Public space for recreation, play, wildlife, solitary reflection, and social interaction remains as important as ever and is increasingly at the overlapping points of carefully crafted urban programmes. This is leading to innovation and experimentation.

Understanding the ways in which public space programmes can respond to overlapping agendas is well illustrated by a project carried out by social entrepreneur David Barrie in Middlesbrough in 2008.[19] One of the remarkable aspects of this project that sought to use vacant space in the city centre for food production was the unit of survey. A square metre is sufficient for a window box or planter, and the study area was analysed on this basis – every square metre had functional potential within the design. Contrast this to the ways in which we designed the precious resource of public space in the city and one can usually see space wasted in ill-considered designs. This is not to argue that every piece of a public space needs to have a specific function – but functionality needs to be considered, even if the result is the sublime simplicity of St Mark's Square in Venice. Great spaces such as this act as an unchoreographed stage upon which different activities can take place throughout the day and the seasons. Good public space offers a platform of possibility (Figure 25.6).

Figure 25.6 Public space as a platform of possibility. (*Source:* Author Peter Bishop.)

---

19 Barrie, D., 2008. https://www.planningresource.co.uk/article/775913/it-urban-farming-taps-regeneration-resolution.

Regulatory frameworks form a foundation for urban living but do not of themselves set a clear vision of a future state, the expectations of values and outcomes, and a process to get there. Planning policies to protect open space are essential (and should be non-negotiable) if this precious resource is not to be slowly eroded. But policies are rarely effective in *creating* new parks and public spaces. This requires agency from city government. Traditionally this was through capital programmes (although there were precious few new public spaces built in UK cities between 1980 and 2005 – and nothing on the scale of the municipal parks of the late nineteenth century). More recently, open spaces have been provided through partnership with landowners or through sophisticated arrangements between government and other stakeholders. In this respect a grasp of finance and a sophisticated understanding of process are essential. Increasingly, incremental strategies and "tactical urbanism" are coming to the fore.

The pace of change in our cities is matched by a crisis of government at local level and a growing demand for local empowerment. The intensification and diversification of the use of public space through incremental strategies and "tactical urbanism" reflects this. The unsolicited use of forgotten or marginal public spaces has been widely explored by writers such as Karen Franck and Quentin Stevens,[20] John Chase, Margaret Crawford, and John Kaliski,[21] and Peter Bishop and Lesley Williams.[22] This is an aspect of appropriation and colonization of city spaces that fits well with the fluid world in which we live. Public space is there to be discovered in places of the city where it lies dormant beneath road junctions, in car parks, and along the acres of tarmac that surround housing estates. This is of particular importance in poorer neighbourhoods where gardens are a rarity and there is little access to outdoor space. Temporary interventions to provide new public spaces are easy to realize, they are cheap to build, and they can be the seeds of community activism and confidence. Ideas can be prototyped and risks that would often be deemed too risky under traditional procurement programmes may be taken. Finally, temporary urbanism, if built on local dialogue, can build community capacity. Contrasting examples include the work of Design for London and Muf Architecture and Design in their project "Making Space for Dalston". Here a wide range of incremental projects were developed with the community to create new community spaces in a deprived town centre in inner north London. Completed in 2009, many of these "temporary" projects are still there and valued by local people. An alternative approach was pioneered in New York by Traffic Commissioner Janet Sadik-Khan where she implemented overnight closures of busy city streets, including Broadway, to create temporary open spaces. The programme was instantly popular, and many have since become permanent.

20 Franck, K., Stevens, Q., 2007. *Loose Space: Possibility and Diversity in Urban Life*. Routledge, Oxford.

21 Chase, J., Crawford, M., Kaliski, J. (eds), 1999. *Everyday Urbanism*. Monacelli Press, New York. Crawford, M., 2008a The current state of everyday urbanism. In: Chase, J., Crawford, M., Kaliski, J. (eds), *Everyday Urbanism*, 2nd edn. Monacelli Press, New York. Crawford, M., 2008b. Blurring the boundaries: public space and private life. In: Chase, J., Crawford, M., Kaliski, J. (eds), *Everyday Urbanism*, 2nd edn. Monacelli Press, New York.

22 Bishop, P., Williams, L., 2012. *The Temporary City*. Routledge, Oxford.

Conclusion

You can judge the health of a city by its parks and open spaces. It is not just the great showpieces where a city can show off, it is the wider mosaic of parks, squares, and streets that provide the framework for civic life to flourish. Public space is not a commodity and the market will not provide it (except under very limited conditions). It is *public* – that is, communally owned and maintained for the use and enjoyment of all. It needs to be protected, managed, and cared for. Where it is lacking it needs to be provided, not as a luxury but as a necessity for urban living. At the time of writing, a global pandemic is causing city regions to be put into quarantine with restrictions on travel, major reductions in economic activity, and extraordinary restrictions on our private lives. Many individuals have relearnt the value of public services and community spirit and the basic qualities of urban living such as clean air, parks and open spaces, gardens and balconies, and the absence of the daily commute.

Let us hope that this continues.

26

The Natural City – Introduction

The global pandemic is causing many individuals to relearn the value of public service and community spirit alongside that of clean air, parks, open spaces, and gardens. In Chapter 25 Peter traced the theory and practice of providing public spaces in the city as an essential ingredient of the richness and messiness of the twenty-first century city.

Carlo, in this chapter, *The Natural City*, considers a range of disparate issues – asking how an artificial phenomenon like increasing urbanization, on a planetary scale, can ever be compatible with the natural world. Architectural trends, natural geomorphological forms, and planning issues are seen under the microscope of whether they, in reality, help or hinder cities becoming more natural. Parallels are drawn between human migrations and the associated social diversity these migrations bring to our cities, and how these compare with recent biodiversity winners and losers, as well as how terms such as "unwelcome visitors" enter our language, discussed alongside questions about land-use and food growth.

We are sharing our planet with many others. In Figure 26.1 we see that animal life accounts for 0.4% of life on Earth. Humans account for 0.01%. We wield a tremendous responsibility to preserve and maintain habitats for all living things. As we move towards

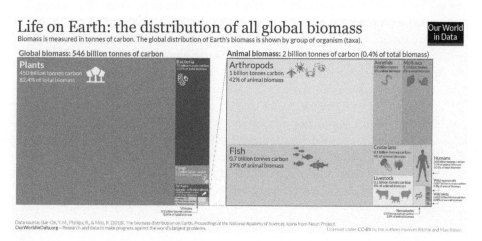

Figure 26.1 Life on Earth: the distribution of all global biomass. (*Source:* Humans make up just 0.01% of Earth's life – what's the rest?, by Hannah Ritchie, Our World in Data, April 24, 2019. Retrieved from: https://ourworldindata.org/life-on-earth. Licensed under CC BY 4.0.)

The Climate City, First Edition. Edited by Martin Powell.
© 2022 John Wiley & Sons Ltd. Published 2022 by John Wiley & Sons Ltd.

2050 and our consumption of resources increases, cities offer the place to house and contain our population's impact on these spaces. We can turn cities into these sustainable ecosystems and allow the vast spaces of the planet to thrive. That is the importance of achieving "The Climate City" – it is already an efficient way of living; we just have to keep going.

In *The World Without Us*, Alan Weisman writes about what would happen to the natural and built environment if humans suddenly disappeared.[1] Written largely as a thought experiment, it outlines how cities and houses would deteriorate, how long man-made artefacts would last, and how remaining lifeforms would evolve. Weisman concludes that residential neighbourhoods would become forests within 500 years, and that radioactive waste, bronze statues, plastics, and Mount Rushmore would be among the longest-lasting evidence of human presence on Earth. It does draw on evidence – flora previously considered extinct being discovered in the demilitarized zone between North and South Korea – demonstrating that humans are pushing many species to the brink of extinction, but without us they can flourish and thrive.

Carlo interrogates the apparent oxymoron of "the natural city" and makes it clear that it is not an oxymoron at all. Highlighting the connections between nature and the city, Carlo looks at what makes a city natural. Using London as a canvas, he not only compares the need for human diversity in cities to our need for biodiversity, but also examines how cities can provide a neutral zone of tolerance and acceptance and mirror a new way for the natural world to follow.

He acknowledges the dangers posed by house pets in urban settings, the education lapse caused by the loss of sight in our cities of farm animals, and the disconnect between what we eat and where it comes from. Carlo also looks at how the UK's obsession with house prices has led to our housing stock becoming less natural, our rising consciousness for the need of broader climate management, and the prioritization of ambitious architectural green projects over preserving existing ones. It is clear that our collective action must prioritize enhancement over mitigation.

As we have seen in other chapters, Carlo highlights the importance of a sustainable public transport infrastructure and the challenges associated with retrofitting, building anew, or missing the boat altogether. He examines how we can tackle other familiar problems such as air pollution through including a greater numbers of trees in our cities and removing asphalt surfaces. He also looks at water contamination, biodiversity, flood resilience, and ongoing sustainable urban drainage (SUD) schemes and discusses the interesting link between improved mental health and promotion of the natural world.

Carlo has provided a fascinating case study on the Woodbury Wetlands in Hackney, London. With help from the Heritage Lottery Fund and Thames Water, the site was transformed from an unattractive functional reservoir to a beautiful nature reserve and tourist destination. By preserving and enhancing, this project proves what is possible for both people and wildlife in urban settings and gives us new insight into how we can envision and create a natural city.

---

1 Weisman, A., 2007. *The World Without Us*. Thomas Dunne Books/St. Martin's Press, New York.

26

The Natural City
Carlo Laurenzi

Official statistics put London's population at around 9 million people, but when one considers the full ambit of living things in the UK's capital, and Europe's largest city, it would probably run into billions if not trillions – it's impossible to quantify such an important thing accurately. As a result, public policy is primarily directed towards human beings rather than what might be good for the planet as a whole. The opposite is also true and just as erroneous: we should not make policy decisions about the so-called natural world without considering the human component, even if its contribution is both negative and seemingly out of control. Sadly, this latter trend is often reflected in, otherwise marvellous, nature programmes on TV by referring to the natural world as that part of our planet which exists without people, and by extension without towns and cities!

The term natural city must surely be an oxymoron. How could something so inextricably artificial like urbanization really be considered as natural? Maybe a better term would be to describe a city as having certain *naturalistic* features. Most cities are not designed, or created, with nature in mind; it is therefore somewhat ironic that most of the current proposals on the table – to improve human life and to adapt for a future, changed climate – require us to be moving away from the artificial towards something more natural! So where does that leave the search for a natural city?

I have heard it often quoted that Venice is a natural city as it was built to sit inside a defensive lagoon, but as beautiful and unique as it is, it has almost no green space, so I contend that it is far from being very natural. Amsterdam with its myriad canals, similar to Venice, is widely recognized as one of Europe's greenest cities,[2] but is it natural? Modern cities carved out of jungles, like Singapore, or the expansion of Bangkok into its surrounding wilderness, whilst photogenic and interesting, does not go to make them natural; indeed these urban developments are replacing what naturally existed with something really quite artificial. China's stalled attempt to create a zero-carbon city close to Shanghai, Dongtan, and Korea's green futuristic city, Songdo, are a step forward in terms of twenty-first-century thinking, but, like similar attempts in the past to "create green", they fail to put biodiversity at their core. Again, a multiplicity of naturalistic features does not make it natural.

---

2 According to the publication *Culture Trip*, 2020. https://theculturetrip.com/asia/articles/culture-trips-guide-to-multi-destination-routes-for-2020/.

The Climate City, First Edition. Edited by Martin Powell.
© 2022 John Wiley & Sons Ltd. Published 2022 by John Wiley & Sons Ltd.

London, as a city, is perhaps the best-known example of human diversity, in that it is said that 250 or more different languages are spoken here every day.[3] As a major urban area, it has drawn in many of the people who otherwise would have felt themselves to be on the fringes of broader UK society. These incoming communities, within the metropolis, feel safer and are able to more easily access services, forms of association, and relationships they might otherwise struggle to achieve in the countryside. London, perhaps uniquely within the UK, is also an overpowering drain on its hinterland, to the point that even some major cities within England also fall into its magnetic pull: for example, the conurbations of both the East and West Midlands. London dominates both England and the UK in the way few other capital cities do. It creates a contradictory dynamic; it draws in resources (e.g. food, talent, and finance), and at the same time creates a haven for groups who feel themselves on the margins of UK society.

Great cities like London can also become a refuge for biodiversity, for example, the water vole, peregrine falcon, and, of course, the controversial fox (Figure 26.2). If biodiversity feels under threat in one place, given time, it moves, if it can, to another. Clearly it's easier for a red kite to move its nest compared to a mature fungal mycelium, which would struggle to move very far at all! The growing biodiversity in our city is in part due to nature being increasingly under threat in the countryside by farmers, transport infrastructure, and estate managers. It is shocking that in the twenty-first century, we tolerate the continuing practice of poisoning raptors, gassing badgers, hunting foxes, and now even the shooting of recently reintroduced beaver!

Figure 26.2 The controversial urban fox. (*Source:* London Wildlife Trust, www.wildlondon.org.uk.)

---

3 http://projectbritain.com/regions/languages.htm.

A city operates to a different set of rules. These creatures are not seen as a threat to our livelihoods; in fact, many new jobs and voluntary initiatives are created on the back of this influx: for example, witness the growth of local nature groups since the 1980s and the expansion of larger wildlife bodies, along with their showpiece facilities, such as the Wildfowl & Wetlands Trust (WWT) Wetland Centre in southwest London. Alongside this *forced migration*, there are new economic and social activities associated with these "flights" from the rural to the urban: for example, conservation volunteering; manufacturers of merchandizing and kit for those NGOs; the numerous refreshment providers in or close to these "wildlife" centres like local nature reserves; and training and skills learning. Think of the thousands of people just in London who are employed in these and related industries. The other side of this push–pull formula is centred on the opportunities some selected fauna have found in making our city their home. Consider the abundance of available food close to or even on our streets. It's estimated that around 10 million tonnes of all food bought in the UK is thrown away;[4] little wonder that fox and rat populations have proliferated. Peregrines have done well nesting on tall buildings, enjoying a veritable feast of pigeon each day, which, in turn, are in town for the food provided through human negligence and waste!

Not all species that could migrate to the city would be welcome. Badgers are naturally more aggressive so would be less appreciated by the public; weasels and stoats are too timid to enter; and deer, which are today found in peri-urban areas, are unlikely to come into the centre. One might be forgiven for thinking that Richmond Park with its gorgeous multitude of deer is actually a country estate, but it's only around 10 km from central London. I have observed the rapid growth of corvids in town – not just crows and rooks, but the more recent ubiquity of the very clever magpie. These relative newcomers must also be contributing to a tiny loss of biodiversity. Peregrines like pigeons; foxes will eat voles; and magpies enjoy other birds' fledglings!

Let us consider the parallels between the thinking around what is indigenous and what is a more recent *alien* arrival, within both the natural and social worlds. Take the example of the east London district of Spitalfields, which since the seventeenth century has been a home to Huguenots escaping France, followed by Jewish refugees from a series of pogroms in Europe, and then various Bangladeshi communities; and today it's a trendy boho district, too expensive for ordinary folk to live there! At which point do these incomers become seen as indigenous? Similarly, some of the other alien species to visit our shores, such as Himalayan Balsam and Japanese Knotweed, continue to be, quite understandably, met with certain negativity, whereas the American grey squirrel (Figure 26.3) and the Himalayan green ring-necked parakeet (Figure 26.4) generally receive a positive press by the public at large. Like the earlier point about human societies, it is difficult to reach a consensus – at a societal level – around what it means to naturally belong in any one town or city? After all, no one seriously talks about deporting anyone of Viking or Saxon ancestry from the UK; likewise, policy-makers are not promoting the idea of exterminating every animal arriving in the UK since the Roman occupation, like the rabbit, edible dormouse, pheasant, or half of our deer species. Alien or not, intended or serendipity, it is often a matter of perspective; after all, one person's wildflower is another person's weed!

---

4 https://wrap.org.uk/sites/files/wrap/food-waste-reduction_roadmap_progress-report-2019.pdf.

Figure 26.3 The American grey squirrel is perceived positively by the public. (*Source:* London Wildlife Trust, www.wildlondon.org.uk.)

Figure 26.4 The Himalayan green ring-necked parakeet is also perceived positively by the public. (*Source:* London Wildlife Trust, www.wildlondon.org.uk.)

The city can create a neutral zone of tolerance and acceptance, where most things can at least attempt to thrive, compared to the monocultures of farmland or the relative human homogeneity of rural areas. Nature in this way imitates human society, rather than the other way round. An impressive example of the opposite must surely be the determination shown by New Zealand conservationists in and around the capital city, Wellington, where they have built a Berlin Wall-style barrier, keeping the newcomers (e.g. cats, rats, and pigs) apart from the now-vulnerable indigenous fauna, in order to protect the latter.

One of the biggest threats to urban wildlife has to be household pets – yes, moggy and pooch (Figure 26.5) In the UK, owning a pet is seen almost as an inalienable right – like freedom of expression. The damage I have seen with my own eyes on nature reserves or even back gardens is horrifying. I can see why the New Zealand government made its biosecurity fence cat-proof. I remain concerned by the impact of faecal toxicity and local authorities' reluctance to confront pet owners. Dogs have a disruptive impact whilst being walked, even on those rare occasions when the animal is on a leash! Cats too have their own unique hunt/catch/kill instinct, which becomes exacerbated when the animal occupies one of the city's micro greenspaces, like a private garden, which later on I will show are essential to the health of the metropolis – thus creating an unnecessary threat to populations of birds, shrews, amphibians, and so on. In Hal Herzog's book,[5] he estimates that cats, globally, kill somewhere between 1 and 5 billion birds each year!

The right to own a pet must be met with the responsibility not to harm other living things, even by accident. There are many examples of the media stoking up fear of urban foxes coming into people's home and so potentially hurting children. There are only a tiny number of verifiable records of fox attacks; compare that to the fact that serious dog attacks in England amount to 250,000 each year.[6] Moggy and Pooch don't always "love" their owners, plus they

Figure 26.5 The moggy and the pooch. (*Source of cat image:* Paul de Zylva, Friends of the Earth. Source of dog image: mlkvous/Adobe Stock.)

5 Herzog, H., 2010. *Some We Love, Some We Hate, Some We Eat: Why It's So Hard to Think Straight about Animals.* Harper Collins, New York.

6 http://sciencesearch.defra.gov.uk.

might also kill something/someone if they perceive it to be a potential threat. Pets are not natural in terms of breed, yet they are ubiquitous in urban settings. Again, Herzog found that compared to a wolf in the wild, a thoroughbred dog was considered to be 70% natural by owners, despite millennia of breeding interventions!

Ownership of a pet does indeed bring joy, companionship, and even a sense of security to many owners; that does not mean, in the future, cities will tolerate pets either on moral or on environmental grounds. In a totally counter-intuitive way, it may well be more acceptable to have a virtual pet or even a robot pet than a real one. The latter already exists – Aibo made by Sony – yet whilst it could be said to be falling into the phenomenological trap identified by the late Umberto Eco and others, in this case the artefact might well be seen as preferable to the highly inbred and destructive nature of today's urban pet!

Domesticated animals of the sort we often see at city farms, like goats, sheep, and chickens, are also facsimiles of their former selves. The sweet calf children like to stroke bears little resemblance to the mighty auroch from which it was bred. Many domesticated animals have been altered either to produce more meat or milk or to be more manageable in terms of size or temperament, thereby facilitating entry to and from our towns and cities in bygone eras. Once upon a time herds were driven on foot into markets across London, but trains and later refrigeration put an end to the practice. Indeed, one can still "technically" drive a gaggle of geese across London Bridge, according to the Corporation of the City of London by-laws. The loss of the sight of farm animals in our cities creates a dislocation for many children about where food comes from; couple this with the virtual disappearance of local butcher's shops on our high streets and it means that children no longer see animal carcasses hanging in shop windows, thereby further loosening the links between animals on the farm and food on the plate. I am reminded of the amusing sketch by Sacha Baron Cohen, as the eponymous Ali G, on discovering the provenance of eggs: *"You mean they come out of the chicken's bum?"*!

Food security and food poverty in towns across the globe have been strong motivators for the creation of numerous urban food-growing initiatives, including large-scale and well-known examples like the impressive array coordinated by Keep Growing Detroit in the USA. Closer to home, the Incredible Edible initiative which began in Yorkshire now has projects across the nation, including various in south London (Figure 26.6).

Whilst horticulture is not nature, many of the processes used by growers can help the general public to gain a better understanding of key concepts essential to the lives of both flora and fauna. An extension of this logic can be seen in the provision of environmental education in our primary schools. Lessons are often accompanied by trips to remarkable facsimiles of the natural world, like Camley Street Natural Park in central London.[7] The vivid nature of these visits means that it's more likely to make an impact upon a young child. Sadly, the scant research that exists does not confirm that these lessons are reflected in later, adult behaviours. This is not an argument for ceasing this aspect of the curriculum; indeed, this author would like to see the bulk of all education delivered outdoors; to which I would forcefully add a caveat to cease the whimsical style that many lessons adopt through their naive descriptions of nature. Chimps don't just eat bananas, they really enjoy baby monkeys!

---

7 Run by the London Wildlife Trust since its inception; opening in 1985.

Figure 26.6 The Incredible Edible London initiative. (*Source:* London Wildlife Trust, www. wildlondon.org.uk.)

In attempting to define a natural city, I am reminded that a large urban area will have a huge diversity of opinions on almost every issue; this is often a direct result of its very heterogeneity. One such fault-line of views is around so-called personal freedoms regarding our own properties. I remember when I was doing some work around disappearing front gardens in London, which in the past were routinely paved over (Figure 26.7), many people got very angry at being told what they could or could not do with their own homes. This work was later eclipsed by the ground-breaking (and painstaking) research published as "*London: Garden City*".[8] It's a complex issue. Let's consider the transformation of people's home and gardens, using London as our example.

When one person paves over their front lawn, data from estate agents tell us that it increases the value of that home over its neighbours; but if everyone on that same street does it, it actually reduces everyone's property value. Quite a conundrum! People want nature close by but not necessarily inside or too close to their homes. This is partly borne out of research by CABE Space,[9] which found that having a good-quality piece of so-called natural greenspace close by is said to increase the value of a London home by around 15%. This only goes to add fuel to the fire which is the UK's obsession with house prices and DIY/property improvement programmes. I contend that this has directly led to our housing stock becoming less *natural* – through roof eaves being blocked up, hedges removed,

8 Smith, C., Dawson, D., Archer, J., et al., 2011. *London: Garden City?* London Wildlife Trust, Greenspace Information for Greater London, and Greater London Authority, London.

9 Commission for Architecture and the Built Environment Space, 2005. *Does Money Grow on Trees?* CABE Space, London.

Figure 26.7 Front gardens paved over. (*Source:* London Wildlife Trust, www.wildlondon.org.uk.)

and green surfaces replaced with impervious concrete and asphalt, thereby removing, at a stroke, nesting sites for birds like house sparrows, let alone the heavy blow to insects and other species. Figure 26.8 shows the reversal of paved spaces in Lambeth.

The impact this practice might also have on broader climate management is almost too obvious to dwell on: increased water run-off leading to localized flooding; diffuse pollution; increased temperatures through the heat island effect; and further losses of habitat. These are the hallmarks of how not to manage climate change at a city-wide level!

Figure 26.8 Depaved front garden in three stages, Lambeth. (*Source:* London Wildlife Trust, www. wildlondon.org.uk.)

Figure 26.9   Wildlife friendly garden. (*Source:* London Wildlife Trust, www.wildlondon.org.uk.)

London has almost 4 million garden plots and about 3,000 parks of varying sizes.[10] The micro is as important as the large-scale site to nature. These small sites act as an essential set of stepping-stones. Each small park or back garden has the potential to become a micro haven for animals on their route between larger, higher-quality green spaces (Figure 26.9). Even if the urban garden has little obvious wildlife, its very existence could well be essential to the city's overall biodiversity and environmental functioning.[11] Think of the additional biodiversity benefit we would gain if each micro greenspace had a small pond?

The micro greenspace, which we have seen is found in abundance in London, has another important function – and this is as true for a rear garden or balcony or a simple window box – that is, to illuminate, educate, and hopefully inspire children. Think of a child growing a single flowering indigenous plant in a window box; the child could learn about the key processes underpinning nature, like the carbon and water cycles, and growth, development, and death. Imagine what a child could learn if they were free to create a haven for insects, put up bird boxes, and plant trees and shrubs? The call for children to be taught more subjects outside will soon become deafening!

Recent trends in architectural forays into nature often feel contrived and tangential to solving the real problems we face in the city. My current bugbear is skyscraper rooftop beehives. In themselves there is nothing amiss, as long as there's plenty of food for the insects; otherwise individual bees could literally starve to death in the process of foraging. Green roofs and walls are appealing (Figure 26.10); sadly, designers and planners often are more excited to be putting a greenspace on a rooftop or on a wall, rather than preserving existing ones on the ground. Yet the ones on the ground are often accessible to all, but rooftops are generally not. Criticisms about elitism and exclusion come to mind. My own (and failed) foray into trying to put a green roof on bus-stop shelters in central London can

10 Smith et al., 2011. Op. cit.

11 Sir Lawton, J., 2010. Making space for nature. Department for Environment, Food & Rural Affairs, London.

Figure 26.10 The importance of a green roof for enhancing biodiversity. (*Source:* London Wildlife Trust, www.wildlondon.org.uk.)

be seen as part of the wider trend to create micro, localized, and somewhat overstylized greenery in the city. Sadly, it is often at the expense of losing existing, quality greenspace. Maybe our collective action should be to preserve and enhance, rather than mitigate and imitate no matter how creative?

Planners and developers within the city often opt for facsimile rather than the object in true form. The impressive metallic "trees of Singapore" might well turn out to be a biodiversity success, though I doubt it. Nearer to home, almost outside of City Hall, a development called More London was brave in creating an attractive water feature, which, whilst an obvious trip hazard, does help cool the black surface at ground level. These same developers chose to erect lollipop-shaped artificial trees rather than plant the real thing. Yes, plastic and metal require very little maintenance, but it provides no biodiversity or climate benefit. The late Umberto Eco, literary genius and semiologist, from Bologna university, warned us about unwittingly *Disneyfying* our world, in various essays introducing readers to the concept of hyper-reality.[12] Developers seem determined to create the "artefact", i.e. facsimile, rather than deal with the more complex real thing. In other words, it's easier to erect plastic trees than to plant real ones. The risk to the city lies in the consequences of these acts of myopic short termism; people will become accustomed to the artefact and may eventually prefer it to the real thing. In an urban setting, this could include artificial turf in gardens and public spaces; wave machines in swimming pools rather than teaching

12 Eco, U., 1986. *Travels in Hyperreality: Essays*. Harcourt, San Diego.

children to swim in rivers or lakes; decking over soil; and so on. Children need to experience the real thing, and to learn that humans cannot control everything and nor should they try. I contend that it would make learning about nature far more interesting.

The challenge for most progressive cities also comes from transport infrastructure. In Europe we rightly cherish what the Americans refer to as mass transit. On the one hand, it has a major social and environmental benefit – like being a low local carbon emitting, speedy, and efficient transport system, thereby keeping thousands of cars off our streets. Yet on the other hand, transport planners are unlikely to knock down part of an existing housing or office development, and most of the world is less keen on tunnelling due to cost, thus leaving greenspace as the obvious target when needing to expand or develop transport infrastructure. Look at High Speed Two, a rail development in England, set to destroy ancient woodland plus a large tract of west London's greenspace, so as to be able to arrive in Birmingham 20 minutes faster.[13] Likewise, the continued expansion of Heathrow Airport, where the argument for it is often represented as air travel is an essential activity for future economic growth, but is our collective, future prosperity really pegged to air travel? It's no good destroying what is good, i.e. ancient woodlands, with an offer to mitigate with an alternative which is new, untested, and often falling far short of the biodiversity value that was unnecessarily lost in the first place! It makes no environmental sense at all.[14]

One of the challenges of designing and creating green infrastructure is that it is invariably tacked on retrospectively. It finds itself in the "it-would-be-nice-to-have" category of various investment programmes. Having said that, some of London's *green pearls* are well located and of a high standard and designation. Take the redefining of the thousands of greenspaces into a single overarching strategy, like the Greater London Authority's (GLA's) *All London Green Grid*.[15] This is an outstanding exercise, consistent with Lawtonian principles of connectivity, which, rather than attempting to create something new, has joined up what already exists. It goes on to prioritize specific pearls, and in so doing it sends a strong message to planners as well as funders around which sites the GLA would most prioritize. The Green Grid not only sets out to define a new connectivity but also has a strong biodiversity benefit, plus an associated social component, wherein the various strands of the *necklace* help connect schools and places of work, thereby aiding the development of routes for walkers and cyclists.

Some cities have serious and major challenges in retrofitting greenspaces and/or green infrastructure. Take an ancient and beautiful city like Rome, where even transport infrastructure takes an age to agree, let alone deliver, due to its precious and ancient foundations. Old cities do not quite so readily make themselves amenable to retrofitting, and maybe we should not expect them to do so. Having said that, Rome is home to flocks of sparrows and starlings that we, in London, rightly envy. The shape of the city can also have an impact on its potential for green infrastructure. Take the German capital Berlin (Figure 26.11), where elements of the countryside enter the city, quite deeply, like fingers of green from its hinterland. Similarly, England's "second city", Birmingham, is located on a series of waterways and transport infrastructure directly into and through the city centre.

---

13 Department for Transport, 2020. Travel times down from 71 minutes to 52! Department for Transport, London.

14 http://www.hs2actionalliance.org/case-against-hs2/environment.

15 https://www.london.gov.uk/what-we-do/environment/parks-green-spaces-and-biodiversity/all-london-green-grid.

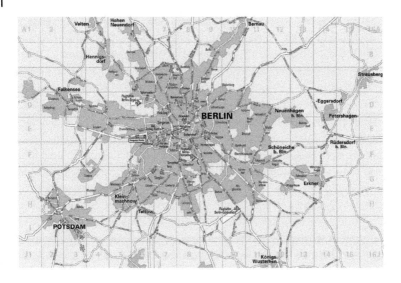

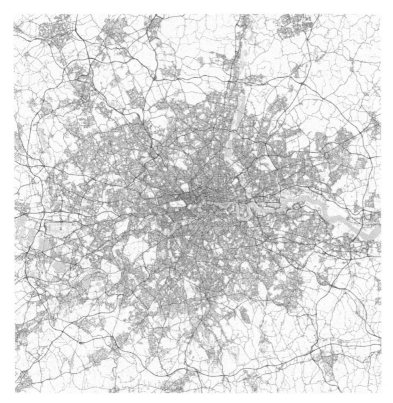

Figure 26.11 Plan of Berlin with fingers of green and Plan of London showing the peripheral transport network. *Source of Berlin image*: Detailed hi-res maps of Berlin, OrangeSmile. Retrieved from: https://www.orangesmile.com/travelguide/berlin/high-resolution-maps.htm. Source of London image: Anton Shahrai/Adobe Stock.)

Compare these two examples with Paris or London, where since the 1800s transport infrastructure has, in effect, cut off the growing city from its greener hinterland. This is further exacerbated by the trend to employ increasingly wider concentric orbital road schemes. It means that larger mammals, like wild boar and badgers, are less likely to find comfortable routes into their respective city centres. It also helps explain why the fox, using railway lines, has conquered an otherwise impenetrable metropolis like London. Wild boar are seen as both interesting and a pest, in equal measure, in the German capital. I doubt Londoners would much relish boar rummaging in their parks and gardens! Outer London boroughs like Croydon and Richmond do have badgers visiting their parks and gardens, and their presence is not always appreciated. Rivers and estuaries within or close to major cities provide another route into central areas. Think of the numerous cetaceans observed entering London as far up as Tower Bridge. Sadly, it's not just due to the improved water quality, but often as a result of marine noise pollution, confusing these wonderful creatures. Having said that, increased numbers of fish have been found in the Thames, right across London, which now hosts a reported 125 individual species.

The world has finally begun to find mental health an acceptable subject for polite conversation. Authors like Richard Mabey[16] have long been outspoken in their promotion of the natural world both as a means to prevent mental ill health and being the best remedy for its improvement. The challenge for the city is how to maximize the biodiversity/green features which could help and/or prevent mental ill health, when there are so many competing demands on healthcare budgets. Interacting with nature is free but only if it is actually present![17] Volunteering on greenspace like a nature reserve or simply strolling through an urban woodland is said to have a dramatic and immediate effect, so why isn't its preservation or even expansion on the agenda?

The recent restrictions surrounding the COVID-19 pandemic have done more for the improvement in air quality than 50 years of campaigning. Londoners have seen an improvement in urban air quality, fewer road accidents involving pedestrians, increased use of bicycles, and, more surprisingly, improved local cohesion. The downside is the associated nervousness of the unseen, the microscopic, and the generally unwanted, such as bugs. London Wildlife Trust (LWT) was successful in the establishment of a water-based project in London's Herne Hill area, called Lost Effra (Figure 26.12).[18] We intentionally chose an area that was both water stressed and flood prone. An exciting and multi-award winning project I helped to design, one of the early issues we had to contend with during the proposed creation of numerous, small, localized sustainable urban drainage (SUD) schemes, within the wider project area, was an objection on the grounds that these would attract litter and insects! Insects are often the cornerstone of a whole series of essential environmental processes, including food for birds and other animals, pollination, and breaking down matter, whether in the wildness or an inner city. We can't do without bugs, yet households appear to spend a good deal of time and money trying to eradicate them. Nature is not tidy-looking and should not be treated as an extension of one's home!

---

16 Mabey, R., 2008. *Nature Cure*. Vintage, New York.

17 https://www.wildlifetrusts.org/nature-health-andwild-wellbeing.

18 https://www.wildlondon.org.uk/projects.

Figure 26.12 Lost Effra water solutions booklet. (*Source:* https://www.lbhf.gov.uk/emergencies-and-safety/floods/living-rainwater.)

The role of SUD schemes in a city like London, where many of its rivers have not seen the light of day for centuries, cannot be overstated (Figure 26.13). The early and pioneering scheme conducted by the Environment Agency (EA) and its then-NGO partner WWT in the Pymmes Brook area of London has set the standard for others to follow. Likewise, work on the River Quaggy's reprofiling shows what can actually be achieved in an urban context, creating a leisure resource, greater flood resilience, and a boon to biodiversity – a living exemplar to demonstrate what's possible.[19]

Pollution affects non-human species as well. Poor air quality and toxic food sources hurt all living things; if we add to this picture the constant sound and light pollution affecting wildlife in our cities, then it's a miracle that our major cities have any wildlife at

19 http://nwrm.eu/sites/default/files/case_studies_ressources/cs-uk-02-final_version.pdf.

Figure 26.13 Rain garden concept using SUDS as the water source. (*Source:* susDrain.)

all. The very recent fall in air pollution in our cities, along with the near absence of cars and trucks on our streets have been a real boost to our urban wildlife, but what will happen to these animals once lockdown begins to end? Fears around SARS, COVID-19, bird flu, swine flu, etc., which are all seen as emanating from animal sources, could create widespread concern among parents, so that in the future they might prevent their children petting a lost baby hedgehog or even rescuing a small bird if they associate these serious diseases with naturally occurring fauna. I doubt many parents will take the risk, thereby depriving the child of an invaluable experience and, possibly, the animal of its life. One of the unexpected debates we have begun to have in the UK is what will happen to the currently empty office and retail spaces in cities like London. Will we head for a "doughnut" entity a little like downtown Portland, Oregon, or convert these spaces into homes for people? If we are serious about welcoming up to 3 million British National Overseas (BNO) people from Hong Kong, then why not use what may well become a near-empty financial district.

We know that cutting the lower branches of urban trees to the 2-m mark, especially in parks, can create the optimal circumstances for reducing ambient overheating.[20] Letting grass grow won't kill anyone, so why do park managers continue to cut grass when it's already desiccated? I challenged a contractor doing exactly this in 2006; his reply was that they were legally obliged through their local authority contract. In marked contrast, a wonderful example of effective management of nature with transport infrastructure, and presumably saving a good deal of money, is where the Highways Agency, which is responsible for main arterial roads in England, has left generous-sized verges to become de facto

---

20 Advice given to the Lost Effra project by Professor Ceda Maksimovic, from his "Blue Green Dream" solutions on urban water solutions. https://www.imperial.ac.uk/people/c.maksimovic.

wildlife meadows. Drivers along the M1 motorway, leaving north London, are treated to a spectacular palette of colour.

Planners, at least since the eighteenth century, have sought to include trees along our streets and in public parks, principally for aesthetic reasons. Some street trees play an invaluable role in moderating temperatures, capturing water, holding onto pollutants, creating nesting opportunities, and even providing food sources. Trees for Cities estimates that in London alone, street trees capture 2.4 million tonnes of airborne pollutants. Our changing climate will no doubt see our street tree strategy replacing iconic but *limited* species like the London Plane with something that might tick more environmental boxes, and possibly not even be indigenous – a difficult discussion to be had.

Initiatives that encourage the removal of hard, dark surfaces such as asphalt and replacing these with something more naturalistic are very likely to have climate-positive benefits, in terms of both mitigation by reducing urban heat island effects and putting us ahead of the curve in terms of adaptation. These simple actions could result in moderating humidity levels, capturing rainfall, creating habitats for wildlife, and so on. The answer is too obvious to dwell on; the issue is one of political choice: do we continue to push economic interests above those of the majority of living things? It's a simple matter. Nature provides us with a broad range of measurable goods: by improving air quality; production of oxygen; absorption of carbon dioxide; pollination; waste management; acoustic management; water control and thereby reducing run-off; and so on. The list is endless, but we continue to not value nature either in itself or for the economic lifeline it provides us, all for free (Figure 26.14)!

Recent pandemic planning suggests that the majority of politicians want to restore the status quo; however, I feel strongly that many people, especially in urban areas, want to see a radical shift in priorities away from private wealth back into local communities. One fork in the road ahead will help us to achieve biodiversity gains and greater local social cohesion, and diverse ownership and control of greenspaces, and to begin to get ready for a changed climate; the other route is back to where we were. It is somewhat ironic that the American economist Galbraith warned us way back in the 1950s that wealth was moving

Figure 26.14 Lost Effra project removing hard surfaces, Southwell Road, Loughborough Junction: before and after shots. (*Source:* susDrain.)

Figure 26.15   The community coming together to reinvigorate public spaces. (*Source:* London Wildlife Trust, www.wildlondon.org.uk.)

out of the public domain into private hands.[21] Maybe now is the time to buck that trend in our cities and towns by returning real wealth and resources to the people and things that actually live in the modern city. The choice about quality of life, improved biodiversity, and better preparedness for a future and changed climate requires political will to make it happen. Time will tell if we choose to improve the planet or fill the coffers of the global elites – we can't do both. Figure 26.15 shows how a community garden can bring people together.

Case Study: Woodberry Wetlands, Hackney, London

Woodberry Wetlands is a beautiful nature reserve in the east London borough of Hackney. Looking at the site today (Figure 26.16), one would be forgiven for thinking that something so *naturally* attractive was always thus. It couldn't be further from the truth.

Background

The site is managed by the LWT and is run in partnership with the private sector utility Thames Water PLC. Much of the detail of this case study comes from LWT itself, who wrote the story used below rather succinctly in their 2017 submission to the CIEEM awards.

Ariel photos show just how grim and functional the site looked when it was a fully operational reservoir (Figure 26.17). When I acquired the site on behalf of LWT in 2007, it was known simply as the East Reservoir, as it was the twin to West Reservoir across the road. The opportunity to acquire East Reservoir almost never happened. At the time, our tenancy at West Reservoir had become untenable for a whole host of reasons, not least of which was that Thames Water, the owner of the twin site, needed to sink a major borehole so as to link the reservoirs with the rest of the water supply network for London. The most feasible spot was LWT's wildlife garden. In return for our *understanding*, Thames Water granted us a limited access tenancy at East Reservoir, and financed a classroom, made up from three shipping containers joined together, at the eastern end of the site. The classroom continues on the same spot and is ever popular with visitors.

---

21 Galbraith, J.K., 1958. *The Affluent Society*. Houghton Mifflin, Boston.

Figure 26.16 Image of Woodberry Wetlands today. (*Source:* London Wildlife Trust, www.wildlondon. org.uk.)

Figure 26.17 Aerial view of the site, circa 1952. (*Source:* Historic England.)

History

"In 1833 the two Stoke Newington reservoirs were built to help supply water, via the New River, to Londoners. London Wildlife Trust first identified the reservoirs' nature potential in 1985, despite both being biologically dead, but subsequent threats to build on them led to a long-running, but successful, local campaign to save them. However, East Reservoir remained closed off behind fencing. To generate interest and support for the site as a potential nature reserve, the Trust established a small community initiative alongside the reservoir in 2007, with support from land owners Thames Water. The Trust's activities at East Reservoir involved many residents of the adjacent Woodberry

Down estate, by then subject to a major regeneration scheme to build 5,500 new homes (from the existing 1,900 homes) by 2031. This led to a close working relationship with Hackney Council, the owners of Woodberry Down, and its developer, Berkeley Homes. In 2010 the Trust drew up an initial vision for transforming the reservoir and integrating more naturally resilient green infrastructure into the estate's regeneration, through the design of swales and "natural green fingers" – the Woodberry Wetlands. Support from Thames Water, the Council, Berkeley Homes and others enabled the design and funding to be secured over 2011–14."

In the early years of our occupation, visitors were restricted to the eastern entrance area and classroom. Under supervision, staff and volunteers worked on conservation enhancements on the northern bank, close to the New River. Within a couple of years, huge areas of reed bed had been built up, making it an attractive site for waders and migratory birds alike.

Associated infrastructure like the Old Coalhouse, with its somewhat toxic history, posed challenges to the redesign exercise (Figure 26.18). Photos of the building today show how it very quickly became a popular café and resting spot on the west bank between the reservoir and the dividing road (Figure 26.19). Building a path around the perimeter was an obvious element to incorporate into the master plan; today it's popular with both joggers and wildlife watchers alike.

A wonderful rethinking of the wetland board walk was developed, whereby the platform supported by pontoons would sweep in an arc-like shape, joining the northwest corner to the southwest. Visitors feel like they are floating above the vibrant wetlands.

The newly established wetlands were designed, in part, to integrate with a large housing development of the same name, delivered by Barclays Homes, one of the main

Figure 26.18   The Old Coalhouse before it was redeveloped. (*Source:* Mathew Frith.)

Figure 26.19 The Old Coalhouse café, boardwalk, and long view of site – 2016 opening. (*Source:* London Wildlife Trust, www.wildlondon.org.uk.)

Figure 26.20 Transformation of the reservoir, now teeming with life. (*Source:* London Wildlife Trust, www.wildlondon.org.uk.)

sources of funding. The other source of funding was the Heritage Lottery Fund. Between them and Thames Water, a total of £1.2 million was found to transform an unattractive, functional reservoir into the go-to visitor destination it is today (Figure 26.20).

Visitors are treated to an array of wildlife – again taken from LWT's documents:

> "At least 50 species of bird now regularly breed on the reserve, including little grebe, common pochard, lesser black-backed gull, sedge warbler, and reed bunting. Seven pairs of breeding reed warbler – a target species – were recorded in the season Woodberry opened, following expansion of the reedbeds. Cetti's warbler bred at the site in 2016, a first for the borough of Hackney, and bittern returned in January 2017."

Figure 26.21 Woodberry Wetlands. (*Source:* London Wildlife Trust, www.wildlondon.org.uk.)

Since opening its doors to the public, Woodberry, hosted 45,000 visitors in its first eight months alone.

The broader environmental benefits of the site have been summarized by LWT as follows:

> "The project embodies and advances the principles of sustainable development. The Wetlands used materials recycled from the site wherever possible, for example silt dredged from the reservoir was used to create the reedbeds (none was taken off-site). The materials used in creating the chestnut and hazel fencing were from sustainable sources and, where possible, recycled from London Wildlife Trust nature reserves across the capital. The volunteer 'hub', an earth rammed green roofed structure, is built from ash felled for chalk grassland restoration at LWT's Chapel Bank nature reserve. Bicycle parking has been installed at each of the two site entrances, and maps indicate travel to/from stations (less than 1 km distant). The site is powered using green energy."

Woodberry shows what can be done with patience, imagination, and a little funding, if the right partners can come together to deliver a radical urban vision for both people and wildlife, in a major city, in the process of adjusting to a changing climate and uncertain global circumstances (Figure 26.21).

27

The Climate-Resilient City – Introduction

In Chapter 26 Carlo raises controversial subjects like children's education and the individual's right to keep pets, along with issues around health, and mental health in particular, as seen under the prism of achieving a natural city. London is used as a canvas to paint strategies and to examine what works and why; and hopefully some of these ideas will have relevance beyond the UK capital. The barriers to achieving a natural city are not centred, for once, around money or technology, but the political and social will to make it happen.

We now begin to pull the story of *The Climate City* together with examples of bold action and progress in Quito, and also we return to the story of Copenhagen told by the Lord Mayor.

In this chapter, Mauricio, the former Mayor of Quito, expands further on Colin and James' chapters, *The Invested City* and *The Financed City*, and the importance of city-friendly access to finance and how cities can finance their climate-resilient infrastructure. He tells the story of the Quito Metro, Quito's first metro, the largest infrastructure project the city has ever developed, and how it almost failed to become a reality because of political rivalries. This speaks to a wider problem that cities face in their lack of direct access to international finance and their dependence on national governments for support to develop urban climate-resilient infrastructure projects. Cities are responsible for 70% of global greenhouse gas emissions, and by 2050 they will hold over half of the world's population. They are consequently best placed to deal with the problem and take effective action. If the international financial landscape for cities is not improved, however, it will be extremely difficult for cities to finance the climate-resilient infrastructure necessary to avert further climate change.

Mauricio calls for a "comprehensive paradigmatic shift", as he takes us through the myriad of finance options available to cities to cover financial gaps and the steps being taken to make them more accessible. Multilateral development banks (MDBs) and other global development finance institutions (DFIs), which have historically been the primary source of infrastructure funding, are taking steps to channel more funding directly to cities. He examines how public financing has a crucial role in leveraging private investment for low-carbon and resilient infrastructure, as well as looking at local government funding

The Climate City, First Edition. Edited by Martin Powell.
© 2022 John Wiley & Sons Ltd. Published 2022 by John Wiley & Sons Ltd.

agencies (LGFAs), and specific instruments such as green bonds, which are fast gaining popularity, with investments reaching US$47.9 billion in the first quarter of 2019.

However, the demand for financing climate-resilient infrastructure must also be there, with cities and local authorities, particularly small cities in low-income countries, in need of a strong business case to secure the necessary funding. These cities often struggle to fund basic services, pushing climate-resilient infrastructure to the back of the queue; moreover, in many countries, public–private partnerships' regulatory frameworks are extremely confusing and complex, preventing cities from receiving investments. As Mauricio says of his experience as mayor in dealing with various city networks, "you get out what you put in". Cities need to employ active involvement and make climate change a priority, whilst local government must enthusiastically embrace the challenge.

Mauricio shows us that innovation is key to strengthening the capacity in cities around financing demand. His detailed examination of the Quito Metro is evidence that a good financial model can turn a highly complex infrastructure project into a landmark, as well as meeting exacting climate standards.

27

The Climate-Resilient City

Mauricio Rodas

The Quito Metro, the first one in Ecuador and the largest infrastructure project the city has ever developed, almost failed to become a reality because of political rivalries with the back-then president (Rafael Correa). The lack of cities' direct access to international finance and their dependence on needing the support of their national governments for that purpose frequently hamper the investment needed to cover the financing gap to develop urban climate-resilient infrastructure projects. Even though there have been advancements in recognizing the importance of timely and adequate finance for cities, a paradigm shift of the international financial architecture, which was designed for nations and not for cities, is urgently needed. Cities are responsible for 70% of global greenhouse gas (GHG) emissions, and by 2050 cities will hold seven out of ten of the world's population, with almost half living in city slums. The UN warned that the clock is ticking as we only have limited years left to take effective action and impede irremediable damage from climate change. Cities play a crucial role and their substantial contribution is now widely acknowledged in the international arena, but if the international finance landscape for cities is not improved, it will be extremely difficult for them to play such a role.

By and large, cities have gained a seat at the global table either through city networks or on their own, acknowledging the fact that the mission of cities is evolving and changing rapidly. City diplomacy's relevance goes hand in hand with the increasingly leading role cities have in tackling global challenges such as climate change, income inequality, migration, and pandemics like COVID-19. Cities are first responders to the world's most pressing issues, and they are the most innovative in adapting to change. Cities have an ever-growing need to partner, learn from each other, advocate and act collectively, make their voices heard, and influence decision-making in the international arena. City networks like the C40 Cities Climate Leadership Group, United Cities and Local Governments (UCLG), Global Covenant of Mayors (GCoM), and ICLEI – Local Governments for Sustainability play an important function in delineating and implementing some of the main global agendas. "Today, their involvement in the COP, the United Nations Conference on Climate Change, their success in adding a territorial dimension to the UN

The Climate City, First Edition. Edited by Martin Powell.
© 2022 John Wiley & Sons Ltd. Published 2022 by John Wiley & Sons Ltd.

2030 Agenda, and their participation in the Steering Committee of the Global Partnership for Effective Development Co-operation, are good examples of how city networks are making their voice heard."[1]

Cities are where international agendas get implemented and where challenges and solutions are more evident, and their proximity and direct engagement with citizens validates the policymaking process; partnership building is more effective and a certain administrative flexibility allows them to act faster than national governments. Cities have new responsibilities derived from global international agendas approved in the last few years that go beyond their day-to-day dealings, like the 2030 Agenda Sustainable Development Goals (SDGs), the Paris Agreement, and the New Urban Agenda, but these international agreements did not contemplate the access for cities to financial resources that will allow their local implementation.

The New Urban Agenda, which was approved in Quito during the Habitat III conference, and which I as mayor had the privilege of hosting, with the participation of more than 500 mayors and local leaders, marked the first milestone in having city officials participating in crafting a UN declaration on urban sustainable development – a key step in the direction of cities raising their voice on the global stage.

The SDGs constitute the international community's effort to safeguard the planet, reduce inequality, and fight poverty. Such a comprehensive agenda needs multistakeholder coordination and involvement, and cities have an essential function in translating and localizing SDG objectives through implementable local policies. Cities are in a unique position and are capable of meeting the SDGs given their proximity to citizens and knowing their territories' strengths and shortcomings; however, a vast majority of cities need to overcome difficulties such as ineffective governance, lack of adequate regulatory frameworks, and data-related issues, among others.

The Paris Agreement brought the world together in the fight against climate change and has the ultimate goal to limit global average temperatures from rising well below 2°C above pre-industrial levels. Nationally determined contributions (NDCs) are at the centre of the Agreement and embrace each country's efforts to reduce carbon emissions. However, countries' best attempts will fall short to deliver on their NDC targets if cities don't take effective action.

According to the Global Commission on Economy and Climate, under a low-carbon scenario, US$93 trillion needs to be invested worldwide in climate-resilient infrastructure by 2030; 70% of this estimate relates to urban areas.[2] To ensure that cities contribute to their countries' abilities to meet the Paris Agreement commitments, they will need to invest a vast amount of money in renewable energy, transportation, water and waste management, green buildings, sustainable public spaces, and other climate-resilient initiatives. An important part of these resources must be channelled to develop resilient

---

1 Fernández de Losada, A., 2019. Rethinking the ecosystem of international city networks. CIDOB, Barcelona, p. 11. https://www.metropolis.org/sites/default/files/resources/Rethinking%20the%20ecosystem%20%20of%20international%20city%20networks.pdf.

2 Alexander, J., Nassiry, D., Barnard, S., et al., 2019. Financing the sustainable urban future. ODI, London. https://odi.org/en/publications/financing-the-sustainable-urban-future-scoping-a-green-cities-development-bank/.

Figure 27.1 The Quito Metro tunnel, which was internationally recognized for its construction quality. (*Source:* Camilo Cevallos.)

infrastructure and programmes to help cities and their communities cope with challenges like increasingly severe natural disasters and extreme heating events. Figure 27.1 is an example of one such project, the Quito Metro (at the early stages of construction), highlighted in more detail throughout the chapter.

In order to be met, all these global agendas require major participation from cities. However, for that to happen, cities need to have direct access to new financial mechanisms without depending on the willingness of their national governments or international finance institutions (IFIs) that have nation states as their stakeholders. It is urgent to make the international financial landscape more city-friendly.

While there is a growing supply of financing mechanisms potentially available for cities, most of them are chained to sovereign debt and creditworthiness and are highly politicized, which heavily delays the financial flows cities need. "Estimates suggest that less than 4% of the largest 500 cities in developing countries are considered creditworthy in international markets, and less than 20% are considered creditworthy in local markets."[3] To finance high-cost infrastructure projects cities need to be creditworthy, but this is a complex challenge for local governments in the developing world. Most countries in the developing world don't even have a credit rating, and private investors and banks

3 OECD, 2018. *Financing Climate Futures.* OECD, Paris. https://www.oecd.org/environment/financing-climate-futures-9789264308114-en.htm.

consider investing in these economies as too financially risky. "One of the most important criteria for good credit is to make sure city revenues predictably exceed costs. So as a first step, the municipal authorities had to know where all the city's money comes from and how fast it might grow, as well as how tax rates are set and taxes collected."[4] For cities to be creditworthy they need technical assistance and capacity building to develop innovative strategies to strengthen municipal finances and attract investment.

The demand for financing climate-resilient infrastructure continues to escalate, as the world becomes more urban.[5] Cities have gradually started to modernize urban policy frameworks, build their capacity for developing "bankable" climate infrastructure projects, incorporate sustainable development planning, and raise their concerns in the international arena. But time is of the essence, and much more remains to be done.

To address the dissonance between the supply and growing demand for financing climate-resilient infrastructure for cities, a comprehensive paradigmatic shift away from the international financial ecosystem and boosting local capacity for receiving financing and executing projects are eminently needed.

Supply of Climate-Resilient Infrastructure Financing

Cities are key to addressing climate change; however, the current financing architecture does not offer cities affordable financing for climate-resilient infrastructure, because the system was intended to finance national governments and not subnational ones.

Historically, global development finance institutions (DFIs) in the form of multilateral development banks (MDBs) have been the primary source of infrastructure funding worldwide. Although they are cooperative banks for a group of countries, the provision of technical assistance for advancing development is still relevant; their contribution to cover cities' financing gap depends on their ability to adapt the way they do business in a world where the decisions cities make today will determine the planet's future.

In order to fill cities' financing gaps, MDBs and other DFIs (such as bilateral cooperation agencies) are taking important steps to channel more funding directly to them. However, subnational governments still face obstacles in accessing these resources. In several countries, cities are not allowed to access international borrowing directly. In many others, cities are required to obtain a national sovereign guarantee or authorization from legislative bodies to borrow, which poses a great risk since political disputes often impede or delay access to credit. Figure 27.2 provides an overview of available finance mechanisms.

Furthermore, public financing has a crucial role in leveraging private investment for low-carbon and resilient infrastructure; and more so in developing countries, where

---

4 Bernasconi, L., 2017. Cities need good credit scores too. Here's why. World Economic Forum, Cologny. https://www.weforum.org/agenda/2017/06/cities-need-good-credit-scores-too.

5 United Nations, 2018. Department of Economic and Social Affairs. https://www.un.org/development/desa/en/news/population/2018-revision-of-world-urbanization-prospects.html.

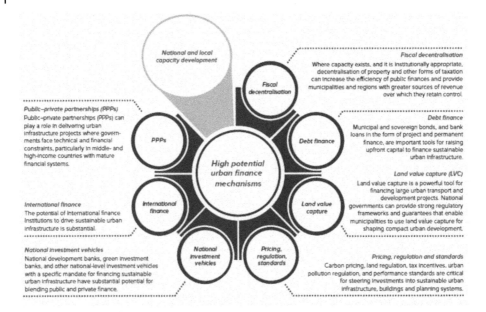

National and local
capacity development

Fiscal
decentralisation

High potential
urban finance
mechanisms

PPPs

International
finance

National
investment
vehicles

Pricing,
regulation,
standards

Land value
capture

Debt finance

Fiscal decentralisation
Where capacity exists, and it is institutionally appropriate, decentralisation of property and other forms of taxation can increase the efficiency of public finances and provide municipalities and regions with greater sources of revenue over which they retain control.

Public–private partnerships (PPPs)
Public–private partnerships (PPPs) can play a role in delivering urban infrastructure projects where governments face technical and financial constraints, particularly in middle- and high-income countries with mature financial systems.

Debt finance
Municipal and sovereign bonds, and bank loans in the form of project and permanent finance, are important tools for raising upfront capital to finance sustainable urban infrastructure.

International finance
The potential of international finance institutions to drive sustainable urban infrastructure is substantial.

Land value capture (LVC)
Land value capture is a powerful tool for financing large urban transport and development projects. National governments can provide strong regulatory frameworks and guarantees that enable municipalities to use land value capture for shaping compact urban development.

National investment vehicles
National development banks, green investment banks, and other national-level investment vehicles with a specific mandate for financing sustainable urban infrastructure have substantial potential for blending public and private finance.

Pricing, regulation and standards
Carbon pricing, land regulation, tax incentives, urban pollution regulation, and performance standards are critical for steering investments into sustainable urban infrastructure, buildings and planning systems.

Figure 27.2 Finance mechanisms with potential to raise, steer, and blend finance. (*Source:* Floater, G., Dowling, D., Chan, D., Ulterino, M., Braunstein, J., McMinn, T. Financing the Urban Transition: Policymakers' Summary. Coalition for Urban Transitions. London and Washington, DC. Available at: http://newclimateeconomy.net/content/cities-working-papers.)

60–65% of infrastructure projects are financed by public resources, while in advanced economies it is 40%.[6]

At the national level, national development banks (NDBs) and funds provide credit (or a mix of grants and loans) to local governments and other institutions investing in local infrastructure. Besides providing credit at lower than market prices, their goal is to improve the effectiveness of local investment, build capacity, and set the stage for independent municipal credit systems. Although this model has worked in some countries, the lack of institutional capacity and financial reliability has often prevented cities from accessing these resources. Similar to the challenges cities face regarding sovereign guarantees, the strong link to national governments has made borrowing from NDBs and funds highly politicized.

Cities can also access financing from local government funding agencies (LGFAs), which are jointly owned by member cities and local governments, whose primary mission is to pool their borrowing needs and issue bonds in capital markets. This model is thriving in the Netherlands and Scandinavia, but it has had very limited acceptance elsewhere.[7] LGFAs advocate for the building of local creditworthiness, help create local markets, and increase transparency in local decision-making.

---

6 Global Commission on the Economy and Climate, 2016. The New Climate Economy Report. https://www.un.org/en/climatechange/reports.

7 Kim, J., 2016. Handbook on urban infrastructure finance. New Cities Foundation. https://newcities.org/wp-content/uploads/2016/03/PDF-Handbook-on-Urban-Infrastructure-Finance-Julie-Kim.pdf

As cities pool their lending needs, investors aim to pool their funds through subnational pooled financing mechanisms (SPFMs), which provide joint access to private capital markets and public sector funding with favourable terms for local and regional government borrowers. Yet, in order for SPFMs to be feasible, they need adequate legal frameworks, upfront payment of project structuring costs, and strong political support.

DFIs and other financial institutions are developing specific instruments for climate-resilient infrastructure projects. Green bonds, in particular, are gaining popularity. Investment in green bonds reached US$47.9 billion in the first quarter of 2019, surpassing the first quarter of 2018 volume of US$33.8 billion by 42%. For 2020, the "credit-rating agency Moody's predicted a record-setting year for green bonds from the outset but had to revise its projection to US$250 billion from US$200 billion after growth exceeded expectations".[8] This suggests investors see these instruments as an alluring option. Nevertheless, local governments encounter cumbersome conditions on borrowing rights and creditworthiness coming from national law which are required to attract private funding.

A smart way to attract private investment, particularly from corporations, is through social impact bonds (SIB). The SIB concept is still new but gaining significant interest in countries like Canada, the USA, the UK, and Australia. "The use of SIBs to finance 'hard' infrastructure projects have [sic] been limited, but they can potentially become an interesting financing source for climate-resilient projects."[9]

Within the myriad of financing possibilities it is important to mention green banks. They offer low interest rates and finance "bankable" initiatives in which debt capital is not available due to market failures. Their success though depends on the savviness of their public and private sector experts. For example, since 1999, the Global Environment Facility (GEF) has invested US$580 million in grants and has leveraged an additional US$7.23 billion in co-financing from the private sector and other sources in 110 cities across 60 countries.

Disruptive proposals are in the making to solve cities' limitations to financing climate-compatible projects. The Green Cities Development Bank (GCDB)[10] is a promising idea being developed by the C40 Cities Climate Leadership Group and the Overseas Development Institute. It aims to be flexible, timely, and entirely focused on financing climate infrastructure in cities and will adapt to fast-changing technology and new lending practices. Nowadays, GCDB is still being discussed, as these models will require further development to become a reality.

Demand for Financing Climate-Resilient Infrastructure

Many cities have shown leadership on climate change, implementing pioneering actions and setting ambitious goals to reduce emissions. However, local authorities need a strong business case to secure funding, particularly intermediary and small cities in low-income countries, where the financing reality and capacities differ greatly from those in developed countries.

---

8 Nauman, B., 2020. Green bonds set to keep flying off shelves in 2020. *Financial Times*, 2 January 2020. https://www.ft.com/content/61631c2c-1a65-11ea-9186-7348c2f183af..

9 Kim, 2016. Op. cit., p. 15.

10 Alexander et al., 2019. Op. cit.

These cities often struggle to fund basic services and have limited fiscal resources to cover infrastructure costs; therefore demand for financing climate-resilient infrastructure is often overlooked. Lack of project preparation capacity and management, limited financial knowledge, lack of creditworthiness, poor access to information, regulatory uncertainty, and corruption impede necessary capital flows from coming.

Furthermore, in many countries, public–private partnerships' regulatory frameworks are extremely confusing and complex, preventing cities from receiving private investment. A comprehensive international benchmark analysis is necessary to develop policy recommendations and legal reforms in this field; international organizations can play a pivotal role in this regard.

Several cities have taken advantage of being active participants in city networks and practise city diplomacy as a way to further build their capacities and make their voices heard regarding city financing for climate infrastructure. As mayor, I learned that you get out what you put into different networks; contributing and having an active involvement will bring great benefits to your city. Thus, the increasingly important role cities have in solving some of the world's most pressing issues makes it impossible for mayors to stay indifferent from global city platforms.

On the local side, cities need to turn climate change into a "politically sexy" agenda for mayors, so they can embrace it enthusiastically. This can be accomplished by strategically communicating about climate change goals and achievements, the urgency for effective action, and how better service provision will improve the constituents' quality of life.

The Quito Metro: A Latin American Case Study

Cities face the challenge to find ways to access financing for infrastructure projects, something in which innovation is key, taking advantage of a much more open perspective from different players to implement new financial models. It was thanks to innovation that during my mayoral term in Quito, the municipality managed to launch and successfully build the first metro line in the city and country (Figure 27.3). The project was financed by the World Bank, the Inter-American Development Bank, the Latin American Development Bank, and the European Investment Bank. These IFIs not only provided loans, but also played a major oversight and assessment role during the entire cycle of the project. The regulations applicable to the project were mandated by the four IFIs, which made the compliance requirements much stricter than what was required by national law. This innovative model and rigorous framework allowed our administration to successfully develop the project, meeting the schedule and budget, something difficult to attain on large infrastructure projects.

The Quito Metro is a fully underground 23-km line with 15 stations, six of which are physically integrated with Metrobus-Q (the bus rapid transit, or BRT, network). It is expected to transport 400,000 people per day, and it will become the backbone of the public transportation system in Quito, integrating five BRT corridors and the network of conventional bus lines. The construction of the metro created 5,000 direct and 15,000 indirect jobs and in the near future will improve access to other employment opportunities, as an estimated 760,000 jobs are located in the zone of the Quito Metro

Figure 27.3 Mayor Rodas reviewing Quito Metro's blueprints when construction began. (*Source:* Camilo Cevallos.)

area of influence.[11] When I left office in May of 2019, this infrastructure project had a 94% completion, and in March of the same year we had the first trials with moving trains, only three years after beginning the construction of the system. The project's main objective is to improve urban mobility to encourage productivity in Quito, serving the growing demand for public transport. "According to experts, the Metro will substantially improve the quality of life of the population, especially the most vulnerable groups, while bringing new sources of economic development around stations."[12] Figure 27.4 shows the drilling machines that made the 23 km of lines.

The Quito Metro is the first in the world being built under a gender-equity perspective; this characteristic is relevant for the city, as one-third of women who use public transportation in Quito have reported some form of abuse. Gender-sensitive design entails appropriate lighting in stations and platforms and child-friendly access for strollers, among others. "Additionally, instead of creating 'segregated' carriages for women and children as done in some other countries, the Quito Metro will have open carriages allowing any person who feels unsafe to walk through to another car. This is expected to reduce the number of abuses by the mere fact that the open space will intimidate an attacker, and

11 Wahba, S., Muhimpundu, G., 2016. World Bank Blogs. Visiting Ecuador's very first metro. https://blogs.worldbank.org/transport/visiting-ecuador-s-very-first-metro.

12 Development Bank of Latin America, 2019. The subway that will change Quito. https://www.caf.com/en/currently/news/2019/04/why-the-new-quito-subway-will-change-life-in-the-city.

Figure 27.4 Drilling machines finishing Quito Metro's tunnel construction, which was internationally recognized for high quality standards. (*Source:* Camilo Cevallos.)

will make it easier for anyone feeling endangered to move cars or solicit help from other passengers."[13] Moreover, the metro system's stations have universal access to people with disabilities; all stops are fully equipped with elevators and ramps and have signs in Braille and tactile surfaces.

The Quito Metro will substantially reduce travel time, reduce traffic congestion, decrease operational costs of the transport service, improve connectivity, security, and comfort of the current system, and reduce emissions of pollutants and GHG. It will capture more than 135,000 tCO_2 per year, becoming an important climate change mitigation action.

"Climate change is one of the most challenging social, environmental and economic issues Ecuador and its capital, the Metropolitan District of Quito, are [*sic*] facing. It is estimated that in the last 100 years, the average temperature in Quito increased by 1.2–1.4°C, causing a significant change in weather patterns and directly and indirectly impacting ecosystems, infrastructure, water availability, human health, food security and hydroelectric generation, among others. In response to current and projected impacts, the municipality of Quito adopted Quito's Climate Change Strategy (QCCS). Quito is one of the few Latin American cities that have [*sic*] a local policy instrument such as the Climate Change Strategy, which together with the use of the city's ecological footprint as a planning tool, placed the city at a vanguard in local responses to climate

---

13 Wahba and Muhimpundu, 2016. Op. cit.

change."[14] The Quito Metro met rigorous environmental standards; for the first time, environmental remediation on aquifers contaminated by gasoline in a major Latin American city was conducted. More than 1 million tons of soil material from the metro excavation was used to shape the landscape of a park located in Quito's former airport, and an old mine was filled and converted into a park using soil waste materials from the excavations. Figure 27.5 shows the trains arriving in Quito and Figure 27.6 the metro trains in-situ.

The total cost of the project was US$2 billion; 68% of the total investment came from the municipality and 32% from the national government. The Quito Metro has the most municipal funding in Latin America. Cost wise, it is the most inexpensive per-kilometre metro in all Latin America. Concessional credits were granted for 25–30 years with a 2–3% interest rate and a grace period of 20-plus years. The conditions for the city's borrowing were extremely beneficial, and they did not compromise investments in other public infrastructure areas. Moreover, the manufacturing of the trains for the metro line was financed by the Fondo para la internacionalizatión de la Empresa (FIEM) fund from the Spanish government, at 0.75% interest rate.

The success on the way we built the first metro line is the result of the innovative way we financed and supervised it, as well as how it became a priority for the municipality. As

Figure 27.5   Quito Metro trains arriving in the city. (*Source:* Camilo Cevallos.)

---

14 Zambrano-Barragán, C., Zevallos, O., Villacis, M., Enriquez, D., 2011. Quito's climate change strategy: A response to climate change in the metropolitan district of Quito, Ecuador. In: *Resilient Cities.* Springer, Heidelberg. doi: 10.1007/978-94-007-0785-6_51.

Figure 27.6   Quito Metro trains. (*Source:* Camilo Cevallos.)

Figure 27.7   An historical day when the Quito Metro made its first trip with passengers (18 March 2019). (*Source:* Camilo Cevallos.)

mayor, I devoted a significant portion of my time to personally orchestrating the project, working hand in hand with different municipal departments, and that sent a powerful message to all my team. It was worth it. The Quito Metro will transform the city forever. It became a flagship project for the banks that financed it, evidence that a good financial model, along with strict international regulations, can turn even a highly complex infrastructure project into a landmark (Figure 27.7).

Conclusion

Cities' shortcomings affect the potentially available supply of financing mechanisms for climate infrastructure, and we can no longer allow this disconnection to happen.

During the past few years, cities have made progress regarding securing new players and financing opportunities, including new facilities, lines of credit, funds, and bonds. However, most cities in the developing world do not have access to these financial mechanisms due to the reasons mentioned in the previous sections. Essentially there is a need to strengthen capacity in cities around financing demand, as well as to foster political and economic inclusiveness for them to access and deploy scale-adequate financing. Innovation is key.

In most cases, mayors and city officials – too busy with their daily tasks – are not even aware of the rapidly changing developments in the financing landscape. Moreover, the existence of a disparity between the financiers' highly technical rationale and language and city halls' political or institutional constraints prevents the transition of ideas into "bankable" projects, updated policy frameworks, enhanced social and economic planning, and improved financial performance.

The COVID-19 crisis has triggered the creation of multi-billion-dollar recovery packages from national governments and new lines of credit from different DFIs and MDBs across the world. A consensus is growing to make these investments climate resilient. Since much of these resources should be disbursed to urban areas because of their relevance to meet with the Paris Agreement and other international agendas, this can become a historic turning point to recognize the need for cities to better access financing for fostering the infrastructure transformation required to cope with the climate change and resilience building challenges. How much of those stimulus packages would be geared towards cities still remains to be seen. Therefore, active advocacy work from mayors and city networks across the world must be strengthened. This can also become an opportunity to elevate the discussion regarding much-needed long-term structural reforms to the international financial system in order to develop more direct financing facilities for cities.

However, if the increasing supply of financial flows available is not matched with the growing demand capacity from cities and their communities, this can turn into a wasted opportunity. We are living in unprecedented times, and now it is more important than ever to support cities, mainly in developing countries, to build institutional, legal, and financial capabilities, as well as to guide them on how to access the most suitable financing options.

Baby steps are no longer an option. The clock is ticking, and we have only a few years left to take effective action and impede irremediable damage from climate change.

28

The Green City – Introduction

In Chapter 27 Mauricio showed how a city such as Quito can combine its governance and leadership not only with finance but with strong environmental performance criteria hard-wired into the project outcomes – a great example of progress to help a city get more people in circulation, safely and affordably, as well as ensuring it does not adversely contribute to global CO_2 emissions.

We now move to Copenhagen, where The Lord Mayor of Copenhagen talks us through "*The Green City*". This chapter is a culmination of many of the other chapters in this book. There are a cluster of Scandinavian cities that are leading the way in achieving *The Climate City*, but Copenhagen remains tenacious in achieving carbon neutrality by 2025.

The Mayor describes the history of this incredible initiative. Beginning in 2009 when the goal was first set – Copenhagen was to become the first carbon neutral capital in the world by 2025 – the bill found support across the political spectrum – unusual, in comparison to challenges faced by other initiatives we have seen, such as the Quito Metro discussed in Chapter 27, *The Climate-Resilient City*. The Mayor looks at how the 2009 Climate Plan sought to implement data, research, and stakeholders, and to include businesses, universities, and research institutes in the planning of action on climate. The city also began developing business plans for 13 focus areas relating to climate change. This groundwork ensured a strong foundation for the current Climate Plan from 2012 which is accompanied by a Climate Adaptation Plan. The latter is to ensure resilience towards current and future consequences of climate change, the necessity of which was made clear by the cloudburst of 2011 which cost Copenhagen €1.4 billion in damage.

The results, challenges, and opportunities posed by these steps have been numerous. Copenhagen has reduced its carbon emissions by close to 65.5% from 2.3 million tCO_2e in 2005 down to 736,000 tCO_2e in 2020. At the same time the population has grown by approximately 10,000 citizens per year, proving once and for all that a sustainable and climate-friendly city does not have to compromise growth, welfare, or liveability.

The Mayor looks in detail at some important factors in maintaining this balance. How the city has tackled energy consumption by the promotion of energy saving through renovation and new construction by formulating partnerships with private real-estate owners to increase the city's control over building stock. The transformation of energy

The Climate City, First Edition. Edited by Martin Powell.
© 2022 John Wiley & Sons Ltd. Published 2022 by John Wiley & Sons Ltd.

production to green and sustainable alternatives, as well as the investments in renewable modes of transport such as cycling pathways, which we read about in Chapter 6, *The Governed City*, and Chapter 12, *The Agile City*. The Mayor explores city administration initiatives and the work they do to inspire and empower Copenhagen's citizens, and, going into the specifics of climate adaptation, and shows how Copenhagen has inspired other cities with its cross-municipality approach and close cooperation with suppliers and companies to implement its projects.

Reaching the 2025 goal is not enough, however, and, having caught us up to date, The Mayor shares the city's plans for the future – the challenges posed by consumption, plans to counter it, and the need to support development of new technical solutions, to name a few. Copenhagen has proved that a city can work across its many sectors to pioneer a climate neutral network and remain liveable. It is a shining example to the world, but in order for there to be global change, international cooperation and networks are essential. Perhaps the most significant point in this green growth agenda is that it will not increase inequality and all Copenhageners will be part of this journey.

28

The Green City

Sophie Hæstorp Andersen

The history of climate action in the city of Copenhagen really took off in 2009 when Copenhagen began its journey towards the goal to become a climate neutral city. Denmark was the host country of the COP15 in Copenhagen, and the City Council in Copenhagen was eager to seize this momentum to accelerate a green transition. The result was the adoption of a political goal that was both ambitious and long-term: Copenhagen would become the first carbon neutral capital in the world by 2025.

Despite being one of the most ambitious climate goals at the time, the bill found support from across the political spectrum from the beginning. Led by the former Lord Mayor, my party colleague Ritt Bjerregaard (Social Democrat) and the other six mayors in the City Council, this was not easy work. The aim of the Climate Plan from 2009 was to reduce carbon emissions by 20% from 2005 to 2015. The plan also formulated a vision for carbon neutrality in 2025, and in this way the first steps to become carbon neutral by 2025 were taken. By "carbon neutrality" we mean that the carbon emissions from our city match our activities that reduce the emissions through sustainable and green energy production and afforestation. We measure carbon emissions as direct combustion (scope 1) or consumption of grid-supplied electricity heating and cooling (scope 2) in the plan.

The Climate Plan of 2009 followed previous political processes with several analyses on the future impact of climate changes, possible solutions, and economic analyses of these. This comprehensive analysis provided solid foundations for making informed decisions and thereby increased the possibility of reaching the set climate targets.

The concrete initiatives have been refined over the years, but the principle remains the same: in the city of Copenhagen we consciously seek to incorporate data, research, analysis, and stakeholders in the planning of action on climate. We work with businesses, universities, and research institutes and have done so from the very beginning. This approach ensures that our Climate Plan and actions enjoy support from key players in the city and that our initiatives work to reduce carbon emissions and make a significant difference. All of this takes Copenhagen a step closer to the goal of carbon neutrality each and every day.

The year 2009 was not only a remarkable year for Copenhagen, it was also the year in which cities across the globe strengthened their role in the climate fight. In cooperation with C40 and ICLEI – Local Governments for Sustainability, Copenhagen organized a parallel city summit to COP15, attended by mayors from all over the world. While negotiations between the national governments were failing, the cities discussed implementation

The Climate City, First Edition. Edited by Martin Powell.
© 2022 John Wiley & Sons Ltd. Published 2022 by John Wiley & Sons Ltd.

of concrete climate solutions and actions. Thus, COP15 was the beginning of what was proven at the C40 World Mayors Summit in Copenhagen 10 years later in 2019, that *nations talked, but cities acted.*

In 2010, the city had already taken its first steps towards being carbon neutral by 2025, but there was still a need for further actions to stay on track to reach that goal.

That same year, the city began developing business plans in 13 focus areas in relation to climate actions. The business plans contained suggestions of goals and milestones for the potential for reducing carbon emissions, the need for investments and funding, the different relevant actors and their roles, the perspectives for green growth, and a risk assessment. The focus areas covered from existing and new constructions, smart city, solar cells, and windmills, to biomass and waste, public transport, bicycling and mobility planning, and the city's own climate initiatives. The plans were developed in close partnership with more than 200 stakeholders from businesses, universities, research institutes, and other organizations, as well as the national government. These stakeholders were invited in to participate in developing analyses, business strategies, and policies, which ensured an all-round perspective and sharing of valuable knowledge, while at the same time strengthening the private–public network and cooperation within the city of Copenhagen. The work ensured a strong foundation for the current Climate Plan from 2012, which replaced the 2009 plan. I am still proud that this new plan was unanimously adopted by the City Council in August 2012 and widely supported by partners and stakeholders across the city. The plan is built on four pillars with concrete political targets: (1) energy consumption, (2) energy production, (3) mobility, and (4) city administration initiatives.

The Climate Plan from 2012 does not stand alone but is accompanied by a Climate Adaptation Plan. Whereas the Climate Plan is focused on reducing carbon emissions, the aim of the Climate Adaptation Plan is to ensure resilience towards current and future consequences of climate change. In 2010 and 2011 Copenhagen experienced massive rainfalls, with more than 100 mm of rain falling in just 1 hour. The cloudburst in July 2011 was the most expensive insurance event in Europe that year, and costs were estimated to exceed €1.4 billion in Copenhagen alone. In addition, it greatly affected many Copenhageners whose homes and belongings were flooded and damaged, which was highly disturbing to witness.

From then on, it was clear that the city needed to invest more heavily in climate adaptation initiatives to secure the city in case of future heavy rainfalls and storm surge flooding. Together with rest of the City Council we adopted a Climate Adaptation Plan in 2011, followed by a Cloudburst Plan in 2012. The vision behind the plan is to invest in climate adaptation and make the city safe to live in, instead of running the risk of having to recover from severe damages. We quickly learned that this approach is also the most beneficial financially.

Both the Climate Plan, the Climate Adaptation Plan, and the Cloudburst Plan required difficult large-scale political decisions as we faced the massive challenges of climate change. But Copenhagen has shown that it is possible for cities to reduce carbon emissions while at the same time enjoying green growth, and without extra costs for the average citizen. By working towards a carbon neutrality goal, we have created new solutions, new green jobs, green investments, and new habits among citizens. The result is a city that is greener and more "liveable" than ever before. The approach has made Copenhagen world-renowned for its green ambitions and front-runner position, just as it has made me proud of my city. Figure 28.1 illustrates the liveable city.

Results, Challenges, and Opportunities

How Far Have We Come and How Did We Get There?

Today, cities consume over two-thirds of the world's energy and account for more than 70% of global carbon emissions. By 2050, nearly two-thirds of the world's population will live in cities. In Copenhagen, it is the city administration that is responsible for the city's development and planning. As the city government, we are closer to its citizens and responsible for public transportation, public meals, public schools, etc. Therefore, cities like the City of Copenhagen have an essential role and a responsibility to act and fight climate change.

In Copenhagen we have reduced our carbon emissions by close to 65.5% from 2.3 million tCO_2e in 2005 down to 736,000 tCO_2e in 2020. At the same time, the population of Copenhagen has grown by about 10,000 citizens each year. This means that the emissions for each Copenhagener in the period were reduced by 73%. Copenhagen is an example of how a sustainable and climate-friendly city does not have to compromise growth, welfare, or liveability. Copenhagen has several times been ranked as one of the most liveable cities,

Figure 28.1 The Liveable City. (*Source:* Sergii Figurnyi/Adobe Stock.)

and our social economic index has increased. At the same time, we have won several C40 prizes for our Climate Plan, our Climate Adaptation Plan, and our energy monitoring system. This work has not the least been championed since 2010 by my predecessor, former Lord Mayor Frank Jensen, with the result that today we have an extremely strong foundation to build our further ambitions upon.

As previously described, these foundations are the result of an extensive, coordinated, and long-term effort working in close partnership with city stakeholders. But Copenhageners themselves also have to be considered as important stakeholders in this matter. They use bicycles to get to and from work and schools, they save the energy in their homes and workplaces, and they sort their waste. They are concerned about climate change and demand actions. A survey from Rambøll in 2019 showed that the majority of Copenhageners are more worried about climate change than they were 5 years before. They demand political actions and are willing to pay more for services such as waste collection and driving their cars in the inner city. They are also happy to invest in local energy solutions such as solar cells and offshore wind-farms close to Copenhagen. The green transition is a shared ambition that requires collective effort. It is a journey that we are on together.

The four pillars in the climate plan (energy consumption, energy production, mobility, and the city's own initiatives) are the key elements of the green transition happening in Copenhagen. In addition, our political initiatives are continuously evaluated and optimized as we gain more knowledge and develop new technologies that push the limits of what we can do.

Energy Consumption

It is not possible to reduce our carbon emissions without talking about energy. In our thinking, "energy" includes energy consumption (the need to reduce energy use) and energy production (the need for sustainable and green production of energy).

The overall objective for energy consumption is simple: to save energy and thereby reduce carbon emissions. This is happening through efficient operation of district heating and promotion of energy savings in renovation and new construction. In Copenhagen, we own around 5% of the building stock. Moreover, we support private renovation through building renewal funds and renovation of social housing. Social housing is estimated to account for about 10% of the total building stock.

But this is not enough. It is essential and important that we cooperate with private real-estate owners and developers to reduce energy use. Thus, in 2016 the city established a partnership called *"Energy Leap"* with the city's largest private real-estate owners. This partnership is built on voluntary agreement between 43 partners and counts on reductions of energy consumption through knowledge sharing, benchmarking, and branding. In 2020, the partnership represented 9.5 million m^2, corresponding to about 26% of the total building stock in Copenhagen. This area includes all types of properties, and a part belongs to a large number of smaller cooperatives and owners' associations. The goal is to reduce the energy in the buildings by 3% the first year, 6% the second year, and 9% the third year after the buildings become a part of the partnership. The goal is to expand the partnership to 750 buildings in the benchmark.

Energy Production

We have also forged partnerships to strengthen green and sustainable energy production. Nordhavn, a brand-new district by the Øresund strait being built from scratch, is an example of how we continuously seek to explore sustainable ways of energy production in Copenhagen. It represents a unique opportunity for us to create a full-scale urban lab. We have asked ourselves questions such as, is it possible to store renewable energy? Is it possible to heat homes with surplus heat from a supermarket? Is it possible to charge electric vehicles while the price of electricity is low? How can we store energy? Through partnerships with relevant stakeholders[1] we have addressed these questions and now seek to make Nordhavn a sustainable district where every resource is exploited to the absolute maximum.

The transformation of energy production from fossil fuels to green and sustainable alternatives and a better use of the city's resources such as waste management are central elements for us to reach carbon neutrality in 2025. Our citizens sort bio-waste, and the resources are used "smarter" as the bio-waste is then gasified and turned into energy or used to fertilize organic agriculture. It is absolutely critical that we replace coal, oil, and natural gas with wind power, sustainable biomass, geothermal energy, and solar power. Green energy must be as inexpensive as possible in order to keep costs down for our citizens and at the same time maintain security of supply.

One key feature of the change we have seen is that 99% of households are now connected to the carbon neutral district heating system. Another is the strategic effort to expand the use of energy from wind and solar cells with HOFOR, Greater Copenhagen Utility. HOFOR is owned by the city of Copenhagen and the other municipalities around Copenhagen. The goal is to build at least 560 MW of wind power before 2025. Efforts to reduce carbon emissions from energy production will account for 80% of the total reduction in 2025. In 2010 the production of electricity and heat in Copenhagen was the largest source of carbon emissions. It accounted for 1.6 million tCO_2e, 76% of the overall carbon emissions in the city. I am proud to say that with reductions in energy consumption and investments in renewable energy production, the emissions from energy consumption were reduced by 1,35 million tCO_2e in 2020, equal to 38% of the overall carbon emissions in Copenhagen.

Mobility

Green transport is key to our green transition. Our goal is that 75% of all trips in 2025 should be by foot, bicycle, or public transport. Since 2009 almost €270 million has been invested in bicycle infrastructure, so that Copenhagen now has 382 km of bicycle lanes which the people of Copenhagen use every day. To put this into perspective, Copenhageners now complete the combined equivalent of 400 Tour de France stages every day. In Copenhagen, 73% of parents of school children report that their children cycle or walk to school, and 42% of all trips to work or education in Copenhagen is by bike. The investments in bicycle infrastructure have meant that it is safer to cycle around our city. Today, three out of four Copenhageners feel it is safe to bike around Copenhagen. This is reinforced by the fact that cycling is the fastest, healthiest, and most climate-friendly choice.

---

1 The stakeholders are (besides the city of Copenhagen) the Technical University of Denmark (DTU), CPH City & Port Development, HOFOR (Greater Copenhagen Energy Utility), Radius, ABB, Danfoss, Balslev, CleanCharge, Glen DImplex, METRO THERM, and PowerLabDK.

Green mobility also means investments in sustainable public transport, and the ambition here goes beyond reducing carbon emissions. As Lord Mayor of Copenhagen, I strive towards a city with air that is as clean as possible, so that our citizens can live and breathe freely. Each year around 440 citizens die prematurely from the effects of air pollution in the city of Copenhagen. That is 440 too many. To measure pollution in the city, we established a partnership with Google. From 2019–2021, a Google car has measured the air pollution while mapping the streets as a supplement to our stationary meters. The data and the footage from the car can collectively teach us more about the sources of the pollution. In this way, we can identify the most effective efforts to reduce air pollution in the city. This has created a more detailed overview and met the tremendous need of citizens, researchers, and city planners to obtain hyperlocal air pollution data.

To reduce air pollution in the city and reduce carbon emissions we invest in sustainable and climate-friendly public transport. Since 2010, we have invested in buses that run on sustainable biogas and a new metro system covering the whole city. The most popular bus route, 5C, has been powered by sustainable biogas since 2017. In June 2020 we introduced the first water buses driven by electricity to routes in our harbour and now all water buses are driven by emission free fuels.

In 2019 the new metro circle line opened. When citizens choose the metro instead of their cars, it makes the city greener and the air cleaner. It brings Copenhageners closer to each other and saves time spent travelling. In January 2022, 1 billion passengers had travelled with The Copenhagen Metro since its opening in 2002.

Moreover, together with 26 other cities in Denmark, we have signed a climate agreement with the national government. For Copenhagen, and many of the other cities, this means, that all new buses will be run on sustainable electricity and hydrogen from 2021. From 2025 it will not only be the new buses – but all buses – that will run on electricity. This will result in a reduction of particle pollution in the cities and make the air cleaner. I hope that

Figure 28.2 Bicycles in Copenhagen. (*Source:* Sergii Figurnyi/Adobe Stock.)

this initiative can inspire other cities to do the same. More sustainable transport will support even more active transport in the city.

City Administration Own Initiatives

As the city administration, it is essential that we set a good example and show that a green transition is possible. Our vehicles, buildings, and consumption should be as green and sustainable as possible to inspire private actors to make green choices as well. When a public health visitor visits a family or the social and health assistant visits an older resident, they drive on green fuel. Today, 93% of the city's owned cars are powered by electricity or hydrogen. When we as the city procure for €1.2 billion each year, issue tenders, and gives permissions to use the city, this is caveated with climate, environmental and social demands. It is, for example, illegal for public events in the city to use disposable plastic cups. Children and young people in our schools are educated in climate knowledge and learn action skills with the goal to inspire to more sustainable behaviour. Since 2009 we have trained around 550 young pupils to be climate ambassadors in our city. They have gained extensive skills and knowledge on sustainability and climate, with the goal that they will act as climate ambassadors both in schools and within their families and so encourage green behaviour. Our climate ambassadors were also an important part of our Youth Summit at the City Hall held during the C40 World Mayors Summit in October 2019. We gathered more than 100 young climate activists from around the world to discuss how young people can participate in cities' climate actions.

We also set a good example by saving energy in our public buildings. In 2017, Copenhagen won a C40 award for our strategic energy monitoring, management, and renovation system in our schools, nurseries, care homes, office buildings, and other public buildings. We are the first city in the world to have developed a monitoring system, where we can read the consumption of electricity, water, and heat on an hourly basis and continuously adjust and improve energy consumption in our buildings. This is key to reaching our goal of being a carbon neutral city in 2025. It is also a big focus in our close cooperation and partnerships with both Beijing and Buenos Aires. Box 28.1 highlights both the scale and opportunity for buildings in Chinese cities.

Box 28.1 Key facts about energy consumption in buildings in Chinese cities

The potential for reducing energy consumption in buildings in Chinese cities:

- China's building sector energy consumption accounted for 46% of the country's total primary energy consumption and 51.3% of carbon emissions in 2019 according to the China Building Energy Consumption Annual Research Report issued in 2021.
- Over 63% of China's building stock is in cities; this is predicted to reach 80% by 2050.
- With continuous urbanization, China's annual new construction area is about 2 billion square meters, which means GHG emissions in the construction sector will continue to rise in the coming 10 years.
- If the existing building energy efficiency standards and technologies remain unchanged, the carbon peak time in the building sector will be around 2038, which will pose a huge challenge to China's 2060 carbon neutrality goal.

As the city of Copenhagen, we also participate in a range of innovation and development projects with businesses, universities, and research institutions to develop new green solutions that are essential to reaching our goal of becoming carbon neutral. One example is our partnership with the global light and traffic company Citelum on LED street-lights in the city. Copenhagen was one of the first cities to use intelligent and energy-efficient LED street-lighting. Connected to a wireless control system, our street-lighting automatically adjusts brightness to traffic flow, and the system is ready for integration with future communication. This contributes to our carbon neutrality aims and at the same time makes our city traffic safer.

Climate Adaptation

As mentioned, our climate action policies are bilateral. They focus both on reducing our carbon emissions and ensuring that the city is prepared for the consequences of climate change. Sankt Kjelds Plads and Enghaveparken in Copenhagen are two examples of how we work on climate adaptation. The aim of the new climate neighbourhood Sankt Kjelds Plads in Østerbro is to protect the area in the event of massive rainfalls and at the same time be a social and green urban space where local people can hang out. The same approach is applied at Enghaveparken in Vesterbro, which the city, together with the city's energy utility HOFOR, the Danish philanthropic association Realdania, and the architect company Tredje Natur, have modernized. The park is designed so that water is collected after heavy rain and cloudbursts, rather than ending up in the citizens' basements. The renovation has also meant that the park is now seen by local citizens as more safe, welcoming, and fun. This way of working to adapt to climate change serves as an inspiration and blueprint for initiatives in megacities around the world. The city administration of New York has been inspired by how we in the Copenhagen have developed a plan for the whole city. They have been inspired by the way we work across the municipality and in close cooperation with suppliers and companies to implement the projects. We also assist Buenos Aires with technical advice to develop a climate adaptation plan through our three-year partnership with Buenos Aires.

The Climate Adaptation Plan was developed in close cooperation with our national government. Copenhagen prepared detailed analysis on the financial gains from investing in a Cloudburst Plan. This thorough preparatory work helped convince our national government to change the national legislation and helped finance the city's Cloudburst Plan. This is an example of how cities can successfully further the green agenda as well as push for governments to take responsibility and form and adopt ambitious new legislation on climate action. It also shows that cities should not wait for national governments to become proactive on climate actions, but they can and should lead the way as green front-runners. Through our membership of the global city climate network C40, we use our collective voice to engage in political discussions at both regional and national level, ensuring ambitious climate action and at least ensuring that we as mayors have the proper legislative framework to implement effective solutions.

When in 2009 we agreed the goal of being carbon neutral by 2025, we were one of the first cities in the world to work with such an ambitious and long-term goal. But we were also setting the stage for green and sustainable transition in the cities. This has made Copenhagen a leader in the field. But we have not reached our goal yet, and the reality of climate change calls for further actions. We have the potential to reduce carbon emissions in relation to transport and energy consumption in buildings, and we have funded two reviews on how we can reduce the carbon emissions from transportation in the city and one on how we can expand our effort on wind turbines.

Conclusion

What Are Our Next Steps? What Will Happen after 2025?

In 2025, Copenhagen will be a city with cleaner air, less noise, energy-efficient buildings, and green transportation. But the green journey is not over. Our activities and the development of new initiatives will not stop in 2025. The 2025 goal is merely a stepping-stone in the development of the city as we head towards 2050. In the energy sector we will work to transform and develop our large-scale utilities and invest in energy plants with a lifespan of 30–50 years. Our political vision is to remain one of the greenest cities in the world. This means that myself and my colleagues in the City Council will face difficult political decisions, as we must all change our behaviour and adapt to an even more climate-friendly lifestyle.

We will need to consider consumption in our way of thinking about carbon emissions. This includes carbon emissions from what urban businesses and citizens use, eat, and wear. It goes further than the traditional way of measuring carbon emissions (scope 1 and 2) and shows that cities, including Copenhagen, have an even more important role to play in the green transition. Carbon emissions, including consumption, will be much higher than measured today in a city like Copenhagen. We need to reduce emissions from the city's procurement. When we buy nappies, pens and food it must be green and sustainable. It should not only be the price, which is essential – but also the climate impact. This spring (2022), the city council therefore adopted that we need to have a climate budget. It is essential that the yearly budget not only has an economic bottom line but also has a climate bottom line and that we know about the climate impacts from the decisions of the budget negotiations. The economic bottom line can change year from year – but the climate bottom line needs to be reduced.

Last year, the city council agreed that the new climate plan after 2025 should aim towards including a goal on reducing the emissions from the citizens comsumption with 50% in 2035 and reducing the emissions from municipal consumption with 50% in 2030.

We have already taken the first steps to include consumption in our work to reduce the carbon emissions in the city. With the introduction of our Food Strategy adopted by the City Council in 2019, we are making public meals greener and more organic for our citizens. In Copenhagen we serve around 70,000 public meals every day in schools, day-care centres, and care homes. The goal of our strategy is to reduce carbon emissions from our public meals by 25% by 2025, reduce food waste, and continue working to make our meals healthy and 90% organic. Our ambition is to inspire our citizens to buy and eat more green, healthy, and local food. For example, we currently educate pupils in 15 food schools around the city on how to cook plant-based and healthy meals.

Today, 30% of global emissions come from food systems. Without substantial changes, greenhouse gas emissions from the food sector will increase by 38% by 2050. Most of the world's food is consumed in cities (80% of all food is expected to be consumed there by

2050) and cities are therefore a major player when it comes to reducing the emissions from our food system.

We also need to look at consumption from construction. Together with Oslo and other cities, we have the ambition to reduce GHG emissions and air pollution from construction sites and civil works. Through our regulations and policies, we began to purchase biofuels and emission-free machinery for our own use in our cities in 2020. Moreover, we as cities demand fossil- and emissions-free solutions in public procurement and city-supported projects. We now demand it in pilot projects and we hope to demand it as a general demand from 2025. Not only will this together reduce carbon emissions, but also it will improve air quality, noice reduction, and create opportunities for new jobs for our citizens. Research shows that the construction industry has the potential to cut emissions generated from buildings and infrastructure by 44% by 2050. In the new climate plan after 2025 we are striving towards including a goal on maintaining and sharpening the goal of climate neutrality and a goal of being climate positive in 2035 including phasing out biomass in 2035. This means that the city in 2035 will remove more CO_2 than we produce.

We in Copenhagen will continue to support the development of new technical solutions. This includes solutions to capture and store carbon (carbon capture) and solutions to use wind-power in the production of sustainable fuels (power-to-X). At our waste treatment utility Amager Ressourcecenter (ARC) in Copenhagen we have now developed and launched a demonstration project with the aim of capturing and storing carbon in the future. This technology is called carbon capture. The ambition is to have a full scale carbon capture facility in operation in 2025. In a large facility we expect that waste incineration will be rendered carbon neutral by capturing the carbon from the fossil fraction of the waste. This will supplement other current initiatives on sorting fossil waste fractions.

Copenhagen needs to continue to be a responsible and pioneer city and be a green role model for cities around the world. Increased globalization means that cities will have a more important role to play as global actors in the climate fight. As cities, we will need to incorporate more with other cities and businesses to transform specific sectors, to adopt to more green energy, and to develop future sustainable solutions. International cooperation and networks are essential if we want to accelerate the green transition. Copenhagen cannot do it alone. We need cooperation with other cities to learn what we need to continue being one of the most sustainable and green cities in the world.

So, the journey is not over. We will continue to move towards a more green and sustainable city and inspire cities around the world to join us on this journey. And we will continue to work in a fair and socially inclusive way. The green transformation will continue to include all Copenhageners and not increase inequality in our city. Copenhagen will continue to be a green city and a liveable city in green growth.

29

The Powerful City – Introduction

What is amazing about the story of Copenhagen in Chapter 28 is that in achieving the carbon neutral target by 2025 it will have delivered a way of life to Copenhageners that will begin to top the rankings for health and wellbeing indices, and its drive to whole-system approaches to energy and waste infrastructure will also make its way of life one of the most affordable. These are the consequences of solving the climate crisis that should be motivation enough for all cities.

The story concludes with Mark and Sarah in this chapter highlighting the achievements of modern cities and how collective action amongst some of the world's biggest cities is leading to planning reforms, congestion pricing, and low emission zones, the electrification of transport, cutting our emissions in the built environment and from construction, and this shift to sustainable finance. The chapter also looks at Kate Raworth's "doughnut" economic model as a tool to ensure that humanity lives within planetary boundaries.

We know that cities account for more than half the world's population and most of its economic activity. Consequently, the potential for cities to drive action on emission reduction is huge, with cities poised to feel the first impacts of rising global temperatures whilst also, simultaneously, accounting for an estimated 75% of global carbon emissions. Mark and Sarah look at how mayors are uniquely placed to act on climate, and, using evidence from the C40 network of the world's 100 most influential mayors, they consider how cities can work together to keep global heating below 1.5°C, to achieve the future we want.

Mark and Sarah tackle transport and mobility in cities and how the internal combustion engine, for so long associated with freedom, has instead locked us into inevitable temperature rise, premature deaths, and increased air pollution. They show examples from cities such as Mexico City, Shenzhen, London, and New York of how local governance has sought to reduce emissions and successfully diversify public transport.

They look in more detail at the environmental benefits of the electrification of transport in cities. The strongest shifts to electric vehicles are happening in the bus sector, with municipal buses expected to lead the growth in the electrical vehicle market, with Shenzhen already converting its entire bus fleet and most other Chinese cities expected to do so by 2025.

Cutting emissions from energy use is a matter of urgency, and Mark and Sarah explore how cities across the world are solving this problem through investment in solar power, wind power, and other forms of renewable energy. They also show examples of how city governments are well placed to reduce emissions from buildings, which account for

The Climate City, First Edition. Edited by Martin Powell.
© 2022 John Wiley & Sons Ltd. Published 2022 by John Wiley & Sons Ltd.

approximately half of total city emissions, and explore more fully the topic of construction emissions, which account for a quarter of the world's GHG emissions.

Mark and Sarah pinpoint the obstruction cities face to sustainable finance, as we have seen highlighted in Chapter 22, *The Invested City*, Chapter 23, *The Financed City*, and Chapter 27, *The Climate-Resilient City*. Ending investment in fossil fuels and stepping up green investments is now a major priority for cities around the world, and Mark and Sarah examine the ways in which cities have taken action. They also look at the challenges facing cities and post-pandemic recovery, such as jobs and an inclusive economy; resilience and equality; and health and wellbeing.

Cities are stepping up and modelling effective governance on emission reduction, taking bold actions based on science, diversifying urban mobility, energy use, and buildings, and improving liveability. Cities are innovating new technologies whilst acting to support the growth of green jobs and sustainable finance. More importantly, they are advancing the process by collaborating and sharing expertise. But rapid emission cuts are needed at all levels of government, and not only in urban settings but in energy systems, agriculture, and land-use. As Mark and Sarah write, "cities have shown the way forward"; but they are not all-powerful. The challenge to achieve the goals set by the Paris Agreement still remains, and action by cities alone is not enough to put the world on the right trajectory.

29

The Powerful City

Mark Watts and Sarah Lewis

The COVID-19 pandemic has exposed how ill-prepared human civilization is to manage major global shocks. This is not a one-off problem, because after the dust settles on the coronavirus crisis an even bigger threat will still be looming: climate breakdown. More than five years after the apparent breakthrough of the 2015 Paris Agreement, we have yet to see the level of action required from national governments, despite an increase in commitments from some of the world's highest emitting countries. It is in this space that mayors emerge as leading political players, offering a model for climate action in line with what the science demands and hope for global collaboration to build the future we want. Working in partnership with mission-driven business and an increasingly dynamic civil society, cities are becoming the arenas for a new era of economic opportunity based on sustainability and social equity.

The potential for cities to drive action on emissions cuts is huge; they are home to more than half[1] of the world's population and most of its economic activity.[2] Cities account for an estimated 75% of global CO_2 emissions;[3] without action at the urban level there will be no limiting global heating to 1.5°C.[4] Moreover, cities are the first to feel the impacts of rising global temperatures; by 2050 over 570[5] low-lying coastal cities will face sea level rises of at least 1.5 m. By the same year almost 90% of the world's additional 2.5 billion urban citizens will be in Asia and Africa,[6] two of the regions that are most vulnerable to the effects of climate breakdown.[7]

Megacity mayors are uniquely placed to act on climate; they have local powers over two of the biggest contributors to global CO_2 emissions, transport and buildings, while their size and economic clout positions them to influence national policy. When cities

---

1 https://www.un.org/development/desa/en/news/population/2018-revision-of-world-urbanization-prospects.html.

2 https://www.ipcc.ch/site/assets/uploads/sites/2/2018/12/SPM-for-cities.pdf.

3 https://www.unep.org/explore-topics/resource-efficiency/what-we-do/cities/cities-and-climate-change.

4 https://www.ipcc.ch/site/assets/uploads/sites/2/2018/12/SPM-for-cities.pdf.

5 https://www.c40.org/other/the-future-we-don-t-want-staying-afloat-the-urban-response-to-sea-level-rise.

6 https://www.un.org/en/climatechange/climatesolutions/cities-pollution.

7 https://reliefweb.int/report/world/climate-change-vulnerability-index-2017.

The Climate City, First Edition. Edited by Martin Powell.
© 2022 John Wiley & Sons Ltd. Published 2022 by John Wiley & Sons Ltd.

collaborate to scale up action on climate they create and shape global markets. In the pantheon of governments, cities are agile and inventive, employing direct investment, regulatory powers, and market measures to drive the shift to a green economy and stimulate behaviour change in citizens.

City governments are implementing ambitious science-based policies to tackle climate breakdown that are often far in advance of national efforts to cut emissions.[8] While the Australian federal government supported coal-powered electricity,[9] the cities of Melbourne[10] and Sydney[11] used their procurement powers to increase the demand for renewables by switching to 100% clean power. Los Angeles launched its Green New Deal[12] while the US federal government injected huge stimulus into the most polluting sectors of the economy.

Part of this is down to the objective conditions of major cities, as diverse, cosmopolitan melting-pots where millions of creative people buzz off each other's ideas. But it is also a product of the subjective character of mayors, who tend to show a far greater trust in science and data, along with a stronger inclination towards collaboration and knowledge-sharing than national leaders. Reaching out, rather than withdrawing in, means that good ideas spread fast between mayors and a global pool of knowledge tends to overcome national exceptionalism (just look at how many mayors acted to lockdown while their federal governments vacillated). This is not exclusive to mayors, and there are many brilliant, internationalist leaders of nation states, but data-driven government and multilateralism are the norm among big city mayors.

Globally, over 200 formal city networks[13] exist in which local governments mobilize resources, engage in global advocacy, and connect urban innovation. C40 Cities is a network of nearly 100 of the world's largest and most influential cities collaborating to keep global heating below 1.5°C. Created and led by cities, it represents more than 700 million citizens and a quarter of the global economy. The group aims to collectively halve greenhouse gas emissions by 2030, increase resilience, and improve quality of life for citizens.

Ending the Reign of the Car and Redesigning Cities for People

Tackling climate change has to be about more than just cutting pollution if it is to be successful. Citizens need to be convinced that slashing emissions is synonymous with building the future they want. Most immediately, this means building back better from the COVID crisis.

---

8 https://www.c40knowledgehub.org

9 https://www.bloomberg.com/news/articles/2020-02-08/australia-to-fund-study-into-new-coal-fired-power-station?sref=pxdaafso.

10 https://www.theguardian.com/australia-news/2019/jan/17/melbourne-becomes-first-city-with-all-council-infrastructure-powered-by-renewables.

11 https://www.abc.net.au/news/2020-07-02/net-zero-by-2050-push-on-whether-government-follows-or-not/12414426.

12 https://www.lamayor.org/mayor-garcetti-launcheslas-green-new-deal.

13 https://www.thechicagocouncil.org/research/report/conducting-city-diplomacy-survey-international-engagement-47-cities.

That is perhaps most true in relation to transport and mobility. The internal combustion engine was once mistakenly equated with freedom, and it has been a long, slow journey to change that narrative, even though reliance on private cars often excludes large numbers of people from mobility and vehicle exhausts were linked to 385,000 premature deaths worldwide in 2015.[14] Moreover, congestion and pollution are now understood as a drag on economic development. In Beijing,[15] congestion reduces the city's GDP by 15%, while traffic jams in Nairobi[16] cost the national economy approximately US$1 billion per year. Mayors know that to cut emissions and make streets work for people means ending the era of cities designed for cars.

Globally, city governments are united in redefining mobility on the basis of public transport (mass transit), cycling, and walking. International collaboration has shifted mayors' perceptions of what is possible; overcoming local limitations and transforming brave innovation have rapidly become accepted wisdom. There has been a powerful interplay between city government policy and commercial entrepreneurialism.

Mobility is an area where mayors have strongly intervened to correct market failure and used policy and public investment to drive innovation. Regulatory powers have been used creatively in order to reclaim the streets from a minority of car users, to benefit everyone – from parking regulations, to vehicle bans, to people-centred urban planning.

Mexico City reformed its planning code;[17] where previously it had required a minimum number of parking spaces in housing and commercial developments, the reforms set an upper limit on the space reserved for parking as well as mandatory bicycle parking spaces. Developers who reach the upper limit are required to contribute to a fund to improve mass transit. San Francisco followed suit,[18] removing minimum parking requirements altogether to allow more space for walking and cycling and to reduce housing costs.

To cut air pollution many cities are prioritizing the removal of polluting vehicles. Seoul has implemented a Green Transport Zone[19] in the city centre which excludes the dirtiest vehicles, while the mayors of Milan and Paris have plans to take diesel cars and vans off the roads by the years 2030 and 2024 respectively.[20] Amsterdam has started the process of making the entire city a zero-emission area for transport by 2030. The city of Shenzhen has electrified its entire fleet of 16,000 buses.[21] Such commitments impact markets; in 2018 Carlos Ghosn, who heads the global alliance of automakers Renault, Nissan, and Mitsubishi Motors, complained that the Paris mayor's intention

---

14 https://theicct.org/news/health-impacts-transport-sector-pr-20190227.

15 https://urbantransitions.global/wp-content/uploads/2020/03/Towards-Sustainable-Mobility-and-Improved-Public-Health-Lessons-from-bike-sharing-in-Shanghai-China-final.pdf.

16 https://weetracker.com/2019/09/24/nairobis-traffic-congestion-cost-kenya-usd-1-billion-in-a-year/.

17 https://www.itdp.org/2017/07/26/mexico-city-became-leader-parking-reform/.

18 https://www.livablecity.org/time-san-francisco-say-goodbye-minimum-parking-requirements.

19 http://english.seoul.go.kr/level-5-vehicles-to-be-limited-in-green-transport-zone-in-the-city-center/.

20 https://www.theguardian.com/environment/2016/dec/02/four-of-worlds-biggest-cities-to-ban-diesel-cars-from-their-centres.

21 https://www.wired.co.uk/article/shenzhen-electric-buses-public-transport.

to ban diesel cars was influencing consumer choices.[22] Manufacturers of electric and hydrogen vehicles derive a major part of their market opportunity from city government policy.

Perhaps one of the most game-changing city policies on car use of the twenty-first century has been the London Congestion Charge (a policy that dominated my life for three years as a young mayoral adviser). In a great example of a public–private strategic partnership, it was implemented first in 2003 by a radical mayor leveraging off a powerful lobbying campaign by the business group London First, when time lost to congestion was costing the economy £4 million per week.[23] Hugely controversial while in the planning stages, the charge quickly achieved majority public support on the back of a reduction in traffic congestion of 30% in its first year. Drivers were required to pay a flat fee to drive within central London, and by 2018 the scheme had raised over £2 billion in revenue to reinvest in a huge expansion of bus services, cheaper fares, and cycle lanes. In 2019 the city introduced strict emissions standards for all vehicles entering the congestion charge zone; by 2020 it had reduced exhaust emissions of nitrogen oxides by 45% and roadside nitrogen dioxide within the zone by 44%.[24]

Longer term, cities have the power to either lock in car-dependent development or avert it. This is going to be even more important while the COVID-19 risk ensures mass-transit usage remains low. Local governments send clear signals to developers and citizens when they set targets for locating jobs, housing, services, and amenities near public transport hubs.

New York City has a target for 95% of new housing to be built within half a mile of transit stations,[25] which should maintain the city's relatively low per capita emission from transport. In Jakarta,[26] where congestion is estimated to cost over US$7 billion per year in economic losses, the City Governor Dr Anies Baswedan set an ambition for 95% of Jakarta's residents to be within 500 m of public transport[27] (Figure 29.1), leaning on European city best practice. The city's bus rapid transit system, Transjakarta, integrated with other local transport operators to serve a larger region and has a long-term strategy to transition to electric buses.[28] In addition to reducing congestion and emissions, transit-orientated development improves quality of life for the majority of residents who can't afford to drive by shortening commute times and reducing transport costs.

---

22 https://www.ft.com/content/b4f72ccc-c59b-11e8-8167-bea19d5dd52e.

23 https://www.c40knowledgehub.org.

24 https://www.london.gov.uk/what-we-do/environment/pollution-and-air-quality/mayors-ultra-low-emission-zone-london.

25 http://www.nyc.gov/html/planyc/downloads/pdf/publications/planyc_2011_planyc_full_report.pdf.

26 https://www.itdp.org/2019/07/15/transjakarta-study-success/.

27 https://www.spf.org/en/spfnews/information/20190530.html.

28 https://www.c40cff.org/news-and-events/jakarta-is-spearheading-indonesias-e-bus-transition-cff-supports-transjakartas-pilot-for-electric-buses.

Figure 29.1 Jakarta's City Governor has pledged for 95% of residents to be within 500 m of public transport. (*Source:* JOKO SL / Getty Images.)

Reducing Emissions with the Electrification of Transport

Key to cutting emissions from transport is shifting the remainder of motorized vehicles to electric.[29] One 2019 analysis found that lifecycle electric vehicle emissions were around three times lower than a car with an internal combustion engine,[30] and as electricity grids decarbonize, emissions from the use of electric vehicles will reduce further.

The strongest shifts to electric vehicles are happening in the bus sector first, driven by action in C40 Cities. According to Bloomberg New Energy Finance, municipal buses are expected to lead the growth in the electric vehicle market until at least 2030.[31] As noted above, Shenzhen has already converted its entire bus fleet to electric vehicles, and most other Chinese cities will have also done so by 2025. In 2015, frustrated at their individual inability to speed up the pace at which bus manufacturers were shifting to production of clean vehicles, 20 global cities worked with C40 to aggregate their purchasing power through a Clean Bus Declaration of Intent.[32] The market received the message that if they

---

29 https://www.c40knowledgehub.org.

30 https://www.carbonbrief.org/factcheck-how-electric-vehicles-help-to-tackle-climate-change.

31 https://about.bnef.com/electric-vehicle-outlook/.

32 https://www.c40.org/news/c40-clean-bus-declaration-urges-cities-and-manufacturers-to-adopt-innovative-clean-bus-technologies/.

produced electric buses at an affordable price there would be buyers for them, and by 2017 the average price for a hybrid bus had declined by 10%.

Following this success, the C40 Green and Healthy Streets Declaration[33] signalled another shift in policy, with 35 global cities committing to a future in which walking, cycling, and shared transport are the primary means of mobility for the majority of citizens. The declaration committed signatories to procuring only zero-emission buses from 2025 at the latest, and by 2021, eight cities were already doing so. With an average renewal rate of 10% per year, the declaration signalled significant opportunities to world markets.

At the time of writing, Latin American cities are now setting the pace outside of China. Mexico City, Medellin, Quito, Rio de Janeiro, and Santiago are all signatories to the Green and Healthy Streets Declaration, and in 2018 the city of Sao Paolo amended its climate law[34] to require that, within 20 years, transit operators must reduce particulate matter and nitrogen dioxide emissions by 95% and eliminate CO_2 emissions entirely from their bus fleets. By 2022, the city of Bogota will have the largest number of electric buses outside China, with a fleet of 1,485.[35] Barriers such as the high upfront costs of the vehicles and charging infrastructure have been overcome with new business models, such as the separation of asset ownership;[36] in Santiago, the city guarantees payment to third parties who purchase electric vehicles and lease them to operators.

Cutting Emissions from Energy Use

Around 30% of global GHG emissions can be attributed to the production of electricity and heat,[37] and as consumers of around two-thirds of global energy, cities are critical to accelerating progress towards net-zero in the sector. Globally, cities are actively shaping energy markets by investing in the supply of clean energy and finding innovative ways to increase demand, while making use of their powers over buildings and planning to reduce energy consumption.

The city of Seoul is installing 1 GW of solar power by 2022 as part of its efforts to lead the country-wide transition away from dependence on fossil fuels for power generation.[38] After the national government abolished feed-in-tariffs in 2012 the city government set up its own, providing US$2.4 million to power generators by the end of 2018. Seoul launched its Solar City project in 2017 by deploying solar PV panels on private households and municipal sites. In 2018, the project avoided 109 tCO_2 in GHG emissions and 27.6 tonnes of fine particulate matter pollution.

---

33 https://www.c40.org/other/green-and-healthy-streets.

34 https://theicct.org/publication/climate-and-air-pollutant-emissions-benefits-of-bus-technology-options-in-sao-paulo/.

35 https://chargedevs.com/newswire/bogota-orders-596-new-electric-buses-fleet-of-1485-e-buses-will-be-the-largest-outside-china/.

36 https://www.c40knowledgehub.org/

37 https://www.wri.org/insights/4-charts-explain-greenhouse-gas-emissions-countries-and-sectors.

38 https://www.c40knowledgehub.org/

US cities with greater oversight of their energy supply have used this effectively to drive the energy transition. Los Angeles partnered with a solar energy company to develop the largest solar and battery storage system in the country, and in February 2020 Mayor Eric Garcetti announced that the city's Department of Water and Power – the largest municipal utility in the USA – would phase out three natural gas units in order to accelerate the city's transition to 100% clean energy.[39] The city has now partnered with the US Department of Energy to publish a detailed roadmap for how they will get there. Cities with fewer powers over energy supply are shaping the markets by increasing demand; by 2018 the city of San Francisco had saved approximately 258,000 tCO_2 in two years with its Community Choice Aggregation Program, in which the city buys clean power on behalf of residents and businesses.[40]

Even where city governments don't own energy utilities and lack significant investment capital for energy projects, determined mayors can use their convening power to shift markets. As part of its commitment to create a carbon neutral municipality, the city of Melbourne partnered with a group of local governments, universities, and corporations to purchase renewable energy directly from a wind farm owned by a local firm. By 2019 all city-council-owned infrastructure was powered by renewable electricity, and by contracting directly with the generator the project provided the company with the financial support required to develop the wind farm and bring more renewable energy into the market.[41]

City governments are especially well placed to reduce emissions from the use of buildings, which are responsible, on average, for over half of total city emissions. In China, the city of Qingdao used financial incentives[42] to encourage a buildings renovation scheme which by 2019 had abated 298,000 tCO_2e. Barcelona has seen savings of 25,000 MWh energy per year since the city introduced regulations making it compulsory to use solar energy to supply 60% of hot water heating in large new buildings and renovations. In both cases, strong public policy has created the space for mission-driven private companies to prosper. By 2021, 28 cities had signed the C40 Cities' "Net Zero Carbon Buildings Declaration", mandating each signatory to design a roadmap to reach net-zero carbon buildings, develop a suite of supporting initiatives and programmes, and report annually on progress.

These individual city success stories matter more because of the propensity of mayors to share ideas with each other, often through networks like C40. In 2017 Copenhagen won a C40 award for its building energy management programme which uses a central database to monitor the energy consumption of municipal buildings with a granularity down to 1 hour. As a result, the scheme gained attention among Chinese cities and a knowledge-sharing partnership was established via the C40 China Buildings

---

39 https://www.lamayor.org/mayor-garcetti-ladwp-will-phase-out-natural-gas-operations-three-power-plants.

40 https://www.c40knowledgehub.org/

41 https://www.theguardian.com/australia-news/2019/jan/17/melbourne-becomes-first-city-with-all-council-infrastructure-powered-by-renewables.

42 https://www.c40knowledgehub.org

Programme.[43] Copenhagen has a population of 800,000, but with China's urban population expected to reach 1 billion by 2030, the potential impact of Copenhagen's innovation has been dramatically scaled up.

Reducing Emissions from Construction

Construction contributes nearly a quarter of the world's GHG emissions and rising – with the UN estimating that we will reach the level of a new city of more than 1.5 million people being built each week by 2050.[44]

Most construction materials are not produced in cities and their embodied emissions are not typically accounted for in city inventories, but as interest in consumption-based emissions grows so has city action on the lifecycle emissions of construction projects. City policies can mandate a range of interventions to reduce emissions from construction and infrastructure, such as reducing material use, enhancing building utilization, and reducing the embodied carbon of materials. Figure 29.2 highlights the growing attention to policies for embodied carbon in cities. A 2019 study by C40, ARUP, and the University of Leeds found that if the most ambitious targets are reached, cities can achieve a 44% reduction in buildings- and infrastructure-related emissions between 2017 and 2050.[45]

The Mayor of Oslo, Raymond Johansen, has spear-headed a drive within C40 to achieve emission-free construction, using public procurement and city-supported development projects to stimulate market innovation towards zero emission construction vehicles and operations (Figure 29.3). When the mayor first touted the idea, heavy-duty zero emission construction vehicles weren't thought to be viable. But Oslo's procurement muscle stimulated rapid commercial innovation, and the city's first zero emission urban construction site began operation in 2019, using only electrified construction equipment.[46] Oslo is one of four global cities to have launched the C40 Clean Construction Declaration,[47] which sets out how signatory cities will reduce embodied emissions for all new buildings by at least 50% by 2030 and require zero emissions construction sites city-wide by the same date.

Other cities are intervening at all stages of the building lifecycle to reduce emissions. Rotterdam has a target to halve the use of primary raw materials used in the city by 2030;[48] in 2020 the Mayor of Los Angeles issued a directive to reduce embodied carbon in building materials. Mexico City aims to triple the amount of waste it recycles, including

---

43 https://c40.my.salesforce.com/sfc/p/#36000001enhz/a/1q000000kuqv/5nfcloqpkwiwvjcidgzsld15onc9v5ywps0uvtrd0rc.

44 https://www.c40knowledgehub.org

45 https://www.c40knowledgehub.org

46 https://www.c40knowledgehub.org/

47 https://www.c40.org/clean-construction-declaration.

48 https://rotterdamcirculair.nl/wp-content/uploads/2019/05/Rotterdam_Circularity_Programme_2019–2023.pdf.

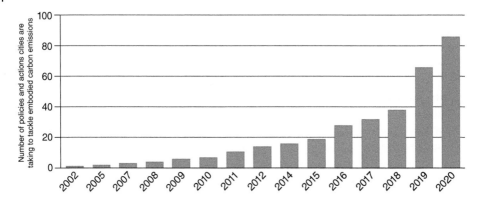

Figure 29.2 Global indication of the growing number of policies and actions cities are taking to tackle embodied carbon emissions. The number is cumulative over the years. (*Source:* JOKO SL / Getty Images.)

Figure 29.3 The Mayor of Oslo – along with other Nordic mayors – as pledged emissions-free construction. (*Source:* anouchka / Getty Images.)

construction and demolition waste,[49] by 2024, while Auckland has made construction and demolition waste a central focus of its waste management plan.[50] Other policies take a whole lifecycle approach; Quezon City encourages material reuse and waste reduction

---

49 https://about.bnef.com/blog/mexico-city-cleans-act-200-million-zero-waste/.

50 https://www.aucklandcouncil.govt.nz/plans-projects-policies-reports-bylaws/our-plans-strategies/topic-based-plans-strategies/environmental-plans-strategies/docswastemanagementplan/auckland-waste-management-minimisation-plan.pdf.

in building projects via a rating system and corresponding tax incentives.[51] London requires whole lifecycle carbon assessments for all its strategic planning applications, including emissions from energy use during the lifetime of the building;[52] applicants must demonstrate that emissions will be reduced. In each case, strong city policy and regulation creates markets for new products and services, giving certainty of opportunity to investors who want to make a return out of helping solve climate breakdown.

The city of Amsterdam has particularly ambitious policies on reducing the impacts of construction; its circular strategy[53] sets targets for reducing the use of raw materials and implementing circular construction principles such as reusing materials and the use of adaptive and modular construction. Also seeking to reduce emissions through better design is the city of Paris, which asks developers to ensure that new builds are designed to accommodate several functions which can change over time; the city has a target of 30% new office space to be reversible by 2030 and 50% by 2050.

Cities and the Shift to Sustainable Finance

Ultimately it won't be possible to stop climate breakdown by simply doing more of the good stuff – clean energy, smarter mobility, more efficient buildings. The biggest obstacle to constraining global heating is ending investment in fossil fuels. Cities around the world are now playing a leading role in the movement to divest from fossil fuel companies, while also stepping up their green investments.

In 2018, the Mayors of London and New York launched the C40 Divest/Invest Forum,[54] a peer-to-peer learning network, to share best practices and tools to accelerate city-led action on divesting city pension funds from fossil fuels and increasing investments in climate solutions (Figure 29.4). Announcing the move, the mayors described city divestment as "a clear message to the fossil fuel industry: change your ways now".[55]

Figure 29.4 London and New York are divesting city pension funds from fossil fuels and into climate solutions. (*Source:* Left: Cavan Images/Adobe Stock Right: Mario Savoia/Adobe Stock.)

---

51 https://quezoncity.gov.ph/green-building-ordinance.

52 https://www.london.gov.uk/sites/default/files/wlc_guidance_april_2020.pdf.

53 https://www.amsterdam.nl/en/policy/sustainability/circular-economy/.

54 https://www.c40.org/press_releases/fossil-fuel-divestment-city-partnership-network.

55 https://www.theguardian.com/commentisfree/2018/sep/10/london-new-york-cities-divest-fossil-fuels-bill-de-blasio-sadiq-khan.

The forum connects leading cities both inside and outside the C40 network, including Pittsburgh, San Francisco, Los Angeles, Vancouver, Oslo, Stockholm, Durban, and Auckland.

This collaborative move came on the back of London and New York's high-profile decisions to divest their own city pension funds. In early 2021, New York City Mayor Bill de Blasio announced that two of the city's pension funds had voted to divest their portfolios of an estimated US$4 billion from securities related to fossil fuel companies, in what is expected to be to be one of the largest divestments in the world.

Although most city pension funds are legally independent from municipal control, mayors have used soft power and influence to call for proper consideration of the climate crisis in investment portfolios. The London Pension Fund Authority (LPFA) developed its Climate Change Policy in collaboration with Mayor Khan, following his commitment to "take all possible steps to divest the pension fund", and the Fund has since reduced its direct holdings in extractive fossil fuel companies from 1% to under 0.2% of its assets. The Fund also assessed its portfolio to ensure that its holdings are compatible with the 1.5°C goal of the Paris Agreement. The mayor, who appoints trustees and the Chair to the LPFA Board, also worked with the LPFA to recruit trustees with climate change experience to help implement and champion their climate policy. Mayor Khan then publicly invited the city's 32 local boroughs to follow the city's lead by divesting from fossil fuels. As of 2021, at least 13 of London's boroughs had agreed to take action on divestment.

The impact of a divestment decision extends far beyond city boundaries; bold steps by leading cities can facilitate changes in the investment landscape and enable further action. The city of Berlin not only divested the equity portfolio of its €1 billion city pension fund, but also initiated the creation of a fossil-fuel-free equity index,[56] not only enabling itself to divest but also providing an opportunity for other municipalities and states to do the same. Berlin's initiative has allowed the federal state of Schleswig-Holstein to divest. After the city of Copenhagen made the political decision to divest from all fossil fuels, pressure from municipalities in Denmark helped persuade Danish banks to offer fossil-free investment funds. The new funds abided by the criteria set by the Copenhagen municipality. Cities can leverage the moral authority gained through local divest/invest action to put pressure on the wider investment sector; New York City's pension funds, already committed to ambitious action on their own portfolios, issued a strong call on multinational insurance companies to sever all ties with the coal industry.

Cities and the Post-Pandemic Recovery

Cities have long been centres for innovation and fresh ideas; they are "where the future happens first".[57] They also hold significant economic power; Seoul accounts for over half of South Korea's GDP,[58] while London, Paris, and Tokyo make up one-third of national

---

56 https://c40.org/wp-content/uploads/2021/07/2602_D_I_CITIES_TEMPLATE_2209.original.pdf.

57 https://www.c40.org/what-we-do/raising-climate-ambition/green-just-recovery-agenda/.

58 https://www.weforum.org/agenda/2020/01/cities-mayors-not-nation-states-challenges-climate.

GDP in their respective countries. The New York metropolitan area alone generates more economic output than Canada[59] – itself a G7 country.

Global mayors are harnessing the progressive energy of cities and combining it with their economic power to drive the transition to a greener, more equitable future following the COVID-19 pandemic. In July 2020, a task force of global city mayors released the C40 Mayors' Agenda for a Green and Just Recovery.[60] Forty city leaders signed up to its principles, which expand on the ideas of the Green New Deal and emphasize public investment, green jobs, and social equity. A further iteration in October 2020[61] set out how a large, rapid investment in a green recovery would not only create 50–80 million good, new green jobs across C40 Cities but also achieve far greater health and environmental outcomes than sliding back to the old, polluting economy.

Jobs and an Inclusive Economy

In the context of a global recession, mayors committed to acting as entrepreneurial governments are setting out to turn a crisis into a new start, accelerating the creation of new, good green jobs, supporting essential workers, and massively expanding training to facilitate a just transition. C40's research during the pandemic showed that the nexus between job creation and emissions reduction is strongest in the building sector – both retrofitting existing building to be more energy efficient and steering the sector more rapidly towards zero emission new build. But there is also significant job creation potential in sustainable transport, low-carbon clean infrastructure and, most notably in Africa and south Asia, nature-based solutions such as parks, green roofs, blue infrastructure, and green walls.

The city of Medellín in Colombia will create 20,000 jobs in industries associated with the digital revolution, such as its Software Valley Centres, as well as a new metro line with the potential to generate around 21,000 jobs during construction and 900 permanent jobs after completion. The city is investing in training for 25,000 people in science, technology, and innovation, with an emphasis on women, youth, and older people. Hong Kong launched the Green Employment Scheme to create more than 1,000 jobs in environmental protection; it will subsidize the installation of electric vehicle charging infrastructure and promote the environmental protection of parks and shorelines. In Los Angeles, the city is investing in the GRID Alternatives Solar Powered Homes and Jobs Program, which by 2020 had already installed solar panels at no cost to the homes of almost 2,000 low-income families, helping 200 individuals gain skills and employment after release from incarceration.

Resilience and Equity

As part of the agenda, mayors will take the lead in providing "fundamental public services for all that underpin a fair society and strong economy and are resilient to future shocks", including safe and resilient post-COVID mass transit systems. The city of Lisbon is

---

59 https://www.weforum.org/agenda/2020/01/cities-mayors-not-nation-states-challenges-climate.

60 https://www.c40knowledgehub.org/s/article/c40-mayors-agenda-for-a-green-and-justrecovery?language=en_us.

61 https://www.c40.org/green-and-just-recovery-benefits.

implementing daily deep-cleaning of buses and trams and creating new dedicated bus lanes to increase frequency and capacity. It has also invested in active-travel, low emission zones and increased the length of its cycle network. In New Orleans, Mayor LaToya Cantrell is overseeing the Climate Action Equity Project, which ensures equity considerations in policymaking and programming. In Freetown, Sierra Leone, Mayor Yvonne Aki-Sawyerr is increasing water provision in informal settlements by using a rainwater harvesting system to collect and store water.

Health and Wellbeing

Cities will "give space back to people and nature, rethink and reclaim our streets, clean our air and create liveable, local communities". Key to delivering this commitment is the implementation of the "15-minute city",[62] a concept to transform urban spaces into connected and self-sufficient neighbourhoods in which residents can access most of their needs within a short walk or bike ride from their home. As well as prioritizing cycle lanes and public transport, the vision involves planning and zoning to ensure that critical public services and infrastructure are easily accessible to all residents. It also represents a rethink of how we view buildings, encouraging repurposing, adaptive reuse, modular design, and reversible buildings, such as the transition of housing to offices.

The 15-minute city concept was a re-election promise of Paris Mayor Anne Hidalgo, and the city has plans to expand the use of existing amenities as well as adding co-working hubs, encouraging remote working, and providing new green space. Milan is also basing its recovery on the 15-minute city framework (Figure 29.5); having already created 35 km of new cycle lanes, the city plans to pedestrianize several school streets. Originally designed with European cities in mind, local governments elsewhere have adapted the idea to their local context. Chinese cities Shanghai and Guangzhou both reference the "15-Minute Community Life Circles" in their masterplans, while the US city of Portland's "Complete Neighborhoods" sets a goal for 90% of residents to access their daily non-work needs by foot or bike.

Amsterdam and the Doughnut Economic Model

A standout example of a city pioneering new ideas is Amsterdam's adoption of the "doughnut" economic model. Designed by British economist Kate Raworth, the model is a tool to ensure that humanity lives within planetary boundaries (the outside ring of the doughnut) without falling short of essentials (the inside ring) (Figure 29.6).

Beginning with a vision to be a "thriving, regenerative and inclusive city for all citizens, while respecting planetary boundaries", the city used a self-portrait[63] to view itself through four lenses: social, ecological, local, and global. A series of workshops were held in diverse

---

62 https://www.theguardian.com/world/2020/feb/07/paris-mayor-unveils-15-minute-city-plan-in-re-election-campaign.

63 https://www.kateraworth.com/doughnut

Figure 29.5 Blending the old with the new in Porta Nuova business district in Milan. (*Source:* RossHelen / Getty Images.)

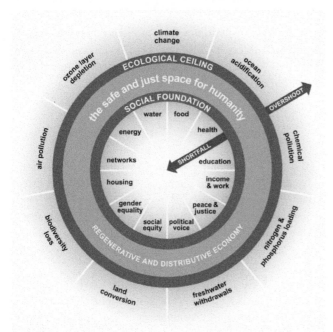

Figure 29.6 Doughnut of social and planetary boundaries. (*Source:* The Doughnut of social and planetary boundaries (2017), Kate Raworth.)

Figure 29.7 Amsterdam living the 'doughnut' economic model. (*Source:* SolStock / Getty Images.)

neighbourhoods to understand people's priorities for a thriving city, and urban designers have begun integrating biomimetic designs into buildings such as creating habitats for species and incorporating green roofs. The city plans to reduce air pollution by completely banning petrol and diesel cars and motorbikes by 2030 and to move towards a circular economy as soon as possible (Figure 29.7). The municipality has also adopted responsible procurement guidelines aligned with the city's social and ecological targets.

Conclusion

Across the globe cities are stepping up and modelling effective governance on emissions reduction. They are taking bold actions based on the science, driving the shifts in urban mobility, energy use, and buildings that cut emissions and improve quality of life for citizens. City governments are innovating and working to overcome barriers to the uptake of clean technologies. They are acting to support the growth of local green jobs and leading on the shift to sustainable finance. They are scaling up ambition through collaboration and advancing progress by sharing ideas and expertise with other cities.

But the challenge remains. Action by cities alone will not put the world on a 1.5°C trajectory. To achieve the goals of the Paris Agreement urgent structural changes are needed, not only in urban areas but also in energy systems, land-use, and agriculture. Rapid cuts in emissions will need all levels of government, the private sector, and civil society on board. Cities have shown the way forward.

30

Epilogue

Martin Powell

I'll tell you how this all ends ...

... We solved it.

We were late and we had to adapt our cities with new canopy and modified greenspaces. Buildings were modified, renewable energy projects that were started were completed. We began to scale up carbon sequestration, direct air capture technology and geo-engineering solutions. Species were lost, numbers dwindling, others however thrived. Some communities were relocated, many more people having to adjust to an altered environment, lives were lost.

We compromised. Our urgency did not ramp up efforts fast enough. Our policymaking did not go far enough. Not enough people believed soon enough to create the momentum needed for the pace and scale of change.

It's spring 2022, and that future is unwritten. Whatever we are locked into does not define how we will grow our cities over the coming decades. How we build our buildings, how we handle our waste, our water, our natural resources, or how we decide we want to live our lives.

We are emerging from a pandemic that has taught us about compromise and restriction, and we have used policy, science, data, innovation, and cooperation to emerge largely intact as a society. There has been a tremendous loss of life, and we would be remiss not to take a long hard look at how we would handle this if it ever happened again. But we also have an opportunity to use this experience to enact the changes that will make our lives better every day.

The future has long presented artists and innovators alike with a great sense of possibility. The unknown has become attractive. Consequently, we are inundated with a variety of depictions and real-life enactments of what the future might be.

Daniel Grataloup's art-piece (Figure 30.1) shows a futuristic view of the city, which captures the imagination, as we fly in and out of our high-rise centres in our capsule homes. Grataloup prides himself on creating "the architecture of the free man today", and this large architectural promise certainly would be something.[1]

Innovation is inherently ambitious, and this is a good thing. But we must also seek to ensure what we have already established and evaluate what is in line with and benefits

---

1 https://communedesign.tumblr.com/post/121302775475/daniel-grataloup.

The Climate City, First Edition. Edited by Martin Powell.
© 2022 John Wiley & Sons Ltd. Published 2022 by John Wiley & Sons Ltd.

Figure 30.1 Daniel Grataloup's Urban Proposal with Multi Thin-Shell Capsules (1974). (*Source*: Editor's photo from Museum of Modern Art (MoMA), New York.)

the city inhabitants, instead of getting too far ahead of ourselves, as we have seen through the problems faced by eco-cities.

Tom Clohosy Cole offers a more recent depiction of a sustainable city, with its greens-pace, pedestrian pathways, cycle lanes, wind turbines in the distance, and solar panels, and not a car in sight (Figure 30.2)! It seems a simpler and more realistic reality to obtain in comparison to Grataloup's vision. Many of the technologies we see in this picture are already starting to be implemented in cities across the world today.

Bryant Park in New York City is a real-life example of how we can transform our cities. As a hidden oasis nestled in a canyon of skyscrapers, it is a lush green respite for New Yorkers and tourists alike (Figure 30.3). The space is adapted throughout the year to attract newcomers and allow residents to enjoy it. The buildings that surround it are con-suming less energy every year and that energy is greener every year. In parallel to my walk along the embankment in London, this image offers me hope that we are conquering this path to the Climate City, but we must accelerate our journey.

Fiction, too, likes to imagine the future and present alternative realities to what actually is, and given the chapter titles in this book it would be wrong of me not to mention the Emerald City. In both the film version of *The Wizard of Oz* and in the original books by L. Frank Baum, Dorothy, a young girl from Kansas, begins a long journey to the Emerald City in search of the mayor, here a wizard, who she believes will use his powers to help

Figure 30.2 Tom Clohosy Cole's depiction of a sustainable city. (*Source*: Image ©Tom Clohosy Cole, *Source*: Destination: Planet Earth, ©Wide Eyed Editions.)

Figure 30.3 Bryant Park, New York City, 2020 could be the real image of Tom Clohosy Cole's depiction. (*Source*: Boogich/Getty Images.)

her return home. When Dorothy and her friends finally reach the Emerald City, they discover that the Wizard is not a wizard at all but just a person – and with limited powers.

Dorothy's return home was a result of collaboration – a leader enforced a clear set of instructions, experts were given a platform and listened to, and the citizen was inspired into action and, most important of all, to take responsibility. These components resulted

Figure 30.4 How could this book not mention "the Emerald City" – perhaps fiction defining reality. (*Source*: stuart/Adobe Stock.)

in success on multiple fronts – Oz was freed from the Wicked Witch, the Scarecrow got a brain, the Tin Man a heart, and the lion courage, Dorothy got to go home, and the Wizard looks like a competent leader.

The Climate City is the story of component parts, each with a message of what, and how, to achieve the right outcomes and some examples of how we can, and have, achieved them.

The Ambitious City sets targets that force historic emissions to be accounted for – incentives to protect the forests through cooperation rather than through tokenistic tree planting.

The Civilized City goes back in time and applies its perpetual learning to improve the everyday and empower the governing entities across the city that have been shaped through time to act decisively and justly while maintaining this finely balanced ecosystem.

In line with these historic lessons, *The Emerging City* grows responsibly, fully rectifying its environmental blunders.

The Sustainable City follows its "North Star" and adheres to the principles of the UN SDGs, putting itself in constant question about the impacts of new policy on these core objectives.

The Vocal City advocates on behalf of all cities and presses national governments to support and fund change.

The Governed City commits only to pre-aligned budgets that have priced the full carbon abatement costs of its plan, ensuring healthy tension in the budgeting process but ensuring that our future is assured for everything new that we do.

The Decoupled City ensures that economic success does not come at the cost of the environment and that we keep cities as places where jobs can be found and lives can be lived.

The Responsible City highlights how governments must use their powers to shape market conditions and adopt circular economy principles in their own supply chains

The Energized City shows how each city must find its own path to curbing emissions, as each starts out from a different and unique starting point.

The Agile City gives us ways to drive diversity of choice for citizens to move around the city and focus on satisfying everyone's purpose to travel, as well as how we must integrate active transport as part of a bigger transport policy.

The Habitable City gives us a way to build affordable housing, and densify our housing whilst retaining high quality, and shows how we must build responsibly, from sourcing materials to construction, and restrict new development within an "emissions envelope".

Both *The Resourceful City* and *The Zero Waste City* show how we can achieve "circularity" and make cities responsible for all waste. This will immediately drive new infrastructure.

The Resilient City shows how we use new metrics to measure climate risk, and for this we need better measurement and good positioning of the data.

The Fragile City furthers this idea and ensures a systems-based approach that identifies factors that interact to shape patterns of risk and resilience. This is essential to effectively tackle the multiple dimensions of urban fragility and address deficits in the urban economy, infrastructure, and services, in social and economic inequalities, and in governance by enhancing political inclusion, social protections, and access to essential services.

The Data City embraces the ability to extract infrastructure data from across disparate systems and shows how using analytics we can find business cases for new business models, enhanced performance, and optimized infrastructure. The time must be spent looking hard at the problem we are trying to solve, to innovate, and to accept new ways of doing things for the greater good of the city to make our data efforts pay off.

The Measured City hones in on our data efforts and gives us the simple conclusion that we must start to measure the same thing in the same way. We will learn faster by comparison and understanding of what works well and what could be improved.

The Smart City, however, extends beyond a digital and connected city. It is a future city that uses technologies to improve the lives of its citizens by modernizing infrastructure (for example, the generation and transmission of energy, creating more efficient and livable buildings, and optimizing transportation) and by using information technology to enhance city operations and services, irrespective of its geography or economic size.

The Just City acknowledges that air pollution is a daily threat that will continue to worsen. Consequently, the just city measures air pollution better, targets its measures, and puts in congestion pricing and low emission zones to fund this effort. Every city should calculate the number of premature deaths per year targeted by location and use that as a starting point to reduce it every year.

The Invested City shows that greater environmental, social, and governance (ESG) transparency will drive new investment in areas of critical societal needs such as building, transport, food, and energy.

The Financed City sees a net-zero carbon city as insufficient by itself. Cities also need to reduce consumption-based emissions by adopting circular economy principles so they can eliminate their contribution to climate change. A concerted and collective effort can accelerate the transition to low carbon circular cities and provide opportunities for capital and investors. Optimizing the funding and financing of the cities of the future is urgent and important.

The Adaptive City shows that one of the greatest opportunities to tackle climate change in cities is to prepare for an increasing urban heat island effect. More greenspaces put less pressure on sewer systems, they provide important cooling, and they help reduce impacts of poor air quality. Once we calculate these costs, we can begin to see new investment in greening our cities.

As we begin the process of greening our cities, *The Open City* shows us that public space is not a commodity and the market will not provide it.

The Natural City reduces our impact on the natural world. In the same way as a new development completed within an emissions envelope, we need to accurately calculate the impact on natural habitats to ensure the natural world thrives within our urban centres.

The Climate-Resilient City uses public financing to leverage private investment for low carbon and resilient infrastructure.

The Green City shows that from the governance and budget setting process, to policy and bringing the citizens along on the journey, we have seen working models for what our cities need to move towards.

Finally, *The Powerful City* embodies the huge potential for cities to drive action on emissions cuts. Cities are home to more than half of the world's population and most of its economic activity, they account for an estimated 75% of global CO_2 emissions, and without action at the urban level there will be no limiting global heating to 1.5°C.

The Powerful City also emphasizes that action by cities alone will not be enough. To achieve the goals of the Paris Agreement, urgent structural changes are needed, not only in urban areas but also in energy systems, land–use, and agriculture. Rapid cuts in emissions will need all levels of government, the private sector, and civil society on board. All cities can do is show the way forward.

With half the world's populations currently living in cities, it is safe to say that a city without people is not a city. In *The Civilized City* I wrote about how a city is often successful because it can service the needs of its people. A city is an epicentre where worlds collide and combine, and every problem we have across society, everything, is inherently tied to the city and reflected back through its inhabitants. Cities house the world's problems and, as a consequence, they must also carry the solutions.

In the Introduction I showed you the population is at risk if we continue on our current path. In Figure 30.5, C40 and Arup show that same trajectory, but they show us close to achieving the 1.5°C scenario trajectory with actions from the C40 Cities. If all of the other cities embark on the journey of *The Climate City*, we will make it.

As the world moves towards these sustainable principles, we can assess how deep they really run as we look to build cities (and houses!) on the moon or on Mars (Figure 30.6). Can they be conceived and built without the need for constant deliveries from Earth of key resources to sustain them? Will they start from the beginning to pile their trash in the next neighbourhood or eject it into space? We must move our thinking to ensure we limit our impacts – it's a new play book and it is one we must absorb quickly.

The Emerald City in *The Wizard of Oz* was secured by collaboration, and the same can also be said for us. Climate change is a collective problem with a collective solution. As we strive to reverse the environmental damage we have done, we need to be better at measuring it. Our currency for improvement has to become the longevity of the action we take. The city must measure its impact, not just within the city boundary, but also the resources it draws into the city, the resources it expels from the city, and the emissions it generates

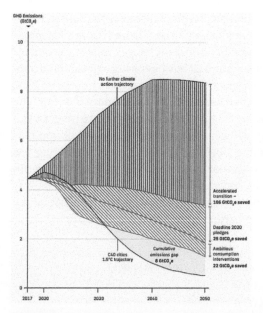

Figure 30.5 Emissions reduction potential of Deadline 2020 commitments and consumption interventions under a future scenario with an accelerated transition of production. (*Source*: https://www.arup.com/-/media/arup/files/publications/c/arup-c40-the-future-of-urban-consumption-in-a-1-5c-world.pdf.)

Figure 30.6 A new home on the moon. Can we apply sustainable principles and can we apply them back on Earth? (*Source*: homegrowngraphics/Getty Images.)

anywhere on the planet as a result of its very existence. The city must negate those impacts. The city must mitigate and remediate as a single solution with full sight of all species that inhabit the Earth.

We need to measure the air we breathe, the water we drink; we need to count our seal-ife, our oceanlife, our wildlife; our insect populations; our trees; all of our flora and fauna – so that everything we do to grow, to build, to expand, and to change can be measured for its impact. We need to embrace all governments and all businesses to progressively decarbonize, accepting the economic realities of a long transition period away from fossil fuels and balance the equation with innovations that can sequester and remove CO_2 from the atmosphere. This will ensure everyone has access to affordable energy as we progress as a society. Our cities will lead the way, they will continue to innovate, they will find new ways to provide finance, and they will shine a bright light for all to see and bring this recovery to a positive conclusion. This collection of stories, told by these wonderful authors, has never been about "solving" the climate crisis; it is a pathway to a better way of life for everyone, leaving no one behind and ensuring only a positive impact on the environment around us as we continue to expand our footprint on the planet. That's worth exploring.

Index

The Climate City, First Edition. Edited by Martin Powell.
© 2022 John Wiley & Sons Ltd. Published 2022 by John Wiley & Sons Ltd.